Matrix Metalloproteinases and TIMPs

Series editor

Peter Sheterline, *School of Biological Sciences, University of Wales Swansea, Singleton Park, Swansea SA2 8PP, Wales*

Advisory Editor

Martin Humphries, *School of Biological Sciences, University of Manchester, Stopford Building, Manchester M13 9PT, UK*

Editorial Board

Alan Barrett, *Peptidase Laboratory, The Babraham Institute, Babraham Hall, Cambridge, CB2 4AT, UK.*

Julio E. Celis, *Aarhus University, Department of Medical Biochemistry, Ole Worms Alle, Bldg 170, University Park, DK-8000, Aarhus, Denmark.*

Benny Geiger, *Department of Chemical Immunology, Weizmann Institute of Science, Rehovot 76100, Israel.*

Keith Gull, *School of Biological Sciences, University of Manchester, Oxford Road, Manchester, UK.*

Tony Hunter, *The Salk Institute, 10010 North Torrey Pines Road, La Jolla, CA 92037, USA.*

Katsuhiko Mikoshiba, *Department of Molecular Neurobiology, Institute of Medical Sciences, University of Tokyo, 4-6-1 Shirokanedai, Minato-ku Tokyo 108, Japan.*

Thomas D. Pollard, *Department of Cell Biology/Anatomy, Johns Hopkins University School of Medicine, 725 North Wolfe St, Baltimore, MD 21205-2105, USA.*

Erkki Ruoslahti, *Burnham Institute, 10901 North Torrey Pines Road, La Jolla, CA 92037-1062, USA.*

Urs Rutishauser, *Department of Genetics, Case Western Reserve University School of Medicine, 2119 Abington Road, Cleveland, OH 44106-2333, USA.*

Solomon H. Snyder, *Department of Neuroscience, Pharmacology and Molecular Sciences, Johns Hopkins University School of Medicine, 725 North Wolfe St, Baltimore, MD 21205, USA.*

Rupert Timpl, *Abteilung Proteinchemie, Max-Planck-Institut für Biochemie, D82152 Martinsried, Germany.*

Published Titles

Tyrosine phosphoprotein phosphates (second edition). *Barry Goldstein*

Lysosomal cysteine proteinases (second edition). *Heidrun Kirschke, Alan J. Barrett and Neil D. Rawlings*

Helix–loop–helix transcription factors (third edition). *Trevor Littlewood and Gerard Evan*

Peptidyl–prolyl cis/trans isomerases. *Andrzej Galat and Sylvie Rivière*

Actins (fourth edition). *Peter Sheterline, Jon Clayton and John C. Sparrow*

Myosins (second edition). *Jim Sellers*

Matrix metalloproteinases and TIMPs. *J. Frederick Woessner and Hideaki Nagase*

Unconventional collagens. *Sylvie Ricard-Blum, Bernard Dublet and Michel van der Rest*

Other Forthcoming Titles

Actin monomer-binding proteins. *Uno Lindberg and Clarence Schutt*

Intermediate filament proteins (third edition). *Roy Quinlan, Chris Hutchison and Birgit Lane*

Gelsolin family. *Horst Hinssen*

Kinesins (third edition). *Viki Allan and Rob Cross*

Cadherins. *Sasha Bershadsky and Benny Geiger*

Integrins. *Martin Humphries*

IgCAMs (third edition). *Fritz Rathjen and Thomas Brümmendorf*

Nuclear receptors (second edition). *Hinrich Gronemeyer and Vincent Laudet*

EF-hand calcium-binding proteins (third edition). *Hiroshi Kawasaki, Susumu Nakayama, and Bob Kretsinger*

Heterotrimeric G proteins (third edition). *Steve Pennington*

Serine phosphoprotein phosphatases. *Patricia Cohen and David Barford*

Signalling domains. *Bruce Mayer and Matthias Wilmann*

Thermolysins. *Rob Beynon and Ann Beaumont*

Colicins. *Richard James*

For more information on the Book Series see
http://www.oup.co.uk/protein_profile

Matrix Metalloproteinases and TIMPs

J. Frederick Woessner

Department of Biochemistry and Molecular Biology, University of Miami School of Medicine, PO Box 106960, Miami, Florida 33101, USA

Telephone: 305 243 6510
Fax: 305 243 3955
Email: fwoessne@med.miami.edu

Hideaki Nagase

The Kennedy Institute of Rheumatology, Imperial College School of Medicine, 1 Aspenlea Road, London W6 8LH, UK

This book has been printed digitally in order to ensure its continuing availability

OXFORD
UNIVERSITY PRESS

Great Clarendon Street, Oxford OX2 6DP

Oxford University Press is a department of the University of Oxford.
It furthers the University's objective of excellence in research, scholarship,
and education by publishing worldwide in

Oxford New York

Auckland Bangkok Buenos Aires Cape Town Chennai
Dar es Salaam Delhi Hong Kong Istanbul Karachi Kolkata
Kuala Lumpur Madrid Melbourne Mexico City Mumbai Nairobi
São Paulo Shanghai Singapore Taipei Tokyo Toronto

with an associated company in Berlin

Oxford is a registered trade mark of Oxford University Press
in the UK and in certain other countries

Published in the United States
by Oxford University Press Inc., New York

© Oxford University Press, 2000

The moral rights of the author have been asserted
Database right Oxford University Press (maker)

First printed 2000
Reprinted 2002

A catalogue record for this book is available from the British Library

Library of Congress Cataloging in Publication Data
Woessner, J. F.
Matrix metalloproteinases and TIMPs/J. Frederick Woessner and Hideaki Nagase.
(Protein profile)
1. Metalloproteinases. 2. Metalloproteinases—inhibitors. 3. Extracellular matrix
proteins. I. Nagase, Hideaki. II. Title. III. Protein profile (Oxford, England)
QP601.7.W38 2000 572'.76—dc21 99- 049877

ISBN 0- 19- 850268- 0

Series preface

The Protein Profile *series has been developed from a recognition that individuals find it increasingly difficult to access readily the enormous amount of information accumulated by the international research community; information which is crucial for the efficiency and quality of their activities.*

The *Protein Profile* series aims to provide a practical, comprehensive and accessible information source on all major families of proteins. Each volume of *Protein Profile* is focused on a single family or sub-family of proteins, and contains tables and figures presenting as comprehensive an accumulation of structural, kinetic and biochemical information available on that particular protein group, coupled to an extensive bibliography. Every volume will be refined and updated to provide users with a practical up-to-date single source of information by the publication of new editions approximately every two years.

Content

The text provides a brief overview of the biological context of the function of the protein group followed by an overview of available information on:

- function
- kinetic and biochemical properties
- sequences, sequence relationships and sequence features
- domain structure
- mutations
- 3-D structure
- binding sites of protein
- ligand binding sites and interactions with drugs
- derivatization sites
- proteolytic cleavage sites

Each volume follows the same format but with different emphases depending on the protein family. The text is extensively supported by tables and figures listing key information gathered from the literature with comprehensive reference to primary sources.

Available protein sequences, or where there are very large numbers, representatives from each subgroup, are aligned in an Appendix at the back of each issue.

Available references pertinent to the properties, structure and function of the proteins are listed in a full bibliography. References are numbered for access in the text, but are also arranged alphabetically under headings to allow browsing. Reviews are listed at the beginning and key papers are identified.

For more information on the Protein Profile Series, including access to online resources, please see our web page at

`http://www.oup.co.uk/protein_profile`

Series preface

The Protein Profile *series has been developed from a recognition that individuals find it increasingly difficult to access readily the enormous amount of information accumulated by the international research community; information which is crucial for the efficiency and quality of their activities.*

The *Protein Profile* series aims to provide a practical, comprehensive and accessible information source on all major families of proteins. Each volume of *Protein Profile* is focused on a single family or sub-family of proteins, and contains tables and figures presenting as comprehensive an accumulation of structural, kinetic and biochemical information available on that particular protein group, coupled to an extensive bibliography. Every volume will be refined and updated to provide users with a practical up-to-date single source of information by the publication of new editions approximately every two years.

Content

The text provides a brief overview of the biological context of the function of the protein group followed by an overview of available information on:

- function
- kinetic and biochemical properties
- sequences, sequence relationships and sequence features
- domain structure
- mutations
- 3-D structure
- binding sites of protein
- ligand binding sites and interactions with drugs
- derivatization sites
- proteolytic cleavage sites

Each volume follows the same format but with different emphases depending on the protein family. The text is extensively supported by tables and figures listing key information gathered from the literature with comprehensive reference to primary sources.

Available protein sequences, or where there are very large numbers, representatives from each subgroup, are aligned in an Appendix at the back of each issue.

Available references pertinent to the properties, structure and function of the proteins are listed in a full bibliography. References are numbered for access in the text, but are also arranged alphabetically under headings to allow browsing. Reviews are listed at the beginning and key papers are identified.

For more information on the Protein Profile Series, including access to online resources, please see our web page at

`http://www.oup.co.uk/protein_profile`

Preface

It is appropriate to publish a review of the MMPs and TIMPs at this time because of the increasing interest of a large number of researchers. Yearly publications of papers in this area approached 1200 in 1998 and is increasing by about 10% per year. While there have been a number of brief review articles in recent years, and even several books, there has not been a venue like that offered by Protein Profiles to accommodate an in-depth overview of the field since its beginnings. Even in this case there is a limit in that the sequence, structure, and function of the proteins are given in detail, while aspects of biological regulation and disease processes must be given short shift.

An attractive feature of the Protein Profiles is that they are continually updated in subsequent years and that they will be available online so that even more frequent updates are possible. We would plead with the readers to send along any corrections or missing information so that these updates will be as useful as possible. We would particularly appreciate information on newly-discovered MMPs. Attendees at the last two Gordon Research Conferences on Matrix Metalloproteinases have designated the authors as an informal board to oversee the numbering of MMPs. Since this Profile was submitted, MMP-21 through MMP-23 have been added (see Nagase, H. and Woessner, J. F., Jr., *J. Biol. Chem.* **274**: 21491–21494, 1999). MMP-24 will appear shortly as an MT6-MMP. If new enzymes were referred to the authors for numbering prior to publication it would avoid confusion such as that arising over MMP-18/MMP-19 and MT2-MMP/MT3-MMP. The numbering system is used only for vertebrate enzymes.

Finally, the authors would like to thank the Editors, Dr. Peter Sheterline, University of Wales, Swansea, and Dr. Beth Knight, Oxford University Press, for their help in organizing the book and seeing it through the press.

Miami   J. F. W.
London   H. N.

Acknowledgements

We thank Dr Klaus Maskos and Professor Wolfram Bode of Max-Planck-Institut für Biochemie at Martinsried, Germany for providing us with diagrams of superimposed MMP structures, TIMP-1 and the TIMP-2–MT1-MMP complex, and Dr Deendayal Dinakarpandian at the University of Kansas Medical Center for preparation of structural diagrams of MMPs and TIMPs. Dr Ken-ichi Shimokawa and Ms Linda Chung at the University of Kansas Medical Center are also acknowledged for preparation of illustrations. Dr. Neil Rawlings, Babraham, England, prepared dendrograms of MMPs and TIMPs.

Sequence data were retrieved from SwissProt, revised and collated into tables and alignments in a collaboration between the European Bioinformatics Institute at Hinxton, Cambridge, UK and Protein Profile by curators Steffi Kappus, Fiona Lang, Michele Magrane, Youla Karavidopoulous and Wolfgang Fleischmann. The project was coordinated at EBI by Rolf Apweiler. Data resulting from this collaboration are available at www3.ebi.ac.uk/Services/protein_profiles.

Contents

Protein substrates of the MMPs 87

Specificity requirements of the MMPs 98

Inhibition of the MMPs 109

List of figures

List of tables

Glossary

APMA, aminophenylmercuric acetate
CAP, cartilage activating protease, MMP-3
dnp, dinitrophenol
Dpa, N-3-(2,4-dinitrophenyl)-ʟ-2,3-diaminopropionic
ECM, extracellular matrix
EDANS, 5-[(2-aminoethyl)amino]naphthalene-1-sulfonic acid
EDTA, ethylene diamine tetra-acetate
EPF, erythroid potentiating factor
ESAF, endothelial cell stimulating angiogenesis factor
IGFB, insulin-like growth factor binding protein
IL, interleukin
kDa, kilodaltons
α_2**M**, α_2-macroglobulin
Matrixin, the subfamily of family M10, containing the MMPs

Mca, (7-methoxycoumarin-4-yl)acetyl
MMP, matrix metalloproteinase
MMP-1 etc., individual MMPs are listed in Table 1
NMR, nuclear magnetic resonance
N-TIMP, truncated TIMP lacking C-terminal domain
Pump-1, putative metalloproteinase, an early name for matrilysin
SDS–PAGE, sodium dodecyl sulfate–polyacrylamide gel electrophoresis
TC, tropocollagen
TIMP, tissue inhibitor of metalloproteinases
Transin, an early name for stromelysin
uPA, urokinase

Introduction

Introduction to the matrix metalloproteinases (MMPs)

It is a simple matter to trace the roots of this field. Although it had long been recognized that connective tissue collagen was rapidly broken down in many physiological situations, such as bone remodelling, postpartum involution of the uterus and embryogenesis, and in pathological process, such as decubitus ulcer, cancer and arthritis [143], there was no evidence identifying the degradative factor(s) involved. Then in 1962, Gross and Lapière [672] demonstrated that skin from the resorbing tail of the metamorphosing frog, when cultured on collagen gels, released an enzyme that could attack the triple-helix of collagen. This was a true collagenase that could digest collagen at neutral pH and 27 °C (well below the denaturation temperature of collagen). The same report also showed a similar activity produced by cultures of embryo skin, postpartum uterus and bone. In retrospect, these last three activities were probably due to MMP-13.

Following that first report, there has been a rapid development of the field. Collagenases have been demonstrated in human tissues (MMP-1) and in neutrophils (MMP-8) and the activity has been shown to be metal-dependent and to cut collagen at a single site. Over the intervening years a number of additional MMPs have been found, starting with gelatinase A (MMP-2) and stromelysin (MMP-3) and continuing through to the recent identification of enamelysin (MMP-20). There has been a tremendous burgeoning of the field during this time, so that today the literature of the field includes more than 10 000 citations with approximately 1000 being added every year.

General review articles that have appeared in the past 10 years include [4, 37, 42, 71, 72, 83, 87, 144, 145] and particularly [17]. There are also a large number of specialized reviews that will be cited in appropriate sections below. An invaluable book covering the field up to 1991 evolved from proceedings of the Destin Beach conference on MMPs [1]. The principles of naming the MMPs and their grouping into a matrixin subfamily were presented in 1992 [91] and a number of chapters covering seven individual MMPs were published in 1995 [8]; although these were largely methodological. Two important books were published in 1998—*Matrix metalloproteinases* [2] and the *Handbook of proteolytic enzymes* [3], both contain individual chapters on the major MMPs.

A list of currently known MMPs and related enzymes is presented in Table 1. Each of these entries will be taken up in an introductory paragraph, prior to examining the sequence relationships. There is also a sketch of the domain structures of the MMPs (Fig. 1), which will be important to the subsequent understanding of the sequence interrelationships of the MMPs. It can be seen that the simplest enzyme (matrilysin) has only a signal peptide, a propeptide and a catalytic domain; while the most complex (gelatinase B) has in addition the common hemopexin domain as well as fibronectin-like repeats and a collagen type V-like domain. Other features may include furin cleavage site, hinge region and transmembrane domain. These will all be discussed in a section below devoted to the domains.

The relationship of MMPs to the broad class of metalloproteinases has been discussed recently by one of us [147]. There are more than 200 known metalloproteinases, almost all of them dependent on zinc at the active centre for their catalytic function. Rawlings and Barrett [104] have proposed dividing the metalloproteinases into clans based on three-dimensional protein folding and into

Table 1 The matrix metalloproteinases

MMP-No.	Name	EC number	M_r latent/active*	Other names, notes
MMP-1	Collagenase 1	3.4.24.7	52 000/43 000	Interstititial collagenase
MMP-2	Gelatinase A	3.4.24.24	71 000/62 000	72 kDa gelatinase Type IV collagenase
MMP-3	Stromelysin 1	3.4.24.17	52 000/43 000	Transin, proteoglycanase, CAP
(MMP-4	Procollagen peptidase	--		Discontinued = MMP-3)
(MMP-5	¾-Collagenase	--		Discontinued = MMP-2)
(MMP-6	Acid metalloproteinase	--		Discontinued = MMP-3)
MMP-7	Matrilysin	3.4.24.23	28 000/19 000	Pump-1
MMP-8	Collagenase 2	3.4.24.34	51 000/42 000	Neutrophil collagenase
MMP-9	Gelatinase B	3.4.24.35	76 000/67 000	92 kDa gelatinase Type V collagenase
MMP-10	Stromelysin 2	3.4.24.22	52 000/44 000	Transin 2
MMP-11	Stromelysin 3		51 000/46 000	Furin motif
MMP-12	Macrophage elastase	3.4.24.65	52 000/20 000	Metalloelastase
MMP-13	Collagenase 3		52 000/42 000	Rat intersititial collagenase
MMP-14	Membrane-type matrix metalloproteinase 1		64 000/54 000	MT1-MMP, furin motif
MMP-15	Membrane-type matrix metalloproteinase 2		71 000/61 000	MT2-MMP, furin motif
MMP-16	Membrane-type matrix metalloproteinase 3		66 000/56 000	MT3-MMP, furin motif
MMP-17	Membrane-type matrix metalloproteinase 4		62 000/51 000	MT4-MMP, furin motif
MMP-18	Collagenase 4		53 000/42 000	*Xenopus*
MMP-19	No trivial name		54 000/45 000	RASI-1
MMP-20	Enamelysin		54 000/22 000	
MMP-21	XMMP		70 000/53 000	*Xenopus*, furin motif
MMP-22	CMMP		51 000/42 000	Chick embryo
–	MMP-C31			*Caenorhabditis elegans*
–	MMP-H19			*C. elegans*, furin motif
–	MMP-Y19			*C. elegans*, furin motif
–	Envelysin	3.4.24.12	63 000/48 000	Sea urchin
–	Soybean MMP		31 000/19 000	*Glycine max*
–	Fragilysin	3.4.24.74	43 000/21 000	*Bacteroides fragilis*

*M_r is based on cDNA sequence excluding signal peptide; some MMPs may have glycosylated forms of higher M_r. Thus, MMP-8 in neutrophils but not other tissues has M_r = 85 000/64 000; MMP-9 has M_r = 92 000/84 000; stromelysin 3 has M_r = 64 000/46 000; human MMP-13 has a latent form of 65 000. Strictly speaking, fragilysin should be assigned to subfamily C of M10, but shows up in affinity searches.

families based on evolutionary relationships of the primary sequences. Using these criteria, the metalloproteinase class currently contains eight clans and 40 families [3]. A common theme in the sequence of metalloproteinases is the signature sequence HELGH, which contains two of the three zinc-binding histidine residues and a catalytic glutamic acid residue. Several clans contain this signature; here we are concerned with clan MB which has the so-called metzincin fold [19] and the signature HEXXHXXGXXH with the third histidine also bound to the zinc atom. However, while the fold is consistent, the sequences of the members of the clan are sufficiently diverse as to require division into families that cannot be shown to have common ancestry. Two families clearly possess the metzincin fold: M10, the interstitial collagenases, and M12, the astacins. Each family is

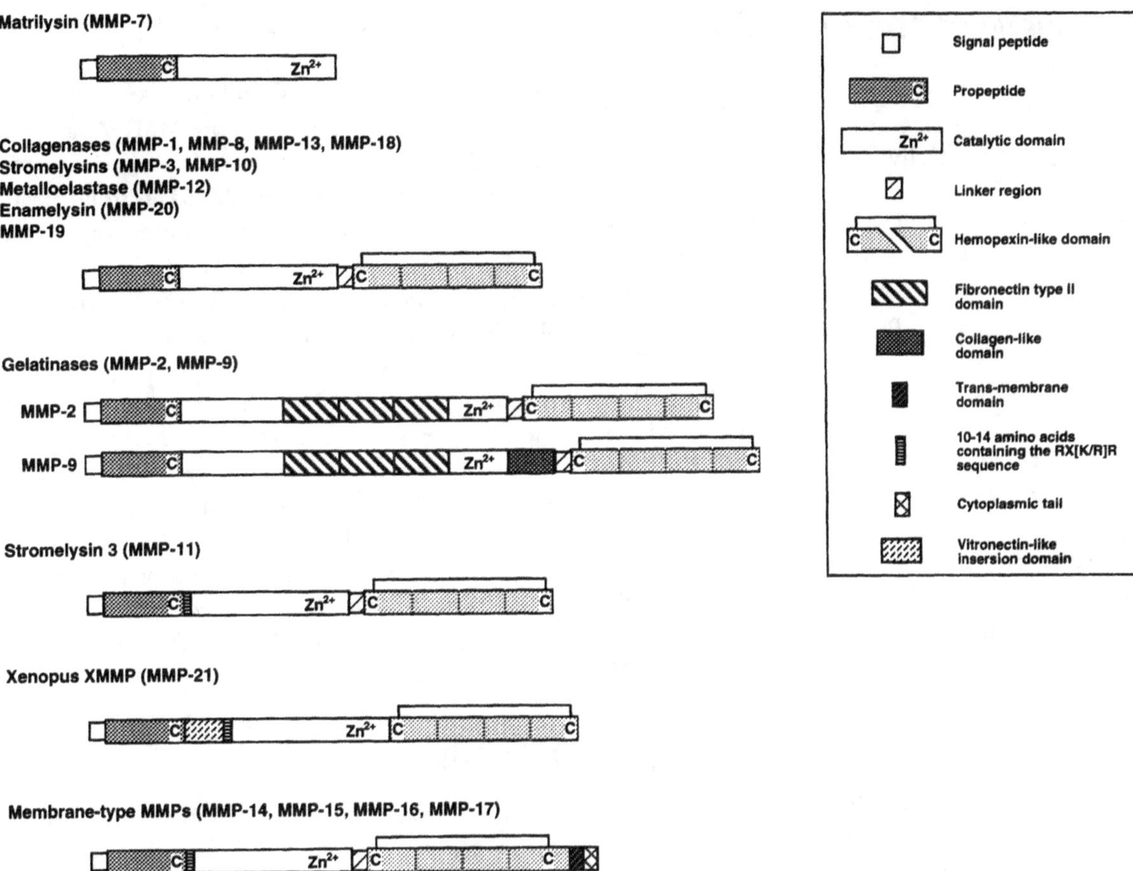

Figure 1

Arrangement of the domain structures of vertebrate matrixins. The propeptide region contains the 'cysteine switch' sequence of PRCG[V/N]PD whose cysteine residue coordinates to the catalytic Zn^{2+} ion in the proenzyme. MMP-11 and MT-MMPs have extra residues containing the RX[K/R]R furin-recognition sequence. MMP-2 and MMP-9 have three repeats of a 58-residue motif homologous to the fibronectin type II domain (F2) before the catalytic zinc binding site. MMP-9 has a collagen type V-like domain before the linker region. MT-MMPs have a hydrophobic transmembrane domain and a cytoplasmic tail. *Xenopus* XMMP contains a 37-residue-long ID sequence in the propeptide. C indicates a conserved cysteine.

further subdivided in two: the collagenase subfamily and the serralysin subfamily of M10, and the astacin and reprolysin subfamilies of M12 [3].

A consequence of the close similarities in the active sites of these two families is that it has become more difficult to define what an MMP is. An early definition was that an MMP was blocked by chelators, had a latent form activated by organomercurials, was inhibited by TIMP and acted on at least one component of the extracellular matrix [91]. The first criterion is too broad, the second and fourth are not valid for all MMPs, so

only inhibition by TIMP remains from this list. Further criteria advanced later, such as possession of a cysteine switch and an extracellular site of function [145], are also not sufficiently specific. The best current criterion would be sequence similarity to collagenase, establishing an evolutionary relationship [147]. On the other hand, inhibition by synthetic inhibitors, such as hydroxamates, is no longer useful because the reprolysins, particularly the ADAMs such as tumour necrosis factor-α convertase, are effectively inhibited by the same compounds [641, 714].

Overview of the individual MMPs

Collagenase 1 (MMP-1)

The first discovery of a vertebrate collagenase activity by Gross and Lapière [672] involved an enzyme from frog, but activity was also shown in cultured tissues of rat, pig, chick and mouse. It is unlikely that all of these enzymes were MMP-1; thus the frog enzyme was possibly MMP-18, the chick, MMP-2 and the rat and mouse enzymes, MMP-13. However, the pig enzyme was probably MMP-1. The use of electron microscopy to study segment long spacing (SLS) forms of reconstituted collagen, demonstrated that the frog enzyme cleaves the collagen molecule into a ¾-length piece from the N-terminus and a ¼-length piece from the C-terminus [673]. Early studies demonstrated collagenase in human periodontal tissue [637, 660], skin [661], bone [662] and rheumatoid synovium [658]. A complete purification of human skin MMP-1 was first achieved in 1970 [634]. However, it was not until 1971 that it was noticed that the collagenases occurred as zymogens [678] that could be activated by trypsin [16, 639] or organomercurials [66, 796]. The first complete sequence, based on human cDNA, was published in 1986 [669]. MMP-1 enzyme is a typical MMP, having signal, propeptide, catalytic and hemopexin domains (see Fig. 1). The names **fibroblast collagenase** and **interstitial collagenase** (the official EC name) have been used extensively. Recent book chapters dealing in detail with collagenase 1 are found in [28, 36, 60].

Gelatinase A (MMP-2)

Harris and Krane in 1972 [679] observed gelatinase activity in rheumatoid synovial tissue, that, in retrospect, was possibly MMP-2. A gelatinase activity was separated from collagenase 1 and stromelysin 1 in the culture medium from rabbit bone in 1978 by Sellers *et al.* [763]. The following year, Liotta *et al.* [701] found a similar enzyme in mouse

tumour that digested type IV collagen of basement membrane. This led to the early designation of the enzyme as **type IV collagenase**, a designation that grew less useful as more MMPs digesting this substrate were discovered. Another name commonly encountered is **72 kDa gelatinase**. MMP-2 was purified from cultures of mouse tumour [755], rabbit bone [719], human skin [765] and gingival tissue [723]. The enzyme was also discovered independently as a gelatin-binding protein [787, 789]. Collier *et al.* [650] reported the complete sequence of the human enzyme except for the signal peptide. The sequence includes a triple repeat of fibronectin type II domains inserted in the catalytic domain; these contribute to the binding of the enzyme to gelatin substrates [649]; this makes gelatinase A one of the longest of the MMPs (see Fig. 1). Recent overviews of gelatinase A are found in [80, 81, 150].

Stromelysin 1 (MMP-3)

A neutral proteinase activity distinct from collagenase was first detected in 1974 in human cartilage by Sapolsky *et al.* [756] and in rabbit bone fibroblast culture by Werb and Reynolds [799]. Rabbit chondrocytes [655] and mouse bone [784] showed similar activity. The enzyme was metal-dependent and activation of the latent form caused a 20 kDa reduction in mass [797]. The rabbit bone enzyme was purified, characterized and referred to as **proteoglycanase** [663]. The name 'stromelysin' was introduced by Chin *et al.* [648], but other names have included **transin** [707] and **collagenase activating protein** [782]. The human cartilage enzyme is one of the few to have been completely purified directly from tissue extracts [675]; this helped to establish its role in tissue and to complete the identification of an activity with acid pH optimum as being MMP-3 [628]. The cDNA sequence for rat transin, obtained in 1985 [707] was the first for any MMP, although the proteinase nature of the protein was not immediately recognized. The enzyme has the typical MMP domains of signal, propeptide, catalytic and hemopexin (see Fig. 1). Recent chapters describing this enzyme are found in [86, 89, 90].

Vacant MMP designations (MMP-4, MMP-5, MMP-6)

Many readers may be puzzled as to why three of the numbers are vacant. This is because numbers were assigned early to MMPs that subsequently proved to be one of the known enzymes. Thus, collagen telopeptidase activity was described in 1983 in pig and human gingival extracts [760, 761]; Nakano *et al.* [724] identified an enzyme cleaving C-telopeptides from type I collagen; and the name MMP-4 was proposed by Overall and Sodek [737]. However, it was later shown that stromelysin can cleave the N-telopeptide, and probably the C-telopeptide as well, from type II collagen [809] indicating (along with other similarities) that MMP-4 is probably stromelysin 1. However, MMP-13 may have also played a role in the original observations [691].

A 3/4-collagenase that digests collagen following the initial cleavage by collagenase was described by Overall and Sodek [737] and named MMP-5. However, this enzyme was subsequently shown to be MMP-2 [722]. Finally, an aggrecan-degrading proteinase with an acid pH optimum was isolated from human articular cartilage [757, 804], purified [628] and shown to increase in osteoarthritis [742]. However, it was later shown that the pure enzyme was identical to stromelysin 1 [675, 802], which in fact has an acid pH optimum [680].

Matrilysin (MMP-7)

Matrilysin is the smallest of the MMPs; it has only signal, propeptide and catalytic domains (see Fig. 1). Sellers and Woessner [764], searching for stromelysin in the postpartum involuting uterus of the rat, found a similar but much smaller enzyme. Finally, in 1988 the enzyme was purified from rat uterus and characterized [805]. At about the same time a human homologue was cloned, sequenced and named **pump-1** for putative metalloproteinase-1 [716]. Expression of the human cDNA established that this enzyme was an actual, not a putative, proteinase with properties similar to the rat enzyme [745]. The enzyme was discovered independently as the urokinase activator of human kidney [706]. It was also purified from a human carcinoma cell line and given the name **matrin** [711]. An extensive review of matrilysin was recently published [140] and three chapters have appeared [141, 142, 146].

Collagenase 2 (MMP-8)

Lazarus *et al.* in 1968 [695, 696] were the first to report a collagenase activity from human neutrophils. This was the second MMP, after interstitial collagenase, to be identified. For many years it was referred to as **neutrophil collagenase** to distinguish it from fibroblast collagenase (MMP-1). There is a factor in synovial fluid capable of activating this enzyme [692]. The enzyme is localized to the specific granules [720] and is released following phagocytic events [735]. Purification of the enzyme [704] was followed by immunological demonstration that the enzyme is distinct from MMP-1 [681]. The enzyme was sequenced in 1990 [682] and found to be of similar size to MMP-1 (Fig. 1). The higher M_r usually seen *in vivo* is due to glycosylation. There was a series of papers by Ohlsson in the 1970s describing a neutrophil collagenolytic activity that was subsequently shown to be a serine enzyme that did not, in fact, digest collagen [733]. MMP-8 was originally thought to be confined to neutrophils/polymorphonuclear leukocytes, but recent developments indicate it may be expressed in other cells, such as osteoarthritic chondrocytes [769], synovial fibroblasts and endothelial cells [677]. Review chapters dealing with MMP-8 are found in [60, 129, 130].

Gelatinase B (MMP-9)

A gelatinase from human polymorphonuclear leukocytes was described by Sopata *et al.* [771]. This was in a latent form activated by mercurials [772].

A similar enzyme found in rabbit macrophages is able to digest type V collagen [705], leading to its earlier designation as **type V collagenase**. The collagenase and gelatinase activities of neutrophils were separated [689, 717]. The enzyme was purified and characterized [748], followed by sequencing of the cDNA [801]. Gelatinase B is the largest of the MMPs, it includes three fibronectin-like domains and a collagen type V-like domain (Fig. 1). Its molecular weight of about 92 000 daltons has led to another common designation: **92 kDa gelatinase.** TIMP-1 binds to proMMP-9, producing a complex that serves to regulate the activation and subsequent activity of the enzyme [668]. The human neutrophil form of gelatinase B is usually found complexed to lipocalin through an intermolecular disulfide bond [688]. Several papers concerning a 95 kDa gelatin-binding protein of human plasma concluded by identifying this protein as MMP-9 [786, 788]. Gelatinase B has been reviewed in three recent chapters [34, 81, 135].

Stromelysin 2 (MMP-10)

Stromelysin 2 was the first of the MMPs to be discovered by cloning and sequencing, in this case as rat **transin 2** [645], which revealed an enzyme 71% identical to transin 1 (MMP-3). The human proteinase was cloned soon after [716]. The enzyme has the typical composition of signal, propeptide, catalytic and hemopexin domains (Fig. 1). Expression of both rat and human cDNA showed that they have proteolytic activity, are inhibited by TIMP and can activate latent MMP-1 [727]. Activity is prominent in human tumours of head, neck and lung [715]. Reviews may be found in [73, 86, 90].

Stromelysin 3 (MMP-11)

Basset *et al.* in 1990 [633] found a cDNA clone that was identified as a putative MMP. Several years later it was demonstrated that the expressed protein had weak proteolytic activity [721]. This may have been a substrate problem because Pei *et*

al. [740] showed that the enzyme had significant digestive action on α1-proteinase inhibitor as well as α_2-macroglobulin. While this MMP has a typical domain structure (Fig. 1) it does not fit well with the other stromelysins because it contains an RKRR sequence at the C-end of its propeptide. It was established by Pei and Weiss [741] that, as expected from this sequence, there is indeed intracellular activation of the proenzyme by a furin-like proteinase. It has been implicated in rheumatoid synovial inflammation [725], cycling human endometrium [750] and embryonic mouse development [698]. There are recent chapters summarizing current knowledge of this enzyme [100] and its role in cancer progression [10].

Macrophage elastase (MMP-12)

Mouse macrophages have both serine and metalloproteinases capable of digesting elastin [798]. Banda and Werb [629] established that the proteinase activity of **metalloelastase** was not blocked by α1-proteinase inhibitor, which was in fact a substrate. The mouse enzyme was purified [630] and specificity studies [686] were the earliest reported for any MMP except MMP-1. The mouse enzyme was cloned, sequenced and assigned to chromosome 9 [766]; this was followed in a few years by sequencing and chromosome assignment (11q22.2–22.3) of the human enzyme [636]. Recent reviews appear in [115, 117].

Collagenase 3 (MMP-13)

In the first paper of Gross and Lapière [672], collagenase was detected in cultures of involuting rat uterus and in mouse and rat bone. At the time there was no reason to think that there might be more than one collagenase. However, when rat uterine collagenase was sequenced [746] homology to human MMP-1 was surprisingly weak. Still later when human MMP-13 had been sequenced [659] it became clear that the rat enzyme was more closely related to human MMP-13 than to MMP-1. A retrospective look suggests that many of the reports of rat/mouse collagenase, such as its purification from bone [623, 753, 768] and uterus [676, 684, 751], actually described MMP-13. The uterine enzyme is one of the earlier MMPs to have been extracted directly from tissue [794]. However, one cannot automatically identify all early rat and mouse reports as due to MMP-13 in view of the recent demonstration of MMP-8 (collagenase 2) in these species [694]. Recent reviews may be found in [59, 60].

Membrane-type matrix metalloproteinase 1 (MMP-14)

A novel MMP with a transmembrane domain close to its C-terminus and lying beyond the hemopexin domain was cloned and sequenced by Sato *et al.* in 1994 [758] (see Fig. 1); this proteinase can activate proMMP-2 bound to the cell surface [627, 631]. Additional cloning or partial protein sequencing of the human enzyme soon followed in three laboratories [734, 778, 780]. These first reports introduced the abbreviation MT-MMP-1 for membrane-type matrix metalloproteinase 1. However, MMP-1 (i.e. interstitial collagenase) does not insert into the membrane bilayer. The recommended abbreviation is MT1-MMP to avoid this confusion. The enzyme is detected in carcinomas [730], developing embryos [624, 687], placenta [640] and cartilage [646]. Recent chapters discussing this enzyme are found in [63, 114].

Membrane-type matrix metalloproteinases 2, 3 and 4 (MMP-15, MMP-16, MMP-17)

There is very little literature concerning these recently discovered enzymes. An enzyme cloned and sequenced by Takino *et al.* in 1995 [779] was named MT2-MMP, but this group was unaware that Will and Hinzmann [803] had already cloned an enzyme that they named MT2-MMP. Therefore, the Will enzyme should be considered as MT2-MMP and the Takino enzyme should be moved down one position to MT3-MMP. The expressed protein from the cDNA for MT2-MMP

showed proteolytic activity [652]. The mRNA is found in human endometrium [754] and breast cancer [783]. MT3-MMP mRNA undergoes alternative splicing, resulting in enzymes of two sizes [709]. The enzyme has been found in breast tumours [783], brain and placenta [770]. The latter matches the brain enzyme sequence by only 94%. MT4-MMP has been sequenced by Puente *et al.* [744]. These additional MT-MMPs are reviewed briefly in the MT1-MMP chapters cited above.

MMP-18 and higher

Collagenase 4 (MMP-18) is described in a single paper where the cDNA was isolated from a *Xenopus* library and is the first amphibian MMP to be sequenced [776], in spite of the early discovery of tadpole collagenase in 1962. The expressed protein is a collagenase. A brief recapitulation may be found in [118].

The sequence of MMP-19 was published in 1996 [651]. The authors were unaware that MMP-18 had been used for the *Xenopus* enzyme and so re-used the abbreviation MMP-18. The enzyme should be called MMP-19 and there is, as yet, no trivial name. Two further sequences have been reported [690, 743]. The former group uses the shorthand RASI-1 (rheumatoid arthritis synovial inflammation) for their enzyme. The protein has been identified on activated lymphocytes and in rheumatoid plasma [762] but there is very little information on its proteolytic activity [743]. The enzyme is similar to the other 55 kDa MMPs, but has a long hinge region between the catalytic and hemopexin domains (Fig. 1). There are no reviews available.

Enamelysin (MMP-20) was first seen as a poorly characterized metalloproteinase activity in the developing enamel of pig [654], rat [781] and cow [713]. Pig enamelysin was cloned and sequenced in 1996 [632], followed shortly by the human enzyme [703]. The most recent MMP to be discovered is XMMP (MMP-21), a novel proteinase from *Xenopus* [811]; this has a long 37-residue insert between the propeptide and the catalytic domain, including an RRKR motif that suggests processing by furin. A second MMP found by the same group has been named CMMP (MMP-22) because it was derived from the chicken. It is similar to XMMP and MMP-19 in having an unusual insertion of cysteine into the catalytic domain [810].

MMPs from invertebrates, plants and bacteria

Envelysin of the sea urchin is a metalloproteinase responsible for the embryo hatching process. Proteinases were long known to participate in hatching, but it was 1989 before an enzyme was purified and characterized [699]. It was sequenced and shown to be homologous to MMP-1 and to have the same domain structure [700]. The propeptide has an unusual length of 148 amino acids, whereas the catalytic and hemopexin domains span a more typical 421 residues. Envelysin was the name introduced by Nomura *et al.* [731] who sequenced the enzyme from a second species of urchin. For a recent review see [44].

There are at least seven MMP homologues in the nematode *Caenorhabditis elegans*. However, only one paper demonstrates that three of these, when expressed in *E. coli* show proteolytic activity; these are named MMP-C31, MMP-H19 and MMP-Y19 [792]. The first two are inhibited by TIMP-1, while the last two have the RXKR motif suggesting furin processing. It should be noted that the practice of assigning MMP numbers has not yet been extended below the vertebrates.

A metalloproteinase from soybean leaf was detected [747] and purified in 1991 [670]. Classical Edman degradation provided a sequence of the active form of the enzyme; this was a 20 kDa domain with homology to MMP-1 [710]. It is inhibited by TIMP and hydroxamate compounds. More recently the full cDNA has been sequenced and the presence of a propeptide and cysteine switch have been shown [738]. A recent update can be found in [74]. A number of sequences resembling MMPs have been found in *Arabidopsis* (see Table 3, below), but there is no evidence yet that any of these functions as a proteinase.

Finally, mention may be made of the intestinal bacterium *Bacteroides fragilis*. This organism exports an enterotoxin that is believed to be causative in animal and human disease. The enzyme, **fragilysin**, was purified [785], sequenced [712] and shown to be related to MMP-1. However, this relationship is not very close and N. D. Rawlings (personal communication) favours a separate subfamily 10C for this bacterial enzyme. The expressed protein produced disease symptoms in lamb, rat and rabbit [732]. The enzyme is reviewed in [95].

Introduction to the tissue inhibitors of metalloproteinases (TIMPs)

How does the cell control the MMPs once they have been released into the extracellular space? There would seem to be a considerable risk in exposing the surrounding matrix to uncontrolled degradation. In fact, the cell does not lose control of the MMPs after they are secreted. Firstly, the cells do not release MMPs in an unrestrained manner, but only in response to well-defined signals arising at times when degradation is appropriate. Secondly, many of the MMPs remain in the vicinity of the cell, bound to the cell surface or to components of the matrix. Thirdly, the enzymes are frequently in a latent proform that requires activation. These processes are controlled by the cell, e.g. by cell-surface MMPs such as MMP-14, which activates MMP-2 at the surface, or by binding urokinase and activating plasminogen to plasmin, which in turn activates proMMP. Finally, there are general inhibitors of MMPs that rapidly block the action of MMPs. Among these are the α_2-macroglobulin family members and the much more specific TIMPs. The latter are produced by the local cells, often the same cells that release MMPs. Thus, the cell can 'titrate' the activity of the MMPs.

The TIMPs were first identified in 1975 and since then at least four genetic types of TIMP have been characterized. The reasons for this multiplicity are not yet clear. The general properties of

Table 2 Properties of the TIMPs

Property	TIMP-1	TIMP-2	TIMP-3	TIMP-4
Molecular mass, seq.	20.6 kDa	21.5 kDa	21.6 kDa	22.3 kDa
Glycosylated mass	28.5 kDa	–	27 kDa	–
Human gene locus	Cp11.23–11.4	17q2.3–2.5	22q12.1–13.2	3p25
mRNA transcript size Kb	0.9	3.5, 1.0	5.0, 2.6, 2.4	1.4
Tissue location	Ovary, bone	Kidney, placenta	Spleen, ovary	Heart, brain
ProMMP complex	MMP-9	MMP-2	–	MMP-2
Soluble	Yes	Yes	No	Yes
Inhibition of MMPs	All	All	All	1, 2, 3, 7, 9
Erythroid potentiation	Yes	Yes	–	–
Mitogenicity	Yes	Yes	–	–

the four TIMPs are summarized in Table 2. The minimum TIMP appears to be the N-terminal domain with an N-terminal Cys residue, but such a TIMP, while inhibiting the MMPs, would fail to produce all the effects expected of TIMPs. In addition to maintaining control over the activity of MMPs, TIMP-1 and TIMP-2 are able to bind directly to the hemopexin domain of MMP-9 and MMP-2, respectively, exerting further control over the activation process. There are also other roles of TIMPs that do not seem to be directly attributable to proteinase inhibition, such as growth factor activity, steroidogenesis, and cell morphology modulation. There is a growing interest in the TIMPs; of the 1000 references each year dealing with the field of matrix metalloproteinases about 25% deal with TIMPs. An important source of information about TIMPs and other MMP inhibitors is the entire Vol. 732 of the *Annals of the New York Academy of Sciences*, NY, 1994 [51]. A recent overview of TIMPs is presented by Gomez *et al.* [49].

Overview of the individual TIMPs

Tissue inhibitor of metalloproteinases 1 (TIMP-1)

The production of a collagenase-inhibitory protein in the medium of cultured human fibroblasts was first reported in 1975 [635]. A serum inhibitor named β_1-anticollagenase was reported in the same year [806, 807]. Extracts of bovine cartilage also showed the presence of such an inhibitor [693] and the activity was partially purified [752]. Rabbit bone and uterus produced an inhibitor [718] as did smooth muscle cells of the pig [728] and human synovium [790] and tendon [791]. The inhibitor was purified from skin fibroblasts [795] and from rabbit bone [647] and cow aorta [729]. The resultant inhibitor was resistant to denaturation by high temperature (even boiling), but was quite sensitive to reduction and alkylation [729]. The amino acid sequence deduced in 1985 [656] was surprising in that the protein proved to be already known in a completely unrelated context.

It was shown in 1980 that human T-lymphoblasts produce a factor that is able to potentiate the action of erythroid factor [665]. Cloning and sequencing of this **erythroid potentiating factor** in 1985 [664] was complete before the identical sequence of TIMP-1 was established [656]. There was initial resistance to the idea that one protein exhibited two distinct activities [777]; the idea was slow to catch on in the TIMP field until later studies of TIMP growth effects [55]. However, a similar theme kept repeating itself. Thus, fibroblasts stimulated to undergo cell division by serum expressed a gene with the sequence of EPF/TIMP [657]. Mouse cells stimulated by phorbol esters produced a **tumour promoting factor TPA-S1**; again the sequence matched that of TIMP-1 [685]. This protein was dubbed **phorbin**

[674]. Newcastle virus disease induces a gene identical to TIMP-1 [666], as is that of **embryo-genin-1** produced by bovine granulosa cells [759]. **Endometriosis protein II** [767], that of a stimulator of steroidogenesis produced by Sertoli cells, proves to be a complex of TIMP-1 and pro-cathepsin L [644]. Reviews of TIMP activity as a growth factor are presented by Gomez *et al.* [49]. Another unrelated feature of TIMP-1 is its ability to bind the hemopexin domain of proMMP-9 [801]; in this case the presence of the propeptide prevents binding to the active centre. The complex is believed to regulate the activation and subsequent function of gelatinase B. A review of TIMP-1 is presented by Murphy [84] and Gomez *et al.* [49].

Tissue inhibitor of metalloproteinases 2 (TIMP-2)

The existence of more than one form of TIMP was recognized when the technique of reverse zymography was developed [683]. Both a 28 kDa form (TIMP-1) and a 20 kDa form of TIMP were found in sheep cervix [749]. The two forms were purified from bovine endothelial cell culture and found to be 51% identical by N-terminal sequencing [653]. A partial sequence was also obtained by Goldberg *et al.* [667] who noted that TIMP-2 and proMMP-2 form a noncovalent complex similar to that mentioned above for TIMP-1 and proMMP-9. This complex may play a regulatory role in the activation of proMMP-2 by cell-surface MMP-14 [778]. A complete sequence was first provided by Edman degradation [775] and then by cDNA sequencing [774]. There have been several rediscoveries of TIMP-2 based on its growth effects: e.g. a mouse **endocrine cell survival factor** [708], an autocrine factor that stimulates SV-40 transformed human fibroblasts [726] and a

factor that stimulates osteosarcoma cells [736]. As in the case of TIMP-1, it is unlikely that these effects can all be explained on the basis of blocking proteolysis. A review of assay procedures for TIMP-1 and TIMP-2 is presented in [84] and [49].

Tissue inhibitor of metalloproteinase 3 (TIMP-3)

Transformed chick embryo fibroblast cells produced a protein in the extracellular matrix [642]. This protein had an M_r of 21 000 and bound tightly to the matrix [643]. Identification of the protein as TIMP-3 did not occur until later [773]. The first name applied to the protein was **ChiMP-3** [739]. Mouse and human sequences of TIMP-3 soon followed [625, 626]. Human fibroblasts display a mitogen-inducible gene, *mig-5*, that is the gene for TIMP-3 [800]. TIMP-3 has excited considerable interest due to the finding that Sorsby's fundus dystrophy involves mutations in this protein [793]. A brief review may be found in [49].

Tissue inhibitor of metalloproteinases 4 (TIMP-4)

Human TIMP-4 was found first by cloning [671]; its mRNA was prominent in heart but low in testes, colon, kidney and placenta and was not found in many other tissues. TIMP-4 resembles TIMP-2 in its ability to bind to proMMP-2 [638]. Rat and mouse TIMP-4 have also been cloned and sequenced [697, 808]. Expressed TIMP-4 has been shown to inhibit at least MMP-1, -2, -3, -7 and -9 [702]. In general, most TIMPs appear to be able to inhibit most active MMPs, but there are considerable differences in binding affinities. A brief review is given in [49].

MMP sequences

Sequence alignments of the MMPs

There are 96 amino acid alignments presented at the end of the book. These are linked to a more complete description of each sequence presented in Table 3. For each item in Table 3 there is a corresponding number giving the line number in the alignment table. The two tables are in the same order (with a few additional inserts in Table 3) so that matching the items is simplified. In this first section, the major features of the alignments—domains, invariant amino acids, etc.—will be pointed out. The roles of these various regions and domains will then be discussed in detail below in the section on domains and inserts.

The inclusion of a number of invertebrate and plant MMPs results in a number of inserts in the sequences of the first group of MMP-1 through MMP-20. Thus, while a typical MMP may have 500–600 amino acids, the entire width of the table is 1032 residues to accommodate all of these inserts. Features will be pointed out on the first line (COG1_BOVIN) which can generally be followed down the length of the table and the position with respect to the 1032 residues will be given. Some of these features are essentially invariant through all 96 members, others will vary slightly to left or right of the indicated position and some features are lacking altogether in the lower members of the group.

The first feature, starting from the N-terminus, is the signal peptide which begins with Met and is of variable length. In COG1_BOVIN this peptide is residues 1–18 (positions 42–59 on the Sequence Alignment table). Cleavage of the signal peptide almost always leaves XaaPro at the N-terminus of the propeptide (where Xaa is hydrophobic). Thus, most of the N-termini of propeptides are found at position 60, with some variations detectable by the fact that the Pro residue is slightly displaced. The propeptide is typically about 80 residues long. For COG1_BOVIN it includes residues 19–99 (positions 60–182). The most striking feature of the propeptide is the cysteine switch sequence PRCGVPD (positions 173–179), which is highly conserved. The C-end of the propeptide is not sharply defined, there may be two or three residues grouped together where cleavage can occur and the resultant differences of 1–3 residues can have marked effects on the enzyme activity of the resultant catalytic domain (see section on activation).

The catalytic domain begins at about position 216 (Phe residue 100 in the case of COG1_BOVIN) and continues to the position where fibronectin-like repeats occur in the gelatinases (i.e. position 376); it typically spans about 110 residues. Several of the MMPs contain a furin-sensitive sequence RXKR and activation occurs immediately to the carboxyl-side of the further arginine. This typically displaces the end of the catalytic domain about two residues to the right on the alignment. It may be noted that the catalytic zinc does not appear in the portion of the catalytic domain found ahead of the fibronectin repeats. However, this domain does contain a structural zinc atom and two or three calcium ions. The zinc ligands include His325 and His365 and the last calcium-binding residue is Glu379. Within this stretch from 325 to 379 are found 12 or more residues that form bonds with one or more of these structural cations.

Fibronectin type II domain inserts are found in MMP-2 and MMP-9, the gelatinases. These are found at positions 378–554. The start of each of the three repeats is generally taken to be the sequence GN(A/S, E/D)G found at positions 385,

Table 3 Matrix metalloproteinases (MMP)

Line	Name	AccNumber*	Description	GeneName	Organism	Citation		Additional database information
MMP-1 Interstitial collagenase								
A1	SWISSPROT: COG1_BOVIN	P28053	INTERSTITIAL COLLAGENASE PRECURSOR (EC 3.4.24.7) (MATRIX METALLOPROTEINASE-1) (MMP-1) (FIBROBLAST COLLAGENASE)	MMP1 or CLG	Bos taurus (bovine)	TAMURA M.:	[1127]	EMBL; X58256; G260; - PIR; S14654; KCBOI; HSSP; P03956; 1HFC; PROSITE; PS00024; HEMOPEXIN; 1 PROSITE; PS00142; ZINC_PROTEASE; 1 PROSITE; PS00546; CYSTEINE_SWITCH; 1 PFAM; PF00045; hemopexin; 4 PFAM; PF00099; zn-protease; 1 PFAM; PF00413; matrixin; 1
A2	SWISSPROT: COG1_HUMAN	P03956 P08156	INTERSTITIAL COLLAGENASE PRECURSOR (EC 3.4.24.7) (MATRIX METALLOPROTEINASE-1) (MMP-1) (FIBROBLAST COLLAGENASE)	MMP1 or CLG	Homo sapiens (human)	TEMPLETON N.S.: [1130] WHITHAM S.E.: [1138] GOLDBERG G.I.: [1073] LIN D.: Submitted (Dec 1996) ANGEL P.: [1043] BRINCKERHOFF C.E.: [1052] CLARK I.M.: [1231] MCKERROW J.H.: J. Biol. Chem. 262:5943-5943(1987) BORKAKOTI N.: [1349] LOVEJOY B.: [1387] LOVEJOY B.: [1386] SPURLINO J.C.: [1410] MOY F.J.: [1395]		EMBL; X05231; G38267; - EMBL; M13509; G180665; - EMBL; M16567; G180669; - EMBL; U78045; G1688258; - EMBL; M15996; G180667; - EMBL; X54925; G30126; - PIR; A37308; KCHUI; PDB; 1CGE; 31-MAR-95; PDB; 1CGF; 31-MAR-95; PDB; 1CGL; 27-FEB-95; PDB; 1HFC; 26-JAN-95; PDB; 2TCL; 08-MAR-96; PDB; 1AYK; 25-FEB-98; PDB; 2AYK; 25-FEB-98; MIM; 120353; -; PROSITE; PS00024; HEMOPEXIN; 1 PROSITE; PS00142; ZINC_PROTEASE; 1 PROSITE; PS00546; CYSTEINE_SWITCH; 1 PFAM; PF00045; hemopexin; 4 PFAM; PF00099; zn-protease; 1 PFAM; PF00413; matrixin; 1

A3	SWISSPROT: COG1_RABIT	P13943	INTERSTITIAL COLLAGENASE PRECURSOR (EC 3.4.24.7) (MATRIX METALLOPROTEINASE-1) (MMP-1)	MMP1	*Oryctolagus cuniculus* (rabbit)	FINI M.E.: FINI M.E.:	[1067] Coll. Relat. Res. 6:239-248(1986)	EMBL; M17823; G164888; - EMBL; M17820; G164888; JOINED EMBL; M17821; G164888; JOINED EMBL; M17822; G164888; JOINED EMBL; M19240; G164888; JOINED EMBL; M25663; G531212; - PIR; A27500; KCRBI; HSSP; P03956; 1HFC; PROSITE; PS00024; HEMOPEXIN; 1 PROSITE; PS00142; ZINC_PROTEASE; 1 PROSITE; PS00546; CYSTEINE_SWITCH; 1 PFAM; PF00045; hemopexin; 4 PFAM; PF00099; zn-protease; 1 PFAM; PF00413; matrixin; 1
A4	SWISSPROT: COG1_RANCA	Q11133	INTERSTITIAL COLLAGENASE PRECURSOR (EC 3.4.24.7) (MATRIX METALLOPROTEINASE-1) (MMP-1) (TC1)		*Rana catesbeiana* (bull frog)	OOFUSA K.:	[1104]	EMBL; S75623; G913071; - PROSITE; PS00024; HEMOPEXIN; FALSE_NEG PROSITE; PS00142; ZINC_PROTEASE; 1 PFAM; PF00045; hemopexin; 1 PFAM; PF00099; zn-protease; 1 PFAM; PF00413; matrixin; 2
A5	SWISSPROT: COG1_PIG	P21692	INTERSTITIAL COLLAGENASE PRECURSOR (EC 3.4.24.7) (MATRIX METALLOPROTEINASE-1) (MMP-1)	MMP1	*Sus scrofa* (pig)	RICHARDS C.D.: CLARKE N.J.: LI J.: CLARK I.M.:	[1115] [1057] [1382] Arch. Biochem. Biophys. 316:123-127(1995)	EMBL; X54724; G930269; - PIR; S15986; KCPGI; PDB; 1FBL; 29-JAN-96; PROSITE; PS00024; HEMOPEXIN; 1 PROSITE; PS00142; ZINC_PROTEASE; 1 PROSITE; PS00546; CYSTEINE_SWITCH; 1 PFAM; PF00045; hemopexin; 4 PFAM; PF00099; zn-protease; 1 PFAM; PF00413; matrixin; 1

MMP-2 Gelatinase A

A6	SWISSPROT: COG2_CHICK	Q90611	72 KD TYPE IV COLLAGENASE PRECURSOR (EC 3.4.24.24) (72 KD GELATINASE) (MATRIX METALLOPROTEINASE-2) (MMP-2) (GELATINASE A)	MMP2	*Gallus gallus* (chicken)	AIMES R.T.: CHEN J.-M.:	[1042] [1055]	EMBL; U07775; G504476; - PROSITE; PS00023; FIBRONECTIN_2; 3 PROSITE; PS00024; HEMOPEXIN; 1 PROSITE; PS00142; ZINC_PROTEASE; 1 PROSITE; PS00546; CYSTEINE_SWITCH; 1 PFAM; PF00040; fn2; 3 PFAM; PF00045; hemopexin; 4 PFAM; PF00099; zn-protease; 1 PFAM; PF00413; matrixin; 1

Continued

Table 3   Continued

Line	Name	AccNumber	Description	GeneName	Organism	Citation		Additional database information
A7	SWISSPROT: COG2_HUMAN	P08253	72 KD TYPE IV COLLAGENASE PRECURSOR (EC 3.4.24.24) (72 KD GELATINASE) (MATRIX METALLOPROTEINASE-2) (MMP-2) (GELATINASE A) (TBE-1)	MMP2 or CLG4A	*Homo sapiens* (human)	COLLIER I.E.: COLLIER I.E.: HUHTALA P.: HUHTALA P.: LIBSON A.M.: GOHLKE U.:	[1058] Genomics 9:429-434(1991) [1078] [1079] [1383] [1367]	EMBL; J03210; G180671; - EMBL; M33789; G180601; - EMBL; M55593; G180616; - EMBL; M58552; G180616; JOINED EMBL; M55582; G180616; JOINED EMBL; M55583; G180616; JOINED EMBL; M55584; G180616; JOINED EMBL; M55585; G180616; JOINED EMBL; M55586; G180616; JOINED EMBL; M55587; G180616; JOINED EMBL; M55589; G180616; JOINED EMBL; M55590; G180616; JOINED EMBL; M55591; G180616; JOINED EMBL; M55592; G180616; JOINED PIR; A28153; A28153; PDB; 1RTG; 10-JUN-96; PDB; 1GEN; 17-AUG-96; MIM; 120360; -; PROSITE; PS00023; FIBRONECTIN_2; 3 PROSITE; PS00024; HEMOPEXIN; 1 PROSITE; PS00142; ZINC_PROTEASE; 1 PROSITE; PS00546; CYSTEINE_SWITCH; 1 PFAM; PF00040; fn2; 3 PFAM; PF00045; hemopexin; 4 PFAM; PF00099; zn-protease; 1 PFAM; PF00413; matrixin; 1
A8	SWISSPROT: COG2_MOUSE	P33434	72 KD TYPE IV COLLAGENASE PRECURSOR (EC 3.4.24.24) (72 KD GELATINASE) (MATRIX METALLOPROTEINASE-2) (MMP-2) (GELATINASE A)	MMP2	*Mus musculus* (mouse)	REPONEN P.:	[1113]	EMBL; M84324; G198466; - PIR; A42496; A42496; HSSP; P08254; 2SRT; MGD; MGI:97009; MMP2; PROSITE; PS00023; FIBRONECTIN_2; 3 PROSITE; PS00024; HEMOPEXIN; 1 PROSITE; PS00142; ZINC_PROTEASE; 1 PROSITE; PS00546; CYSTEINE_SWITCH; 1 PFAM; PF00040; fn2; 3 PFAM; PF00045; hemopexin; 4 PFAM; PF00099; zn-protease; 1 PFAM; PF00413; matrixin; 1

A9	SWISSPROT: COG2_RABIT	P50757	72 KD TYPE IV COLLAGENASE PRECURSOR (EC 3.4.24.24) (72 KD GELATINASE) (MATRIX METALLOPROTEINASE-2) (MMP-2) (GELATINASE A)	MMP2	*Oryctolagus cuniculus* (rabbit)	MATSUMOTO S.:	[1093]	EMBL; D63579; G944817; - PROSITE; PS00023; FIBRONECTIN_2; 3 PROSITE; PS00024; HEMOPEXIN; 1 PROSITE; PS00142; ZINC_PROTEASE; 1 PROSITE; PS00546; CYSTEINE_SWITCH; 1 PFAM; PF00040; fn2; 3 PFAM; PF00045; hemopexin; 4 PFAM; PF00099; zn-protease; 1 PFAM; PF00413; matrixin; 1
A10	SWISSPROT: COG2_RAT	P33436	72 KD TYPE IV COLLAGENASE PRECURSOR (EC 3.4.24.24) (72 KD GELATINASE) (MATRIX METALLOPROTEINASE-2) (MMP-2) (GELATINASE A)	MMP2	*Rattus norvegicus* (rat)	MARTI H.P.:	[1089]	EMBL; X71466; G854415; - PIR; S34780; S34780; HSSP; P08254; 2SRT; PROSITE; PS00023; FIBRONECTIN_2; 3 PROSITE; PS00024; HEMOPEXIN; 1 PROSITE; PS00142; ZINC_PROTEASE; 1 PROSITE; PS00546; CYSTEINE_SWITCH; 1 PFAM; PF00040; fn2; 3 PFAM; PF00045; hemopexin; 4 PFAM; PF00099; zn-protease; 1 PFAM; PF00413; matrixin; 1

MMP-3 Stromelysin 1

A11	TREMBLNEW: G257061	G257061	PROSTROMELYSIN {N-TERMINAL}		*Canis familiaris*	BAYNE E.K.:	[1050]	
A12	TREMBL: Q98857	Q98857	STROMELYSIN-1/2-A		*Cynops pyrrhogaster* (japanese common newt)	MIYAZAKI K.:	[1096]	EMBL; D82053; G1514963; - PFAM; PF00045; hemopexin; 4 PFAM; PF00099; zn-protease; 1 PFAM; PF00413; matrixin; 1
A13	TREMBL: Q98858	Q98858	STROMELYSIN-1/2-B		*Cynops pyrrhogaster* (japanese common newt)	MIYAZAKI K.:	[1096]	EMBL; D82054; G1514965; - PFAM; PF00045; hemopexin; 4 PFAM; PF00099; zn-protease; 1 PFAM; PF00413; matrixin; 1
A14	TREMBL: Q28397	Q28397	MATRIX METALLOPROTEINASE 3		*Equus caballus* (horse)	RICHARDSON D.W.:	Submitted (Jun 1996)	EMBL; U62529; G1480746; - PROSITE; PS00024; HEMOPEXIN; 1 PFAM; PF00045; hemopexin; 4 PFAM; PF00099; zn-protease; 1 PFAM; PF00413; matrixin; 1

Continued

Table 3 Continued

Line	Name	AccNumber	Description	GeneName	Organism	Citation		Additional database information
A15	SWISSPROT: COG3_HUMAN	P08254	STROMELYSIN-1 PRECURSOR (EC 3.4.24.17) (MATRIX METALLOPROTEINASE-3) (MMP-3) (TRANSIN-1) (SL-1)	MMP3 or STMY1	*Homo sapiens* (human)	SAUS J.: WHITHAM S.E.: WILHELM S.M.: LIN D.: NAGASE H.: GOOLEY P.R.: BECKER J.W.: DHANARAJ V.: ESSER C.K.:	[1118] [1138] [1139] Submitted (Dec 1996) [1288] [1372] [1343] [1361] [1363]	EMBL; X05232; G36633; - EMBL; J03209; G188619; - EMBL; U78045; G1688259; - PIR; A28156; KCHUS1; PDB; 2SRT; 10-JUL-95; PDB; 1SLM; 17-DEC-96; PDB; 1SLN; 17-DEC-96; PDB; 1UMS; 08-MAR-96; PDB; 1UMT; 08-MAR-96; PDB; 1HFS; 18-FEB-98; MIM; 185250; -; PROSITE; PS00024; HEMOPEXIN; 1 PROSITE; PS00142; ZINC_PROTEASE; 1 PROSITE; PS00546; CYSTEINE_SWITCH; 1 PFAM; PF00045; hemopexin; 4 PFAM; PF00099; zn-protease; 1 PFAM; PF00413; matrixin; 1
A16	SWISSPROT: COG3_MOUSE	P28862	STROMELYSIN-1 PRECURSOR (EC 3.4.24.17) (MATRIX METALLOPROTEINASE-3) (MMP-3) (TRANSIN-1) (SL-1)	MMP3	*Mus musculus* (mouse)	HAMMANI K.: LI F.:	[1075] Submitted (Aug 1991)	EMBL; X66402; G296168; - EMBL; X63162; G54872; ALT_INIT PIR; JC1476; KCMSS1; PIR; S33139; S33139; HSSP; P08254; 2SRT; MGD; MGI:97010; MMP3; PROSITE; PS00024; HEMOPEXIN; 1 PROSITE; PS00142; ZINC_PROTEASE; 1 PROSITE; PS00546; CYSTEINE_SWITCH; 1 PFAM; PF00045; hemopexin; 4 PFAM; PF00099; zn-protease; 1 PFAM; PF00413; matrixin; 1
A17	SWISSPROT: COG3_RABIT	P28863	STROMELYSIN-1 PRECURSOR (EC 3.4.24.17) (MATRIX METALLOPROTEINASE-3) (MMP-3) (TRANSIN-1) (SL-1)	MMP3	*Oryctolagus cuniculus* (rabbit)	FINI M.E.: WHITHAM S.E.:	[1066] [1138]	EMBL; M25664; G165710; - PIR; A37306; KCRBS1; HSSP; P08254; 2SRT; PROSITE; PS00024; HEMOPEXIN; 1 PROSITE; PS00142; ZINC_PROTEASE; 1 PROSITE; PS00546; CYSTEINE_SWITCH; 1 PFAM; PF00045; hemopexin; 4 PFAM; PF00099; zn-protease; 1 PFAM; PF00413; matrixin; 1

| A18 | SWISSPROT:
COG3_RAT | P03957 | STROMELYSIN-1 PRECURSOR (EC 3.4.24.17) (MATRIX METALLOPROTEINASE-3) (MMP-3) (TRANSIN-1) (SL-1) (PTR1 PROTEIN) | MMP3 | *Rattus norvegicus* (rat) | MATRISIAN L.M.:
MATRISIAN L.M.:

UMENISHI F.: | [1092]
Mol. Cell. Biol. 6:1679-1686(1986)
[1133] | EMBL; X02601; G57461; -
PIR; A00997; KCRTIH;
HSSP; P08254; 2SRT;
PROSITE; PS00024; HEMOPEXIN; 1
PROSITE; PS00142; ZINC_PROTEASE; 1
PROSITE; PS00546;
CYSTEINE_SWITCH; 1
PFAM; PF00045; hemopexin; 4
PFAM; PF00099; zn-protease; 1
PFAM; PF00413; matrixin; 1 |

MMP-7 Matrilysin

| A19 | SWISSPROT:
COG7_FELCA | P55032 | MATRILYSIN PRECURSOR (EC 3.4.24.23) (PUMP-1 PROTEASE) (UTERINE METALLOPROTEINASE) (MATRIX METALLOPROTEINASE-7) (MMP-7) (MATRIN) (FRAGMENT) | MMP7 | *Felis silvestris catus* (cat) | SCALZO C.M.: | [1119] | EMBL; U04444; G436482; -
PROSITE; PS00142; ZINC_PROTEASE; 1
PROSITE; PS00546;
CYSTEINE_SWITCH; FALSENEG
PFAM; PF00099; zn-protease; 1
PFAM; PF00413; matrixin; 1 |
| A20 | SWISSPROT:
COG7_HUMAN | P09237 | MATRILYSIN PRECURSOR (EC 3.4.24.23) (PUMP-1 PROTEASE) (UTERINE METALLOPROTEINASE) (MATRIX METALLOPROTEINASE-7) (MMP-7) (MATRIN) | MMP7
or
MPSL1
or
PUMP1 | *Homo sapiens* (human) | MULLER D.:
MARTI H.P.:
GAIRE M.:
MIYAZAKI K.:
QUANTIN B.:
BROWNER M.F.: | [1098]
[1090]
[1071]
[1595]
[475]
[520] | EMBL; X07819; G35799; -
EMBL; Z11887; G35803; -
EMBL; L22524; G348021; -
EMBL; L22519; G348021; JOINED
EMBL; L22520; G348021; JOINED
EMBL; L22521; G348021; JOINED
EMBL; L22522; G348021; JOINED
EMBL; L22523; G348021; JOINED
PIR; B28816; KCHUM;
PIR; S24324; S24324;
PDB; 1MMP; 03-APR-96;
PDB; 1MMQ; 03-APR-96;
PDB; 1MMR; 03-APR-96;
MIM; 178990; -;
PROSITE; PS00142; ZINC_PROTEASE; 1
PROSITE; PS00546;
CYSTEINE_SWITCH; 1
PFAM; PF00099; zn-protease; 1
PFAM; PF00413; matrixin; 1 |

Continued

Table 3   Continued

Line	Name	AccNumber	Description	GeneName	Organism	Citation		Additional database information
A21	SWISSPROT: COG7_MOUSE	Q10738	MATRILYSIN PRECURSOR (EC 3.4.24.23) (PUMP-1 PROTEASE) (UTERINE METALLOPROTEINASE) (MATRIX METALLOPROTEINASE-7) (MMP-7) (MATRIN)	MMP7	*Mus musculus* (mouse)	WILSON C.L.:	[1143]	EMBL; L36238; G548182; - EMBL; L36243; G548182; JOINED EMBL; L36242; G548182; JOINED EMBL; L36241; G548182; JOINED EMBL; L36240; G548182; JOINED EMBL; L36239; G548182; JOINED EMBL; L36244; G537929; - MGD; MGI:103189; MMP7; PROSITE; PS00142; ZINC_PROTEASE; 1 PROSITE; PS00546; CYSTEINE_SWITCH; 1 PFAM; PF00099; zn-protease; 1 PFAM; PF00413; matrixin; 1
A22	SWISSPROT: COG7_RAT	P50280	MATRILYSIN PRECURSOR (EC 3.4.24.23) (PUMP-1 PROTEASE) (UTERINE METALLOPROTEINASE) (MATRIX METALLOPROTEINASE-7) (MMP-7) (MATRIN)	MMP7 or MMP-7	*Rattus norregicus* (rat)	ABRAMSON S.R.:	[1041]	EMBL; L24374; G402493; - PROSITE; PS00142; ZINC_PROTEASE; 1 PROSITE; PS00546; CYSTEINE_SWITCH; 1 PFAM; PF00099; zn-protease; 1 PFAM; PF00413; matrixin; 1

MMP-8 Neutrophil collagenase

Line	Name	AccNumber	Description	GeneName	Organism	Citation		Additional database information
A23	SWISSPROT: COG8_HUMAN	P22894	NEUTROPHIL COLLAGENASE PRECURSOR (EC 3.4.24.34) (MATRIX METALLOPROTEINASE-8) (MMP-8) (PMNL COLLAGENASE) (PMNL-CL)	MMP8 or CLG1	*Homo sapiens* (human)	HASTY K.A.: KNAEUPER V.: BLAESER J.: MALLYA S.K.: KNAEUPER V.: KNAEUPER V.: BLAESER J.: BODE W.: REINEMER P.: STAMS T.: BETZ M.:	[1076] [227] [166] [1275] [1081] Biol. Chem. HOPPE-SEYLER 371:733-733(1990) [167] [1348] [1406] [1411] [834]	EMBL; J05556; G180618; - PIR; A37073; KCHUN; PIR; S09680; S09680; PIR; S19576; S19576; PIR; S27225; S27225; PDB; 1JAN; 11-JUL-96; PDB; 1JAO; 11-JUL-96; PDB; 1JAP; 11-JUL-96; PDB; 1JAQ; 11-JUL-96; PDB; 1KBC; 12-AUG-97; PDB; 1MMB; 14-OCT-96; PDB; 1MNC; 07-FEB-95; MIM; 120355; -; PROSITE; PS00024; HEMOPEXIN; 1 PROSITE; PS00142; ZINC_PROTEASE; 1 PROSITE; PS00546; CYSTEINE_SWITCH; 1 PFAM; PF00045; hemopexin; 4 PFAM; PF00099; zn-protease; 1 PFAM; PF00413; matrixin; 1
A24	TREMBL: O70138	O70138	MATRIX METALLOPROTEINASE 8 (NEUTROPHIL COLLAGENASE)	MMP8	*Mus musculus* (mouse)	LAWSON N.D.:	[1082]	EMBL; U96696; G3025475; - MGD; MGI:1202395; MMP8; PROSITE; PS00546; CYSTEINE_SWITCH; 1 PROSITE; PS00024; HEMOPEXIN; 1

| A25 | TREMBL: O88733 | O88733 | NEUTROPHIL COLLAGENASE PRECURSOR | MMP-8 | *Mus musculus* (mouse) | BALBIN M.: | [1045] | EMBL; Y13342; E1311491; - PROSITE; PS00024; HEMOPEXIN; 1 PROSITE; PS00546; CYSTEINE_SWITCH; 1 |
| A26 | TREMBL: O88766 | O88766 | COLLAGENASE PRECURSOR (EC 3.4.24.34) | MMP-8 | *Rattus norvegicus* (rat) | OVERALL C.M.: | Submitted (Jun 1998) | EMBL; AJ007288; E1300704; - PROSITE; PS00024; HEMOPEXIN; 1 PROSITE; PS00546; CYSTEINE_SWITCH; 1 |

MMP-9 Gelatinase B

A27	SWISSPROT: COG9_BOVIN	P52176	92 KD TYPE IV COLLAGENASE PRECURSOR (EC 3.4.24.35) (92 KD GELATINASE) (MATRIX METALLOPROTEINASE-9) (MMP-9) (GELATINASE B)	MMP9	*Bos taurus* (bovine)	BAYLIS H.A.:	[1049]	EMBL; X78324; G467621; - PROSITE; PS00023; FIBRONECTIN_2; 3 PROSITE; PS00024; HEMOPEXIN; FALSE_NEG PROSITE; PS00142; ZINC_PROTEASE; 1 PROSITE; PS00546; CYSTEINE_SWITCH; 1 PFAM; PF00040; fn2; 3 PFAM; PF00045; hemopexin; 4 PFAM; PF00099; zn-protease; 1 PFAM; PF00413; matrixin; 1
A28	SWISSPROT: COG9_CANFA	O18733	92 KD TYPE IV COLLAGENASE PRECURSOR (EC 3.4.24.35) (92 KD GELATINASE) (MATRIX METALLOPROTEINASE-9) (MMP-9) (GELATINASE B)	MMP9	*Canis familiaris* (dog)	YOKOTA H.:	Submitted (Sep 1997)	EMBL; AB006421; D1022946; - PROSITE; PS00142; ZINC_PROTEASE; 1 PROSITE; PS00546; CYSTEINE_SWITCH; FALSENEG PROSITE; PS00023; FIBRONECTIN_2; 3 PROSITE; PS00024; HEMOPEXIN; FALSE_NEG PFAM; PF00040; fn2; 3 PFAM; PF00045; hemopexin; 4 PFAM; PF00099; zn-protease; 1 PFAM; PF00413; matrixin; 1
A29	TREMBL: Q95166	Q95166	METALLOPROTEINASE 9 (FRAGMENT)		*Canis familiaris* (dog)	PARKER H.R.:	Submitted (Aug 1996)	EMBL; U68533; G1546822; - PROSITE; PS00023; FIBRONECTIN_2; 1 PFAM; PF00040; fn2; 2
A30	TREMBL: O19130	O19130	GELATINASE B		*Canis familiaris* (dog)	FANG K.C.:	[1063]	EMBL; U89842; G2564101; - PROSITE; PS00023; FIBRONECTIN_2; 3 PROSITE; PS00024; HEMOPEXIN; 1 PROSITE; PS00546; CYSTEINE_SWITCH; 1 PFAM; PF00040; fn2; 3 PFAM; PF00045; hemopexin; 4 PFAM; PF00099; zn-protease; 1 PFAM; PF00413; matrixin; 1

Continued

Table 3   Continued

Line	Name	AccNumber	Description	GeneName	Organism	Citation		Additional database information
A31	TREMBL: Q98856	Q98856	GELATINASE-B		*Cynops pyrrhogaster* (japanese common newt)	MIYAZAKI K.:	[1096]	EMBL; D82052; G1514961; - PROSITE; PS00023; FIBRONECTIN_2; 3 PROSITE; PS00024; HEMOPEXIN; 1 PROSITE; PS00546; CYSTEINE_SWITCH; 1 PFAM; PF00040; fn2; 3 PFAM; PF00045; hemopexin; 4 PFAM; PF00099; zn-protease; 1 PFAM; PF00413; matrixin; 1
A32	SWISSPROT: COG9_HUMAN	P14780	92 KD TYPE IV COLLAGENASE PRECURSOR (EC 3.4.24.35) (92 KD GELATINASE) (MATRIX METALLOPROTEINASE-9) (MMP-9) (GELATINASE B)	MMP9 or CLG4B	*Homo sapiens* (human)	WILHELM S.M.: HUHTALA P.: SATO H.: VAN RANST M.: MASURE S.: LEE D.H.:	[1140] [1080] Oncogene 8:395-405(1993) [1207] Eur. J. Biochem. 198:391-398(1991) Exp. Mol. Med. 28:161-165(1996)	EMBL; J05070; G177205; - EMBL; D10051; G219892; - PIR; A34458; A34458; PIR; S16097; S16097; HSSP; P02462; 1BBE; MIM; 120361; -; PROSITE; PS00142; ZINC_PROTEASE; 1 PROSITE; PS00023; FIBRONECTIN_2; 3 PROSITE; PS00024; HEMOPEXIN; 1 PROSITE; PS00546; CYSTEINE_SWITCH; 1 PFAM; PF00040; fn2; 3 PFAM; PF00045; hemopexin; 4 PFAM; PF00099; zn-protease; 1 PFAM; PF00413; matrixin; 1
A33	SWISSPROT: COG9_MOUSE	P41245 Q06788	92 KD TYPE IV COLLAGENASE PRECURSOR (EC 3.4.24.35) (92 KD GELATINASE) (MATRIX METALLOPROTEINASE-9) (MMP-9) (GELATINASE B)	MMP9 or CLG4B	*Mus musculus* (mouse)	REPONEN P.: TANAKA H.: MASURE S.: GRAUBERT T.:	[1114] [1128] [1091] [1074]	EMBL; Z27231; G415981; - EMBL; D12712; G286080; - EMBL; X72794; G433433; - EMBL; X72795; G433435; - EMBL; S67830; G460864; - PIR; S38654; S38654; PIR; JC1456; JC1456; MGD; MGI:97011; MMP9; PROSITE; PS00023; FIBRONECTIN_2; 3 PROSITE; PS00024; HEMOPEXIN; 1 PROSITE; PS00142; ZINC_PROTEASE; 1 PROSITE; PS00546; CYSTEINE_SWITCH; 1 PFAM; PF00040; fn2; 3 PFAM; PF00045; hemopexin; 4 PFAM; PF00099; zn-protease; 1 PFAM; PF00413; matrixin; 1

A34	SWISSPROT: COG9_RABIT	P41246	92 KD TYPE IV COLLAGENASE PRECURSOR (EC 3.4.24.35) (92 KD GELATINASE) (MATRIX METALLOPROTEINASE-9) (MMP-9) (GELATINASE B)	MMP9	*Oryctolagus cuniculus* (rabbit)	TEZUKA K.I.: FINI M.E.:	[1131] [1064]	EMBL; D26514; G499373; - EMBL; L36050; G535715; - PROSITE; PS00023; FIBRONECTIN_2; 3 PROSITE; PS00024; HEMOPEXIN; 1 PROSITE; PS00142; ZINC_PROTEASE; 1 PROSITE; PS00546; CYSTEINE_SWITCH; 1 PFAM; PF00040; fn2; 3 PFAM; PF00045; hemopexin; 4 PFAM; PF00099; zn-protease; 1 PFAM; PF00413; matrixin; 1
A35	SWISSPROT: COG9_RAT	P50282	92 KD TYPE IV COLLAGENASE PRECURSOR (EC 3.4.24.35) (92 KD GELATINASE) (MATRIX METALLOPROTEINASE-9) (MMP-9) (GELATINASE B)	MMP9	*Rattus norvegicus* (rat)	XIA Y.: OKADA A.:	[1144] [1102]	EMBL; U36476; G1022784; - EMBL; U24441; G1173506; - PROSITE; PS00023; FIBRONECTIN_2; 2 PROSITE; PS00024; HEMOPEXIN; 1 PROSITE; PS00142; ZINC_PROTEASE; 1 PROSITE; PS00546; CYSTEINE_SWITCH; 1 PFAM; PF00040; fn2; 3 PFAM; PF00045; hemopexin; 4 PFAM; PF00099; zn-protease; 1 PFAM; PF00413; matrixin; 1

MMP-10 Stromelysin 2

A36	SWISSPROT: COGX_HUMAN	P09238	STROMELYSIN-2 PRECURSOR (EC 3.4.24.22) (MATRIX METALLOPROTEINASE-10) (MMP-10) (TRANSIN-2) (SL-2)	MMP10 or STMY2	*Homo sapiens* (human)	MULLER D.:	[1098]	EMBL; X07820; G36629; - PIR; A28816; KCHUS2; HSSP; P08254; 2SRT; MIM; 185260; -; PROSITE; PS00024; HEMOPEXIN; 1 PROSITE; PS00142; ZINC_PROTEASE; 1 PROSITE; PS00546; CYSTEINE_SWITCH; 1 PFAM; PF00045; hemopexin; 4 PFAM; PF00099; zn-protease; 1 PFAM; PF00413; matrixin; 1
A37	TREMBL: COGX_MOUSE	O55123	STROMELYSIN-2 PRECURSOR (EC 3.4.24.22) (MATRIX METALLOPROTEINASE-10) (MMP-10) (TRANSIN-2) (SL-2)	MMP10	*Mus musculus* (mouse)	MADLENER M.:	[1088]	EMBL; Y13185; E321434; - PROSITE; PS00024; HEMOPEXIN; 1 PROSITE; PS00142; ZINC_PROTEASE; 1 PROSITE; PS00546; CYSTEINE_SWITCH; 1

Continued

Table 3   Continued

Line	Name	AccNumber	Description	GeneName	Organism	Citation		Additional database information
A38	SWISSPROT: COGX_RAT	P07152	STROMELYSIN-2 PRECURSOR (EC 3.4.24.22) (MATRIX METALLOPROTEINASE-10) (MMP-10) (TRANSIN-2) (SL-2) (TRANSFORMATION- ASSOCIATED PROTEIN 34A)	MMP10	*Rattus norvegicus* (rat)	BREATHNACH R.: CHAN J.C.:	[1051] [1054]	EMBL; X05083; G57389; - EMBL; M65253; G207151; - PIR; B26403; KCRTS2; HSSP; P08254; 2SRT; PROSITE; PS00024; HEMOPEXIN; 1 PROSITE; PS00142; ZINC_PROTEASE; 1 PROSITE; PS00546; CYSTEINE_SWITCH; 1 PFAM; PF00045; hemopexin; 4 PFAM; PF00099; zn-protease; 1 PFAM; PF00413; matrixin; 1
MMP-11 Stromelysin 3								
A39	SWISSPROT: COGY_HUMAN	P24347	STROMELYSIN-3 PRECURSOR (EC 3.4.24.-) (MATRIX METALLOPROTEINASE-11) (MMP-11) (ST3) (SL-3)	MMP11 or STMY3	*Homo sapiens* (human)	BASSET P.:	[1048]	EMBL; X57766; G456257; ALT_SEQ PIR; S13423; S13423; HSSP; P03956; 1HFC; MIM; 185261; -; PROSITE; PS00024; HEMOPEXIN; 1 PROSITE; PS00142; ZINC_PROTEASE; 1 PROSITE; PS00546; CYSTEINE_SWITCH; 1 PFAM; PF00045; hemopexin; 4 PFAM; PF00099; zn-protease; 1 PFAM; PF00413; matrixin; 1
A40	SWISSPROT: COGY_MOUSE	Q02853	STROMELYSIN-3 PRECURSOR (EC 3.4.24.-) (MATRIX METALLOPROTEINASE-11) (MMP-11) (ST3) (SL-3)	MMP11	*Mus musculus* (mouse)	LEFEBVRE O.: LEFEBVRE O.:	[1083] Submitted (Jun 1994)	EMBL; Z12604; G467683; - PIR; A44399; A44399; HSSP; P03956; 1HFC; MGD; MGI:97008; MMP11; PROSITE; PS00024; HEMOPEXIN; 1 PROSITE; PS00142; ZINC_PROTEASE; 1 PROSITE; PS00546; CYSTEINE_SWITCH; FALSE.NEG PFAM; PF00045; hemopexin; 4 PFAM; PF00099; zn-protease; 1 PFAM; PF00413; matrixin; 1
A41	TREMBL: P97568	P97568	STROMELYSIN 3		*Rattus norvegicus* (rat)	OKADA A.:	[1101]	EMBL; U46034; G1762128; - PROSITE; PS00024; HEMOPEXIN; 1 PFAM; PF00045; hemopexin; 4 PFAM; PF00099; zn-protease; 1 PFAM; PF00413; matrixin; 1

| A42 | SWISSPROT: COGY_XENLA | Q11005 | STROMELYSIN-3 PRECURSOR (EC 3.4.24.-) (MATRIX METALLOPROTEINASE-11) (MMP-11) (ST3) | | *Xenopus laevis* (african clawed frog) | PATTERTON D.: | [1107] | EMBL; Z27093; G414918; - PROSITE; PS00024; HEMOPEXIN; 1 PROSITE; PS00142; ZINC_PROTEASE; 1 PROSITE; PS00546; CYSTEINE_SWITCH; 1 PFAM; PF00045; hemopexin; 4 PFAM; PF00099; zn-protease; 1 PFAM; PF00413; matrixin; 1 |

MMP-12 Macrophage elastase

A43	SWISSPROT: COGM_HUMAN	P39900	MACROPHAGE METALLOELASTASE PRECURSOR (EC 3.4.24.65) (HME) (MATRIX METALLOPROTEINASE-12) (MMP-12)	MMP12 or HME	*Homo sapiens* (human)	SHAPIRO S.D.:	[1122]	EMBL; L23808; G435970; ALT_SEQ PIR; A49499; A49499; MIM; 601046; -; PROSITE; PS00024; HEMOPEXIN; 1 PROSITE; PS00142; ZINC_PROTEASE; 1 PROSITE; PS00546; CYSTEINE_SWITCH; 1 PFAM; PF00045; hemopexin; 4 PFAM; PF00099; zn-protease; 1 PFAM; PF00413; matrixin; 1
A44	TREMBL: Q99745	Q99745	METALLOELASTASE (FRAGMENT)		*Homo sapiens* (human)	SHAPIRO S.D.: BORDEN P.: LIN D.:	[1122] Submitted (Dec 1996) Submitted (Nov 1996)	EMBL; U78045; G1688260; - PROSITE; PS00024; HEMOPEXIN; 1 PFAM; PF00045; hemopexin; 4
A45	SWISSPROT: COGM_MOUSE	P34960	MACROPHAGE METALLOELASTASE PRECURSOR (EC 3.4.24.65) (MME) (MATRIX METALLOPROTEINASE-12) (MMP-12)	MMP12 or MMEL or MME	*Mus musculus* (mouse)	SHAPIRO S.D.:	[1121]	EMBL; M82831; G199128; - PIR; A42401; A42401; HSSP; P08254; 2SRT; MGD; MGI:97005; MMEL; PROSITE; PS00024; HEMOPEXIN; 1 PROSITE; PS00142; ZINC_PROTEASE; 1 PROSITE; PS00546; CYSTEINE_SWITCH; 1 PFAM; PF00045; hemopexin; 4 PFAM; PF00099; zn-protease; 1 PFAM; PF00413; matrixin; 1
A46	SWISSPROT: COGM_RABIT	P79227	MACROPHAGE METALLOELASTASE PRECURSOR (EC 3.4.24.65) (MME) (MATRIX METALLOPROTEINASE-12) (MMP-12)	MMP12	*Oryctolagus cuniculus* (rabbit)	HOU P.:	Submitted (Feb 1997)	EMBL; U88652; G1839256; - PROSITE; PS00024; HEMOPEXIN; 1 PROSITE; PS00142; ZINC_PROTEASE; 1 PROSITE; PS00546; CYSTEINE_SWITCH; 1 PFAM; PF00045; hemopexin; 4 PFAM; PF00099; zn-protease; 1 PFAM; PF00413; matrixin; 1

Continued

Table 3   Continued

Line	Name	AccNumber	Description	GeneName	Organism	Citation		Additional database information
A47	TREMBL: Q63341	Q63341	MACROPHAGE METALLOELASTASE PRECURSOR		*Rattus norvegicus* (rat)	COSSINS J.:	Submitted (Jun 1996)	EMBL; X98517; E249424; - PROSITE; PS00024; HEMOPEXIN; 1 PFAM; PF00045; hemopexin; 4 PFAM; PF00099; zn-protease; 1 PFAM; PF00413; matrixin; 1
MMP-13 Collagenase 3								
A48	TREMBL: O77656	O77656	MATRIX METALLOPROTEINASE 13	MMP13	*Bos taurus* (bovine)	WU C.W.:	Submitted (Jun 1998)	EMBL; AF072685; G3264824; - PROSITE; PS00546; CYSTEINE_SWITCH; 1
A49	TREMBL: O77632	O77632	MATRIX METALLOPROTEINASE 13 PRECURSOR (FRAGMENT)	MMP-13	*Bos taurus* (bovine)	FLANNERY C.R.:	Submitted (Jun 1998)	EMBL; AF069644; G3219743; -
A50	TREMBL: Q98859	Q98859	COLLAGENASE 3		*Cynops pyrrhogaster* (japanese common newt)	MIYAZAKI K.:	[1096]	EMBL; D82055; G1514967; - PROSITE; PS00024; HEMOPEXIN; 1 PROSITE; PS00546; CYSTEINE_SWITCH; 1 PFAM; PF00045; hemopexin; 4 PFAM; PF00099; zn-protease; 1 PFAM; PF00413; matrixin; 1
A51	TREMBL: O18927	O18927	MATRIX METALLOPROTEINASE 13	MMP13	*Equus caballus* (horse)	RICHARDSON D.W.:	Submitted (Nov 1997)	EMBL; AF034087; G2641648; - PROSITE; PS00546; CYSTEINE_SWITCH; 1 PFAM; PF00045; hemopexin; 4 PFAM; PF00099; zn-protease; 1 PFAM; PF00413; matrixin; 1
A52	SWISSPROT: COGZ_HUMAN	P45452	COLLAGENASE 3 PRECURSOR (EC 3.4.24.-) (MATRIX METALLOPROTEINASE-13) (MMP-13)	MMP13	*Homo sapiens* (human)	FREIJE J.M.P.: WILLMROTH F.: GOMIS-RUETH F.X.:	[1069] [1142] [1368]	EMBL; X75308; G516386; - EMBL; X81334; E118011; - PDB; 1PEX; 23-DEC-96; MIM; 600108; -; PROSITE; PS00024; HEMOPEXIN; 1 PROSITE; PS00142; ZINC_PROTEASE; 1 PROSITE; PS00546; CYSTEINE_SWITCH; 1 PFAM; PF00045; hemopexin; 4 PFAM; PF00099; zn-protease; 1 PFAM; PF00413; matrixin; 1

	Database	Accession	Description	MMP	Species	Author	Reference	Cross-references
A53	SWISSPROT: COGZ_MOUSE	P33435	COLLAGENASE 3 PRECURSOR (EC 3.4.24.-) (MATRIX METALLOPROTEINASE-13) (MMP-13)	MMP13	*Mus musculus* (mouse)	HENRIET P.:	[1077]	EMBL; X66473; G53604; - PIR; S29243; S29243; HSSP; P22894; 1MNC; PROSITE; PS00024; HEMOPEXIN; 1 PROSITE; PS00142; ZINC_PROTEASE; 1 PROSITE; PS00546; CYSTEINE_SWITCH; 1 PFAM; PF00045; hemopexin; 4 PFAM; PF00099; zn-protease; 1 PFAM; PF00413; matrixin; 1
A54	TREMBL: O73923	O73923	COLLAGENASE 3 (FRAGMENT)		*Oncorhynchus mykiss* (rainbow trout) (*Salmo gairdneri*)	HENRY M.A.:	Submitted (May 1998)	EMBL; AJ006411; E1294593; -
A55	TREMBL: O62806	O62806	COLLAGENASE-3 PRECURSOR		*Oryctolagus cuniculus* (rabbit)	VINCENTI M.P.:	[1134]	EMBL; AF059201; G3089539; - PROSITE; PS00024; HEMOPEXIN; 1 PROSITE; PS00546; CYSTEINE_SWITCH; 1
A56	SWISSPROT: COGZ_RAT	P23097	COLLAGENASE 3 PRECURSOR (EC 3.4.24.-) (MATRIX METALLOPROTEINASE-13) (MMP-13) (UMRCASE) (FRAGMENT)	MMP13	*Rattus norvegicus* (rat)	QUINN C.O.:	[1111]	EMBL; M60616; G203499; - PIR; A23685; A23685; HSSP; P22894; 1MNC; PROSITE; PS00024; HEMOPEXIN; 1 PROSITE; PS00142; ZINC_PROTEASE; 1 PROSITE; PS00546; CYSTEINE_SWITCH; 1 PFAM; PF00045; hemopexin; 4 PFAM; PF00099; zn-protease; 1 PFAM; PF00413; matrixin; 1
A57	TREMBL: O77631	O77631	MATRIX METALLOPROTEINASE 13 PRECURSOR (FRAGMENT)	MMP-13	*Sus scrofa* (pig)	FLANNERY C.R.:	Submitted (Jun 1998)	EMBL; AF069643; G3219741; -
A58	SWISSPROT: COGZ_XENLA	Q10835	COLLAGENASE 3 PRECURSOR (EC 3.4.24.-) (MATRIX METALLOPROTEINASE-13) (MMP-13) (FRAGMENT)	MMP13	*Xenopus laevis* (african clawed frog)	FINI M.E.:	Submitted (Dec 1995)	EMBL; L49412; G1129121; - PROSITE; PS00024; HEMOPEXIN; 1 PROSITE; PS00142; ZINC_PROTEASE; 1 PROSITE; PS00546; CYSTEINE_SWITCH; 1 PFAM; PF00045; hemopexin; 4 PFAM; PF00099; zn-protease; 1 PFAM; PF00413; matrixin; 1
A59	TREMBL: CGZA_XENLA	Q10833	COLLAGENASE 3A PRECURSOR (EC 3.4.24.-) (MATRIX METALLOPROTEINASE-13A) (MMP-13A)	MMP-13A	*Xenopus laevis* (african clawed frog)	BROWN D.D.:	[1053]	EMBL; U41824; G1223974; - PFAM; PF00045; hemopexin; 4 PFAM; PF00099; zn-protease; 1 PFAM; PF00413; matrixin; 1

Continued

Table 3   Continued

MMP-14 Membrane-type matrix metalloproteinase 1 (MT1-MMP)

Line	Name	AccNumber	Description	GeneName	Organism	Citation		Additional database information
A60	SWISSPROT: COGT_HUMAN	P50281 Q92678	MATRIX METALLOPROTEINASE-14 PRECURSOR (EC 3.4.24.-) (MMP-14) (MEMBRANE-TYPE MATRIX METALLOPROTEINASE 1) (MT-MMP 1) (MTMMP1)	MMP14 or MMP-X1	*Homo sapiens* (human)	SATO H.: TAKINO T.: OKADA A.: WILL H.: LUO G.X.: LOHI J.L.:	[1116] [1126] [1100] [1141] Submitted (Nov 1995) [1087]	EMBL; D26512; G793763; - EMBL; X83535; G804994; - EMBL; Z48481; G963054; - EMBL; U41078; G1127837; - EMBL; X90925; E196537; - MIM; 600754; -; PROSITE; PS00024; HEMOPEXIN; 1 PROSITE; PS00142; ZINC_PROTEASE; 1 PROSITE; PS00546; CYSTEINE_SWITCH; 1 PFAM; PF00045; hemopexin; 4 PFAM; PF00099; zn-protease; 1 PFAM; PF00413; matrixin; 2
A61	SWISSPROT: COGT_MOUSE	P53690	MATRIX METALLOPROTEINASE-14 PRECURSOR (EC 3.4.24.-) (MMP-14) (MEMBRANE-TYPE MATRIX METALLOPROTEINASE 1) (MT-MMP 1) (MTMMP1)	MMP14 or MT-MMP	*Mus musculus* (mouse)	OKADA A.:	[1100]	EMBL; X83536; G805000; - MGD; MGI:101900; MMP14; PROSITE; PS00024; HEMOPEXIN; 1 PROSITE; PS00142; ZINC_PROTEASE; 1 PROSITE; PS00546; CYSTEINE_SWITCH; 1 PFAM; PF00045; hemopexin; 4 PFAM; PF00099; zn-protease; 1 PFAM; PF00413; matrixin; 2
A62	TREMBL: O08645	O08645	MEMBRANE-TYPE MATRIX METALLOPROTEINASE 1		*Mus musculus* (mouse)	OTA K.:	[1105]	EMBL; U54984; G1935025; - PROSITE; PS00024; HEMOPEXIN; 1 PROSITE; PS00546; CYSTEINE_SWITCH; 1 PFAM; PF00045; hemopexin; 4 PFAM; PF00099; zn-protease; 1 PFAM; PF00413; matrixin; 2
A63	SWISSPROT: COGT_RABIT	Q95220 P79225	MATRIX METALLOPROTEINASE-14 PRECURSOR (EC 3.4.24.-) (MMP-14) (MEMBRANE-TYPE MATRIX METALLOPROTEINASE 1) (MT-MMP 1) (MTMMP1)	MMP14	*Oryctolagus cuniculus* (rabbit)	WANG H.: SATO T.:	Submitted (Feb 1997) [1117]	EMBL; U83918; G1805295; - EMBL; U73940; G1658112; - PROSITE; PS00024; HEMOPEXIN; 1 PROSITE; PS00142; ZINC_PROTEASE; 1 PROSITE; PS00546; CYSTEINE_SWITCH; 1 PFAM; PF00045; hemopexin; 4 PFAM; PF00099; zn-protease; 1 PFAM; PF00413; matrixin; 2

A64	SWISSPROT: COGT_RAT	Q10739	MATRIX METALLOPROTEINASE-14 PRECURSOR (EC 3.4.24.-) (MMP-14) (MEMBRANE-TYPE MATRIX METALLOPROTEINASE 1) (MT-MMP 1) (MTMMP1)	MMP14 or MT-MMP	*Rattus norvegicus* (rat)	OKADA A.: COSSINS J.:	[1100] Submitted (Sep 1995)	EMBL; X83537; G805013; - EMBL; X91785; G1001927; - PROSITE; PS00024; HEMOPEXIN; 1 PROSITE; PS00142; ZINC_PROTEASE; 1 PROSITE; PS00546; CYSTEINE_SWITCH; 1 PFAM; PF00045; hemopexin; 4 PFAM; PF00099; zn-protease; 1 PFAM; PF00413; matrixin; 2

MMP-15 Membrane-type matrix metalloproteinase 2 (MT2-MMP)

A65	SWISSPROT: COGU_HUMAN	P51511	MATRIX METALLOPROTEINASE-15 PRECURSOR (EC 3.4.24.-) (MMP-15) (MEMBRANE-TYPE MATRIX METALLOPROTEINASE 2) (MT-MMP 2) (MTMMP2)	MMP15	*Homo sapiens* (human)	WILL H.:	[1141]	EMBL; Z48482; G963056; - MIM; 602261; -; PROSITE; PS00024; HEMOPEXIN; 1 PROSITE; PS00142; ZINC_PROTEASE; 1 PROSITE; PS00546; CYSTEINE_SWITCH; 1 PFAM; PF00045; hemopexin; 4 PFAM; PF00099; zn-protease; 1 PFAM; PF00413; matrixin; 2
A66	TREMBL: Q14111	Q14111	MATRIX METALLOPROTEINASE, MT2MMP	MT2-MMP	*Homo sapiens* (human)	SATO H.H.:	Submitted (Jul 1996)	EMBL; D86331; G1418215; - PROSITE; PS00024; HEMOPEXIN; 1 PROSITE; PS00546; CYSTEINE_SWITCH; 1 PFAM; PF00045; hemopexin; 4 PFAM; PF00099; zn-protease; 1 PFAM; PF00413; matrixin; 1
A67	TREMBLNEW: D1023087	D1023087	SMCP-2 (FRAGMENT)		*Homo sapiens* (human)	SHOFUDA K.:	[1123]	EMBL; D85510; D1023087; -
A68	TREMBL: O54732	O54732	MATRIX METALLOPROTEINASE 15 (MEMBRANE TYPE-2 MATRIX METALLOPROTEINASE)	MMP15 or MT2-MMP	*Mus musculus* (mouse)	TANAKA M.:	[1129]	EMBL; D86332; D1024548; - MGD; MGI:109320; MMP15; PROSITE; PS00546; CYSTEINE_SWITCH; 1 PROSITE; PS00024; HEMOPEXIN; 1

MMP-16 Membrane-type matrix metalloproteinase 3 (MT3-MMP)

A69	TREMBL: Q98947	Q98947	MEMBRANE TYPE-MATRIX METALLOPROTEINASE	MT3-MMP	*Gallus gallus* (chicken)	YANG M.:	[1145]	EMBL; U66463; G1519365; - PROSITE; PS00024; HEMOPEXIN; 1 PROSITE; PS00546; CYSTEINE_SWITCH; 1 PFAM; PF00045; hemopexin; 4 PFAM; PF00099; zn-protease; 1 PFAM; PF00413; matrixin; 1

Continued

J. Frederick Woessner and Hideaki Nagase

Table 3   Continued

Line	Name	AccNumber	Description	GeneName	Organism	Citation		Additional database information
A70	SWISSPROT: COGV_HUMAN	P51512	MATRIX METALLOPROTEINASE-16 PRECURSOR (EC 3.4.24.-) (MMP-16) (MEMBRANE-TYPE MATRIX METALLOPROTEINASE 3) (MT-MMP 3) (MTMMP3) (MMP-X2)	MMP16 or MMPX2	*Homo sapiens* (human)	TAKINO T.: MATSUMOTO S.: SHOFUDA K.:	[1125] Submitted (Mar 1996) [1123]	EMBL; D50477; G1197174; - EMBL; D83646; D1012692; - EMBL; D85511; D1023088; - MIM; 602262; -; PROSITE; PS00024; HEMOPEXIN; 1 PROSITE; PS00142; ZINC_PROTEASE; 1 PROSITE; PS00546; CYSTEINE_SWITCH; 1 PFAM; PF00045; hemopexin; 4 PFAM; PF00099; zn-protease; 1 PFAM; PF00413; matrixin; 1
A71	TREMBL: Q14824	Q14824	METALLOPROTEINASE PRECURSOR (EC 3.4.24.-)		*Homo sapiens* (human)	MATSUMOTO S.:	[1094]	EMBL; D83647; D1012693; - PFAM; PF00045; hemopexin; 1 PFAM; PF00099; zn-protease; 1 PFAM; PF00413; matrixin; 1
A72	SWISSPROT: COGV_RAT	O35548	MATRIX METALLOPROTEINASE-16 PRECURSOR (EC 3.4.24.-) (MMP-16) (MEMBRANE-TYPE MATRIX METALLOPROTEINASE 3) (MT-MMP 3) (MTMMP3)	MMP16	*Rattus norregicus* (rat)	SHOFUDA K.:	[1123]	EMBL; D85509; D1023086; - PROSITE; PS00024; HEMOPEXIN; 1 PROSITE; PS00142; ZINC_PROTEASE; 1 PROSITE; PS00546; CYSTEINE_SWITCH; 1 PFAM; PF00045; hemopexin; 4 PFAM; PF00099; zn-protease; 1 PFAM; PF00413; matrixin; 1
A73	TREMBL: O35541	O35541	MT3-MMP-DEL		*Rattus norregicus* (rat)	SHOFUDA K.:	[1123]	EMBL; D63886; D1023085; - PROSITE; PS00024; HEMOPEXIN; 1 PROSITE; PS00546; CYSTEINE_SWITCH; 1 PFAM; PF00045; hemopexin; 4 PFAM; PF00099; zn-protease; 1 PFAM; PF00413; matrixin; 1

MMP-17 Membrane-type matrix metalloproteinase 4 (MT4-MMP)

Line	Name	AccNumber	Description	GeneName	Organism	Citation		Additional database information
A74	TREMBL: Q14850	Q14850	MT4-MMP	MMP-17	*Homo sapiens* (human)	PUENTE X.S.: LOPEZ-OTIN C.: PUENTE X.S.:	[1110] Submitted (Jul 1995) Submitted (Sep 1998)	EMBL; X89576; E1321562; -

MMP-18 Collagenase 4

Line	Name	AccNumber	Description	GeneName	Organism	Citation		Additional database information
A75	TREMBL: O13065	O13065	COLLAGENASE 4 PRECURSOR		*Xenopus laevis* (african clawed frog)	STOLOW M.A.:	[1124]	EMBL; L76275; G2055321; - PFAM; PF00045; hemopexin; 4 PFAM; PF00099; zn-protease; 1 PFAM; PF00413; matrixin; 1

MMP-19 (No trivial name)

A76	TREMBL: Q99542	Q99542	MMP-19 (MATRIX METALLOPROTEINASE)	MMP-19	*Homo sapiens* (human)	PENDAS A.M.: MAUCH S.: SEDLACEK R.: KOLB C.:	[1108] Submitted (Oct 1995) [1120] [690]	EMBL; X92521; E208554; - EMBL; U38321; G2228244; - EMBL; U37791; G2253587; - PFAM; PF00045; hemopexin; 4 PFAM; PF00099; zn-protease; 1 PFAM; PF00413; matrixin; 2
A77	TREMBL: Q99580	Q99580	MATRIX METALLOPROTEINASE PRECURSOR	'MMP-18'	*Homo sapiens* (human)	COSSINS J.:	[1059]	EMBL; Y08622; E274423; - PFAM; PF00045; hemopexin; 4 PFAM; PF00099; zn-protease; 1 PFAM; PF00413; matrixin; 2
A78	TREMBL: O15278	O15278	MATRIX METALLOPROTEINASE RASI-1		*Homo sapiens* (human)	MAUCH S.:	Submitted (Oct 1995)	EMBL; U38322; G2228246; - PFAM; PF00045; hemopexin; 4

MMP-20 Enamelysin

A79	TREMBL: O18767	O18767	ENAMEL METALLOPROTEINASE PRECURSOR		*Bos taurus* (bovine)	DENBESTEN P.K.:	[1060]	EMBL; AF009922; G2326212; - PROSITE; PS00546; CYSTEINE_SWITCH; 1 PFAM; PF00045; hemopexin; 4 PFAM; PF00099; zn-protease; 1 PFAM; PF00413; matrixin; 1
A80	TREMBL: O60882	O60882	ENAMELYSIN	MMP-20	*Homo sapiens* (human)	LLANO E.:	[1086]	EMBL; Y12779; E314926; - PROSITE; PS00546; CYSTEINE_SWITCH; 1
A81	TREMBL: P79287	P79287	MATRIX METALLOPROTEINASE		*Sus scrofa* (pig)	BARTLETT J.D.:	[1047]	EMBL; U54825; G1800213; - PROSITE; PS00546; CYSTEINE_SWITCH; 1 PFAM; PF00045; hemopexin; 4 PFAM; PF00099; zn-protease; 1 PFAM; PF00413; matrixin; 1

Various

A82	TREMBL: O04529	O04529	SIMILAR TO GLYCINE METALLOENDOPROTEINASE	F20P5.11	*Arabidopsis thaliana* (mouse-ear cress)	OSBORNE B.L.:	Submitted (Jun 1997)	EMBL; AC002062; G2194124; - PROSITE; PS00546; CYSTEINE_SWITCH; 1 PFAM; PF00099; zn-protease; 1 PFAM; PF00413; matrixin; 2
A83	TREMBL: O23507	O23507	PROTEINASE HOMOLOG		*Arabidopsis thaliana* (mouse-ear cress)	EU ARABIDOPSIS SEQUENCING PROJECT: BEVAN M.:	Submitted (Jun 1997) Nature 391:485-488(1998)	EMBL; Z97341; E327511; - PFAM; PF00099; zn-protease; 1 PFAM; PF00413; matrixin; 2

Continued

Table 3   Continued

Line	Name	AccNumber	Description	GeneName	Organism	Citation		Additional database information
A84	TREMBL: O48680	O48680	F3I6.6 PROTEIN	F3I' 6	*Arabidopsis thaliana* (mouse-ear cress)	FEDERSPIEL N.A.:	Submitted (Feb 1998)	EMBL; AC002396; G2829864; - PROSITE; PS00546; CYSTEINE_SWITCH; 1
A85	TREMBL: O65340	O65340	METALLOPROTEINASE		*Arabidopsis thaliana* (mouse-ear cress)	PAK J.H.: LIU C.Y.:	[1106] Submitted (May 1998)	EMBL; AF062640; G3128477; -
A86	TREMBL: O16901	O16901	C31B8.8 PROTEIN	C31B8.8	*Caenorhabditis elegans*	WILSON R.: DU Z.: WATERSTON R.:	[1215] Submitted (Sep 1997) Submitted (Sep 1997)	EMBL; AF022971; G2384812; - PFAM; PF00045; hemopexin; 2 PFAM; PF00413; matrixin; 1
A87	TREMBL: O61264	O61264	MATRIX METALLOPROTEINASE		*Caenorhabditis elegans*	WADA K.:	[1136]	EMBL; AB007815; D1029291; -
A88	TREMBL: O61266	O61266	MATRIX METALLOPROTEINASE		*Caenorhabditis elegans*	WADA K.:	[1136]	EMBL; AB007817; D1029293; -
A89	TREMBL: O61265	O61265	MATRIX METALLOPROTEINASE (FRAGMENT)		*Caenorhabditis elegans*	WADA K.:	[1136]	EMBL; AB007816; D1029292; -
A90	TREMBL: O44836	O44836	H19M22.3 PROTEIN	H19M22.3	*Caenorhabditis elegans*	WILSON R.: WILSON R.: WATERSTON R.:	[1215] Submitted (Jan 1998) Submitted (Dec 1997)	EMBL; AF040648; G2746839; -
A91	TREMBL: O55761	O55761	PUTATIVE METALLOPEPTIDASE		*Chilo iridescent virus* (civ) (insect iridescent virus type 6)	BAHR U.:	Virus Genes 15:235-245(1997)	EMBL; AF003534; G2738445; -
A92	SWISSPROT: MEP1_SOYBN	P29136	METALLOENDOPROTEINASE 1 PRECURSOR (EC 3.4.24.-) (SMEP1)		*Glycine max* (soybean)	PAK J.H.: MCGEEHAN G.:	[1106] [1095]	EMBL; U63725; G1679656; - PIR; A41820; A41820; HSSP; P03956; 1HFC; PROSITE; PS00142; ZINC_PROTEASE; 1 PROSITE; PS00546; CYSTEINE_SWITCH; 1 PFAM; PF00099; zn-protease; 1 PFAM; PF00413; matrixin; 2
A93	SWISSPROT: HE_HEMPU	P91953	HATCHING ENZYME PRECURSOR (EC 3.4.24.12) (HE) (HEZ) (ENVELYSIN) (SEA-URCHIN-HATCHING PROTEINASE)		*Hemicentrotus pulcherrimus* (sea urchin)	NOMURA K.: NOMURA K.: NOMURA K.:	[1099] Biochemistry 30:6115-6123(1991) FEBS Lett. 321:84-88(1993)	EMBL; AB000719; G1816431; - PROSITE; PS00024; HEMOPEXIN; 1 PROSITE; PS00142; ZINC_PROTEASE; 1 PROSITE; PS00546; CYSTEINE_SWITCH; 1 PFAM; PF00045; hemopexin; 4 PFAM; PF00099; zn-protease; 1 PFAM; PF00413; matrixin; 1

*Line	*Name	*Acc Number				Authors	Ref	Database cross-references
A94	SWISSPROT: HE_PARLI	P22757	HATCHING ENZYME PRECURSOR (EC 3.4.24.12) (HE) (HEZ) (ENVELYSIN) (SEA-URCHIN-HATCHING PROTEINASE)		*Paracentrotus lividus* (common sea urchin)	LEPAGE T.: GHIGLIONE C.:	[1084] [1072]	EMBL; X53598; G9996; - EMBL; X65722; G416552; - PIR; S12805; S12805; HSSP; P22894; 1MNC; PROSITE; PS00024; HEMOPEXIN; FALSE_NEG PROSITE; PS00142; ZINC_PROTEASE; 1 PROSITE; PS00546; CYSTEINE_SWITCH; 1 PFAM; PF00045; hemopexin; 4 PFAM; PF00099; zn-protease; 1 PFAM; PF00413; matrixin; 1
A95	TREMBL: O93470	O93470	MATRIX METALLOPROTEINASE	XMMP	*Xenopus laevis* (african clawed frog)	YANG M.:	[1147]	EMBL; U82541; G3211705; -
A96	TREMBL: P89294	P89294	MATRIX METALLOPROTEINASE HOMOLOG (FRAGMENT)		*Xestia c-*NIGRUM GRANULOSIS VIRUS	GOTO C.:	Submitted (Sep 1996)	EMBL; U70919; G1835349; - PFAM; PF00099; zn-protease; 1

The table is arranged in order of MMP numbers. Live updated information is available with hot links at http://www3.ebi.ac.uk/Services/protein_profiles/mmp/table.html together with full bibliographic references. Fragments of enzyme, particularly those sequenced by Edman degradation are not included.

*Line, Name and Acc(ession) Number match the first 3 columns of the sequence alignment table, pages 210–220.

443 and 502. Following these inserts is the motif of the catalytic domain that binds the catalytic zinc atom. This highly conserved region starts at position 557 (Leu) and ends at Gly613. The HEXGHXXGXXHS sequence extends from 563–574 and includes all three His ligands for zinc. The Met turn is at 581.

The catalytic region is followed by a highly variable hinge region of 1–80 residues. It follows Gly613 and continues to the residue before Cys697. The hemopexin domain starts at this Cys697 and forms a complete cycle by joining to Cys922 in a disulfide bond. The hemopexin domain has the form of a four-bladed propeller. Each blade starts near the periphery, but the major repeating structure is about six residues further along and takes the form of DAA, DAV, DAI, etc. in which the Asp residue makes a link to the central calcium ion through its carbonyl oxygen (not its carboxyl group). The repeating structure is found at 712, 762, 817 and 876, but each blade begins about six residues before this.

A number of the MMPs, particularly the MT-type, contain additional residues beyond the hemopexin domain and continuing to the C-terminus. For MMP-14, there is a linker of 30 residues from positions 923–988, a 24-residue transmembrane domain (989–1012) and a cytoplasmic domain of 20 residues (1013–1032). MMP-15 is similar but has a much longer linker of 66 residues.

Miscellaneous sequence comments

Mass spectroscopy of human MMP-1 shows several differences from the published sequences [1056]. An early fragment of rabbit MMP-1 was reported [1065] and a long promotor sequence of rabbit MMP-1 was provided by Vincenti *et al.* [1135]. No sequence of MMP-1 has ever been obtained for the rat, although the designation MMP-1 was used in the earlier literature before the recognition of MMP-13. Sequences for human [1148] and horse [1046] MMP-3 are not included in Table 3. A sequence of human MMP-8 [1062] differs in two bases from that of Hasty [1076]. Human fibroblast and PMN MMP-9 were compared, the fibroblast had Arg279 where the PMN had Gln [1109]. The PMN enzyme was also sequenced by Devarajan *et al.* [1061].

A fragment of human MMP-10 DNA [1070] and a sequence of human MMP-13 are not in Table 3 [1137]. There is also a rat MMP-13 fragment [1132]. A genomic sequence for mouse MMP-14 [1044] is known. An entry in TREMBL O43923 is listed as enamelysin (MMP-20), however this enzyme is not enamelysin and is distinct from other MMPs. It is closest to MMP-17 (X89576), matching in 68 of 183 residues.

There are currently four *Arabidopsis* sequences that have both the cysteine switch motif and the zinc-binding motif, but have not been shown to have proteolytic activity and must be considered to be putative proteinases only at this point.

Xenopus produces an MMP referred to as XMMP, which appears on line 95 of the sequence alignment. It is now suggested that this enzyme be numbered MMP-21. It contains a 37-residue insertion similar to vitronectin near the C-end of the propeptide, and an extra cysteine inserted in the catalytic domain in the region where the structural zinc is bound, a feature also seen in MMP-19. A further enzyme from chicken CMMP [1146] also has an extra cysteine in the catalytic domain. It is proposed to name this MMP-22. (Its sequence is not given in the alignment.)

The bacterial enzyme, fragilysin, is not very closely related to the other MMPs and does not possess the cysteine switch [1068, 1097]. Rawlings [1112] proposes that it be placed in a separate subclass of the M10 family, subclass C. Two viruses possess sequences related to the MMPs: chilo iridescent virus O55761 (TREMBL) 264 residues AF003534 (EMBL) and xestia C-nigrum granulosis virus P89294 (TREMBL) 75 residues U70719 (EMBL). The chilo virus is most closely related to *Xenopus* MMP-18 and MMP-11, matching in 58 of 130 positions.

The domains and inserts of MMPs

The three-dimensional structures of the catalytic domain with fibronectin inserts and the hemopexin domain (Fig. 1) are discussed in greater detail in the section on three-dimensional structures (below). In the present section, emphasis is on the function of the domains. There remain also a number of regions of the MMP molecule that have not been amenable to X-ray analysis and these are discussed here.

The signal peptide

All MMPs are synthesized in the cell and secreted to the extracellular space. Thus, their mRNA specifies a hydrophobic signal sequence typically comprising 18–30 residues. This region is cut off during transit from the cell and is not observed in mature enzymes recovered from outside the cell.

The propeptide domain

The propeptide domain extends from the N-terminus created after removal of the signal peptide to the beginning of the catalytic domain, a length typically spanning 80 amino acids. This is considerably longer than most zymogen peptides and the reasons for this are not completely clear. The understanding that the cysteine residue 73 (MMP-1 numbering from propeptide terminus) in a conserved sequence PRCGVPD is positioned directly opposite the zinc atom at the active centre and is coordinated to it through the –SH group was advanced by Springman *et al.* in 1990 [596]. This was termed the '**cysteine switch**' because displacement of the cysteine residue by a wide variety of means (oxidation, proteolytic cleavage, mercurial and gold compounds, etc.) would turn on the enzyme activity. This proposal for human MMP-1 was then extended to MMP-2, -3, -7, -9 and -10 [607] on the basis of their possessing the conserved zinc binding site and the cysteine residue and that all were activated by similar mechanisms.

Another term proposed for this activation is the '**velcro**' mechanism [606].

This concept was then substantiated by mutation of residues in the PRCGVPD sequence [576, 589, 615]; various mutations led to more readily activatable forms of the enzymes (MMP-1 and -3) and MMP-1 expressed with mutated Cys73 became active within the *E. coli* cells and led to their autolysis. Trapping by α_2-macroglobulin [615] proved that there was still some residual latency, indicating that other portions of the propeptide must be involved in protecting the zinc site. Sixteen mutations of rat MMP-3 [576] have indicated that Arg, Cys and Val in PRCGVPD are all important in the binding, Arg forms a salt bridge with Asp of the switch and does not bind to Glu200 in the HELGH sequence. The contact of Cys–SH with zinc was established directly by EXAFS analysis of MMP-3 [543] and later by X-ray crystallography [515]. This X-ray structure (the only one to date for any propeptide) further reveals three α-helices in MMP-3 propeptide, while residues 1–15 and residues 31–39 are not visible due to freedom of motion. The segment with cysteine lies in the active centre with an N to C orientation that is the opposite of that found for synthetic peptide inhibitors. If residues 1–48 are removed from MMP-8 by trypsin action, the enzyme remains latent but activation by stromelysin is very greatly accelerated [554]. A mutation removing the first 20 residues of MMP-3 leads to spontaneous activation, indicating that anchoring of the propeptide at both N- and C-termini is important in maintaining latency. Similarly, a mutant lacking residues 1–34 and the hemopexin domain can still react with TIMP-1 and undergo bimolecular reaction, indicating that peptide association with the enzyme is relatively weak [603]. On the other hand, the propeptide is not important for refolding denatured enzyme, the mature form folds perfectly well without the propeptide. A similar deduction could be made based on the ease of refolding of mature enzyme in zymography after removal of SDS.

Another approach to studying the role of the propeptide has been the preparation of peptides containing the critical zinc-binding cysteine residue. Stetler-Stevenson *et al.* [599] prepared TMRKPRCGNPDVAN and showed that it was a good inhibitor of human MMP-2; replacement of Cys with Ser eliminated activity. They also noted that Cys in the native propeptide could not react with Ellman's reagent, further evidence for anchoring of the propeptide at several points. Smaller peptides were developed for inhibition of MMP-3; Ac-RCGVPD and Ac-RCGVP had an IC_{50} of about 10 μM [540]. Free cysteine is also an inhibitor, but must be present at millimolar levels. Further work with these peptides led to the conclusion that Ac-RCGVPD-NH$_2$ was a good inhibitor and that Arg, Cys and Val were critical for inhibition of MMP-1 and −3 [533], similar to the results noted above with mutants.

Envelysin of the sea urchin contains a long insert of 70 residues in the propeptide, rich in Asp and Glu residues; these are thought to participate in binding Ca^{2+} from seawater, aiding in the activation of this enzyme [570].

Furin-susceptible insert

When the first membrane-inserted MMP (MT1-MMP) was cloned, it was noted that there was an 11-residue insert between the propeptide and catalytic domains that should be susceptible to the action of Kex-2/furin type proteinases. Pei and Weiss [577] showed that the corresponding 10-residue site in stromelysin 3 (MMP-11) GLSARNRQKR~F was indeed cleaved *in vitro* by furin. This was confirmed [590] and it was also shown that mouse MMP-11 was cleaved more effectively by the furin PACE-4. Both human and mouse MMPs have the sequence ARNRQKR, so the enzyme must be affected by other parts of the substrate MMP molecule as well. Cotransfection of COS cells with MMP-11 and furin of rat origin led to activation of the MMP within the cells [574].

Furin activates proMMP-14 at the RRKR~Y113 bond, in turn stimulating the subsequent activation of proMMP-2 by MMP-14 [591]. Mutation in this furin-susceptible region prevents cleavage from 62 to 56 kDa and blocks the action of the enzyme on gelatin [549]. Furthermore, a chimera formed between the propeptide of MMP-3 and the catalytic domain of MMP-14, omitting the furin site, prevents activation by furin. Nonetheless, the full-length enzyme is still capable of activating proMMP-2 [522]. Enzyme on the cell surface is susceptible to cleavage by plasmin at the Arg112 residue; this could provide a mechanism of activation for enzyme not cleaved within the cell [575]. In addition to MMP-11 and MMP-14, this furin insert is found in the other three transmembrane MMPs [583] and in XMMP of *Xenopus* which has an RRKR~F site [619]. This enzyme has an unusually long insert of 37 residues resembling vitronectin but ending in the RRKR motif. MMP-19 lacks a furin site [579].

Catalytic and zinc-binding domain, catalytic mechanism

This domain is C-terminal in matrilysin, but is followed by a hemopexin domain in all other MMPs. It is typically 160–170 residues in length with the catalytic active centre in the C-terminal 50–54 residues. Fibronectin inserts (see below) can break up this domain, separating the last 50 residues from the others; it is this last segment that contains the catalytic zinc. The first catalytic domain to be studied in isolation was that of human collagenase MMP-1. Clark *et al.* [525] observed spontaneous breakdown of MMP-1 to a 22 kDa fragment which had catalytic activity. This proved to be the catalytic domain, free of hemopexin and propeptide. It digested casein and gelatin but not collagen, suggesting it was too short to bind properly to its substrate. The catalytic domain was later expressed in *E. coli* [561]; it had the properties just listed and was also found to denature more readily than full-length enzyme. The rate of cleavage of synthetic peptides by the catalytic domain of MMP-1 and also MMP-3 is not affected by truncation, nor is the inhibition by actinonin [519]. The inability of catalytic domain alone to cleave type I collagen is also reported for collagenase 2/MMP-8 [592] and collagenase 3/MMP-13 [556]. Omission of the hemopexin domain, but with retention of the fibronectin domains which bind substrate, does not compromise the ability of MMP-9 to digest native type IV collagen [555, 618]. A naturally occurring hemopexin-free form of MMP-9 has also been studied [586].

The C-terminal portion of the catalytic domain contains the catalytic zinc, bound in the sequence HELGHXXGXXH by the three His residues. Mutation of each of these residues in turn in MMP-1 leads to an inactive catalytic domain, which can be traced to disrupted folding [616]; mutation of the Glu residue also diminishes or abolishes activity. X-Ray studies indicate that the isolated catalytic domain has two moles of Zn^{2+} and one Ca^{2+} ion [559, 560]. The second 'structural' zinc is ligated by His149, His164, His177 and Asp151 [517] (see Table 6 below). A double mutant of His149/Asp151 is unable to cleave peptide substrates [614].

In the case of stromelysin, the second zinc is bound by His151, His166 and His179 [539]. It is speculated that His166 (of MMP-3) makes a His 'switch', replacing water and permitting Ca^{2+} to dissociate; the second structural zinc atom would prevent this and stabilize the enzyme [610]. However, in the case where the hemopexin domain is attached to the catalytic domain, this second zinc may not be needed [613] (MMP-2 and MMP-3). Against this, Salowe *et al.* [587] find 1.5 molecules of zinc in a full-length molecule. The propeptide has no effect on the zinc content of truncated MMP-3 [595]. The properties of latency, activation and folding of MMP-3 are unaffected by absence of the hemopexin domain and the activity on peptide substrates is also unaffected. [562]. The catalytic domain contains two moles of Ca^{2+} [547] and at least 2 mM calcium is required for stabilization [544].

MMP-8 catalytic domain has been shown by X-ray crystallography to contain two Ca^{2+} ions in addition to the two Zn^{2+} [516]. The Phe79–Gly242 form is shown to have a salt linkage between the ammonium of the Phe79 and the side-chain carboxyl of Asp232. This stabilizes the structure, and also presumably the transition state, explaining the fourfold higher activity of this form versus the form with one less N-terminal residue [584]. A similar effect is seen in human MMP-1 that has been activated by stromelysin; however, effects observed using the large collagen molecule as substrate are not noted with small peptides [536].

Windsor *et al.* [617] have compared mini-enzymes (propeptide plus catalytic domain) of MMP-1, -3 and -7. The cleaved active forms hydrolyse a synthetic Mca-peptide with a much higher k_{cat}/K_m for MMP-7 than for MMP-1 or -3. MMP-1 and MMP-3 can complex to TIMP-1 in a form stable to SDS, whereas MMP-7 cannot. MMP-3 and MMP-7 can fully activate proMMP-1, but MMP-1 cannot. Thus, each of the catalytic domains has its own unique properties and specificities. There have recently been several studies of the expressed catalytic domain of MMP-14, the membrane-type MMP; this form is inhibited by TIMP-2 [612, 622] but not by TIMP-1 [612], and if expressed together with the propeptide, undergoes autoactivation [558].

The **catalytic mechanism** of MMPs has been examined in some detail since the first X-ray structures emerged (see also the section on three-dimensional structures, illustrated by Fig. 13). It was proposed by Spurlino *et al.* [597] and Lovejoy *et al.* [560] that the mechanism of MMP-1 is similar to that of thermolysin; the Glu residue at the active centre HEXXH promotes the nucleophilic attack of water bound to the zinc atom on the carbonyl carbon of the scissile bond. In this case Glu would act as a general base catalyst. However, there are no residues to stabilize the carbonyl oxygen of the tetrahedral intermediate, in contrast to thermolysin, which has Tyr157 and His231 [515]. The same features pertain to the structures of MMP-3 [515] and MMP-7 [520]. The importance of the Glu residues has been demonstrated by mutation analysis. In the case of MMP-1, mutant E200Q was completely inactive and E200D had low activity [616]. In MMP-2 E375D acts on substrates at a low level, but E375A and E375Q do not digest various substrates; however, they do show up on gelatin zymograms at about 1/1000 the level of wild-type enzyme [529]. A similar series of mutants in matrilysin calls into question the role of this Glu residue as a general base catalyst. The mutant E198D has its value of k_{cat}/K_m for synthetic substrate reduced by a factor of 4, for E198C the factor is 160, for E198Q it is 590 and for E198A it is 1900 [524]. These are not considered large changes if the mechanism were really dependent on base catalysis. A careful study of the pH dependence of each mutant indicates that the pKa of 4.3 for the wild-type enzyme is unlikely to be due to the pK of Glu, which is in a hydrophobic environment, and is more likely to be that of the water molecule bound to zinc. This water may act as the nucleophile and the protonated Glu residue may act to stabilize the hemiketal of the transition state, or it may act as a general catalyst following the rate-determining step.

Some further contributions to catalysis are noted for MMP-9, which has a pair of conserved Asp residues 432 and 433 to the C-terminal side of the active centre HEFGH. Mutation of the first of these (D432G) reduces gelatinase activity by 75% and D433G abolishes activity [582]. D433E retains 7% activity. These effects are attributed to stabilizing the conformation of the active centre rather than to calcium binding. However, calcium binding may be critical in the case of MMP-9, which has an N-terminus that is not Phe88. Phe88 is exposed when activation is by trypsin or MMP-3, but other termini such as Gln89 or Glu92 arise by other activation mechanisms. Whereas Phe88 can form a salt bridge to Asp432 to stabilize the active centre, the other residues require calcium to make this link [521]. The Met-turn is critical to forming the base of the active centre binding pocket. A mutation of this Met215 to selenomethionine decreases k_{cat}/K_m for MMP-8 digesting a synthetic peptide and reduces its stability to urea denaturation [580]. Stromelysin is unusual in having an acid pH optimum, an observation that first led to the proposal that there was a distinct enzyme (MMP-6) [510]. However, it was soon established that this acid pH is characteristic of stromelysin [541] although the specific groups involved in determining this optimum are not known.

Fibronectin repeats

Gelatinases A and B (MMP-2 and -9) contain three tandem repeats of about 58 residues each related to fibronectin type II domains. These occur in the catalytic domain, coming shortly before the HEXXH zinc-binding site and contributing about 20 kDa to the mass of the enzymes. Bányai *et al.* in 1991 [511] prepared a β-galactosidase fusion protein with the three tandem repeats of fibronectin II domains from human MMP-2. This protein bound to denatured collagen. Subsequent examination of each domain alone, in pairs or as a triplet showed that each binds to gelatin with cooperative binding as the number of units increased [513]; the triplet could also displace native fibronectin from gelatin. The central unit of the three is stabilized by the other two, and the three units together are postulated to form an extension of the binding cleft of MMP-2 [512]. Assignment of the disulfide bridges in MMP-2 has been made by mass spectrometry [563]. Deletion of the three repeats from MMP-2 (Δ V191-Q364) gave a latent enzyme which was activated normally by APMA or by cell membrane mechanisms (MMP-14) and could bind TIMP-1 and -2 [564]. Proteolysis of dnp-octapeptide was unaffected, but casein and gelatin were digested more slowly than by wild-type enzyme. The enzyme could not bind to type IV collagen. Ye *et al.* [620] also noted digestion of peptides and gelatin by a similar mutant. Expression of the three tandem fibronectin II domains [598] showed that the resultant protein bound to gelatins of type I, IV and V and elastin, but not to native collagen IV or V. There are at least two gelatin-binding sites in this protein, and corresponding binding sites in gelatin are found in peptides CB2, CB7 and CB8 of the $\alpha 1(I)$ chain. The binding of gelatin to the fibronectin domain in full-length MMP-2 protects these domains from cleavage by neutrophil elastase, which would normally produce an active 40 kDa fragment [585]. Binding of MMP-2 to matrix components such as collagen I and IV may prevent the enzyme from diffusing away from the cell vicinity, such binding has been shown to be due to the fibronectin repeats by deletion mutants [508].

Somewhat different results are found with MMP-9. Production of a fusion protein between each of the fibronectin II domains and β-galactosidase [526] showed that only the second domain was critical for binding to gelatin and domains 1 and 3 were less effective. Within domain 2, the residues Arg307, Asp309, Asn319, Tyr320 and Asp323 were critical. Deletion of the entire region (Δ214–395) from recombinant human MMP-9 resulted in an enzyme with no action on gelatin [581]. Such deletion also prevents the enzyme from binding to elastin, resulting in loss of elastase activity [593]; this is also true for MMP-2.

Type V collagen insert

This insert is found only in gelatinase B (MMP-9) where it appears between the active centre and the hemopexin domain; it should be considered as a special case of the hinge region (see next paragraph). It was first observed in 1989 when the enzyme was cloned by Wilhelm *et al.* [611]. It consists of 54 residues, rich in proline and similar to the $\alpha2$ chain of type V collagen. Its role is not clear. A deletion mutant (452–512) digests gelatin, type V and X collagen, producing a typical band pattern of peptides indicating that specificity was not altered [581].

Hinge region

Most of the catalytic domains end in the sequence LY, which is followed shortly by the first cysteine of the hemopexin domain. Between these two domains is a highly variable stretch of amino acids ranging from 2 [619] to 72 (in MMP-9) which is referred to as the linker region or the hinge region. The recent full-length crystal structure of pig collagenase [557] shows that this region, in this case 17 residues long, is not visualized and probably has no secondary structure. The relative

position of the two domains may be an accident of the crystal packing. In the case of pig MMP-1, this region is rich in proline and has the sequence GPSENPVQPSGPQTPQV. De Souza [531] has suggested that this region might adopt a conformation similar to that found in collagen, which could then interact with the triple-helix of the collagen substrate to form a sort of 'proline zipper' that might destabilize the helix to allow a strand to enter the active pocket of the enzyme. Mutation of I270S in the hinge, and of I259L immediately before the hinge, stabilize human MMP-1 and prevent autolytic loss of the hemopexin domain [572].

The hinge region of stromelysin has an additional nine residues (TPLVPTEPV) not found in collagenase 1 or 2 containing mostly hydrophobic amino acids which are also found in MMP-10 [542]. Inserting these nine residues into the hinge sequence of human collagenase 2 (MMP-8) causes loss of collagenolytic action. Extensive alanine scanning mutagenesis of the hinge of MMP-8 shows that this linker region, containing two autolytic sites, can lead to stabilization of the enzyme [552]. Moreover, if four of the five prolines are replaced by alanines then 98.5% of the collagenolytic activity is lost. The hinge of MMP-19 is elongated to 33 residues and has a section with seven Glu or Asp residues with only a single interruption [579]; the role of this motif is not known. The sea urchin envelysin [570] has a long 53-residue hinge, which includes 22 Thr residues.

Hemopexin domain

The hemopexin domain has recently been reviewed by Murphy and Knäuper [82] with respect to its role in MMP-1, -2, -3, -7, -8, -9, -10 and -11. When the first MMP (MMP-1) was sequenced [537] the hemopexin domain had not been identified. However, computer alignment by Hunt [548] indicated that serum hemopexin, rat transin (MMP-3) and human MMP-1 all contained homologous domains. Human S-protein/vitronectin is also in this family [550]. The relationship to vitronectin suggested that the role of the hemopexin domain could be binding to components of the extracellular matrix. Recently, serum hemopexin was shown to have hyaluronidase activity [621]. Both the crystal structures of rabbit hemopexin [532] and pig MMP-1 catalytic/hemopexin domains [557] are now available. Bode [18] commented on the possible role of this domain in MMPs. This four-bladed propeller structure is discussed in the section on three-dimensional structure below. The first and last residues in this domain are cysteines, which link together to close the propeller at its periphery. The intervening length of the domain is just under 200 residues, divided into approximately 48 residues per blade.

The role of the hemopexin domain in MMP catalysis was first noticed by Clark *et al.* [525] who isolated MMP-1 lacking this domain and showed that the enzyme, although still catalytically active, could no longer digest collagen. On the other hand, both latent and active MMP-3 can bind to collagen types I and II through the hemopexin domain [509] even though they are not substrates. This led to the concept of producing chimeras of collagenase and stromelysin in which the catalytic and hemopexin domains were interchanged. The hemopexin domain of MMP-3 could not restore collagenase activity to the catalytic domain of MMP-1, nor could the hemopexin domain of MMP-1 impart such activity to the MMP-3 catalytic domain. In every combination both hemopexin domains could bind collagen [563a]. The requirement for the hemopexin domain for collagenase activity extends also to neutrophil collagenase (MMP-8) [553] and collagenase 3 (MMP-13) [551]. This effect of the hemopexin domain on MMP-8 was explored in more detail [542]. If the last four residues at the C-terminus were removed, including the cysteine that completes the linkage of the propeller blades, 62% of collagenolytic activity was lost. If the first 68 residues of this domain from MMP-8 were retained and the rest of the domain was supplied from the MMP-3 hemopexin domain, 16% of activity was retained. A model has been proposed [538] in which the hemopexin domain swings around on its linker region (which is of critical size, see above) to form

a sandwich after collagen is in the active site. A complete understanding of the hemopexin contribution will probably require a structure of the full-length enzyme bound to collagen substrate lacking the scissile bond.

The hemopexin domain also has the ability to bind heparin. In the case of MMP-2, heparin binding to the proenzyme accelerated the rate of activation eightfold [527]. The effect was lost when the hemopexin domain was deleted. It has been suggested that two molecules of MMP-2 can be aligned by their interactions with heparin to speed the biomolecular autocatalytic step of activation [528]. Heparin also binds to MMP-1 [528]. The binding of MMP-2 to heparin, and also to fibronectin, requires the presence of Ca^{2+} [608]; however, the hemopexin domain of MMP-2 does not bind to collagen. This domain is also required for MMP-2 to bind to the cell surface and to undergo activation by cell membranes [609]; adding excess free hemopexin domain inhibits this activation. Also addition of TIMP-2 abolishes attachment to the cell membrane, suggesting the TIMP-2 might be the link between the enzyme and cell. This effect of hemopexin domain on activation of zymogen is not a general phenomenon, its deletion has no effect on the activation of MMP-13 [551]. The integrin $\alpha_v\beta_3$ binds to the hemopexin domain of MMP-2 and localizes it to the surface of invasive cells [518]. It is possible that the second structural zinc seen in the catalytic domain of many MMPs, may not be required if the hemopexin domain is present. Thus, full-length MMP-3 appears to have only one zinc [613] indicating a stabilizing role for the hemopexin domain that is only taken on by zinc after truncation to the catalytic domain. However, in the case of MMP-1, the X-ray structure of the full-length molecule still shows the presence of this structural zinc.

The hemopexin domain is involved in the binding of TIMP to MMPs. In the case of MMP-3, TIMP-1 binds more tightly than TIMP-2 to the active centre of mature MMP, but the absence of the hemopexin domain does not affect the rate of association [568]. However, Baragi *et al.* [514] report that there is an increase in the K_i for

TIMP-1 from 8.3×10^{-10} to 5.95×10^{-9} M when the hemopexin domain is removed from MMP-3. This is only 14% as tight and is closely similar to the K_i for matrilysin, which naturally lacks the hemopexin domain. In the case of MMP-1, the K_D changes from 0.1 to 2.7 nM upon removal of the hemopexin domain [605]. This effect is attributed to the action of this domain in promoting the initial rapid binding of TIMP-1 to the enzyme. Similarly, with MMP-13, the inhibition by TIMP-1 and TIMP-3 shows an increase in association rates up to 30-fold in the presence of the hemopexin domain, but TIMP-2 action is not altered [551].

The second type of TIMP binding to MMPs is found in the case of complexes between the proenzyme form of gelatinase A and TIMP-2 and gelatinase B and TIMP-1. Howard *et al.* [546] isolated proMMP-2–TIMP-2 complexes from culture medium; the presence of TIMP-2 prevents autocatalytic activation of MMP-2 but not gelatinolysis by enzyme that is partially activated. If the TIMP-2 is first removed by acidification, the enzyme undergoes autoactivation [545]. It was proposed that TIMP binds to the hemopexin domain and acts as a stabilizing factor. If the hemopexin domain is removed from MMP-2 then about ten times more TIMP-2 is required to inhibit the active catalytic domain [534]. The hemopexin domain of MMP-2 is required for the process of membrane activation and TIMP-2 binding to proenzyme [565]. Membrane activation is inhibited when proMMP-2 is complexed with TIMP-2 [602] and the free hemopexin domain competes in this membrane activation process. The latent complex is still able to bind to active MMP-9 [535] and the picture that emerges is that the C-terminal domain of TIMP-2 binds to the hemopexin domain of MMP-2, but the N-terminal domain of TIMP-2 is still exposed and is able to form a complex with the active centre of MMP-9 in a ternary complex.

The case of MMP-9 has not been so thoroughly studied. However, it is known that a complex forms between proMMP-9 and TIMP-1 through binding of TIMP-1 to the hemopexin domain [601]. The inhibition of active MMP-9 by TIMP-1 is much faster when the hemopexin domain is present due to an enhanced on-rate [571]. Chimeras of the catalytic and hemopexin domains of MMP-1 and MMP-10 indicate that all of the activity of MMP-10 resides in the catalytic domain, and collagenolytic activity is not bestowed by adding the MMP-1 hemopexin domain [588]. Oddly, MMP-11 has very little ability to digest protein substrates; however, if 175 residues are removed from the C-terminus of the hemopexin domain, MMP-11 of the mouse develops weak activity against various components of the extracellular matrix [569].

Transmembrane domain

The first of the membrane-inserted MMPs, MT1-MMP or MMP-14, was cloned and sequenced in 1995 [604]. The hemopexin domain ends at residue 508 and there is an extension of 74 residues; at position 539–562 is a highly hydrophobic span suggested to be a transmembrane domain, leaving the final 20 residues within the cytoplasm. Cao *et al.* [523] prepared a deletion mutant lacking residues 535–582; this form was outside the cell but still weakly bound to the cell and it did not activate proMMP-2. If this domain was replaced by the transmembrane domain of the interleukin 2 receptor α-chain, MT1-MMP reappeared in the membrane and was able to activate MMP-2. Attaching the MMP transmembrane domain to TIMP-1 led to the localization of this protein in the cell membrane. When melanoma cells overexpress MMP-14, the enzyme is localized to invadopodia and facilitates cell invasiveness and degradation of the extracellular matrix [567]. Concanavalin A treatment or deletion of the transmembrane domain prevents this localization, and replacement of the transmembrane domain with that from IL-2R fails to restore it. It was shown by Pei and Weiss [578] that the mutant MMP-14 without transmembrane domain can activate proMMP-2, in contrast to the earlier report of Cao *et al.* [523]; this form is soluble and can digest a variety of matrix molecules such as fibronectin,

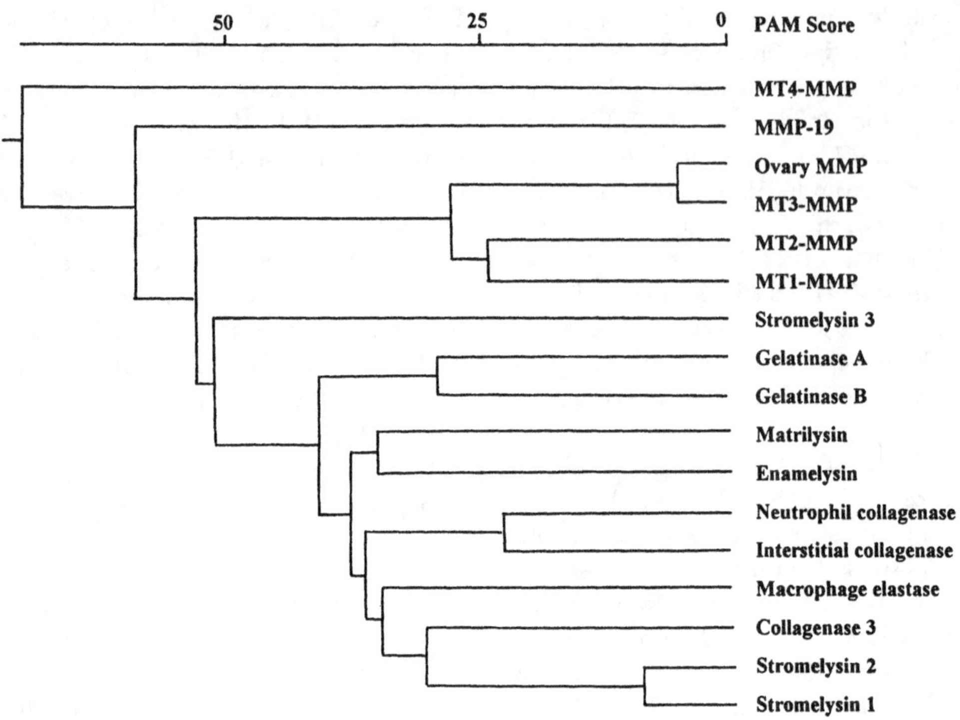

Figure 2

Dendrogram indicating the evolutionary relationships among human MMPs. The sequences were aligned by use of the PILEUP program for the catalytic domains (exclusive of fibronectin repeats). The tree was generated by the KITSCH algorithm. PAM accepted point mutations. The figure was drawn by Dr Neil Rawlings, Babraham, England and is reprinted from [147] by kind permission of Academic Press, Inc., San Diego.

vitronectin, laminin and gelatin. These authors did not observe digestion of collagen, but further studies by Ohuchi *et al.* [573] did show an effect on collagens I, II and III, about 1/5 as much activity as with MMP-1; deletion of the transmembrane domain did not alter this effect on collagen. A naturally occurring deletion of MT3-MMP transmembrane domain in the rat [594] was found to activate membrane-bound proMMP-2 but only to the first step. Autolytic cleavage to the final mature form was not seen due either to rapid reaction of the intermediate with TIMP or to the lack of sufficient concentration of MMP-2 which would normally occur on the cell membrane and facilitate autolysis.

Evolutionary relationships among the MMPs

One of the earliest dendrograms was provided for the MMPs by Murphy *et al.* [566]. They noted the remote relationship between the serralysins (*Serratia, Erwinia, Pseudomonas*) and the MMPs. Of the MMPs, stromelysin 3 was most distant from the remaining ones. de Souza and Brentani [530] presented a similar sized tree, with the addition of the sea urchin envelysin. In their tree, gelatinases A and B are somewhat removed from the others, but this is a consequence of using the entire protein sequence and not just the catalytic portion without the fibronectin inserts. X-Ray studies have led to the recognition that the MMPs are related to a broader group termed the metzincins (due to the importance of a turn at a Met residue in forming the active pocket). These studies are summarized by Stöcker *et al.* [600]. The metzincin group contains four subgroups with almost identical protein fold, but sizeable differences in sequences. The fundamental unit is the zinc binding sequence HEXXHXXGXXHZ, where Z is Glu in the astacins, Ser in the MMPs, Pro in the serralysins and Asp in the adamalysins. Rawlings and Barrett [104] have subdivided the metalloproteinases into some 30 families: astacins and reprolysins (adamalysins) are subfamilies of family M12 and MMPs, serralysins and collagenases are subfamilies of family M10. All of these, plus two small additional families belong to clan MB, which is characterized by the HEXXHXXGXXH binding sequence and the metzincin fold. A dendrogram is presented [103] which suggests that the cysteine switch evolved about 2.5 billion years ago when the serralysins split off, followed by plants acquiring the hemopexin domain about 1 billion years ago. MMP-11 (stromelysin 3) is one of the most ancient MMPs, going back about 800 million years. Matrilysin (MMP-7) subsequently lost its hemopexin domain to become the smallest matrixin.

A similar dendrogram, covering 30 MMPs was presented recently by Sang and Douglas [111]. This was based on the full length of the proteins, so again the gelatinases are separated from the rest. Envelysin, MT1-MMP and MMP-11 are also separated from the bulk of the MMPs. This paper also presents a useful table of sequence identities among 30 MMPs. The recent tree pattern presented in Fig. 2 shows, for purposes of simplification and clarity, only the catalytic domains of the known human enzymes (17 cases). While MMP-11 is still separate from the MMPs below number 13, the more recently discovered membrane-type MMPs and MMP-19 are still more ancient and stromelysins 1 and 2 are recent. Relatively complete trees with 67 enzymes [103] and 64 enzymes [70] from all species have recently been published. The reader is referred to these for a more complete picture of the more distantly related MMPs. In particular, Massova *et al.* [70] analyse the relationships based on the full-length enzyme, upon the catalytic domain only and upon the hemopexin domain. Each gives a somewhat different picture because the addition of domains and inserts and the changes in hemopexin are only loosely associated with the changes in catalytic efficacy that occur in the active part of the enzyme.

TIMP sequences

Alignments of the TIMP sequences

Twenty-five amino acid alignments are presented at the end of the book. These are linked to a more complete description of each sequence presented in Table 4. For each item in Table 4 there is a number relating to the line number in the alignment. The sequences in both tables are in the same order (with a few additional inserts in Table 4) so that matching the items is simplified. The major features of the alignments are the signal peptide, the 12 invariant Cys residues that form six disulfide bridges and the C- and N-terminal domains separated by 1–2 residues between the sixth and seventh Cys.

The number of inserts in the TIMP sequences is relatively small compared to that of the MMPs, so that while a typical TIMP has about 200–210 amino acids, the width of the alignment table is not much larger (234 residues). The signal peptide begins with Met and ends at position 29 (on the alignment) just before the first Cys residue. This Cys1 bridges to the Cys at position 109, the second Cys at position 32 is linked to 139 and the third, at 42, to 164. Residue 164 marks the end of the N-terminal domain, which accounts for most of the inhibition of MMP. There is a join of 1–2 residues from this Cys at position 165 to the first Cys of the C-domain at position 167. This bridges to Cys 214 producing a loop that takes in most of the C-domain. There is a short loop at Cys172–177 and a somewhat larger loop at Cys185–206.

Miscellaneous sequence comments

A few items not yet in SwissProt include fragments of bovine TIMP-1 [1208], human TIMP-1 [1201, 1207], and mouse TIMP-1 [1194] and an almost complete sequence of rat TIMP-1 [1176]. Fragments of bovine [1153, 1163], human [1199] and mouse [1183] TIMP-2 have been reported. Entry Q21265, line C12, for *C. elegans* TIMP-2 is interesting; it matches the N-terminal domain of other TIMPs, but lacks the C-terminal domain. Perhaps the N-terminal domain was the primordial form. An EST fragment of TIMP-2 for zebrafish is in the databanks at AA542611. Fragments of human TIMP-3 have been reported [1174, 1209]. There are two important complete sequences of mouse [1180] and rat [1217] TIMP-4 that have been published, but do not appear in the databanks. Each has 224 residues.

Domain structure of the TIMPs

The TIMP molecules can be readily divided into two regions separated by a two-residue isthmus at residue 125–126 in human TIMP-1 and a single residue at a position between 121 and 128 in the other three TIMP species. It is clear that the N-terminal domain has almost complete inhibitory action on its own, but whether the other portion is a true domain is sometimes questioned. It has not been established that the C-domain can fold on its own. Each of these two domains is held in a relatively rigid conformation by three disulfide bridges, which are highly conserved across all species of TIMPs and organisms. In particular, the first cysteine after the signal peptide, Cys1, is critical for the binding of TIMP to zinc. Cys 1 forms a disulfide bridge to a second Cys at a position between 68 and 73, producing a long loop (the numbering of these residues omits the signal and begins with Cys1). Smaller loops are formed by bridges from Cys3 to a Cys between positions 95

Table 4 Tissue inhibitors of metalloproteinases (TIMP)

Line	Name	AccNumber	Description	GeneName	Organism	Citation		Additional database information
TIMP-1								
C1	SWISSPROT: TIM1_BOVIN	P20414	METALLOPROTEINASE INHIBITOR 1 PRECURSOR (TIMP-1) (EMBRYOGENIN-1) (EG-1)	TIMP1	*Bos taurus* (bovine)	FREUDENSTEIN J.: SATOH T.: DE CLERCK Y.A.:	[1165] [1191] [412]	EMBL; M60073; G163761; - EMBL; S70841; G546974; - PIR; A35685; A35685; PIR; B34468; B34468; PROSITE; PS00288; TIMP; 1 PFAM; PF00965; TIMP; 1
C2	TREMBL: TIM1_CANFA	P81546	METALLOPROTEINASE INHIBITOR 1 PRECURSOR (TIMP-1) (ERYTHROID POTENTIATING ACTIVITY) (EPA) (TISSUE INHIBITOR OF METALLOPROTEINASES) (FIBROBLAST COLLAGENASE INHIBITOR) (COLLAGENASE INHIBITOR)	TIMP1	*Canis familiaris* (dog)	NORITAKE H.: CHOPRA R.:	Submitted (Aug 1998) [1158]	EMBL; AB016817; D1033355; - PROSITE; PS00288; TIMP; 1
C3	SWISSPROT: TIM1_HORSE	O02722	METALLOPROTEINASE INHIBITOR 1 PRECURSOR (TIMP-1)	TIMP1	*Equus caballus* (horse)	RICHARDSON D.W.:	Submitted (Mar 1997)	EMBL; U95039; G2072247; - PROSITE; PS00288; TIMP; 1 PFAM; PF00965; TIMP; 1
C4	SWISSPROT: TIM1_HUMAN	P01033 Q14252	METALLOPROTEINASE INHIBITOR 1 PRECURSOR (TIMP-1) (ERYTHROID POTENTIATING ACTIVITY) (EPA) (TISSUE INHIBITOR OF METALLOPROTEINASES) (FIBROBLAST COLLAGENASE INHIBITOR) (COLLAGENASE INHIBITOR)	TIMF1 or TIMP	*Homo sapiens* (human)	DOCHERTY A.J.P.: GASSON J.C.: CARMICHAEL D.F.: KACZOREK M.: RAPP G.: OPBROEK A.: MATSUDA T.: HARDCASTLE A.J.: WILLIAMSON R.A.: OSTHUES A.: O'SHEA M.:	[1161] [1166] [1157] [1178] [1188] [1186] Submitted (Jul 1992) [1173] [500] FEBS Lett. 296:16-20(1992) [1434]	EMBL; X03124; G37183; - EMBL; M12670; G182483; - EMBL; X02598; G31189; - EMBL; M59906; G189382; - EMBL; S68252; E119406; - EMBL; D11139; G220125; - EMBL; L47361; G994731; - EMBL; A10416; G490094; - PIR; A01269; ZYHUEP; PIR; A23534; A23534; PIR; A35826; A35826; PIR; S20318; S20318; MIM; 305370; -; PROSITE; PS00288; TIMP; 1 PFAM; PF00965; TIMP; 1

Continued

Table 4 Continued

Line	Name	AccNumber	Description	GeneName	Organism	Citation		Additional database information
C5	SWISSPROT: TIM1_MOUSE	P12032 P20064	METALLOPROTEINASE INHIBITOR 1 PRECURSOR (TIMP-1) (ERYTHROID POTENTIATING ACTIVITY) (EPA) (TISSUE INHIBITOR OF METALLOPROTEINASES) (COLLAGENASE INHIBITOR 16C8 FIBROBLAST) (TPA-INDUCED PROTEIN) (TPA-S1)	TIMP1 or TIMP-1 or TIMP	*Mus musculus* (mouse)	GEWERT D.R.: EDWARDS D.R.: JOHNSON M.D.:	[1167] [1162] [1177]	EMBL; M28312; G193042; - EMBL; M28308; G193042; JOINED EMBL; M28309; G193042; JOINED EMBL; M28310; G193042; JOINED EMBL; M28311; G193042; JOINED EMBL; X04684; G49704; - EMBL; M17243; G202112; - EMBL; M28312; G193042; - PIR; A26633; A26633; PIR; A26106; A26106; PIR; A26917; A26917; MGD; MGI:98752; TIMP; PROSITE; PS00288; TIMP; 1 PFAM; PF00965; TIMP; 1
C6	SWISSPROT: TIM1_RABIT	P20614	METALLOPROTEINASE INHIBITOR 1 PRECURSOR (TIMP-1)	TIMP1	*Oryctolagus cuniculus* (rabbit)	HOROWITZ S.:	[1175]	EMBL; J04712; G165743; - PIR; A33350; A33350; PROSITE; PS00288; TIMP; 1 PFAM; PF00965; TIMP; 1
C7	SWISSPROT: TIM1_SHEEP	P50122	METALLOPROTEINASE INHIBITOR 1 PRECURSOR (TIMP-1)	TIMP1	*Oris aries* (sheep)	SMITH G.W.:	[1195]	EMBL; S67450; G456990; - PROSITE; PS00288; TIMP; 1 PFAM; PF00965; TIMP; 1
C8	SWISSPROT: TIM1_PAPCY	P49061	METALLOPROTEINASE INHIBITOR 1 PRECURSOR (TIMP-1)	TIMP1	*Papio cynocephalus* (yellow baboon)	FOROUGH R.:	[1164]	EMBL; L37295; G561546; - PROSITE; PS00288; TIMP; 1 PFAM; PF00965; TIMP; 1
C9	SWISSPROT: TIM1_RAT	P30120	METALLOPROTEINASE INHIBITOR 1 PRECURSOR (TIMP-1)	TIMP1 or TIMP-1	*Rattus norvegicus* (rat)	OKADA A.: GIBBONS K.L.: BOUJRAD N.: ROSWIT W.T.:	[1185] Submitted (Jan 1996) [1155] [1189]	EMBL; U06179; G468058; - EMBL; L31883; G1161234; - EMBL; U16022; G562119; ALT_SEQ PIR; S20326; S20326; PROSITE; PS00288; TIMP; 1 PFAM; PF00965; TIMP; 1
C10	SWISSPROT: TIM1_PIG	P35624	METALLOPROTEINASE INHIBITOR 1 PRECURSOR (TIMP-1)	TIMP1	*Sus scrofa* (pig)	TANAKA T.:	[1205]	EMBL; S96211; G247730; - PROSITE; PS00288; TIMP; 1 PFAM; PF00965; TIMP; 1

TIMP-2

Line	Name	AccNumber	Description	GeneName	Organism	Citation		Additional database information
C11	SWISSPROT: TIM2_BOVIN	P16368	METALLOPROTEINASE INHIBITOR 2 PRECURSOR (TIMP-2) (TISSUE INHIBITOR OF METALLOPROTEINASES-2) (COLLAGENASE INHIBITOR)	TIMP2	*Bos taurus* (bovine)	BOONE T.C.: MURRAY J.B.: DE CLERCK Y.A.:	[1154] J. Biol. Chem. 261:4154-4159(1986) [412]	EMBL; M32303; G163342; - PIR; A25322; A25322; PIR; A35996; A35996; PIR; A34468; A34468; PROSITE; PS00288; TIMP; 1 PFAM; PF00965; TIMP; 1

C12	TREMBL: Q21265	Q21265	SIMILAR TO METALLOPROTEINASE INHIBITOR 2 PRECURSORS	K07C11.5	*Caenorhabditis elegans*	WILSON R.: [1215] WU X.: Submitted (Apr 1996) WATERSTON R.: Submitted (Apr 1996)	EMBL; U53336; G1255824; - PFAM; PF00965; TIMP; 1
C13	SWISSPROT: TIM2_CRILO	Q60453	METALLOPROTEINASE INHIBITOR 2 PRECURSOR (TIMP-2) (TISSUE INHIBITOR OF METALLOPROTEINASES-2) (FRAGMENT)	TIMP2	*Cricetulus longicaudatus* (Long-tailed hamster) (chinese hamster)	SUZUKI Y.: Submitted (Nov 1993)	EMBL; X75924; G414877; - PROSITE; PS00288; TIMP; 1 PFAM; PF00965; TIMP; 1
C14	SWISSPROT: TIM2_CHICK	O42146	METALLOPROTEINASE INHIBITOR 2 PRECURSOR (TIMP-2) (TISSUE INHIBITOR OF METALLOPROTEINASES-2)	TIMP2	*Gallus gallus* (chicken)	AIMES R.T.: [1149]	EMBL; AF004664; G2352473; - PROSITE; PS00288; TIMP; 1 PFAM; PF00965; TIMP; 1
C15	SWISSPROT: TIM2_HUMAN	P16035 Q93006 Q16121	METALLOPROTEINASE INHIBITOR 2 PRECURSOR (TIMP-2) (TISSUE INHIBITOR OF METALLOPROTEINASES-2) (CSC-21K)	TIMP2	*Homo sapiens* (human)	STETLER-STEVENSON W.G.: [1157] BOONE T.C.: [1154] HAMMANI K.: [1171] MALIK K.: Submitted (Aug 1990) STETLER-STEVENSON W.G.: [1198] GOLDBERG G.I.: [1168] OSTHUES A.: FEBS Lett. 296:16-20(1992) DE CLERCK Y.A.: Gene 139:185-191(1994) WILLIAMSON R.A.: [1439]	EMBL; J05593; G339707; - EMBL; S48568; G298202; - EMBL; U44385; G1517893; - EMBL; U44381; G1517893; JOINED EMBL; U44382; G1517893; JOINED EMBL; U44383; G1517893; JOINED EMBL; M32304; G307195; - EMBL; X54533; G37181; - EMBL; S68860; E119468; - PIR; A34415; A34415; PIR; A34464; A34464; PIR; B35996; B35996; PIR; A37128; A37128; PIR; S20319; S20319; MIM; 188825; -; PROSITE; PS00288; TIMP; 1 PFAM; PF00965; TIMP; 1
C16	SWISSPROT: TIM2_MOUSE	P25785	METALLOPROTEINASE INHIBITOR 2 PRECURSOR (TIMP-2) (TISSUE INHIBITOR OF METALLOPROTEINASES-2)	TIMP2 or TIMP-2	*Mus musculus* (mouse)	SHIMIZU S.: [1192] LECO K.J.: [1181] KISHI J.I.: [1179]	EMBL; X62622; G54802; - EMBL; M82858; G202052; - EMBL; M93954; G202054; - PIR; S15987; S15987; PIR; JH0683; JH0683; MGD; MGI:98753; TIMP2; PROSITE; PS00288; TIMP; 1 PFAM; PF00965; TIMP; 1
C17	TREMBLNEW: G1477929	G1477929	TIMP2=TISSUE INHIBITOR OF METALLOPROTEINASE 2		*Oryctolagus cuniculus*	WERTHEIMER S.J.: [1210]	

Continued

Table 4   Continued

Line	Name	AccNumber	Description	GeneName	Organism	Citation		Additional database information
C18	SWISSPROT: TIM2_RAT	P30121	METALLOPROTEINASE INHIBITOR 2 PRECURSOR (TIMP-2) (TISSUE INHIBITOR OF METALLOPROTEINASES-2)	TIMP2 or TIMP-2	*Rattus norvegicus* (rat)	COOK T.F.: GIBBONS K.L.: SANTORO M.: GRIMA J.: ROSWIT W.T.:	[1159] Submitted (Jun 1995) [1190] [1170] [1189]	EMBL; U14526; G540205; - EMBL; L31884; G1141730; - EMBL; S72594; G619233; - EMBL; S82718; G1881814; - PIR; S20325; S20325; PROSITE; PS00288; TIMP; 1 PFAM; PF00965; TIMP; 1
TIMP-3								
C19	SWISSPROT: TIM3_BOVIN	P79121	METALLOPROTEINASE INHIBITOR 3 PRECURSOR (TIMP-3) (TISSUE INHIBITOR OF METALLOPROTEINASES-3)	TIMP3	*Bos taurus* (bovine)	SU S.:	[1202]	EMBL; U77588; G1684818; - PROSITE; PS00288; TIMP; 1 PFAM; PF00965; TIMP; 1
C20	SWISSPROT: TIM3_CHICK	P26652	METALLOPROTEINASE INHIBITOR 3 PRECURSOR (TIMP-3) (TISSUE INHIBITOR OF METALLOPROTEINASES-3) (21 KD PROTEIN OF EXTRACELLULAR MATRIX)	TIMP3 or IMP-3	*Gallus gallus* (chicken)	PAVLOFF N.: STASKUS P.W.:	[1187] [1196]	EMBL; M94531; G211902; - PIR; A39043; A39043; PIR; A43429; A43429; PROSITE; PS00288; TIMP; 1 PFAM; PF00965; TIMP; 1
C21	SWISSPROT: TIM3_HUMAN	P35625	METALLOPROTEINASE INHIBITOR 3 PRECURSOR (TIMP-3) (TISSUE INHIBITOR OF METALLOPROTEINASES-3) (MIG-5 PROTEIN)	TIMP3	*Homo sapiens* (human)	SILBIGER S.M.: URIA J.A.: WICK M.: WILDE C.G.: STOEHR H.: RUIZ A.C.: HAMMANI K.: BURGESS J.: APTE S.S.: WEBER B.H.F.: FELBOR U.: FELBOR U.:	[1193] [1206] [1211] [1212] [1200] Submitted (Sep 1996) Submitted (May 1996) Submitted (Nov 1997) [1151] [1209] [422] [421]	EMBL; U14394; G608129; - EMBL; U02571; G472310; - EMBL; X76227; G495252; - EMBL; Z30183; G520932; ALT_SEQ EMBL; S78453; G998826; - EMBL; U33114; G1215682; - EMBL; U33110; G1215682; JOINED EMBL; U33111; G1215682; JOINED EMBL; U33112; G1215682; JOINED EMBL; U33113; G1215682; JOINED EMBL; U38955; G1304484; - EMBL; U38952; G1304484; JOINED EMBL; U38953; G1304484; JOINED EMBL; U38954; G1304484; JOINED EMBL; U67195; G1519558; - EMBL; Z98256; E1188526; - EMBL; L15078; G407035; - PIR; S45317; S45317; MIM; 188826; -; MIM; 136900; -; PROSITE; PS00288; TIMP; 1 PFAM; PF00965; TIMP; 1

C22	SWISSPROT: TIM3_MOUSE	P39876	METALLOPROTEINASE INHIBITOR 3 PRECURSOR (TIMP-3) (TISSUE INHIBITOR OF METALLOPROTEINASES-3)	TIMP3 or TIMP-3 or SUN	*Mus musculus* (mouse)	LECO K.J.: SUN Y.: APTE S.S.: SUN Y.: APTE S.S.:	[1182] [1203] [1150] [1204] [1152]	EMBL; L27424; G439882; - EMBL; Z30970; E264421; - EMBL; L19622; G438811; - EMBL; U26437; G1167534; - EMBL; U26433; G1167534; JOINED EMBL; U26434; G1167534; JOINED EMBL; U26435; G1167534; JOINED EMBL; U26436; G1167534; JOINED PIR; A53532; A53532; PIR; S43052; S43052; MGD; MGI:98754; TIMP3; PROSITE; PS00288; TIMP; 1 PFAM; PF00965; TIMP; 1
C23	SWISSPROT: TIM3_RAT	P48032	METALLOPROTEINASE INHIBITOR 3 PRECURSOR (TIMP-3) (TISSUE INHIBITOR OF METALLOPROTEINASES-3)	TIMP3 or TIMP-3	*Rattus norvegicus* (rat)	WU I.:	[1216]	EMBL; U27201; G971206; - PROSITE; PS00288; TIMP; 1 PFAM; PF00965; TIMP; 1
C24	SWISSPROT: TIM3_XENLA	O73746	METALLOPROTEINASE INHIBITOR 3 PRECURSOR (TIMP-3) (TISSUE INHIBITOR OF METALLOPROTEINASES-3)	TIMP3	*Xenopus laevis* (african clawed frog)	YANG M.:	[1218]	EMBL; AF042493; G3098488; - PROSITE; PS00288; TIMP; 1
TIMP-4								
C25	SWISSPROT: TIM4_HUMAN	Q99727	METALLOPROTEINASE INHIBITOR 4 PRECURSOR (TIMP-4) (TISSUE INHIBITOR OF METALLOPROTEINASES-4)	TIMP4	*Homo sapiens* (human)	GREENE J.:	[1169]	EMBL; U76456; G1773293; - MIM; 601915; -; PROSITE; PS00288; TIMP; 1 PFAM; PF00965; TIMP; 1

The table is arranged in order of TIMP numbers. Live updated information is available with hot links at http://www3.ebi.ac.uk/Services/protein_profiles/mmp-timp/table.html together with full bibliographic references.

Line, Name and Acc(ession) Number match the first 3 columns of the sequence alignment table, pages 221–223.

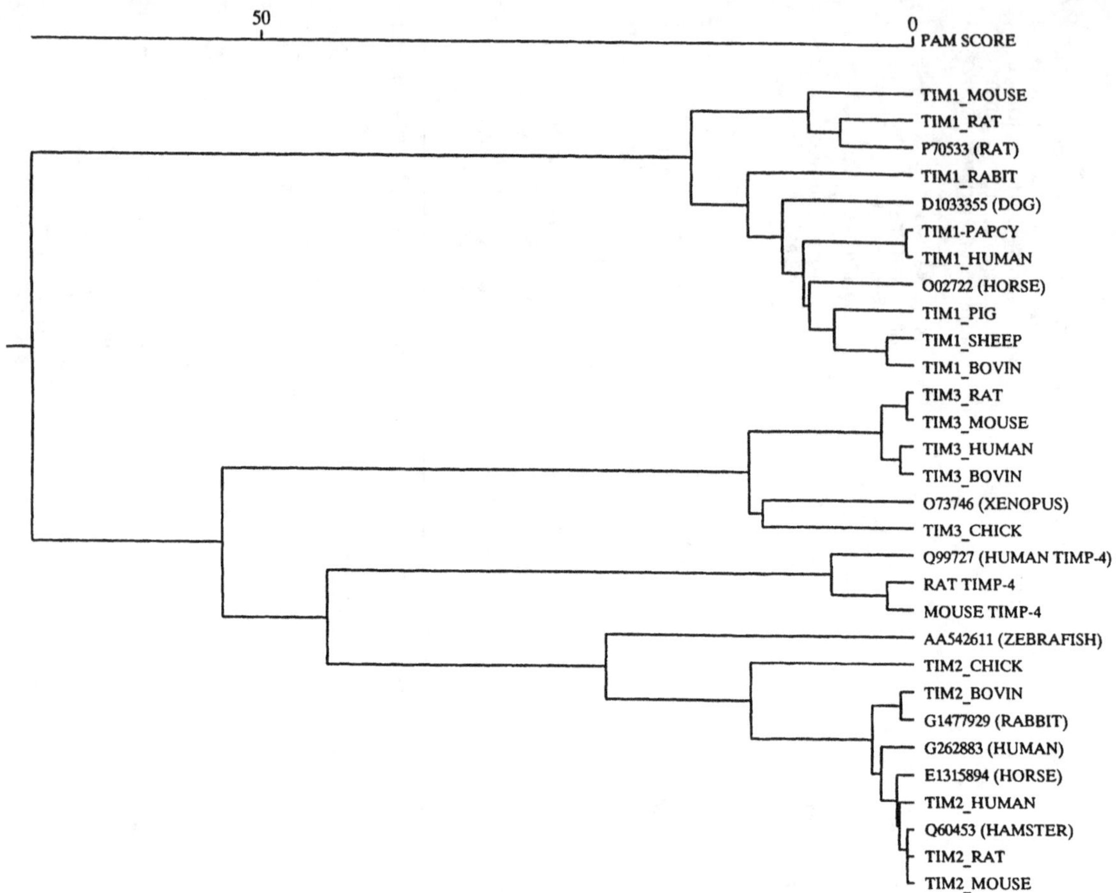

Figure 3

Dendrogram indicating the evolutionary relationships among the known TIMPs. The figure was kindly provided by Dr Neil Rawlings and was prepared as indicated in the legend to Fig. 2.

and 102 and from Cys13 to a Cys at the end of the N-domain (position between 120 and 127). The C-terminal domain has three small loops where a cysteine between 122 and 128 makes a long loop with Cys 169–176, which is pinched into two smaller loops by links at 127–137 joined to a cysteine five residues away and a link between 140–146 and 161–168. The original position of the disulfide bridges in TIMP-1 were deduced by Williamson *et al.* [1214] based on sequencing of proteolytic fragments.

The inhibitory action of the isolated N-domain has been demonstrated for TIMP-1 [1184] and TIMP-2 [1160]. The C-domain has no inhibitory activity, but does enhance the binding, for example, of TIMP-1 to stromelysin, by providing further contact and by helping to orient the TIMP to the active centre [1213]. Further discussion of the domains is found in the section on three-dimensional structures.

Evolutionary aspects

Apte *et al.* [1152] upon working out the gene structure for TIMP-3 suggested that the TIMPs formed a closely related family. Since then a fourth TIMP has been added and the entire group has been analysed in detail by Douglas *et al.* [38] including a dendrogram with 17 proteins. The general order of evolution appears to be the same as the numbering from 1 to 4. A somewhat larger tree of 30 members is presented here in Fig. 3. Here the order appears to be TIMP-1, TIMP-3, TIMP-4 and TIMP-2. Differences in the branch points suggest that there may have been somewhat variable rates of change for the various TIMPs, e.g. TIMP-1 may be evolving at double the rate of TIMP-3.

Three-dimensional structures of the MMPs and TIMPs

The first three-dimensional (3D) structure of a metalloendopeptidase was reported for the thermophilic bacterial metalloproteinase thermolysin in 1972 by Matthews *et al.* [1390]. The 3D structure of a eukaryotic metalloendopeptidase was not resolved until 1992 when Bode *et al.* [1347] reported the crystal structure of astacin from crayfish. The structure of astacin revealed that the topology around the catalytic zinc of the active site in this metalloproteinase can be superimposed on that of thermolysin, although the overall amino acid sequences of the two enzymes do not show a significant similarity [1347]. Recent progress in determining the 3D structures of a number of metalloproteinases has revealed that the overall C_α-backbone structure of astacin [19, 123, 1347] has a high degree of similarity with those of members of the other metalloproteinase families, i.e. snake venom metalloproteinases adamalysin II [1369, 1370] and atrolysin C [1421] in the reprolysin subfamily, MMPs in the matrixin subfamily (see below), and *Pseudomonas aeruginosa* alkaline proteinase [1342] and *Serratia* proteinase [1341] in the serralysin subfamily. Common structural elements are four parallel (sI, sII, sIII, sV) and one antiparallel (sIV) β-strands and three long α-helices (hA, hB, hC), which are arranged in an identical order (Fig. 4).

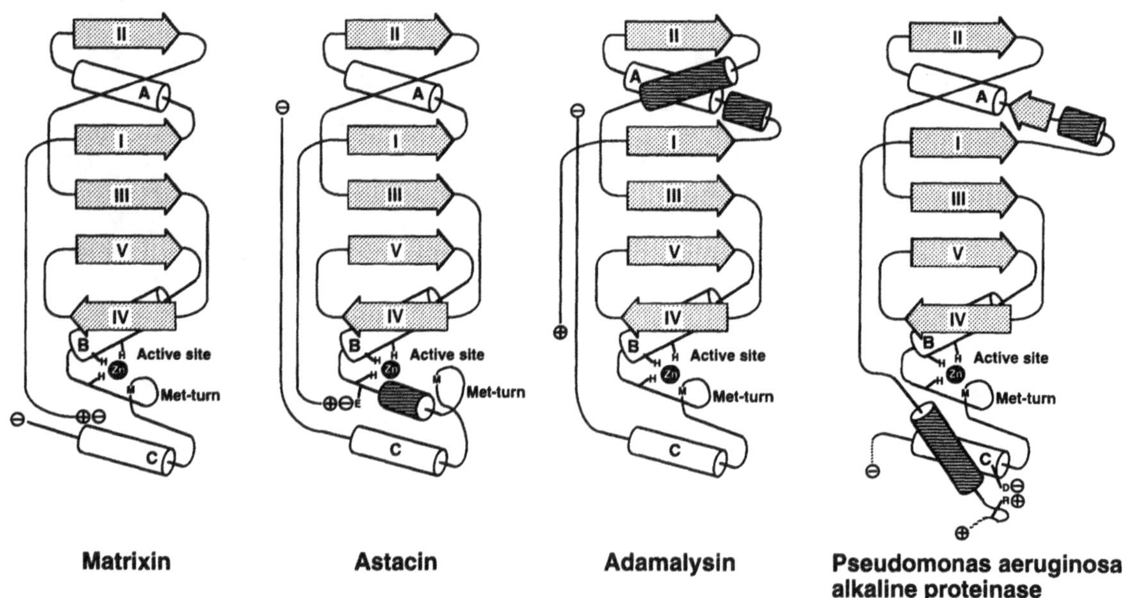

Figure 4

Schematic representation of the secondary structure of the members of the metzincin superfamily. Strands (I–V) and helices (A–C) common to the four families are shaded and white, respectively. Structural elements that are unique to one family are hatched. The figure was redrawn from Stöcker *et al.* [1412] by kind permission of Cambridge University Press, New York.

In addition, all these proteins from four different subfamilies of metalloproteinases (astacins, reprolysins, matrixins and serralysins) exhibit almost identical active-site environments. The main feature is a consensus sequence HEXXHXXGXXH whose histidine side chains ligate the catalytic Zn^{2+}. The zinc binding site is followed by a stretch of random coil that exhibits a divergent structure among members from the four different families, stretches of the polypeptide chains of the four enzymes then come close to the active site zinc by forming a 1,4-tight turn in an essentially identical conformation relative to the zinc. This β-turn contains a conserved methionine in all members of the four proteinase subfamilies. This unique structure is called the 'Met-turn' [19] and its sulfur atom is located about 5 Å away from the zinc. The Met-turn forms a hydrophobic base for the catalytic zinc ion and the three histidine ligands. These four subfamilies of metalloendopeptidases are collectively called 'metzincins' [19]. The 3D structural coordinates for matrixins deposited in the Brookhaven Protein Data Bank are listed in Table 5.

Overall structures of catalytic domains of MMPs

Crystallization of a matrixin was first reported for the 44-kDa porcine synovial MMP-1 in 1989 [1384], but those crystals were not adequate for structural determination. The crystal structures of matrixins were determined for human MMP-1 [1349, 1386, 1387, 1410] and human MMP-8 [1348, 1411] by four independent groups and reported at about the same time. The crystal structures were also determined for the catalytic domain of MMP-3 (MMP-3(ΔC)) [1343, 1361], the C-terminal truncated proMMP-3 [1343], the catalytic domain of MMP-7 [1350], the full-length porcine MMP-1 [1382], the MMP-3(ΔC)–TIMP-1 complex [1371] and the catalytic domain of MT1-MMP (MMP-14)–TIMP-2 complex [1365]. These structures revealed that the catalytic

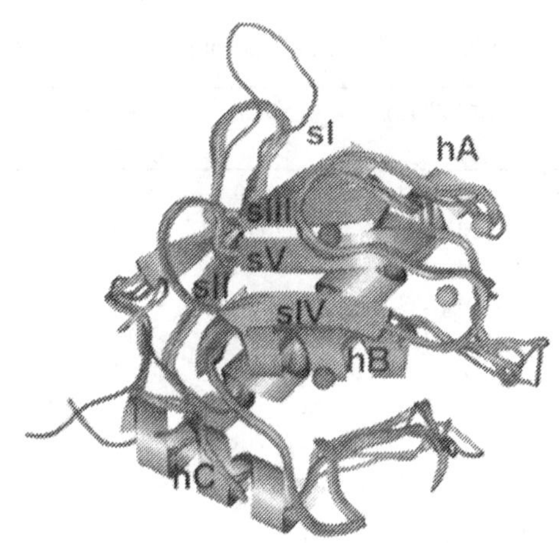

Figure 5

Ribbon diagram of the catalytic domains of matrixins. Catalytic domains of MMP-3, MMP-7, MMP-8, and MMP-14 are superimposed. Strands and helices are labelled as sI–sIV and hA–hC, respectively. The catalytic and the structural zinc (centre and top) and the three calcium ions (flanking) are shown as darker and lighter spheres, respectively. MMP-14 has an extended loop between sII and sIII. Diagram kindly provided by Dr K. Maskos and Prof. W. Bode, Munich.

domains of all these MMPs are essentially superimposable and that all have a five-stranded β-sheet, three α helices, two zinc ions and one to three calcium ions (Fig. 5). The active site cleft is bordered by β-strand IV (sIV), helix B (hB), and a stretch of random coil adjacent to the COOH-terminus of helix B. Of the two zinc ions, one is catalytic and the other is structural. The catalytic zinc is ligated by three conserved histidines in the HEXGHXXGXXH sequence at a distance of 2.1 Å. The structural zinc is coordinated by His149Nε2, Asp151Oδ2, His164Nε2 and His177Nδ1 in a tetrahedral manner in the case of MMP-1. (See Fig. 6 for amino acid numbering.) These four ligands are completely conserved within the matrixin subfamily and the presence of the second zinc is a feature of matrixins. Wetmore and Hardman [1419] propose that the structural zinc of MMP-3 is critical for maintaining the active form of the enzyme by pulling the contiguous loops, which form the zinc- and calcium-binding sites (His151 to His166 in MMP-3), away from the active site and helping to organize

Table 5 Brookhaven Protein Data Bank three-dimensional structures

PDB No.	MMP No.	Domains	Resolution (Å)	Submitted Mo/Da/Yr	Comments	Ref.
1CGL	MMP-1	Catalytic	2.4	11/17/93	Complexed to carboxylate inhibitor	[1386]
1CGF	MMP-1	Catalytic	2.1	02/03/94	Binary complex	[1387]
1CGE	MMP-1	Catalytic	1.9	02/03/94	Complexed with carboxylate inhibitor	[1387]
2TCL	MMP-1	Catalytic	2.2	09/07/94	Complexed to hydroxamate inhibitor	[1349]
1HFC	MMP-1	Catalytic	1.56	09/13/94	Complexed to hydroxamate inhibitor	[1410]
1FBL	MMP-1	Catalytic, hemopexin	2.5	04/24/95	Complexed to hydroxamate inhibitor (This is pig enzyme*)	[1382]
1AYK	MMP-1	Catalytic		11/06/97	No inhibitor, NMR of 30 structures	[1396]
2AYK	MMP-1	Catalytic		11/06/97	NMR minimized average structure	[1396]
1GEN	MMP-2	Hemopexin	2.15	07/19/95		[1383]
1RTG	MMP-2	Hemopexin	2.6	12/21/95		[1367]
2SRT	MMP-3	Catalytic		03/22/95	NMR of 30 structures	[1372]
1SLN	MMP-3	Catalytic	2.27	08/03/95	Complex with carboxyalkyl inhibitor	[1343]
1SLM	MMP-3	Propeptide, catalytic	1.90	08/03/95	Includes part of propeptide	[1343]
1UMS	MMP-3	Catalytic		10/31/95	NMR of 20 hydroxamate complexes	[1416]
1UMT	MMP-3	Catalytic		10/31/95	NMR minimized average structure	[1416]
1UEA	MMP-3	Catalytic	2.80	06/06/97	Complex with TIMP-1 inhibitor	[1371]
1HFS	MMP-3	Catalytic	1.70	02/13/97	Complex with carboxyalkyl inhibitor	[1363]
1USN	MMP-3	Catalytic	1.8	06/09/98	Complex with thiadiazole inhibitor	[1366]
2USN	MMP-3	Catalytic	2.2	06/09/98	Complex with different thiadiazole	[1366]
1MMP	MMP-7	Catalytic	2.3	03/22/95	Complex with carboxylate inhibitor	[1350]
1MMQ	MMP-7	Catalytic	1.9	03/22/95	Complex with hydroxamate inhibitor	[1350]
1MMR	MMP-7	Catalytic	2.3	03/22/95	Complex with sulfodiimine inhibitor	[1350]
1MNC	MMP-8	Catalytic	2.1	01/12/94	Complex with hydroxamate inhibitor	[1411]
1MMB	MMP-8	Catalytic	2.1	08/23/95	Complex with hydroxamate inhibitor	[1375]
1JAN	MMP-8	Catalytic	2.5	03/11/96	Complex with hydroxamate inhibitor	[1406]
1JAO	MMP-8	Catalytic	2.40	03/11/96	Complex with thiolpeptide inhibitor	[1375]
1JAP	MMP-8	Catalytic	1.82	03/11/96	Complex with hydroxamate inhibitor	[1348]
1JAQ	MMP-8	Catalytic	2.25	03/11/96	Complex with hydroxamate inhibitor	[1375]
1KBC	MMP-8	Catalytic	1.81	04/29/97	Complex with hydroxamate inhibitor	[1345]
1PEX	MMP-13	Hemopexin	2.7	05/24/96		[1368]
2TMP	TIMP-2	N-terminal		05/26/98	NMR of 49 structures	[1399]

*All enzymes in this table are of human origin except for pig collagenase.

the active site pocket, particularly the S1′ substrate specificity pocket. One calcium was found in the carboxylate or hydroxamate inhibitor-inhibited human MMP-1 [1386, 1410]. It was ligated in an octahedral manner by Asp156Oδ1, Asp179Oδ1, Glu182Oε2 and the main chain carbonyl oxygens of Gly157, Gly159 and Asn161. The Ca^{2+} and the structural Zn^{2+} are ligated through the β sheet and the highly exposed large S-shaped loop (residues 146 to 162) between strand III and IV; thus they stabilize the loop onto the β sheet.

Lovejoy *et al.* [1387] also reported two other crystal structures of MMP-1 without the carboxylate inhibitor, and one was determined at 1.9 Å resolution. The striking difference between those two crystal structures and the one shown previously for inhibited MMP-1 is the different conformations of the N-terminal four residues (Leu83-Thr-Glu-Gly). In the inhibitor-free crystals, residues 83–86 occupy the active site cleft of a symmetry-related molecule, and due to crystallographic symmetry the interaction of the N-terminal peptide of one MMP-1 molecule and the active site cleft of an adjacent MMP-1 results in the formation of an infinite lattice of MMP-1 molecules. Both the amino groups and the carbonyl oxygen of the N-terminal Leu83 interact with the symmetry-related catalytic zinc, and the N-terminal peptide interacts like a substrate with the active site cleft with Thr84 extended into a deep hydrophobic S1′ pocket in a manner similar to a peptide carboxylate inhibitor. These structures also revealed two additional calcium ions: one interacting with Asp105Oδ1, Asp105Oδ2, Glu180Oε2, Glu180O and two water molecules; and the other with Asp139O, Gly171O, Gly173O, Asp175Oδ1 and two water molecules. The number of Ca^{2+} ions reported in the catalytic domains is one [1386, 1410], two [1348–1350, 1365] or three [1343, 1361, 1382, 1387]. The sites identified for metal coordination are summarized in Table 6. Amino acids involved in metal binding sites are conserved in matrixins (Fig. 6 and Table 6).

Substrate binding site

Structures of matrixins inhibited by a carboxyalkylamide-, thiol- or hydroxamate-based peptide inhibitor have provided information on how the peptide substrate may interact with the catalytic domain of MMPs. Figure 7(A) illustrates the

Table 6 Metal coordination sites

Metal ion	Enzyme				
	MMP-1	MMP-3	MMP-7	MMP-8	MMP-14
Catalytic Zn^{2+}	His199Nε2	His201Nε2	His197Nε2	His197Nε2	His239Nε2
	His203Nε2	His205Nε2	His201Nε2	His201Nε2	His243Nε2
	His209Nε2	His211Nε2	His207Nε2	His207Nε2	His249Nε2
Structural Zn^{2+}	Asp151Oδ1	Asp153Oδ2	Glu149Oε2	Asp149Oδ2	Asp188Oδ2
	His149Nε2	His151Nε2	His147Nε2	His147Nε2	His186Nε2
	His164Nε2	His166Nε2	His162Nε2	His162Nε2	His201Nε2
	His177Nδ1	His179Nδ1	His175Nδ1	His175Nδ1	His214Nδ1
Ca^{2+} (1)	Asp156Oδ1	Asp158Oδ1	Asp153Oδ1	Asp154Oδ1	Asp199Oδ1
	Gly157O	Gly159O	Gly154O	Gly155O	Gly194O
	Gly159O	Gly161O	Gly156O	Asn157O	Gly196O
	Asn161O	Val163O	Thr158O	Ile159O	Phe198O
	Asp179Oδ2	Asp181Oδ2	Asp175Oδ2	Asp177Oδ2	Asp216Oδ2
	Glu182Oε2	Glu184Oε2	Glu179Oε2	Glu180Oε2	Glu199Oε2
Ca^{2+} (2)	Asp139O	Asp141O	Asp136O	Asp137O	Asp176O
	Gly171O	Gly173O	Gly167O	Gly171O	Asn208O
	Gly171O	Asn175O	Gly169O	Gly169O	Gly210O
	Asp175Oδ1	Asp177Oδ1	Asp171Oδ1	Asp173Oδ1	Asp212Oδ1
	water	water	water	water	water
	water	water	water	water	water
Ca^{2+} (3)	Asp105Oδ1	Asp107Oδ1			
	Asp105Oδ2	Asp107Oδ2			
	Gly180Oε2	Asp182Oδ1			
	Gly180O	Glu184O			
	water	water			
	water	water			

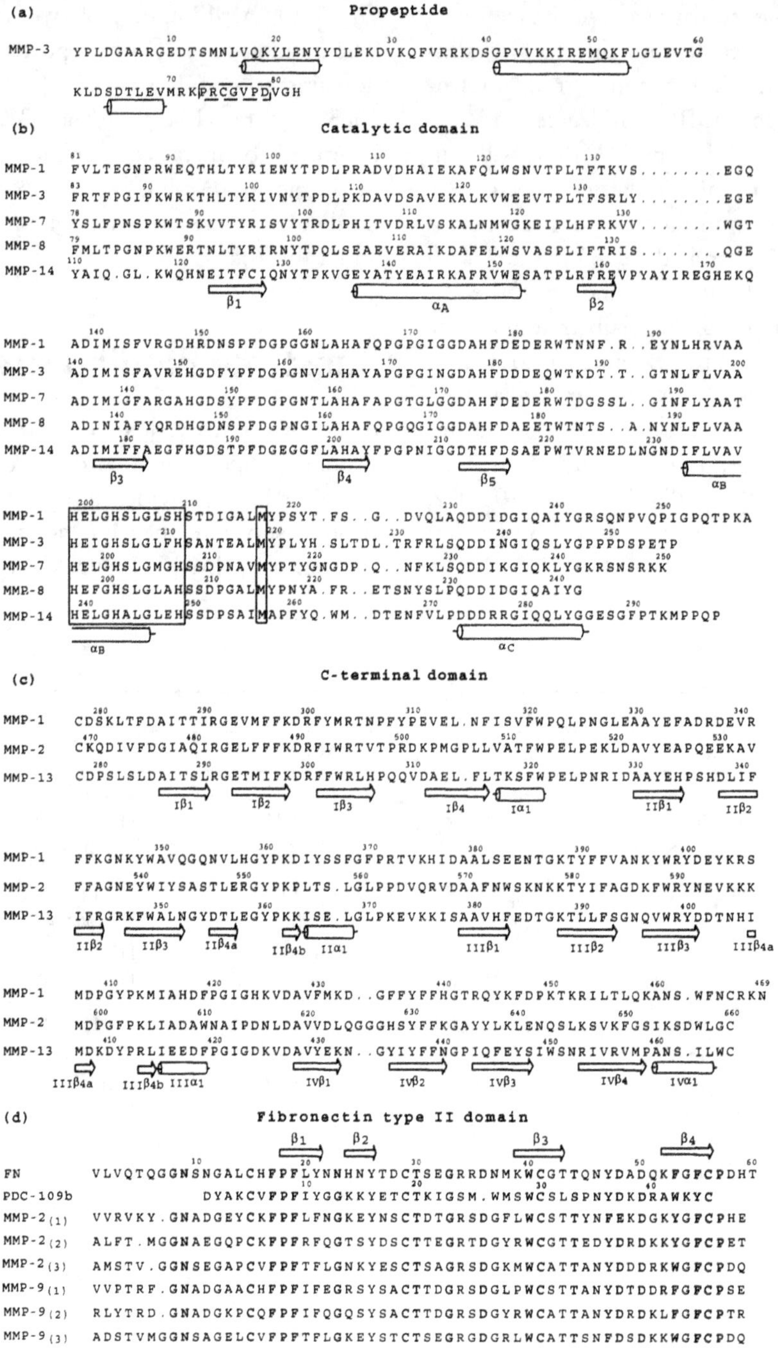

(a)   **Propeptide**

```
                    10                  20                  30                  40                  50                  60
MMP-3   Y P L D G A A R G E D T S M N L V Q K Y L E N Y Y D L E K D V K Q F V R R K D S G P V V K K I R E M Q K F L G L E V T G

                    70                  80
        K L D S D T L E V M R K P R C G V P D V G H
```

(b)   **Catalytic domain**

```
        81            90            100           110           120           130
MMP-1   F V L T E G N P R W E Q T H L T Y R I E N Y T P D L P R A D V D H A I E K A F Q L W S N V T P L T F T K V S . . . . . . . . E G Q
        83            90            100           110           120           130
MMP-3   F R T F P G I P K W R K T H L T Y R I V N Y T P D L P K D A V D S A V E K A L K V W E E V T P L T F S R L Y . . . . . . . E G E
        78            90            100           110           120           130
MMP-7   Y S L F P N S P K W T S K V V T Y R I S V Y T R D L P H I T V D R L V S K A L N M W G K E I P L H F R K V V . . . . . . . W G T
        79            90            100           110           120           130
MMP-8   F M L T P G N P K W E R T N L T Y R I R N Y T P Q L S E A E V E R A I K D A F E L W S V A S P L I F T R I S . . . . . . . Q G E
        110           120           130           140           150           160           170
MMP-14  Y A I Q . G L . K W Q H N E I T F C I Q N Y T P K V G E Y A T Y E A I R K A F R V W E S A T P L R F R E V P Y A Y I R E G H E K Q
                                        β1                    αA                    β2
```

```
        140           150           160           170           180           190
MMP-1   A D I M I S F V R G D H R D N S P F D G P G G N L A H A F Q P G P G I G G D A H F D E D E R W T N N F . R . . E Y N L H R V A A
        140           150           160           170           180           190           200
MMP-3   A D I M I S F A V R E H G D F Y P F D G P G N V L A H A Y A P G P G I N G D A H F D D D E Q W T K D T . T . . G T N L F L V A A
        140           150           160           170           180           190
MMP-7   A D I M I G F A R G A H G D S Y P F D G P G N T L A H A F A P G T G L G G D A H F D E D E R W T D G S S L . . G I N F L Y A A T
        140           150           160           170           180           190
MMP-8   A D I N I A F Y Q R D H G D N S P F D G P N G I L A H A F Q P G Q G I G G D A H F D A E E T W T N T S . A . N Y N L F L V A A
        180           190           200           210           220           230
MMP-14  A D I M I F F A E G F H G D S T P F D G E G G F L A H A Y F P G P N I G G D T H F D S A E P W T V R N E D L N G N D I F L V A V
                β3                    β4                    β5                    αB
```

```
        200                 210                 220                 230           240           250
MMP-1   H E L G H S L G L S H S T D I G A L M Y P S Y T . F S . . G . . D V Q L A Q D D I D G I Q A I Y G R S Q N P V Q P I G P Q T P K A
                            210                 220                 230           240           250
MMP-3   H E I G H S L G L F H S A N T E A L M Y P L Y H . S L T D L . T R F R L S Q D D I N G I Q S L Y G P P P D S P E T P
        200                 210                 220                 230           240           250
MMP-7   H E L G H S L G M G H S S D P N A V M Y P T Y G N G D P . Q . . N F K L S Q D D I K G I Q K L Y G K R S N S R K K
        200                 210                 220                 230           240
MMP-8   H E F G H S L G L A H S S D P G A L M Y P N Y A . F R . . E T S N Y S L P Q D D I D G I Q A I Y G
        240                 250                 260           270           280           290
MMP-14  H E L G H A L G L E H S S D P S A I M A P F Y Q . W M . . D T E N F V L P D D D R R G I Q Q L Y G G E S G F P T K M P P Q P
                αB                                                                            αC
```

(c)   **C-terminal domain**

```
        280           290           300           310           320           330           340
MMP-1   C D S K L T F D A I T T I R G E V M F F K D R F Y M R T N P F Y P E V E L . N F I S V F W P Q L P N G L E A A Y E F A D R D E V R
        470           480           490           500           510           520           530
MMP-2   C K Q D I V F D G I A Q I R G E L F F F K D R F I W R T V T P R D K P M G P L L V A T F W P E L P E K L D A V Y E A P Q E E K A V
        280           290           300           310           320           330           340
MMP-13  C D P S L S L D A I T S L R G E T M I F K D R F F W R L H P Q Q V D A E L . F L T K S F W P E L P N R I D A A Y E H P S H D L I F
                    Iβ1           Iβ2           Iβ3           Iβ4     Iα1                   IIβ1          IIβ2
```

```
        350           360           370           380           390           400
MMP-1   F F K G N K Y W A V Q G Q N V L H G Y P K D I Y S S F G F P R T V K H I D A A L S E E N T G K T Y F F V A N K Y W R Y D E Y K R S
        540           550           560           570           580           590
MMP-2   F F A G N E Y W I Y S A S T L E R G Y P K P L T S . L G L P P D V Q R V D A A F N W S K N K K T Y I F A G D K F W R Y N E V K K K
        350           360           370           380           390           400
MMP-13  I F R G R K F W A L N G Y D T L E G Y P K K I S E . L G L P K E V K K I S A A V H F E D T G K T L L F S G N Q V W R Y D D T N H I
        IIβ2      IIβ3      IIβ4a     IIβ4b  IIα1              IIIβ1         IIIβ2         IIIβ3         IIIβ4a
```

```
        410           420           430           440           450           460           469
MMP-1   M D P G Y P K M I A H D F P G I G H K V D A V F M K D . . G F F Y F F H G T R Q Y K F D P K T K R I L T L Q K A N S . W F N C R K N
        600           610           620           630           640           650           660
MMP-2   M D P G F P K L I A D A W N A I P D N L D A V V D L Q G G G H S Y F F K G A Y Y L K L E N Q S L K S V K F G S I K S D W L G C
        410           420           430           440           450           460
MMP-13  M D K D Y P R L I E E D F P G I G D K V D A V Y E K N . . G Y I Y F F N G P I Q F E Y S I W S N R I V R V M P A N S . I L W C
        IIIβ4a    IIIβ4b  IIIα1               IVβ1          IVβ2          IVβ3          IVβ4      IVα1
```

(d)   **Fibronectin type II domain**

```
                                        β1          β2                    β3                    β4
                    10                  20                  30                  40                  50                  60
FN      V L V Q T Q G G N S N G A L C H F P F L Y N N H N Y T D C T S E G R R D N M K W C G T T Q N Y D A D Q K F G F C P D H T
                            10                  20                  30            40
PDC-109b            D Y A K C V F P F I Y G G K K Y E T C T K I G S M . W M S W C S L S P N Y D K D R A W K Y C
MMP-2(1)  V V R V K Y . G N A D G E Y C K F P F L F N G K E Y N S C T D T G R S D G F L W C S T T Y N F E K D G K Y G F C P H E
MMP-2(2)  A L F T . M G G N A E G Q P C K F P F R F Q G T S Y D S C T T E G R T D G Y R W C G T T E D Y D R D K K Y G F C P E T
MMP-2(3)  A M S T V . G G N S E G A P C V F P F T F L G N K Y E S C T S A G R S D G K M W C A T T A N Y D D D R K W G F C P D Q
MMP-9(1)  V V P T R F . G N A D G A A C H F P F I F E G R S Y S A C T T D G R S D G L P W C S T T A N Y D T D D R F G F C P S E
MMP-9(2)  R L Y T R D . G N A D G K P C Q F P F I F Q G Q S Y S A C T T D G R S D G Y R W C A T T A N Y D R D K L F G F C P T R
MMP-9(3)  A D S T V M G G N S A G E L C V F P F T F L G K E Y S T C T S E G R G D G R L W C A T T S N F D S D K K W G F C P D Q
```

Figure 6

Sequence alignment of matrixin domains whose structures have been determined. Strands and helices are indicated by arrows and cylinders, respectively. (a) MMP-3 propeptide; (b) catalytic domains of MMP-1, MMP-3, MMP-7, MMP-8 and MMP-14; (c) hemopexin domains of MMP-1, MMP-2, and MMP-13; and (d) fibronectin type II domain (F2). Shaded residues are conserved in F2 domains. FN, fibronectin; PDC-109b, bovine seminal plasma protein PDC-109 domain b. Residue numbers in the catalytic domain of MMP-14 are after Fernandez-Catalan *et al.* [1427]. Residue numbers in the hemopexin domain in MMP-1 and MMP-13 are after Li *et al.* [1382] and those in MMP-2 are after Libson *et al.* [1383].

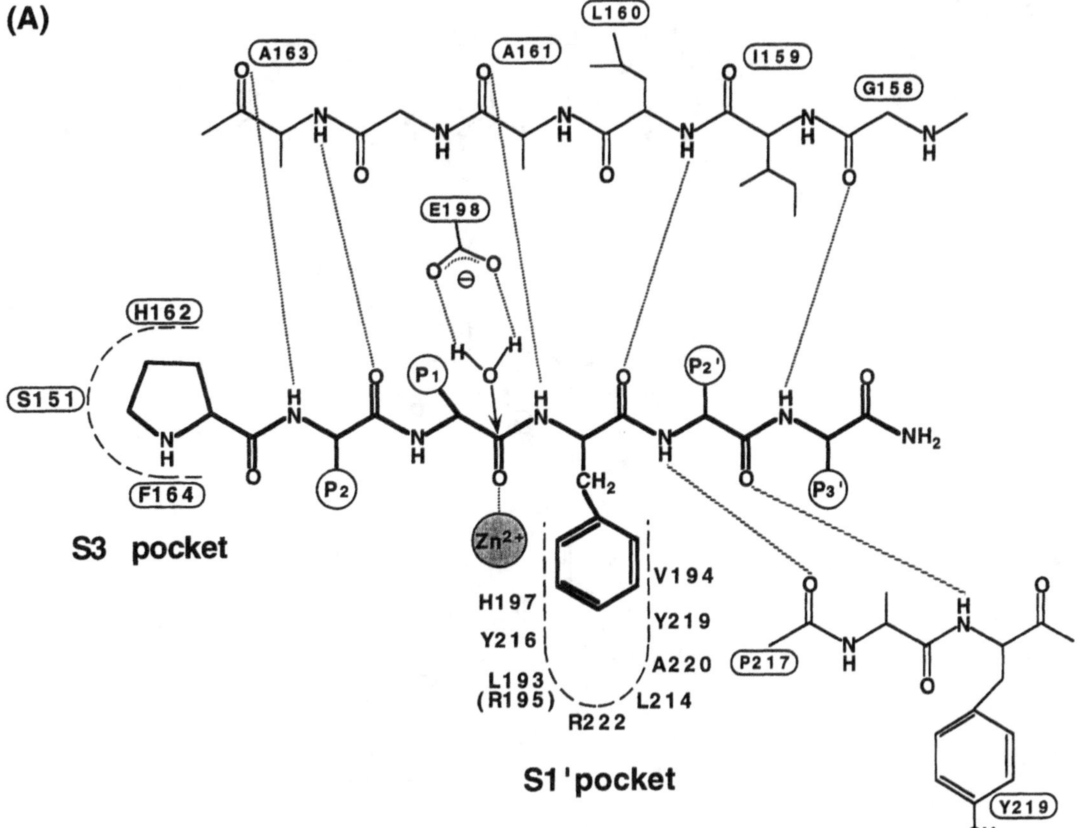

Figure 7

Schematic illustrations of the active site of matrixin with a substrate and the propeptide. (A) A putative binding interaction between a hexapeptide substrate (bold) and the active site of MMP-8 proposed by Grams *et al.* [1375] based on the structural analysis of MMP-8 complexes with Pro-Leu-Gly-NHOH and with SH-BzlPp-Ala-Gly-NH₂. The substrate peptide backbone lies antiparallel to the bulge strand (upper chain) and parallel to the S1′ pocket forming wall segment (Pro217-Asn-Tyr219) by forming five and two interchain hydrogen bonds (dashed lines), respectively. Dominant hydrophobic interactions are made through P1′ and P3 side chains of the substrate and the respective pockets. Residues that participate in forming S3 and S1′ pockets are illustrated. L193 in the S1′ pocket is replaced by R195 in MMP-1 and Y193 in MMP-7. R195 and Y193 significantly influence the specificity of the enzyme [1408, 1418]. Redrawn after Grams *et al.* [1375]. (B) Interaction between the propeptide of MMP-3 and the cognate active site reported by Becker *et al.* [1343]. The cysteine switch segment lies parallel to strand IV of the enzyme, antiparallel to the S1′ pocket wall, but it forms essentially identical interchain hydrogen bonds to those formed with a substrate. Residues in parentheses are corresponding residues in MMP-8. Redrawn after Becker *et al.* [1343].

putative mode of interaction of a hexapeptide and the active site of MMP-8 proposed by Grams *et al.* [1375] that was modelled from the interaction with inhibitors, Pro-Leu-Gly-NHOH and 2-benzyl-3-mercaptopropanoyl-Ala-Gly-NH₂ (SH-BzlPp-Ala-Gly-NH₂). In the structure of MMP-8 inhibited by Pro-Leu-Gly-NHOH [1348] the inhibitor binds to the unprimed subsite of the enzyme. It lies anti-parallel to the edge strand IV by forming two inter-chain hydrogen bonds between the P2 residue (Leu) and Ala163 (see Fig. 7(A)). The N-terminal Pro interacts with the hydrophobic cleft formed by side chains of His162, Ser151 and Phe164, whereas the P2 Leu interacts with a shallow groove lined by His201, Ala206 and His207. Pro at P3 and Leu at P2 are

(B)

preferred residues for matrixins [1401]. The carbonyl and hydroxyl oxygens of the hydroxamic acid moiety coordinate with the catalytic Zn^{2+} with an average metal–ligand distance of 2.1 Å. The SH-inhibitor binds to the primed subsite of the enzyme and its peptidyl moiety mimics binding of P1′–P3′ ('Phe'-Ala-Gly-NH_2) residues of the substrate [1375].

The main chain of the inhibitor runs antiparallel to the bulge segment Gly158-Ile-Leu and is simultaneously parallel to the crossover and S1′ wall-forming segment Pro217-Asn-Tyr by forming four hydrogen bonds (Fig. 7(A)). The sulfur atom of the inhibitor ligates to the catalytic Zn^{2+} with a sulfur–metal distance of 2.2 Å. The dominant interactions between the inhibitor and the enzyme are made by the phenyl side chain of the inhibitor and the central hydrophobic portion of the large S1′ pocket (located to the right of the catalytic zinc in Fig. 7(A)) formed by side chains of His197, Val194, Tyr219 and a main-chain segment

of Pro217-Asn-Tyr219. The S1′ pocket of MMP-8 is much more spacious than any substrate side chains of natural amino acids, and the phenyl group of the inhibitor occupies only 1/3–1/2 of the inner volume of the pocket, leaving space for three ordered water molecules [1375]. The side chain of P2′ Ala of the inhibitor points away from the enzyme and P3′ Gly is located between crossing over segments Gly158–Leu160 and Pro217–Tyr219. The P2′–P3′ residues of a substrate would have to pass through an opening 5 Å wide formed by the segments Asn157–Gly158 (part of the calcium loop) and Asn218–Tyr219 (forming the edge of the S1′ pocket), which provide several amide groups for the main chain–main chain interactions [1348]. The substrate binding mode predicted from the MMP-8 complexes with Pro-Leu-Gly-NHOH and SH-BzlPp-Ala-Gly-NH_2 is illustrated in Fig. 7(A).

The most notable difference among MMP structures is the size of the S1′ specificity pockets.

Comparing those of the two collagenases, MMP-1 and MMP-8, the latter has the larger S1′ pocket. The main difference is due to the side chain of Arg195 in MMP-1, which delimits the S1′ pocket by projecting out towards the catalytic zinc beyond the Arg195Cβ atom, whereas Leu193 is found in the corresponding site in MMP-8 (Fig. 7(A)). Leu193 enters the S1′ pocket but it is orientated away from the catalytic Zn^{2+} beyond the Leu193Cβ atom [1411]. MMP-3 and MT1-MMP also have Leu197 at this position and form a very large, predominantly hydrophobic, S1′ pocket that extends completely through the body of the molecule [1343, 1361,1365]. MMP-7 has Tyr193 in the corresponding position, which limits the size of S1′ pocket [1350]. The enzyme prefers residues with aliphatic to aromatic side chains in the P1′ position [1402]. The mutation of Tyr193 to Leu in MMP-7 altered the P1′ specificity making it similar to that of MMP-3, and reversed results were reported with an MMP-3(L197Y/V194A) mutant [1418]. The S1′ pockets of MMP-2 and MMP-9 are thought to have a similar dimension as that of MMP-8 [1388].

NMR solution structures of MMPs

The secondary structure and zinc ligation of the 19.5-kDa catalytic domain of human MMP-3 were assigned by multidimensional heteronuclear NMR analyses [1372, 1417] prior to the determination of tertiary structures of MMPs. These studies revealed a five-stranded β-sheet with four parallel strands and one antiparallel strand and three α-helices. The earlier observation of the presence of two zinc ions in the catalytic domain of MMP-3 [1407] was confirmed by identifying His201, His205 and His211 as ligands of the catalytic zinc and His151, His166, and His179 as ligands of the structural zinc [1372]. The topological arrangement of the secondary structure elements was remarkably similar to that of astacin [1347]. The initial 3D structure of the catalytic domain

(Phe83–Pro250) of MMP-3 inhibited by an N-carboxyalkyl inhibitor was determined [1374] and its refinement indicated general agreement with the X-ray crystal structure [1373]. However, the root mean square deviation of the backbone atoms showed significant deviations for the regions Phe83–Gly88, His224–Arg231 and Pro249–Pro250. The region, Pro249–Pro250 was not visible in the crystal structure, and it is probably not structurally or functionally significant.

The N-terminal Phe83–Gly88 influences the catalytic activity and specificity of MMP-3 [1344]. An N-terminal truncation or an activation intermediate with an extended N-terminus of MMP-3 is less active than the correctly processed enzyme. Similar results are observed with MMP-1 and MMP-8 [1381, 1413]. The crystal structure of the N-terminal truncated MMP-8 shows a flexible N-terminal hexapeptide and lack of a salt bridge between the ammonium group of the N-terminus and the carboxylate group of the conserved Asp232. This salt bridge is present in the crystal structure of correctly processed MMP-3(ΔC) [1343] but it is not always apparent in the solution structure. The distances between the nitrogen of the N-terminal Phe83 and the two oxygens of the carboxylate of Asp235 are 5.7 $\pm$ 2.0 Å and 5.3 $\pm$ 1.4 Å ranging from 2.7 to 11 Å for the 30 conformers, indicating that the salt bridge is partially disrupted in solution.

The region His224–Arg231 differs significantly between the solution and crystal structures. In the solution structure, this region is poorly characterized, suggesting that the backbone and the side chain of these residues are flexible. Although it is considered that the large root mean square deviation values may be due to a lack of sufficient data rather than flexibility of this region, the temperature factor of residues 224–231 in the crystal structure is sufficiently higher (22.8 $Å^2$) than those of the average protein atoms (11.5 $Å^2$), supporting a flexible structure. It is also notable that the region of Tyr223–Arg231 forms the base and opening of the S1′ pocket of the enzyme [1373].

The solution NMR structures of the inhibitor-free catalytic domains of MMP-1 were also reported [1391, 1395] and their secondary and tertiary structures are identical to those determined by the X-ray crystal structure of inhibited MMP-1. However, a comparison of the high-resolution NMR structure of the inhibitor-free MMP-1 [1395] and the 1.56 Å resolution X-ray structure of the hydroxamate-inhibited MMP-1 [1410] shows structural differences in the vicinity of the active site of the enzyme. This includes the Ca^{2+}-loop (Phe155–Asn161) and residues Pro219–Gly225. The Ca^{2+}-binding loop is 'pushed-up' relative to that in the crystal structure. This region is considered to be flexible, but after incorporation of Ca^{2+} ion into the refinement protocol, this region becomes well-defined [1395]. This indicates that chelation of the Ca^{2+} ion by residues 155–161 is essential for establishing the proper local structure. Because these residues are close to the active site, it is suggested that proper chelation of Ca^{2+} is crucial for enzyme activity, a concept previously developed by Wetmore and Hardman [1419]. Distinct differences are also observed for side-chain torsion angles, particularly the χ_1 for Asn161 of $-60°$ in the NMR structure compared to $-179°$ in the X-ray structure. This indicates that the side chain of Asn161 partially blocks access to the active site of MMP-1. Residues 219–225 participate in forming the S1′ pocket of the enzyme. It is suggested that an inherent flexibility in the protein in this region may be crucial for proper inhibitor/substrate binding [1395]. A *cis*-peptide bond for Glu190–Tyr191 found in the X-ray structure was not identified. Residues 183–191 have been shown to play a critical role in expression of collagenolytic activity in the full-length MMP-1 (L. Chung and H. Nagase, unpublished observation), suggesting that a high level of disorder of this region may account for interaction of MMP-1 with the collagen substrate. The use of ^{15}N-Leu labelled MMP-1 [1391] suggested that Leu162 is also highly mobile in the inhibitor-free MMP-1. The crystal structure of the inhibited MMP-1 demonstrated a hydrogen bond between the Leu162 amide nitrogen and the P1′ carbonyl of the inhibitor [1410].

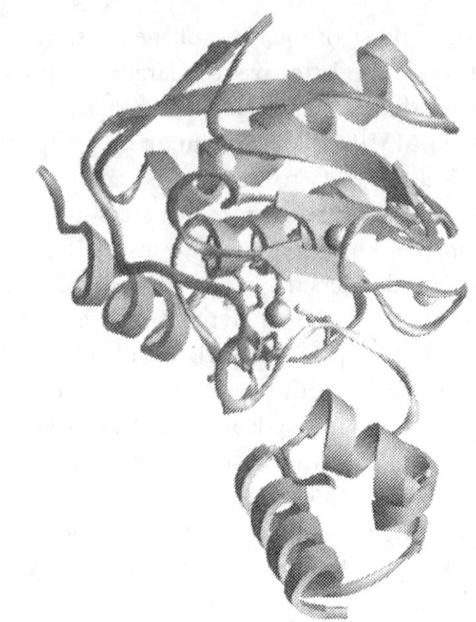

Figure 8

Ribbon diagram of the C-terminal-truncated proMMP-3. The image was prepared from the Brookhaven Protein Data Bank entry (1SLM), deposited by Becker *et al.* [1343], using RIBBONS 2.87 [1351a]. The upper and the lower domains are the catalytic and the propeptide domains, respectively. The catalytic and structural zinc ions are shown as dark grey spheres and the two calcium ions as light grey spheres. The catalytic zinc ligands (His201, His205, His211) and Cys75 in the cysteine switch sequence PRCGVPD (73–79) are shown in ball-and-stick representation.

Propeptide domain

The crystal structure of the C-terminal domain-truncated proMMP-3 revealed that the 82-amino acid pro-domain is a separate folding unit consisting of three α-helices and the extended peptide region, the so-called 'cysteine switch' (Pro73-Arg-Cys-Gly-Val-Pro-Asp) that lies in the active site of the enzyme [1343] (Fig. 8). The N-terminal 15 residues and the proteinase-susceptible 'bait region' (positions 31–40) are not visible in the electron density maps suggesting a flexible nature for these regions. The crystal structure of the N-carboxyalkyl peptide-inhibited catalytic domain of MMP-3 and proMMP-3 indicates that the substrate binding groove of the active site of the enzyme is occupied by extended peptide chains,

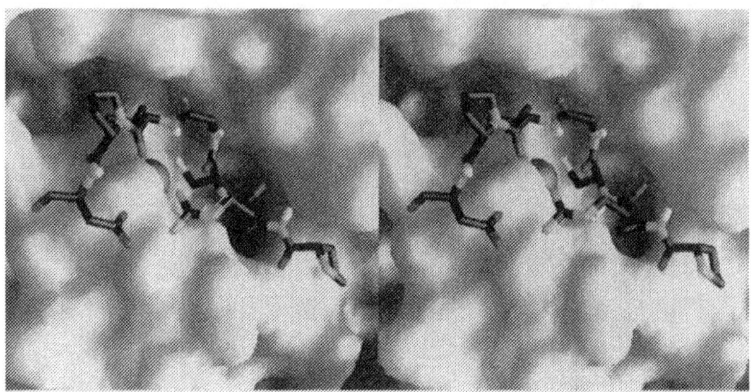

Figure 9

Stereo view of the active site groove of proMM-3. The image was prepared from the Brookhaven Protein Data Bank entry (1SLM) deposited by Becker *et al.* [1343] using GRASP 1.2 [1402a]. The catalytic site of MMP-3 is represented by a solid surface in grey, and PRCGVPD (73–79; from right to left) in the propeptide is shown in stick representation. The catalytic zinc ion is shown as a grey sphere. The deep S1' pocket located below Arg74 is unoccupied, and the side chain of Cys75 acts as a ligand to the catalytic zinc in proMMP-3.

by making several β-structure-like hydrogen bonds with the enzyme in both cases. Thus, the propeptide and the inhibitor make similar interactions with the same groups of the enzyme, but a striking difference is that the direction of the propeptide chains is opposite that of the inhibitor (Figs 7(B) and 9). Similar opposite directional arrangements of the propeptide segment and the peptide substrate in the substrate-binding cleft are also found in procathepsin B [1359, 1415] and procathepsin L [1357]. In proMMP-3, the S1' pocket is empty, but it is occupied by the P1' side chain of the inhibitor in the inhibited MMP-3. The guanidino group of Arg74 and the carboxylate of Asp79 in the cysteine switch region form a salt bridge, and the Cys75 of the propeptide directly interacts with the catalytic zinc ion, indicating that the Zn^{2+}–Cys75 interaction is crucial for maintaining the inactive form of proMMPs.

The structure of proMMP-3 also provides insights into the activation processes of the zymogen. Activation of proMMP-3 proceeds in a stepwise manner which involves initial cleavage of the N-terminal portion of the propeptide and the subsequent removal of the remaining pro-domain by cleavage of the His82–Phe83 bond by a bimolecular reaction of activated MMP-3 intermediates [1400]. The His82–Phe83 bond is sterically hindered and not accessible to MMP-3. Proteolytic attack on the bait region located around residues 34–39, which is flexible and most likely highly exposed to the solvent, probably results in the conformational rearrangement around the His82–Phe83 bond. This allows an efficient cleavage of this bond by the activated MMP-3. The exact activation mechanism of proMMPs by APMA is not clear, but APMA treatment of proMMP-3 results in generation of several intermediates, and the initial cleavage is thought to be due to the intramolecular reaction [1351, 1400, 1403]. Based on the three-dimensional structure of proMMP-3 it is suggested that residues 56–59 in the loop between helices 2 and 3 in the propeptide may be the site of intramolecular cleavage as they are accessible to the active site without substantial conformational changes in the proenzyme [1343].

The structure of the catalytic domain is relatively unchanged upon activation of proMMP-3 except for a rearrangement of residues 83–89. In proMMP-3, residues 83–89 are a part of large loop that links the first β-strand of the catalytic domain and the cysteine switch sequence (residues 73–79). Through the activation process, residues 83–89 move to a completely different location, and the newly generated N-terminal ammonium group of Phe83 forms a salt bridge with the car-

Figure 10

Stereo ribbon diagram of hemopexin domain. The image was prepared from the Brookhaven Protein Data Bank entry (1GEN) deposited by Libson *et al.* [1383] using RIBBONS 2.87 [1351a]. The first β-sheet (blade) starts from the upper left corner, and the disulfide bond is at the top. Ca^{2+}, Cl^- and Na^+ ions are shown as white, grey and black spheres, respectively.

boxylate group of Asp237 in helix C in the catalytic domain, by moving more than 17 Å away from the original position in the proenzyme. The formation of this salt bridge is important for the expression of the full enzymatic activity [1344]. Without it the substrate specificity of MMP-3 alters, but the structural basis for the altered activity and substrate specificity is not clear, since the site of the salt bridge is about 12 Å away from the catalytic zinc ion. Nonetheless, the flexible nature of the N-terminal peptide without the N-terminal Phe83 may interfere with the binding of a substrate in the extended active site. It is notable, however, that the NMR solution structure of MMP-3(ΔC) indicates that the salt bridge is partially disrupted in solution [1373].

Hemopexin domain

The 3D structures of hemopexin/vitronectin-like C-terminal domain have been determined by X-ray crystallography for porcine MMP-1 (determined as the full-length MMP-1 [1382], human MMP-2 [1367, 1383] and human MMP-13 [1368]). Figure 10 shows a stereo ribbon diagram of the hemopexin domain of MMP-2 [1383]. The domain has the shape of an ellipsoidal disk and the overall folds of the hemopexin domains from

three MMPs are very similar, being organized as a β-propeller structure with four β-sheets (blades) I to IV, which are almost symmetrically arranged around a central axis. Each sheet is made up of four anti-parallel β-strands and the four β-sheets have very similar scaffold architectures. The structure is homologous to the C-terminal domain of hemopexin [1364]. The innermost first strands of all four blades run parallel to each other along the propeller axis and form a central channel with a radius of -3.5 Å [1383]. Because of a slight curvature in each strand, the channel opens slightly towards the exit side like a funnel. Within this channel a number of ions and water molecules are found.

In the hemopexin domain of MMP-2, Libson *et al.* [1383] assigned one Ca^{2+} with a temperature factor $B = 15$ Å^2 based on its coordination geometry and scattering power. This Ca^{2+} is at the opening (entrance) of the channel and each inner strand of the channel binds the Ca^{2+} ion through the backbone carbonyl oxygens of Asp476, Asp521, Asp569 and Asp618. The side chains of these Asp residues are not coordinated to the Ca^{2+} ion. In addition to four carbonyl oxygens, Ca^{2+} is coordinated by three water molecules. The binding affinity of this domain for Ca^{2+} is thought to be extremely strong since the protein buffer contained 4 mM EDTA without exogenous Ca^{2+}

[1383]. A Ca^{2+} ion is also assigned in the corresponding sites of MMP-1 [1382] and MMP-13 [1368] structures. Deeper in the channel, about 5–6 Å from Ca^{2+}, is found an Na^+ Cl^- ion pair. This ion pair was not in the MMP-1 structure but was found in the hemopexin structure [1364]. The Na^+ ion is octahedrally coordinated, with four backbone carbonyl oxygens in an equatorial plane about the Na^+ ion. The axial ligands are a water molecule and a Cl^- ion positioned in the channel interior. Libson *et al.* [1383] also noted the presence of a Zn^{2+} ion that was tetrahedrally coordinated between residues of symmetry related molecules. Gohlke *et al.* [1367] also found four ions/solvent molecule within the channel of MMP-2, which were tentatively assigned as two Ca^{2+} ions, a Cl^- ion and a water molecule, based on their scattering power and coordination geometry.

In the hemopexin structure of MMP-13, two Ca^{2+} ions and two Cl^- ions have been assigned based on their scattering power and their coordination geometry [1368]. The first Ca^{2+} corresponds to the one in the MMP-2 structure. Towards the centre of the channel three more electron density spheres were found, which were assigned as Cl^-, a second Ca^{2+} and Cl^- ions, respectively. The second Ca^{2+} ion is octahedrally coordinated with four carbonyl oxygens of residues in each innermost strand (Ile287, Ala331, Ala380 and Val429) and two Cl^- ions. These residues in the MMP-13 structure correspond exactly to the residues in MMP-2 (Ile478, Val523, Ala571 and Val620) that coordinate Na^+ ion in the MMP-2 structure [1383]. The two Cl^--ion axial ligands of the second Ca^{2+} are at distances of 2.9 Å and 3.3 Å, which are within the range for calcium chloride ionic bonds. The distance from the inner Cl^- ion to Ca^{2+} ion at the entrance is 5.6 Å, far beyond typical calcium chloride ionic bond length. The fourth strands of blades II and III are interrupted by characteristic β-bridges consisting of four residues (X-G/D-Y-P). Because those prolines are in *cis* conformation, the fourth strands are kept in phase with the antiparallel strand 3. In all four blades the fourth strand loops around the periphery of the disk and ends with a 1 or 1½ turn α-helix. The chain then turns towards the entrance of the channel and starts the innermost strand 1 of the subsequent blade. The C-terminus of the blade IV is connected to the beginning of the blade I through a disulfide bridge of Cys278–Cys466 (Cys469–Cys660 in MMP-2).

The innermost strands commonly contain smaller hydrophobic residues such as Ala and Val, whereas the aromatic groups (Phe, Tyr, Trp) are found mainly on the second and the third sheets. Hydrophobic residues make up the bulk of the hydrophobic core of the domains. The surface-exposed strands and α-helices contain polar residues. Charged residues are clustered on the surface of the molecule. Although the surface charge distribution patterns of MMP-1 and MMP-13 are similar, they are different from that of MMP-2. A large positively charged surface patch is found on blade II and on blades I and III in collagenases, more acidic residues are found in blade II in MMP-2 [1368]. The exit side surface at blade III is mainly negatively charged in collagenases, but it is positively charged in MMP-2. At the strand 364 loop in this segment MMP-2 has one negatively charged (Glu593) and three positively charged residues (Lys595, Lys596, Lys597), whereas there are two negatively charged residues in collagenases (Asp401, Glu/Asp402).

Full-length MMP-1

The crystal structure for full-length porcine MMP-1 was determined by Li *et al.* [1382]. The structure revealed that the N-terminal catalytic domain and the first C-terminal hemopexin-like domain of the matrixins are joined by highly exposed linker peptide residues 261–277 (numbering from the start of the signal sequence) (Fig. 11). The backbone of the catalytic domain is quite similar to the catalytic domain of collagenases determined as isolates. The structure of the C-terminal hemopexin domain has approximately fourfold symmetry with four blades each containing a sheet of four anti-parallel β-strands. This

Figure 11

Ribbon diagram of full-length porcine MMP-1. The image was prepared from the Brookhaven Protein Data Bank entry (1FBL) deposited by Li *et al.* [1382] using RIBBONS 2.87 [1351a]. The catalytic and structural zinc ions are shown as black spheres, and the three calcium ions in the catalytic domain and one in the hemopexin domain are shown as grey spheres. A collagenase inhibitor, *N*-[3-*N'*-hydroxycarboxamido)-2-(2-methylpropyl)-propananoyl]-*O*-methyl-L-tyrosine-*N*-methylamide, bound to the active site is shown in space filling representation.

four-bladed β-propeller structure is similar to that of the C-terminal domain of rabbit hemopexin reported at the same time [1364] and to those subsequently reported for MMP-2 [1367, 1383] and MMP-13 [1368].

The linker peptide is rich in proline and has no secondary structure. It is most likely flexible and a good target for proteolysis. Upon cleavage of the linker peptide, the catalytic domain loses its ability to cleave the triple helical region of interstitial collagen types I, II and III, although it still retains general proteolytic activity [1346, 1352, 1397]. Similar results were observed with MMP-8 [1380] and MMP-13 [1378]. Those biochemical studies indicate that the hemopexin domain plays a crucial role in collagenolysis. Matrixins bind to substrates in an extended conformation through an opening ~5 Å wide [18, 1348, 1375, 1388]. Attempts to model the binding of the triple helical collagen between the active site and the hemopexin domain did not place a peptide bond of collagen triple helix into the active site because the diameter of the triple helix is too bulky [1348, 1382] and each polypeptide adopts an inappropriate conformation to allow correct interaction with the active site and the substrate binding site of the enzyme. It is therefore postulated that partial relaxation of the supercoiled triple helix is required for collagenolysis [18, 1348, 1382]. Such relaxation may be produced by the concerted binding of the catalytic domain and hemopexin domain. Both domains are folded as individual entities in the crystal and are loosely connected by a flexible 17-residue linker peptide. The correct pairing of the catalytic domain and the hemopexin domain appears to be critical for collagenolysis as the substitution of the MMP-1 hemopexin domain with that of MMP-3 results in little or no activity [1376, 1397]. Replacement of the linker peptide of MMP-8 by the 26-residue linker of MMP-3 abolishes activity against collagen [1376] suggesting a possible involvement of this region. Based on energy-minimized molecular modelling, de Souza *et al.* [1360] proposed that the linker peptide rich in proline forms a collagen-like structure that may interact with interstitial collagens and affect the stability of the triple-helix in the collagenase-sensitive region. Knäuper *et al.* [1379] reported that when four prolines (Pro247, Pro250, Pro253 and Pro256) in the linker peptide of MMP-8 were mutated to alanine the collagenolytic activity dropped to 1.5% of the value detected with the wild-type. However, clearer understanding of how collagenases cleave the triple helical collagen molecule requires further investigation.

Fibronectin type II module

MMP-2 and MMP-9 contain three repeats of fibronectin type II modules [1354, 1420] attached in the loop between strand V and helix B in their catalytic domains. These modules, each consisting of approximately 58 amino acids, are responsible for binding of MMP-2 and -9 to gelatin and collagen [1338–1340, 1353, 1398]. The module is characterized by the presence of highly conserved hydrophobic residues and four cysteines. The proteins that contain the fibronectin type II module include fibronectin [1404], macrophage mannose receptor [1414], phospholipase A2 receptor [1377], factor XII [1392], cation-dependent mannose-6-phosphate receptor [1385, 1394], hepatocyte growth factor activator [1393], bovine seminal plasma proteins PDC-109 [1362] and BSP-A3 [1409]. The 3D structures of a shorter fibronectin type II module of bovine seminal plasma protein PDC-109b [1355, 1356] and the first fibronectin type II module of human fibronectin [1405] (Protein Data Bank ID 2FN2) have been determined by NMR spectroscopy. The 3D structures of the first and second modules of fibronectin and PDC-109b are essentially identical. Because of the sequence similarity among fibronectin type II modules, the overall tertiary structures of fibronectin type II molecules in MMP-2 and MMP-9 are likely to be very similar to those of fibronectin and PDC-109b.

The fibronectin type II module consists of two double-stranded antiparallel β-sheets, which lie roughly perpendicular to each other, and two large irregular loops, one between the two β-sheets and the other between the two strands of the second sheet (Fig. 12). The two disulfide bonds are in close proximity to each other located on the opposite face of the second β-sheet. The strictly conserved Phe19 and Trp40 are buried in the first fibronectin domain structure and constitute the core of the molecule. The highly conserved Phe17, Tyr21, Tyr26, Tyr47, Phe53 and Phe55 cluster around this core and the side chains of Tyr21, Tyr47, Phe53 and Phe55 are on the surface of the molecule forming a solvent-exposed hydrophobic surface. This area is consid-

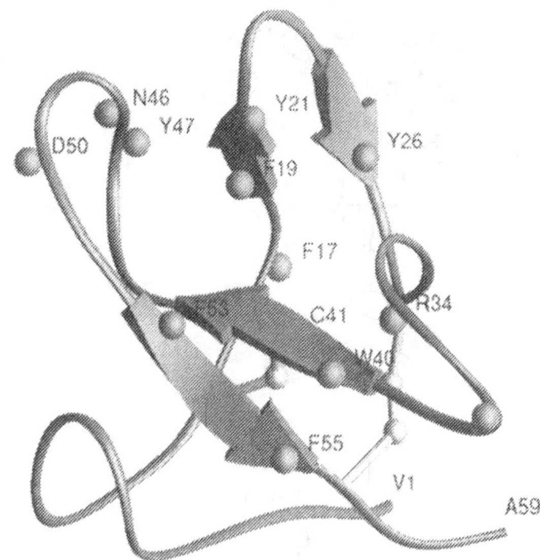

Figure 12

Ribbon diagram of type II fibronectin domain. The image was prepared from the Brookhaven Protein Data Bank entry (2FN2) deposited by Pickford *et al.* [1405] using RIBBONS 2.87 [1351a]. The location of αCs of the indicated residues are shown as grey spheres and disulfide bonds are in ball and stick representation.

ered as a possible ligand (gelatin/collagen) binding pocket with Trp40 at the bottom. Alanine substitution scanning studies of the fibronectin type II module of MMP-9 showed that a single mutation of residues equivalent to Arg34, Asp36, Trp40, Cys41, Asn46, Tyr47 and Asp50 in the first fibronectin type II module greatly reduced (>85%) affinity for gelatin [1353]. Residues Arg34 and Asp36 are conserved in fibronectin type II modules and Arg34 is in close proximity to the putative ligand binding pocket. Since the side chain of Asp36 is oriented towards that of Arg34, it has been suggested that a surface salt bridge is formed between the two residues and stabilizes the functional activity of this module [1405]. These residues are conserved in fibronectin type II modules in MMP-2 and MMP-9. Thus, similar structural arrangements may apply to them.

Mechanism of action

The 3D structures of matrixins revealed that key

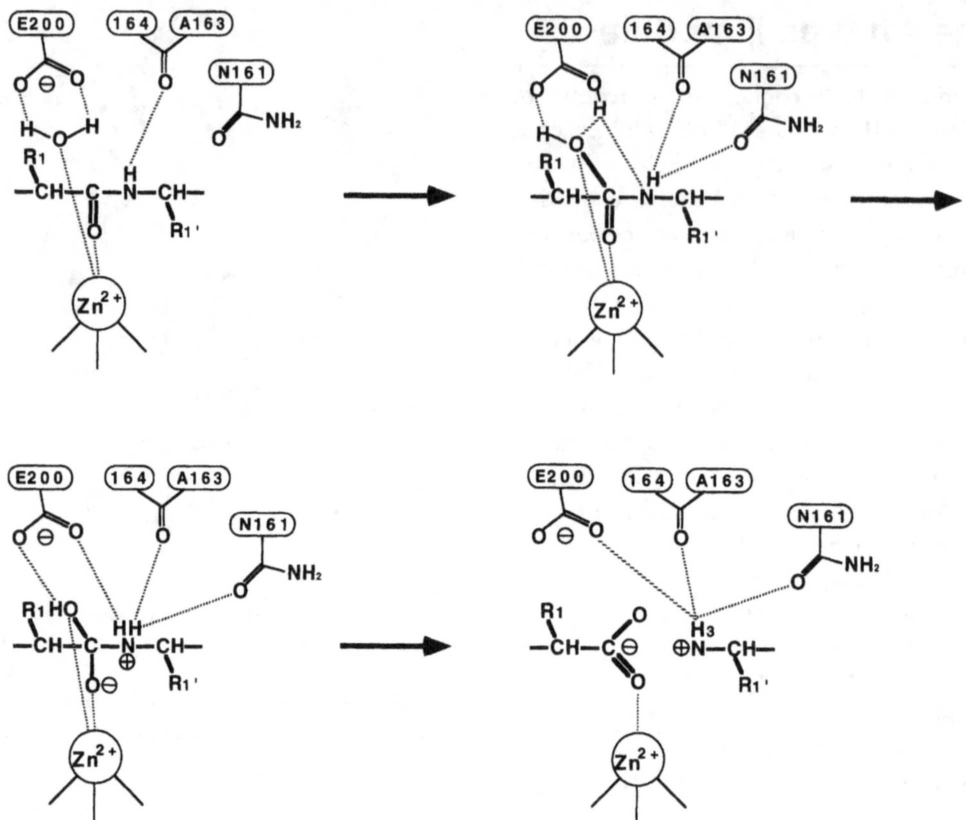

Figure 13

A proposed mechanism of action for matrixins. The substrate is shown in bold. Redrawn after Spurlino *et al.* [1410].

features proposed for hydrolysis of a peptide bond for thermolysin [1389] are also found in the matrixins. Superposition of the active site of matrixins and thermolysin suggests that the catalytic Zn^{2+} in both enzymes and glutamic acids (Glu200 in MMP-1 and Glu143 in thermolysin) in the HEXXH motif are positioned to play equivalent roles in catalysis. Based on the mechanism for thermolysin [1389] a catalytic mechanism has been proposed for matrixins by Spurlino *et al.* [1410] and Becker *et al.* [1343] as illustrated in Fig. 13. As in Glu143 in thermolysin, Glu200 (MMP-1) is thought to promote the nucleophilic attack of the water molecule on the carbonyl of the scissile bond to form a hemiketal. The carbonyl group of the scissile bond is a ligand of the catalytic Zn^{2+} and the water molecule is hydrogen bonded with the side chain of Glu200, which most likely func-

tions as a general acid/base during catalysis. The mutation of the analogous glutamic acid in MMP-2 to aspartic acid reduces the specific activity of the enzyme by 100-fold, and the alanine mutant exhibits only 0.01% of the activity of the wild-type [1358], indicating the proposed role of this residue. It has been proposed that the carbonyl oxygen of Ala163 and the side chain of Asn161 in MMP-1 stabilize the protonated nitrogen of the scissile bond, analogous to residues Ala113 and Asn112 in thermolysin. In MMP-3, however, the space occupied by the side chain of Asn161 is Val63, which does not form a hydrogen bond. In thermolysin, the side chains of Tyr157 and His231 stabilize the oxygenation of the hemiketal [1389], but the corresponding residues that occupy those spaces are not found in MMP-1 and MMP-3 [1343, 1410]. Thus, the matrixins appear to lack

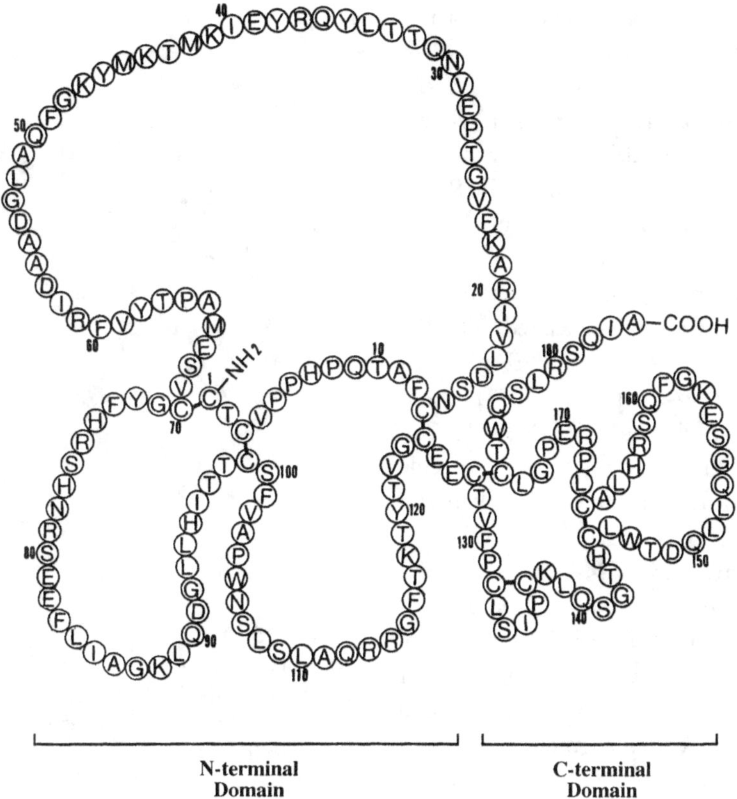

Figure 14

The primary structure of TIMP-1.

the groups that play equivalent roles in stabilizing the oxygenation of the hemiketal. There is further discussion of mechanisms in the section on the catalytic domain of MMP sequences above.

NMR and X-ray structures of TIMPs

TIMPs have 12 conserved cysteines all of which are disulfide bonded [1438]. The disulfide arrangements indicate that the molecule consists of an N-terminal domain (N-TIMP) and C-terminal domain (C-TIMP) (Fig. 14). Crystallization of the N-TIMP-2 [1435] and the unglycosylated TIMP-1 mutated at the N-glycosylation site [1436] have been reported, but the structures were not determined. Williamson *et al.* [1439] reported the sec-

ondary structure and a low resolution tertiary structure of the N-terminal domain of TIMP-2 (N-TIMP-2) determined by two-dimensional and three-dimensional ^{1}H NMR spectroscopy. The structure revealed that N-TIMP-2 contains a five-stranded anti-parallel β-sheet that rolls over on itself to form a closed β-barrel and two short α-helices that are packed close to each other on the same side of the barrel. The β-barrel topology is homologous to that seen in proteins of the oligo-saccharide/oligonucleotide binding (OB)-fold family [1431]. Superposition of the backbone atoms of the N-TIMP-2 β-barrel with those of sta-phylococcal nuclease (SN) and B subunit of *E. coli* heat-labile enterotoxin (LTB) indicated the root mean square deviation values are 2.60 and 1.99, respectively [1439], similar to that reported for SN-LTB comparison. The ligand-binding side of OB-fold proteins is composed of the loops between strands A/B and C/D which protrude outward from the barrel axis and form a shallow groove with the β-sheet at its floor. Because conserved residues in the TIMP family members are clustered on the opposite side of the β-barrel on strands A and B, it was thought that TIMPs are unlikely to share a common ligand binding site with members of the OB fold family [1439].

The crystal structure of the complex formed between the full-length unglycosylated TIMP-1 and the catalytic domain of MMP-3 resolved at 2.8 Å revealed both the structure of TIMP-1 and the mechanism of inhibition of MMPs [1428]. TIMP-1 is an elongated wedge-shaped molecule consisting of contiguously folded N-terminal and C-terminal domains (Fig. 15). As demonstrated in the NMR solution structure of N-TIMP-2, the N-terminal domain exhibits the five-stranded β-pleated sheet rolled into a β-barrel of a conical shape (Figs 15 and 16). The narrow opening of this barrel is covered by the flexible B–C loop and the wider end by the extended segment 'C-connector' between C and D strands (Fig. 16). The N-terminal segment forms a rigid stretch of polypeptide fixed with two disulfide bonds Cys1–Cys70 and Cys3–Cys99 which connect the C-connector loop and the E–F loop, respectively. Three α-helices (residues 7–15, 109–118, 199–124) are closely located together in between the N- and C-terminal domains. The C-terminal domain of TIMP-1 consists of two adjacent β-sheets, one with two-parallel strands (G and H) and the other with two antiparallel strands (I and J), and a short α-helix in between the two β-sheets (Fig. 16). The last three residues of the molecule are not defined, suggesting the flexible nature of this region. The strictly conserved residues His7 and Gln9 of the N-terminal domain, which were once considered to be a potential docking site towards the MMPs [1434], form a part of the interface that interacts with the C-terminal domain.

The high resolution structure of N-TIMP-2 in solution was determined using multidimensional heteronuclear NMR spectroscopy by Muskett *et al.* [1432]. The topology of the protein backbone indicated that the elements of regular secondary

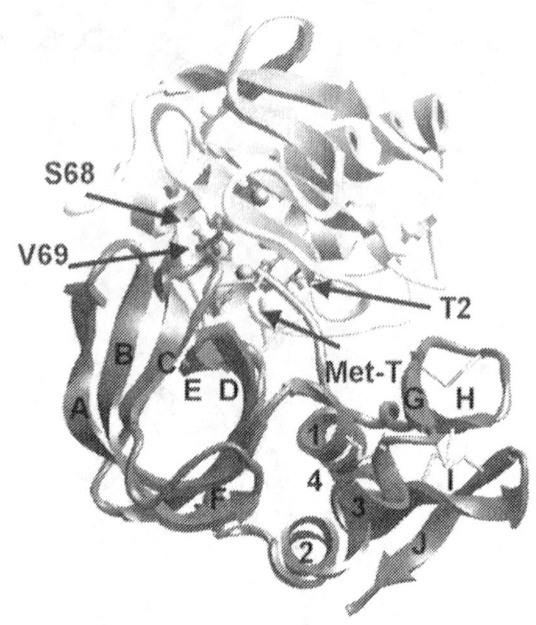

Figure 15

Ribbon diagram of the complex of TIMP-1 and the catalytic domain of MMP-3. The image was prepared from the Brookhaven Protein Data Bank entry (1UEA) deposited by Gomis-Rüth *et al.* [1428] using RIBBONS 2.87 [1351a]. MMP-3 (upper molecule) is shown in light grey and TIMP-1 (lower molecule) is in dark grey. Strands and helices in TIMP-1 are labelled as A–J and 1–4, respectively. Zinc and calcium ions are shown as grey spheres. The zinc ligands (His201, His205, His211) in the catalytic domain of MMP-3 and Thr2, Ser68, and Val69 in TIMP-1 are shown by ball and stick representation. Disulfide bonds in TIMP-1 are shown in light grey. The locations of Thr2, Ser68 and Val69 in TIMP-1 and the 'Met-turn' are indicated.

structure are identical to those described for their previous low-resolution structure [1439]. The structure revealed that the β-hairpin formed with strands A and B in N-TIMP-2 is much longer than that of the corresponding structure of TIMP-1. By determining the chemical shift changes for the backbone nuclei of N-TIMP-2 from the protein backbone upon complex formation with the catalytic domain of MMP-3, these investigators showed that the MMP-binding sites are located on the N-terminus of the inhibitor and the loops between β-strands A and B, C and D, and E and F; the extended A–B strand hairpin loop allows it to make extensive binding interactions with the catalytic domain of MMPs [1432].

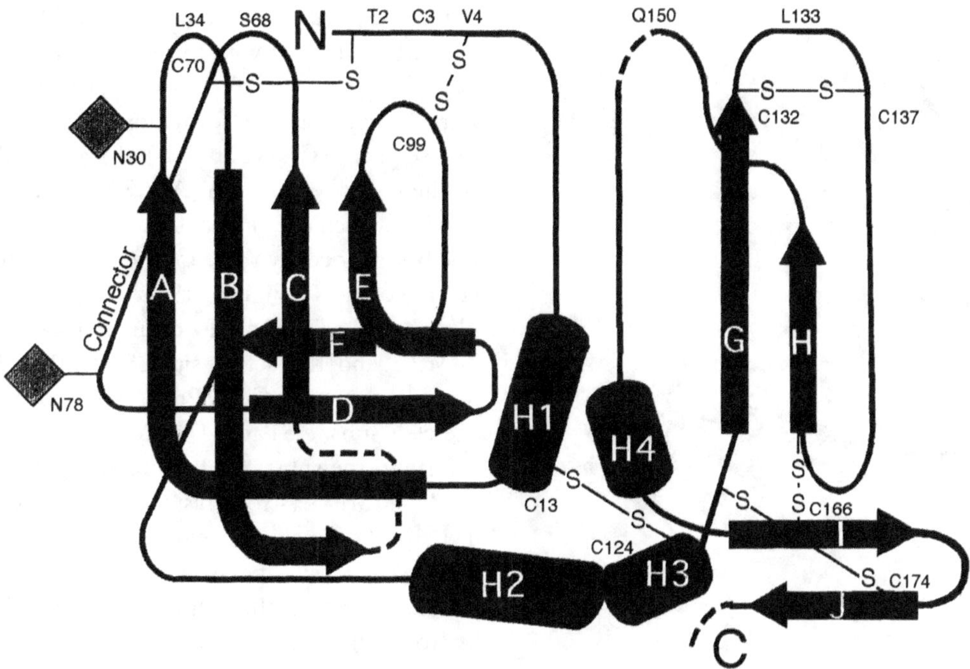

Figure 16

Schematic display of the secondary structure of TIMP-1. The 3D structure of TIMP-1 was determined as a complex with the catalytic domain of MMP-3 [1428]. Strands (A–J) and helices (H1–H4) are illustrated by arrows and cylinders, respectively. The two N-glycosylation sites are indicated by ◆. Diagram kindly provided by Dr. K. Maskos and Prof. W. Bode, Munich.

The 3D structure of the full-length bovine TIMP-2 resolved by X-ray crystallography as a complex of TIMP-2 and the catalytic domain of MT1-MMP [1427] reveals that the full-length TIMP-2 has a similar structure to TIMP-1, but there are several notable differences between the two TIMPs (Fig. 17). As shown by the NMR solution structure, TIMP-2 in the complex exhibits a much longer A–B loop, compared with TIMP-1, due to the insertion of seven additional residues (Figs 17 and 18). This elongated β-hairpin loop does not follow a typical OB-barrel surface, but is twisted and extends away. Other notable differences include the D–E loop in the molecular centre, the less exposed B–C loop, the twisted G–H loop, the twisted and lifted I–J loop, and the protruding C-terminus.

Mechanism of inhibition of MMPs by TIMPs

Several studies aimed at identifying the functional site(s) in TIMP have been reported. Williamson *et al.* [1440] proposed that His95 was essential for MMP inhibition on the basis of chemical modification. O'Shea *et al.* [1434] reported that mutation of His7 and Gln9 to Ala reduced the reactivity with MMP-7, but the effects were small and the binding affinity decreased only about two- to six-fold. Based on peptide and antibody competition experiments, Bodden *et al.* [1423] proposed that the region surrounding the second 'disulfide knot' (Cys13–Cys124 and Cys127–Cys174) located at the junction of the N-terminal and C-terminal domains was important in MMP inhibition. Hanglow *et al.* [1429] reported that a peptide corresponding to residues 70–97 of TIMP-1 inhibited MMP-3 with an IC_{50} value of 47 μM. Based on

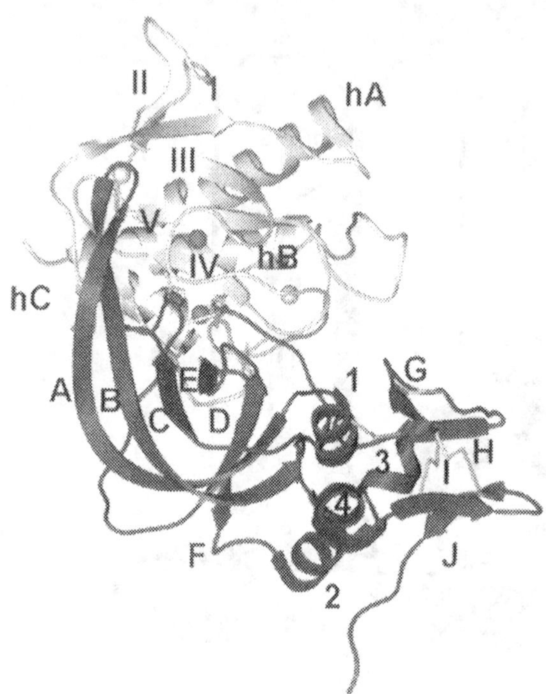

Figure 17

Ribbon diagram of the complex of TIMP-2 and the catalytic domain of MT1-MMP. The catalytic domain of MT1-MMP (upper molecule) and TIMP-2 (lower molecule) are in light and dark grey, respectively. Strands and helices of MT1-MMP are I–V and hA–hC, and strands and helices of TIMP-2 are A–J and 1–4, respectively. The catalytic and structural zinc ions and the two calcium ions in MT1-MMP are shown as dark grey and light grey spheres. The image was prepared using SETOR [1363a] and was kindly provided by Dr. K. Maskos and Prof. W. Bode, Munich. The figure was used by kind permission of Oxford University Press, Oxford.

differential proteinase susceptibility of free TIMP-1 and the MMP-3-bound TIMP-1, Nagase *et al.* [1433] proposed that the region Val69–Cys70 is in close contact with an active MMP. Subsequently, Huang *et al.* [1430] postulated, based on a series of mutagenesis studies, that two sections of polypeptide chain, from Met66 to Cys70 and from Cys1 to Pro5 which are arranged in a continuous surface ridge through a Cys1–Cys70 disulfide bridge, form a part of reactive site of TIMP. The importance of this disulfide bond for TIMP inhibitory activity has been discussed [1424, 1425].

The crystal structure of the complex of TIMP-1 and the catalytic domain of MMP-3 has revealed that six sequentially separate polypeptide segments of TIMP-1 interact with the enzyme [1428]. A surface area of about 1300 Å^2 of each molecule is buried upon complex formation, and the N-terminal segment (Cys1-Thr-Cys-Val-Pro5) and a part of the C-connector loop (Ala65-Met-Glu-Ser-Val-Cys70), which are linked by a Cys1–Cys70 disulfide bridge, occupy three-quarters of all intermolecular contacts. Figure 19 shows a schematic illustration of the mode of interaction between TIMP-1 and the active site of MMP-3. The N-terminal segment (Cys1–Pro5) binds to the active site cleft subsite S1 to S4′ of MMP-3 in a manner similar to the substrate, by interacting antiparallel to the MMP-3 (stromelysin 1) strand AsnS162–TyrS168 (stromelysin residues are indicated with the prefix S) and parallel to segment ProS221–TyrS223 through three and two inter-main-chain hydrogen bonds, respectively. The N-terminal α-amino group and the carbonyl group of Cys1 coordinate the catalytic Zn^{2+} at a distance of about 2.0 Å. The α-amino group could form hydrogen bonds with a neighbouring carbonyl oxygen of Ser68 and with one of the carboxylate oxygens of GluS202 of the active site of MMP-3. The side chain of GluS202 moves closer when TIMP-1 binds, and the zinc-bound water molecule that is normally present in the uninhibited MMP is excluded in the complex. The side chain of Thr2 extends into the large hydrophobic S1′ specificity pocket of MMP-3, with its β-hydroxy group directed towards the more polar neck of the pocket. Likewise, Ser2 of TIMP-2 is directed towards the S1′ pocket of MT1-MMP in the complex, and the electron density indicates that the Oγ of Ser2 is hydrogen bonded to Oε2 of the catalytic Glu240 [1427]. Thus, a similar interaction may take place in the TIMP-1–MMP-3 complex. The cystine bridge, Cys3–Cys99, points away from MMP-3, and the side chains of Val4 and Pro5 slightly interact with MMP-3. The segment of the C-connector loop (Ala65–Cys70) occupies the left hand side of the active site of MMP-3, but Ser68 and Val69 of TIMP-1 are arranged in a nearly opposite orientation to the P3–P2 segment of a substrate. The interactions in this region (made

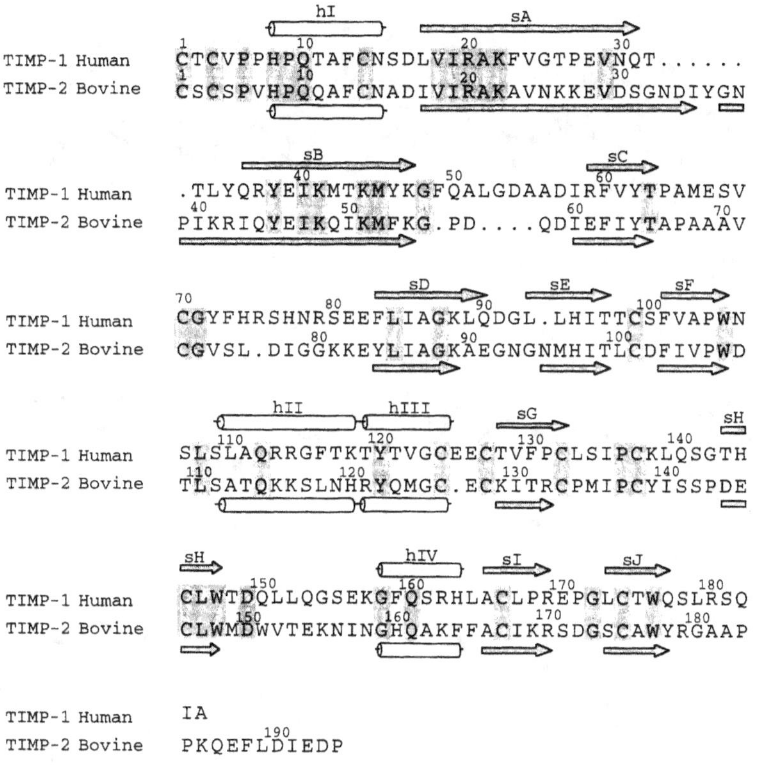

Figure 18

Sequence alignment of TIMP-1 and TIMP-2. Strands and helices are indicated by arrows and cylinders, respectively, after Gomis-Rüth *et al.* [1428] and Fernandez-Catalan *et al.* [1427].

with MMP-3 segments S83–S86, S154–S155, S163–S167, S205 and S210–S211) are primarily hydrophobic. The side chains of Met66 and Val69 in this region extend into a hydrophobic environment created by the N-terminal segment PheS83–ProS90 of MMP-3 as a result of the complete rearrangement of the N-terminal segment with a 15 Å shift by breaking a salt bridge formed between the ammonium group of PheS83 and the carboxylate of AspS237.

Other segments of TIMP-1 that have direct but weaker contacts with MMP-3 are in the A–B loop (Val29, Thr33, Tyr35; corresponding MMP-3 segments, S85–S86, S154–S155) and the E–F loop (Thr98 and Cys99; MMP-3 segments S221–S223) in the N-terminal domains, and the open loop (Leu133 and Ser134; MMP segment S190–S192) and the multiple-turn segment (Gln150–Leu152; MMP-3 segment S223–S228) in the C-terminal domain. These contacts are mostly through hydrophobic side chains. The structure suggests that the sugar chains linked to Asn30 and Asn78 in TIMP-1 (see Fig. 16) should not interact with the inhibited MMP.

The crystal structure of the complex between TIMP-2 and the catalytic domain of MT1-MMP [1427] is similar to that of the TIMP-1–MMP-3 complex. TIMP-2 interacts with the enzyme through six sequentially separate segments, and 3/5 of intermolecular contacts are located in the N-terminal segment Cys1–Pro5 and the disulfide-linked segment Ala66–Cys72 in the C-connector loop, as in the complex of TIMP-1 and MMP-3. However, the elongated A–B loop and the E–F loop make tighter interactions, particularly the A–B loop which folds over the rim of the active-site cleft and reaches to the surface of the β-sheet of MT1-MMP. The exposed side-chain of Tyr36 of

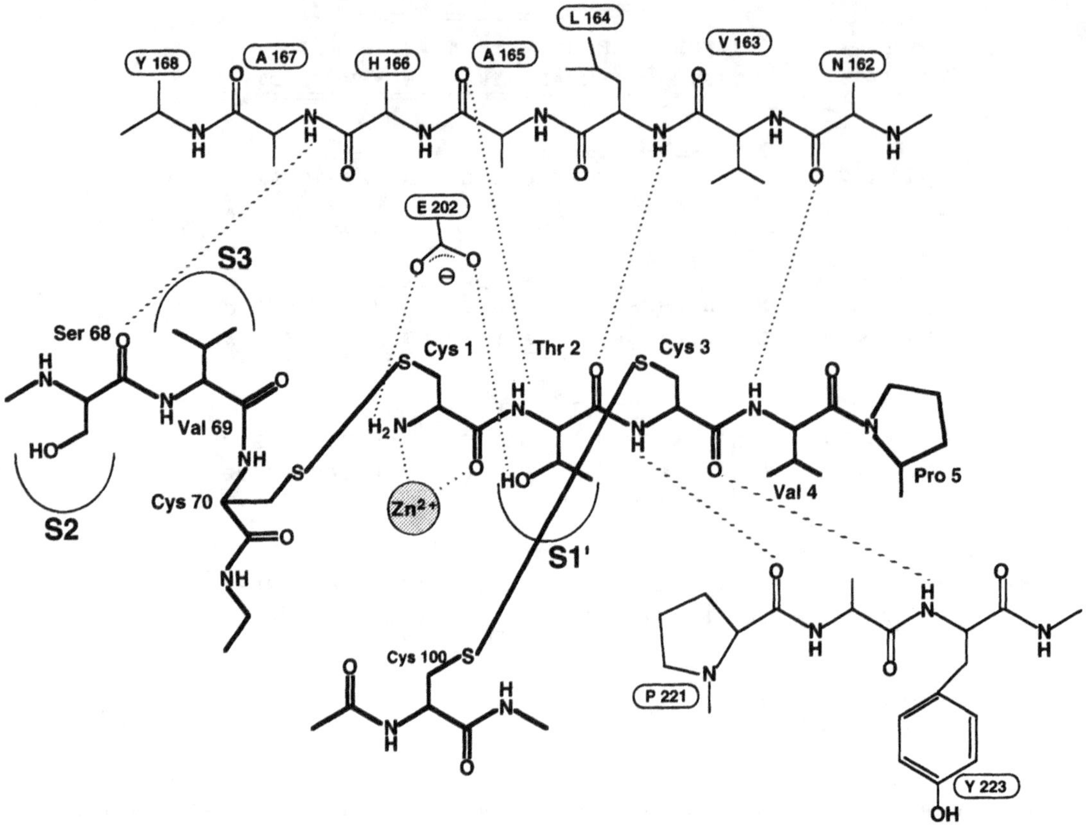

Figure 19

Interaction of TIMP-1 and the active site of MMP-3. The N-terminal α-amino and carbonyl groups of Cys1 bidentately coordinate the catalytic zinc ion. The N-terminal Cys1-Thr-Cys-Val-Pro5 segment of TIMP-1 lies antiparallel to the bulge strand and parallel to the S1′ pocket forming wall segment of MMP-3 by forming three and two hydrogen bonds, respectively [1428]. The residues Val69 and Ser68 of TIMP-1 interact with the S3 and S2 subsites, respectively. Measurement of hydrogen exchange rates [1422] suggests a hydrogen bond may be formed between the amide of Asn162 of MMP-3 and the carbonyl oxygen of Val4 of TIMP-1, but not between the amide of Tyr 223 of MMP-3 and the carbonyl oxygen of Cys3, in solution.

the A–B loop slots into a surface gap bounded by the characteristic MT-loop, side-chains of AspM212 (prefix M indicates residues in MT1-MMP, numbering the first methionine of the signal peptide as 1) in strand V and PheM180 in strand III, and the S-loop of the enzyme. It is anchored there by hydrogen bonds with the carboxylate oxygens of AspM212 and by a number of van der Waals contacts. In addition, two edge loops of the C-terminal domain interact with MT1-MMP, but they do not make major contributions to binding.

Another striking difference in structural arrangement between the MMP-3–TIMP-1 complex and the MT1-MMP–TIMP-2 complex is that when the two MMPs are optimally superimposed, the two inhibitors are tilted with respect to each other by rotations of -20°, 10° and 5° about Ser4 around vertical, perpendicular and horizontal axes, respectively.

The binding sites of N-TIMP-2 and the catalytic domain of MMP-3 were investigated by measuring the NMR chemical shift of backbone amide nuclei of N-TIMP-2 when it forms a complex with the catalytic domain of MMP-3 [1432, 1437]. This identified eight residues whose backbone signals underwent large chemical shift changes on complex formation: Ser32, Tyr36,

Ile40, Lys41, Ala70, Val71, Gly73 and Cys101 [1432]. Those residues fall into three distinct regions of N-TIMP-2:

(1) the end of strand A, the AB loop and the beginning of strand B (Glu28–Lys41);
(2) Ala70–Gly73 in the C-connector loop, which is disulfide-bonded to Cys1 through Cys72; and
(3) the EF loop (His97–Ile104), which is also linked to the N-terminus by the Cys3–Cys101 bridge.

No chemical shift data were obtained for Cys1–Ser4, but the central position of these residues in the NMR-mapped MMP binding site suggests that they play a key role in TIMP–MMP interactions [1432]. These results are consistent with those observed for the crystal structures of TIMP–MMP complexes [1427, 1428].

The sites in the catalytic domain of MMP-3 that interact with N-TIMP-1 in solution were also mapped by protection from amide proton line broadening by paramagnetic Gd-EDTA probing [1422]. Many of the observations made by this approach were consistent with those of the crystallographic studies [1428]. However, measurement of amide proton exchange rates of MMP-3 induced by N-TIMP-1 indicated that the amide of Asn162 became protected from hydrogen exchange with water by at least eightfold [1422]. In addition, the binding of N-TIMP-1 did not protect Tyr223 of MMP-3 from the paramagnetic probe, but it enhanced the rate of the Tyr223 amide exchange with water more than 20-fold [1422]. Those investigators, therefore, suggest that the amide of Asn162 of MMP-3 forms a hydrogen bond with the carbonyl oxygen of Val4 of TIMP-1, but that the amide of Tyr223, in contrast to the crystal structure, lacks hydrogen bonding to the carbonyl oxygen of Cys3 of N-TIMP-1 in solution.

Activation of the zymogen forms of MMPs

Introduction

Early work on collagenase demonstrated that the enzyme was present in tissue or cultured cells largely in a latent form. Activation of the enzyme could be achieved by treating the conditioned medium or the tissue extracts with trypsin [162, 193, 209, 214, 279, 328, 340], mast cell proteinases [163, 307], cathepsin B [193], tissue kallikrein [193] or plasmin [171, 193]. Activation of the latent collagenase also occurred spontaneously during purification or storage [157, 179, 193, 290, 313]. Activation with proteolytic enzymes suggested that the latent form of collagenase was a proenzyme [164, 193, 313, 329, 351]. It had also been proposed that the latent collagenase was an enzyme–inhibitor complex, since non-proteolytic agents such as mercurial compounds [66, 301, 333], NaI [306] and NaSCN [152] can activate the latent collagenase. The enzyme–inhibitor concept was further supported by the observation that latent collagenase from the medium of cultured rabbit bone was eluted at about 40 kDa on gel filtration chromatography [301], but when the medium was treated with 4-aminophenylmercuric acetate (APMA) the activated collagenase was eluted at 28 kDa. Furthermore, when the activated collagenase fraction was mixed with an endogenous inhibitor, the latent collagenase was eluted at 40 kDa. This collagenase–inhibitor complex was activated by treating with APMA [301]. Similar observations were also reported for human rheumatoid synovial collagenase [333], suggesting that APMA and other non-proteolytic agents activated the latent collagenase by dissociating the inhibitor from the collagenase–inhibitor complex.

On the other hand, purified latent collagenase from the medium of cultured skin fibroblasts exhibited a doublet of 55 kDa and 60 kDa on SDS/PAGE [313]. Both forms converted to the 40-kDa and active 45-kDa forms when they were treated with trypsin. These results suggested that collagenase was secreted from the cells as procollagenase of a single polypeptide chain. The synthesis of pre-procollagenase was demonstrated by *in vitro* translation of rabbit synovial collagenase mRNA [261], and in the presence of microsomal membranes, it was processed to procollagenase. In cultured rabbit fibroblasts, procollagenase was secreted from the cell by the constitutive route [258], taking about 35 min [258, 330] from the initiation of pre-procollagenase synthesis to secretion of procollagenase. Trypsin- or APMA-activated chick bone collagenase can also be inhibited by a 25-kDa endogenous inhibitor [291], but inhibition is not reversed by treatment with either trypsin or APMA. Nor can purified collagenase–TIMP complex be activated by an organomercurial compound [180, 314]. These early observations established that the latent collagenase secreted from the cell was a form of proenzyme. Most matrixins are also secreted from the cells as inactive zymogens, but the mechanism by which promatrixins are activated by proteinases and non-proteolytic agents requires further investigation.

In 1990, Van Wart and colleagues proposed the 'cysteine switch' mechanism to explain the activation of the proMMPs by multiple agents [310, 332]. This model proposes that the single cysteine residue found in the conserved PRCG[V/N]PD sequence in the propeptide interacts with the catalytic Zn^{2+} of the enzyme, thereby maintaining the latency of the proMMPs by preventing the association of the Zn^{2+} with a water molecule which is required for the hydrolysis of the peptide. Thus, the disruption of the Cys–Zn^{2+} interaction is essential for the activation of proMMPs. This mechanism has also been called the 'velcro' mechanism [331]. The Cys–Zn^{2+} interaction in the proenzyme is supported by the presence of a single Cys in the propeptide; this Cys could not be modified with an alkylating agent unless proMMPs were first treated with a chelating agent [310, 349]. The cysteine-switch mechanism is also supported by spectroscopic studies of the Co^{2+}-complexed proMMP-3 [292], by extended X-ray absorption fine structure spectroscopy [213] and by the crystal structure of proMMP-3(ΔC) lacking the C-terminal hemopexin-like domain [158].

Many matrixins are secreted from the cell as proenzymes and activated extracellularly in a stepwise manner [88]. However, intracellular activation of the zymogen has been shown for MMP-11 and MT1-MMP (MMP-14) [283, 284, 296, 297]. The latter mechanism is probably applicable to other MT-MMPs which possess a furin-recognition sequence in the C-terminal end of the propeptide. Lee *et al.* [238] also found that proMMP-2 is activated intracellularly when normal skin fibroblasts are cultured in type I collagen lattices. The sites cleaved in the propeptide domains during activation are summarized in Fig. 21 (below).

Physical and chemical activation processes

One of the unique properties of matrixins that distinguishes them from other families of metalloproteinases is that pro-matrixins are not only activated by proteinases, but also by mercurial compounds [66, 301, 308, 309], other thiol reactive reagents such as dithio-*bis*-(2-nitrobenzoic acid) [66], iodoacetamide [308], N-ethylmaleimide [252], oxidized glutathione [242], SDS [165], NaSCN [152, 193, 308], NaI [306], reactive oxygens such as HOCl [246, 270, 285, 310, 341], H$_2$O$_2$ [288], ONOO$^-$ [68, 281], NO$_2$ [275], heat treatment [234] and by brief exposure to acid pH [188, 207]. The cysteine switch model proposes that the Cys–Zn^{2+} interaction at the active site transiently dissociates, thereby allowing Cys to react with SH reagents. The modification of Cys then prevents the reassociation of Cys and Zn^{2+}, i.e. the SH group is permanently dissociated from the Zn^{2+}, and proMMPs autolytically process the propeptide (Fig. 20). Activation of pro-matrixins by the above-mentioned physical and chemical means favours this model. However, the exact mechanisms of proMMP activation by APMA and SH-reactive reagents are not clear. Activation of human skin fibroblast collagenase by various mercurials does not correlate with processing of procollagenase of 55 kDa and 60 kDa (the 60 kDa species is a glycosylated form of the 55 kDa enzyme [258]) to a lower molecular weight active form [314], indicating that mercurials are promoters of autoactivation. The conversion to a low molecular weight active form was due to an induction of autocleavage of collagenase, proposed to be intramolecular since it was independent of the enzyme concentration. Furthermore, trypsin produced an inactive initial product of procollagenase that subsequently underwent activation by a concentration-independent autoprocessing [314]. The initial trypsin cleavage site of procollagenase is the Arg36–Asn37 bond [205], with cleavage resulting in a 46-kDa intermediate that lacks activity towards collagen [314]. This intermediate is then autoprocessed to an active 42-kDa collagenase. The N-terminus of active collagenase is Phe81, Val82 and Leu83. The autoprocessing step requires initial activation by an organomercurial or trypsin. Thus, even with different means of activation, the 52-kDa procollagenase results in the same set of active species of collagenase. Treatment of proMMP-3 (57 kDa) with APMA also results in generation of a 46-kDa initial intermediate by an intramolecular autolysis which then converts to a stable 45-kDa MMP-3 by an intermolecular reaction [271].

The cysteine switch hypothesis postulates that the disruption of Cys–Zn^{2+} interaction is essential for the activation of pro-matrixin. This is most likely the case in activation by chaotropic agents or partial denaturation of the proenzyme, but it is still not clear how mercurial compounds and SH reagents activate pro-matrixins. ProMMP-8 binds to organomercurial-agarose, suggesting that Hg^{2+}

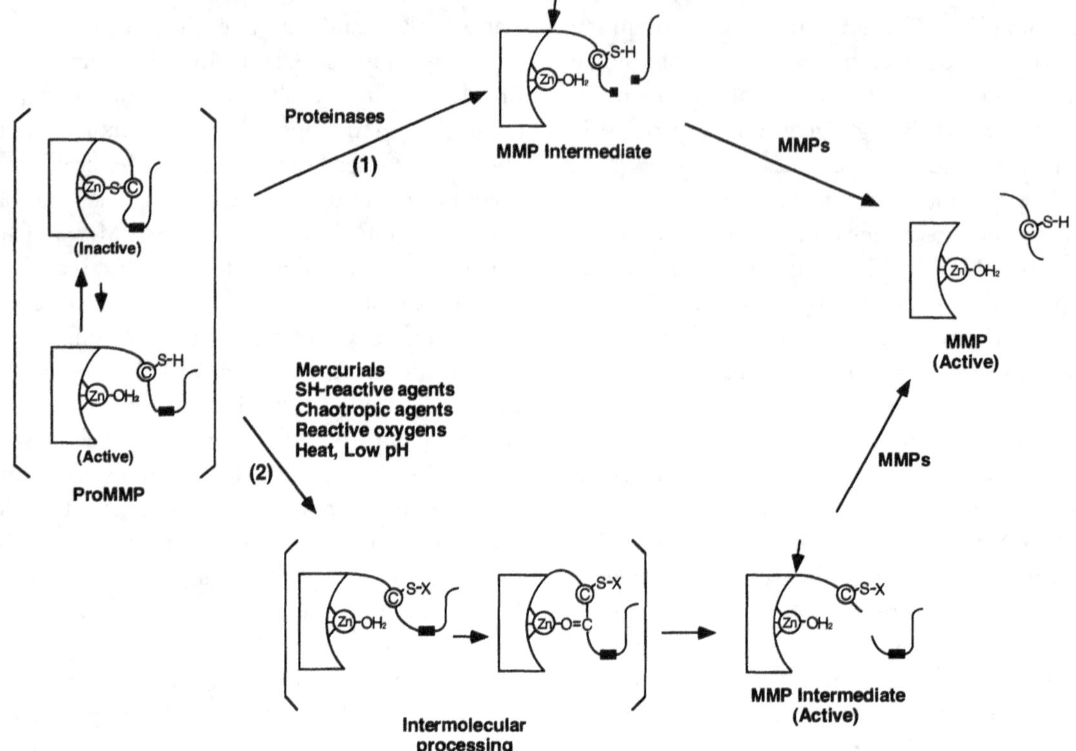

Figure 20

Stepwise activation of pro-matrixin. Pro-matrixins secreted from the cells are largely in an inactive form due to the zinc–cysteine interaction [310, 331, 332]. Proenzymes can be activated by proteinases and non-proteolytic agents. In the proteinase pathway (1), an activator proteinase first attacks the 'bait' region (–■–) in the propeptide, which destabilizes the zinc–cysteine interaction. In the non-proteolytic agent pathway (2), the transiently dissociated cysteine of the propeptide reacts with a SH reacting agent or the pro-matrixin structure is perturbed by denaturants, heat or exposure to an acidic pH. This often results in intramolecular removal of a portion of the propeptide. The complete removal of the propeptide is due to further action of an active MMP or an MMP intermediate in many cases. Some proteinases are, however, capable of removing the entire propeptide from the pro-matrixin (see the text). The figure was redrawn from [88] by kind permission of Walter de Gruyter GmbH and Co., Berlin.

binds to Cys in the propeptide [167]. On the other hand, disruption of the Cys–Zn coordination by chemical modification of the Cys75 in the propeptide by AMPA and other SH reactive agents is not sufficient to activate proMMP-3 [181]. Activation of the Cys75-modified proMMP-3 requires the continuous presence of APMA or treatment with proteinases. These results indicate that activation of proMMP-3 by APMA is initiated by structural perturbation of the proMMP-3 molecule, rather than by its reaction with Cys75. This observation was further substantiated by mutagenesis studies of human proMMP-3 by substituting Ser or His for Cys75 [202]. These mutants were synthesized and secreted from the mutant proMMP-3 cDNA-transfected HeLa S3 cells as latent proenzymes and were also activated by treatment with APMA or trypsin. Mutations of some of the residues in the cysteine switch sequence of rat proMMP-3 resulted in secretion of active MMP-3 from the transfected COS cells, but the mutation of Cys to Ser apparently resulted in degradation of the enzyme [282]. Mutation of Tyr20 or Leu21 to Ala, on the other hand, resulted in spontaneous activation of proMMP-3 [198]. When the proMMP-2–TIMP-2 complex was treated with APMA, proMMP-2 was activated by breaking the Zn^{2+}–Cys interaction, but activation was not caused by the reaction between the SH group of the Cys and APMA [221]. These results suggest that activation of at least some pro-matrixins by non-proteolytic agents, including APMA, is initiated by structural perturbation of the zymogen. ProMMP-2 is activated by concentration-dependent autolysis, which is enhanced by addition of heparin [184]. This may in part be related to the conformational changes induced in proMMP-2 and a trace amount of active MMP-2.

N-terminal sequence analyses of the products generated from proMMP-3 by treatment with APMA indicated that several intermediates are formed initially, which are then processed to a mature MMP-3 of 45 kDa by cleavage of the His82–Phe83 bond [174, 260]. The initial generation of an intermediate is considered to be due to an intramolecular reaction of the zymogen since it follows first-order kinetics [271] and it is not inhibited by an active-site inhibitor of MMPs. The cleavage of the His82–Phe83 bond in the final activation step follows a bimolecular reaction mechanism of MMP-3 [174, 260, 271].

In most cases, the treatment of pro-matrixins with a mercurial compound results in the removal of the propeptide by the induced autoproteolysis. The treatment of proMMP-9 with a mercurial compound, however, generates an active 68-kDa MMP-9 with N-terminal Met75 [270, 294, 323, 344]. This form of MMP-9 still retains the cysteine switch sequence of the propeptide. The enzymatic activity of this species is detected only in the presence of a mercurial compound and removal of Hg(II) from the Hg(II)-activated sample reduces the enzymatic activity to a level similar to that of the original zymogen [294], suggesting that Hg(II) confers activity on this species by binding to the Cys residue in the cysteine switch sequence.

Several investigators reported activation of proMMP-9 by HOCl, but the results have not been conclusive. Neutrophil progelatinase is activated by HOCl to about 30% of the activity produced by mercurial treatment [285]. Also only the partially processed 83-kDa species, but not the 92-kDa proMMP-9 zymogen, was activated by HOCl to the 55-kDa form [270], although other investigators reported negligible activation [246, 294], probably due to destruction of the enzyme.

A naturally occurring non-enzymic low-molecular-mass factor called 'endothelial-cell-stimulating angiogenesis factor' (ESAF) has been reported to be capable of fully activating proMMP-1, proMMP-2 and proMMP-3 [245]. ESAF is a non-protein factor with a molecular mass of approximately 600 Da which stimulates the proliferation of microvascular endothelial cells in culture [300]. ESAF reactivates the MMP-1–TIMP, MMP-2–TIMP and MMP-3–TIMP complexes (TIMP can be either TIMP-1 or TIMP-2), and releases fully active MMPs and functional TIMPs [244, 245]. The released MMPs can no longer be inhibited by TIMPs [245]. The mechanisms by which ESAF activates proMMPs and reactivates the MMP–TIMP complexes are not known.

Activation of proMMPs by SDS, urea and other denaturants is a useful property since it allows many proMMPs to be detected by zymography after electrophoresis on SDS–polyacrylamide gel containing a protein substrate. During SDS/PAGE proMMP is denatured and is in an inactive state. Thus, it migrates to a position corresponding to the molecular mass of the proenzyme. After electrophoresis, washing the gel with a buffer at a neutral pH containing a non-ionic detergent with a small amount of Zn^{2+} and millimolar amount of Ca^{2+} renatures the proenzyme. During this process the proenzyme is activated and digests the substrate in the gel [212, 226]. This provides a useful means of detecting both the precursor and active forms of MMPs.

Mechanisms of cleavage of individual propeptides by proteinases

proMMP-1

Activation pathways for MMP-1 are summarized in Fig. 21(a). Several activators of procollagenase have been reported, including tadpole collagenase

(a)   MMP-1

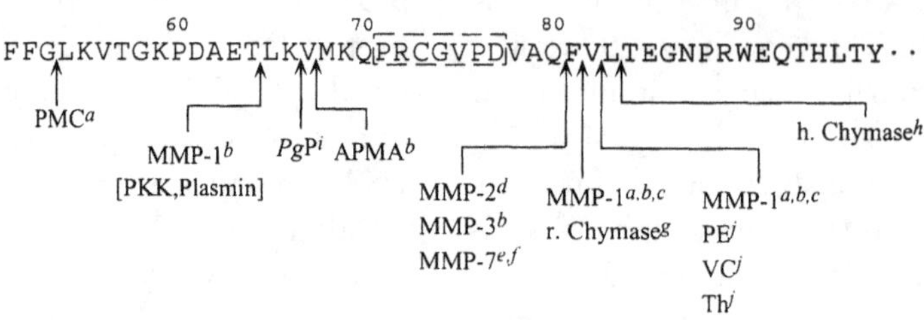

(b)   MMP-2

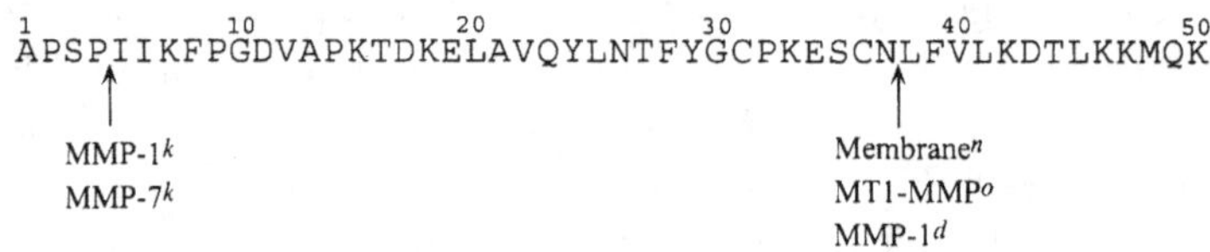

(c)   MMP-3 & MMP-10

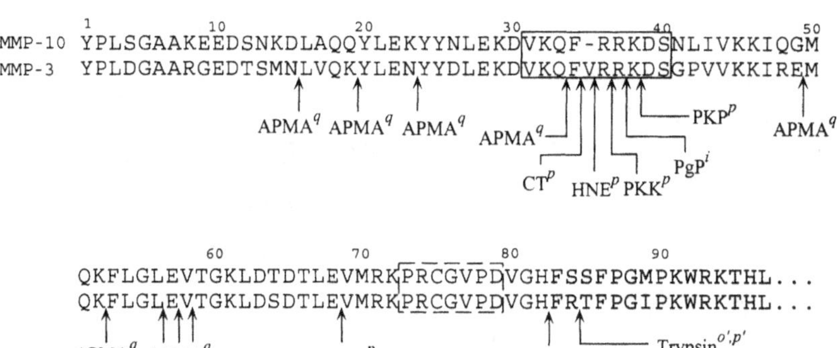

Figure 21 (a)–(c)

(d) MMP-7

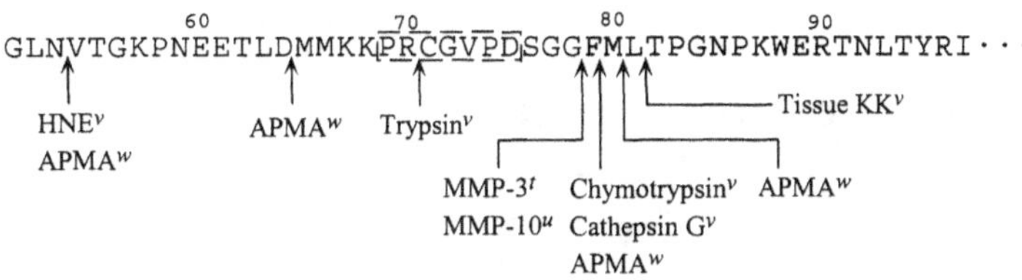

(e) MMP-8

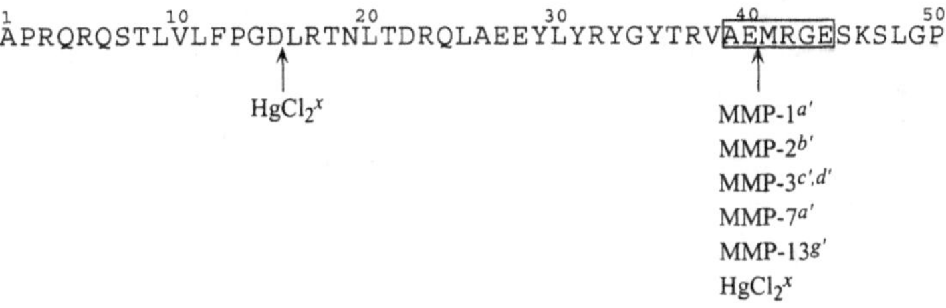

(f) MMP-9

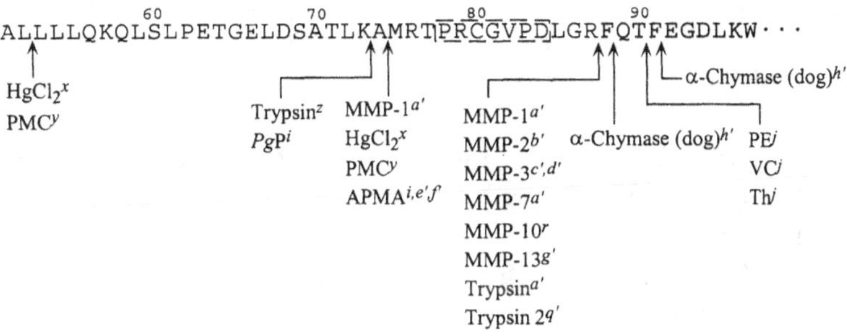

Figure 21 (d)–(f)

(g)   MMP-13

```
 1          10             20                   30              40              50
LPLPSGGDEDDLSEEDLQFAERYLRSYYHPTNLA[GILKEN]AASSMTERLREM
                                          ↑      ↑
                                   MT1-MMP[j′]  Plasmin[i′]
```

```
        60               70          80              90                  100
QSFFGLEVTGKLDDNTLDVMKK[PRCGVPD]VGEYNVFPRTLKYSKMNLTY · · ·
      ↑                          ↑            ↑
   MMP-3[i′]                  Plasmin[i′]   MMP-2[j′]
   APMA[i′]                                 MMP-3/MMP-13[i′]
                                            MT1-MMP[j′]
                                            MMP-13[i′]
```

(h)   MMP-11 & MT1-MMP (MMP-14)

```
                  80              90               100
MMP-11   P[PRCGVPD]PSDGLSARNRQKRFVLS · · ·
                                      ↑
                                    Furin[k′]
```

```
            90              100            110
MMP-14   R[PRCGVPD]KFGAEIKANVRRKRYAIQ · · ·
                                  ↑     ↑
                              Plasmin[n′]  Furin[l′,m′]
                                           Plasmin[n′]
```

Figure 21 (g)

Sites cleaved in the propeptides during activation of pro-matrixins by proteinases and organomercurial compounds. Each MMP is shown in a separate panel. Proteinase susceptible 'bait' regions and the predicted bait regions are boxed with solid lines. The cysteine switch sequences are boxed with broken lines. Furin recognition sites are underlined. The amino acids in the catalytic domains are shaded. PgP, *Porphyromonas gingivalis* proteinase; PKK, plasma kallikrein; PE, *Pseudomonas aeruginosa* elastase; VC, *Vibrio cholerae* proteinase; Th, thermolysin; h, human; r, rat; HNE, human neutrophil elastase; KK, kallikrein; PMC, phenylmercuric chloride. [a]Grant *et al.*, 1987 [205]; [b]Suzuki *et al.*, 1990 [320]; [c]He *et al.*, 1994 [211]; [d]Crabbe *et al.*, 1994 [185]; [e]Abramson *et al.*, 1995 [153]; [f]Imai *et al.*, 1995 [217]; [g]Suzuki *et al.*, 1995 [319]; [h]Saarinen *et al.*, 1994 [289]; [i]DeCarlo *et al.*, 1997 [189]; [j]Okamoto *et al.*, 1997 [276]; [k]Sang *et al.*, 1996 [293]; [l]Stetler-Stevenson *et al.*, 1989 [312]; [m]Okada *et al.*, 1990 [272]; [n]Strongin *et al.*, 1993 [316]; [o]Atkinson *et al.*, 1995 [154]; [p]Nagase *et al.*, 1990 [260]; [q]Cameron *et al.*, 1995 [174]; [r]Nakamura *et al.*, 1998 [264]; [s]Crabbe *et al.*, 1992 [186]; [t]Knäuper *et al.*, 1993 [232]; [u]Knäuper *et al.*, 1996 [229]; [v]Knäuper *et al.*, 1990 [227]; [w]Bläser *et al.*, 1991 [166]; [x]Triebel *et al.*, 1992 [323]; [a′]Wilhelm *et al.*, 1989 [344]; [z]Tschesche *et al.*, 1992 [324]; [a′]Sang *et al.*, 1995 [294]; [b′]Fridman *et al.*, 1995 [200]; [c′]Ogata *et al.*, 1992 [268]; [d′]Goldberg *et al.*, 1992 [204]; [e′]Okada *et al.*, 1992 [270]; [f′]Shapiro *et al.*, 1995 [303]; [g′]Knäuper *et al.*, 1997 [231]; [h′]Fang *et al.*, 1997 [194]; [i′]Knäuper *et al.*, 1996 [228]; [j′]Knäuper *et al.*, 1996 [233]; [k′]Pei and Weiss, 1995 [283]; [l′]Sato *et al.*, 1996 [297]; [m′]Pei and Weiss, 1996 [284]; [n′]Okumura *et al.*, 1997 [277]; [o′]Lark *et al.*, 1990 [237]; [p′]Benbow *et al.*, 1996 [159]; [q′]Sorsa *et al.* [309a].

activator [210], a factor from rheumatoid synovial fluids [236], a mast cell factor [307], factors in the medium of cultured human skin [326], a metalloproteinase activator [218, 334] and cartilage procollagenase activating protein [322]. Other activator proteinases include trypsin [205, 328], plasmin [193, 211, 342], plasma kallikrein [259], chymase [289, 319], tryptase [206, 319], *Pseudomonas* elastase [276], *Vibrio cholerae* proteinase [276] and thermolysin [276]. The treatment of 52-kDa human proMMP-1 with trypsin-like serine proteinases results in initial products of ~53 kDa, which are then converted to 42 kDa [317]. N-Terminal sequence analysis indicates that these enzymes initially attack the region Glu33-Lys-Arg-Arg-Asn37 located in the middle of the propeptide.

The crystal structure of the C-terminal hemopexin domain-deleted proMMP-3(ΔC) [158] indicates that this region in the pro-domain is likely to be exposed on the surface of the molecule. The sequence of this region limits its activators to those that recognize Arg or Lys. Chymotrypsin, leukocyte elastase and cathepsin G do not activate this enzyme [317]. Thus, it is thought that the removal of the first 34–36 residue from the N-terminus is sufficient to destabilize the Cys73–Zn^{2+} coordination in the proMMP-1 site and partially activate proMMP-1. The initial intermediates then undergo autocleavage at Thr64–Leu65, resulting in the 42-kDa MMP-1. This species of MMP-1 has about 20% of the full collagenolytic activity [317]. To fully activate the [Leu65]MMP-1 (the residue in brackets indicates the N-terminus), it must be converted to the 41-kDa species by cleaving the Gln80–Phe81 bond, and this cleavage can be produced by MMP-2 [185], MMP-3 [317], MMP-7 [153, 217], MMP-10 [265, 350] and MMP-11 [256], but they cannot initiate the activation of proMMP-1 effectively [185, 317, 334]. These results explain the observations [334] on the role of rabbit synovial collagenase activator in procollagenase activation.

The collagenase activator has been isolated as a doublet of proenzyme at 58 kDa and 60 kDa. Proactivator can be activated by APMA, trypsin [334] or removal and replenishment of Ca^{2+} [335], and expresses its proteolytic activity, but the activated activator cannot activate procollagenase unless both proactivator and procollagenase are combined and activated together [334]. Enhanced activation of collagenase was also observed when proMMP-1 was activated in the presence of recombinant proMMP-3 [253] or natural proMMP-3 [220, 253]. These studies identified the activator of procollagenase purified from rabbit synovial fibroblasts as MMP-3. The identity of the activator and MMP-3 was also proved by cDNA sequencing [197, 299]. Procollagenase activator was also purified from bovine articular cartilage [322] and rabbit uterine cervical fibroblasts [218] and again is likely to be MMP-3. Thus, proMMP-1 is activated in a stepwise manner requiring at least two proteinases for full activation. Without the activator MMPs, a partially activated MMP-1 undergoes autoprocessing at either the Phe81–Val82 or the Val82–Leu83 bond [205, 317]. The resulting MMP-1 has only 30–40% of the full activity [320].

The importance of MMP-3 in proMMP-1 activation was demonstrated in human rheumatoid synovial fibroblasts co-treated with interleukin 1 and interferon-γ; reduced collagenase activity was detected after trypsin activation of the conditioned medium compared with medium from the interkeukin 1-treated cells [327]. In both media the levels of proMMP-1 were unaltered, but co-treatment of the cells with interleukin 1 and interferon-γ resulted in the inhibition of proMMP-3 production [327]. Human chymase, on the other hand, directly cleaves the Leu84–Thr85 bond [289] and rat chymase cleaves the Phe81–Val82 bond [319]. Rat chymase-generated [Val82]MMP-1, however, has only about 30–40% of the collagenolytic activity of [Phe81]MMP-1 [319]. Bacterial proteinases *Pseudomonas* elastase, *Vibrio* proteinase and thermolysin activate proMMP-1 by cleaving the Val82–Leu83 bond [276]. The resultant MMP-1 probably has a limited collagenolytic activity.

proMMP-2

This proenzyme is readily activated by APMA to a 68-kDa active form by cleavage of the Asn80–Tyr81 bond [272, 312], but is resistant to most endopeptidases which include trypsin, plasmin, plasma kallikrein, thrombin, chymotrypsin, neutrophil elastase, cathepsin G, thermolysin and MMP-3 [251, 255, 272]. Using the human melanoma M24met cell line that produces both urokinase-type plasminogen activator (uPA) and MMPs, the tumour cell-associated activation of proMMP-2 is independent of uPA/plasmin activity [248]. A lack of involvement of the PA/plasmin system in proMMP-2 activation was also demonstrated in the phorbol ester-stimulated HT1080 fibrosarcoma cells [241]. On the other hand, uPA processed the 72-kDa proMMP-2 to 62 kDa, and this processing was not inhibited by a plasmin inhibitor aprotinin, but was inhibited by anti-uPA antibody [224]. Thrombin and plasmin can also process proMMP-2 to the 62 kDa form. However, whether all these 62-kDa forms have proteolytic activity has not been examined. Trypsin 2 cleaves the Arg86–Lys87 bond of proMMP-2 but only limited enzymic activity is detected [309a]. ProMMP-2 can be activated by MMP-1 [185] and MMP-7 [1233, 293]. MMP-1 initially cleaves the Pro4–Ile5 [293] or the Asn37–Leu38 [185] bond of proMMP-2, then MMP-2 autolytically process the Asn80–Tyr81 bond [185, 272, 293, 312]. However, MMP-1 and MMP-7 are not very efficient activators of proMMP-2. In the presence of heparin, however, the activation of proMMP-2 by MMP-1 is greatly enhanced [185]. Heparin acts as a template for the two reactants by interacting through the C-terminal domains of both enzymes. Nonetheless, cell surface activation of proMMP-2 by MT-MMPs is considered to be physiologically more significant (see below). In addition, cell surface-bound proMMP-2 and proMMP-9 were activated by the cell surface-associated uPA/plasmin system [243]. Activation of the two gelatinases takes place only on the cell surface. Inhibition of uPA or plasminogen binding to the cell surface blocked the activation and the soluble-phase plasmin degraded both gelatinases [243]. Thus, under certain conditions the cell surface-associated uPA/plasmin system appears to participate in proMMP-2 and proMMP-9 activation. The activation steps are summarized in Fig. 21(b).

proMMP-3

Activation of proMMP-3 by proteinases follows a similar pattern to that of proMMP-1, but unlike MMP-1, proMMP-3 is activated by a large number of proteinases [189, 239, 271, 319], and this activation does not require other MMPs. Most proteinases attack the 'bait' region in the propeptide of MMP-3 that contains the sequence Phe34-Val-Arg-Arg-Lys-Asp39 (Fig. 21(c)). This promiscuous sequence, exposed to solvent [158], allows many endopeptidases with different specifi-

cities to cleave this region and initiate the activation [260, 262]. The second proteolytic step is catalysed by the intermediates of MMP-3, which cleave the His82–Phe83 bond. This bond is not cleaved by the activator proteinases that can attack the bait region. Furthermore, none of MMPs so far tested can cleave the bait region of MMP-3. Without the cleavage of its bait region MMP-3 cannot cleave the His82–Phe83 bond. Thus, the activation of proMMP-3 requires proteinases other than MMPs, but it can readily occur because many tissue and microbial proteinases can trigger activation. On the other hand, treatment of proMMP-3 or MMP-3 with trypsin results in the cleavage of the Arg84–Thr85 bond [159, 237]. [Thr85]MMP-3 has only about 25% as much activity as the fully active [Phe83]MMP-3 [159].

proMMP-7

Activation of proMMP-7 by trypsin also follows a stepwise process [186] (Fig. 21(d)). Leukocyte elastase and plasmin partially activate proMMP-7 (up to 50% of activity) [217], but the sites of their action are not known. MMP-3 activates proMMP-7 by what appears to be direct cleavage of the Glu77–Tyr78 bond [217].

proMMP-8

ProMMP-8 is activated by tissue kallikrein, leukocyte elastase, trypsin and cathepsin G [177, 227] which cleave various regions of the propeptide (Fig. 21(e)). Since the initial cleavage sites of these enzymes were not investigated by sequencing, it is not known whether the predicted bait region of proMMP-8 is initially attacked by enzymes. This region has a Ser-Thr-Arg-Lys-Asn-Gly sequence and it is likely that this stretch in the propeptide is exposed to the solvent. Thus, some of those serine proteinases may initially attack this region. *Pseudomonas* elastase and *Vibrio* proteinase are also potent activators of proMMP-8, but *Pseudomonas* alkaline proteinase and *Serratia* 56-kDa and 73-kDa proteinases have weak or little activating potential [276].

These bacterial proteinases may not act on the putative bait region of proMMP-8.

proMMP-9

Earlier work of Sopata and Dancewicz [308] showed that the gelatinase in neutrophils was activated by trypsin and chymotrypsin. The latent neutrophil gelatinase was reported to be activated by tissue kallikrein [325] and cathepsin G [252]. Morodomi *et al.* [249] examined eight endopeptidases (trypsin, chymotrypsin, plasmin, plasma kallikrein, thrombin, cathepsin G, neutrophil elastase and thermolysin) for their ability to activate proMMP-9 isolated from U937 cells. Of these, trypsin was the most effective activator and full activation was observed. Only partial activation (10–30%) was detected with plasmin, cathepsin G and chymotrypsin. Although it was suggested that the proMMP-9 released from neutrophils stimulated with fMet-Leu-Phe in the presence of cytochalasin B was likely to be activated by neutrophil elastase [336], some investigators reported that neutrophil elastase was not effective in proMMP-9 activation [223, 249, 270]. On the other hand, proMMP-9 was activated by neutrophil elastase to a 75-kDa form [196]. The active forms generated by trypsin were 80 kDa, 74 kDa and 66 kDa. Trypsin initially cleaves the Lys73–Ala74 bond and the Arg87–Phe88 bond, which results in an 88–82 kDa form [294]. However, trypsin, chymotrypsin and cathepsin G fully activated proMMP-9 by generating 77–65 kDa MMP-9, but longer incubation with those enzymes resulted in degradation of MMP-9 [270]. Trypsin 2 effectively activates proMMP-9 by cleaving at Arg87–Phe88 [309a]. Tissue kallikrein, but not plasma kallikrein, activated proMMP-9 [192]. Dog mast cell α-chymase [194, 195] activated proMMP-9 by cleavage of the Phe88–Gln89 and Phe91–Glu92 bonds of the catalytic domain, whereas *Pseudomonas* elastase and *Vibrio* proteinase activated proMMP-9 by cleaving the Thr90–Phe91 bond [276].

ProMMP-9 is also activated by MMP-1 [294], MMP-2 [200], MMP-3 [204, 268, 270], MMP-7 [217, 294], MMP-10 [264] and MMP-13 [231] by a stepwise mechanism (Fig. 21(f)). N-Terminal sequence analysis of the initial product generated by MMP-1, MMP-2, MMP-3, MMP-7 or MMP-13 indicated that the Glu40–Met41 bond was cleaved first. This intermediate of 86 kDa, however, did not bind to α2-macroglobulin, indicating that it is not proteolytically active [268]. The second cleavage of the Arg87–Phe88 bond generates the fully active 82-kDa MMP-9. Since MMP-9 failed to cleave this bond [268], other activator MMPs must cleave this bond. However, Okada *et al.* reported that MMP-1 failed to activate proMMP-9 [270]. Some discrepancies reported for proMMP-9 activation by proteinases may be due to differences in activation conditions. For example, the concentration of MMP-1 required for the activation of proMMP-9 is 100-fold greater than that for MMP-3 [294].

proMMP-10

MMP-10 and MMP-3 are 78% identical in amino acid sequence, and their propeptides have very similar sequences. ProMMP-3 has the bait region sequence Gln33-Phe-Val-Arg-Arg-Lys-Asp39 in the propeptide but proMMP-10 lacks Val35 in this sequence (Fig. 21(c)). Thus, like proMMP-3, proMMP-10 is activated by plasmin, trypsin and chymotrypsin, but not by neutrophil elastase (K. Suzuki, H. Nagase, G. Murphy and A. J. Docherty, unpublished results). Once the bait region is cleaved, the His81–Phe82 bond is cleaved by intermolecular autoproteolysis.

proMMP-11

The propeptide of MMP-11 contains a subtilisin-like proprotein converting peptidase (e.g. furin) recognition sequence RNRQKR at the C-terminal end (Fig. 21(h)). ProMMP-11 is not autocatalytically processed to an active form by APMA [256, 296]. In COS cells co-transfected with proMMP-11 and furin cDNA, proMMP-11 is activated to a 45-kDa form in the secretory pathway by Golgi-associated furin [283]. Specific cleavage at the furin cleavage site was shown by the lack of cleavage after mutation at this site. The processing of proMMP-11 to an active form was also inhibited by a furin inhibitor, Dec-Arg-Val-Lys-Arg-CH$_2$Cl [296]. The treatment of proMMP-11-transfected MCF7 cells with brefeldin A increased the intracellular accumulation of post-translationally modified proMMP-11, but not the activated enzyme, this indicates that the processing by furin into an active form occurs in the trans-Golgi network [296].

proMMP-12

Native and recombinant proMMP-2 undergo N-terminal and C-terminal processing to an active 22-kDa form during purification [304, 305]. Proteinases that activate proMMP-12 *in vitro* have not been studied. Mice deficient in urokinase-type plasminogen activator (uPA), however, are unable to activate and process proMMP-12 [178], indicating that the uPA–plasminogen system may be involved in activation of this enzyme.

proMMP-13

Plasmin, trypsin, MMP-2, MMP-3 and MT1-MMP (MMP-14) can activate proMMP-13 to the 48-kDa active form with the N-terminus of Tyr85 [228, 233] (Fig. 21(g)). Plasmin generates an intermediate by cleaving the Lys38–Glu39 and Arg76–Cys77 bonds in the propeptide; this is followed by autoproteolysis that produces a 48-kDa active MMP-13 with N-terminal Tyr85 [233]. Activation with trypsin also results in active MMP-13 with the same N-terminus. MT1-MMP initially cleaves the Gly35–Ile36 bond [233] and MMP-3 hydrolyses the Gly57–Leu58 bond [228]. These intermediates are processed by cleavage at the Glu84–Tyr85 bond. The rate of proMMP-13 activation by MT1-MMP is greatly enhanced in the presence of MMP-2 [233]. The treatment of SW1353 human chondrosarcoma cells with interleukin 1 and oncostatin M enhances the production of MT1-MMP, proMMP-2, proMMP-9 and proMMP-13 [183]. When these cells are treated with Con A, active MMP-2, MMP-9 and MMP-13 are found in the culture media; in the presence

of TIMP-2 and TIMP-3, activation of proMMP-13 is completely blocked, and it is strongly inhibited in the presence of TIMP-1 [183]. Since TIMP-1 is an effective inhibitor of MMP-2, but not of MT1-MMP, those observations further support the notion that the cell surface activated MMP-2 plays a major role in proMMP-3 activation. Together these studies suggest that the trimolecular interaction on the plasma membrane greatly enhances pericellular proteolysis of the matrix and cell surface associated proteins.

proMT1-MMP (proMMP-14)

The propeptide of MT1-MMP has a subtilisin-like proprotein converting peptidase (SPCP) recognition sequence RRKR at the C-terminal end, suggesting that the precursor of MT1-MMP is processed to an active form intracellularly (Fig. 21(h)). COS-7 cells transfected with the transmembrane deletion mutants expressed proMT1-MMP, which was than processed to an active MT1-MMP with N-terminal Tyr112 (numbering includes the signal peptide) and secreted from the cell [284]. This processing was inhibited by mutation of the SPCP-recognition site or co-transfection of the cell with the Pittsburgh mutant of α1-proteinase inhibitor that inhibits furin but not PACE 4 [284], suggesting that furin is responsible for activation of proMT1-MMP. The N-terminus of Tyr112 is identical to the MT1-MMP bound to TIMP-2 that was isolated from the plasma membrane as a complex with TIMP-2 [315]. In contrast, Cao *et al.* [175] reported that pro-domain processing was not necessary for MT1-MMP to activate proMMP-2. Co-transfection of COS-1 cells with cDNA encoding MT1-MMP mutated at the furin-recognition site (RRKR → ARAA) and cDNA for proMMP-2 resulted in activation of proMMP-2 to a level similar to that in cells with cDNAs for the wild-type MT1-MMP and proMMP-2. It is not known how the precursor form of MT1-MMP participates in proMMP-2 activation. On the other hand, intracellular processing of the 63-kDa pro-MT1-MMP to an active 60-kDa MT1-MMP is required for proMMP-2 activation [239a]. The C-terminally truncated pro-form of MT1-MMP expressed in *E. coli* was activated by furin [297], and can be also activated by trypsin [345] and plasmin [277]. Furin cleaves the Arg111–Tyr112 bond but plasmin hydrolyses the Arg108–Arg109 and Arg111–Tyr112 bonds. Because all other MT-MMPs, MT2-MMP [346], MT3-MMP [321] and MT4-MMP [286] in humans, chicken MT-MMP [352] and *Xenopus* XMMP [353] have a furin-like proteinase recognition site just before the catalytic domain, these precursors may also be activated intracellularly by a similar mechanism.

Cell-surface activation of proMMP-2: involvement of MT1-MMP and TIMP-2

ProMMP-2 is readily activated by APMA *in vitro* [255, 272], but it is generally considered that activation of proMMP-2 in tissue takes place on the cell surface. The ability of the plasma membrane to activate proMMP-2 was originally observed with normal or neoplastic cells treated with concanavalin A (Con A) [280], phorbol ester [170], or transforming growth factor β [170]. The cells treated with the RGD-containing peptide (FALRGDNP) [295] or cultured on collagen [155] or within collagen gels [208, 302] also show enhanced activation of proMMP-2.

In 1994, Sato *et al.* [298] cloned the cDNA of the first member of the MT-MMPs, now called MT1-MMP (MMP-14), and showed that COS cells transfected with MT1-MMP cDNA activated proMPP-2, but not proMMP-9. Among four MT-MMPs that are currently known, MT1-MMP [240, 284, 298, 346], MT2-MMP [173] and MT3-MMP [321] were shown to activate proMMP-2.

The isolated plasma membranes from the cells stimulated with Con A or a phorbol ester, or from the cells transfected with MT1-MMP activate proMMP-2 in a stepwise manner by initially cleaving the Asn37–Leu38 bond, followed by autocleavage of the Asn80–Tyr81 bond by the processed

MMP-2 [154, 316]. MT1-MMP-generated intermediate can be processed to a mature 62-kDa form by the uPA/plasmin system, but the cleavage site has not been identified [156]. This activation process requires the C-terminal domain from proMMP-2. Deletion of the C-terminal domain from proMMP-2 or addition of an excess C-terminal domain prevents the activation [257, 316, 337]. ProMMP-2 binds to TIMP-2 through the C-terminal domains of the two molecules. Addition of the exogenous TIMP-2 to the system inhibits the activation of proMMP-2 by the membrane [316, 337, 338]. Using the C-terminal domain of MMP-2 as an affinity ligand, the TIMP-2–MT1-MMP complex has been isolated from the plasma membrane of the phorbol ester-treated HT-1080 fibrosarcoma cells [315]. Activation of proMMP-2 by the plasma membrane was enhanced when a small amount of exogenous TIMP-2 was added, but was inhibited by a larger amount of TIMP-2. The polyclonal antibody specific to TIMP-2 blocked the activation of proMMP-2. These observations suggest that MT1-MMP on the cell surface serves as a receptor of TIMP-2, which in turn binds to proMMP-2. The formation of the ternary complex is considered to be critical for binding of proMMP-2 to the cell surface, where the subsequent activation of proMMP-2 by the closely located free MT1-MMP is thought to take place. The requirement of both free MT1-MMP and the TIMP-2–MT1-MMP complex for proMMP-2 activation was demonstrated [172] by TIMP-2 titration of the MT1-MMP expressed on the plasma membrane. Similar observations using MT1-MMP immobilized on agarose beads have been made [225]. On the other hand, TIMP-2 binds to the plasma membrane by two different mechanisms when cells are treated with Con A [222]: one is sensitive to a peptidyl hydroxamate inhibitor of MMPs and the other is insensitive to the inhibitor. The former binding mode of TIMP-2 is presumably mediated by MT1-MMP, but the binding molecule for the latter interaction has not been identified. TIMP-2 bound to the membrane in a hydroxamate-insensitive manner specifically inhibits the cell surface activated MMP-2. This inhibition reaction is thought to be initiated through the C-terminal domains of MMP-2 and TIMP-2 because the TIMP-2 on the cell surface inhibits only the full-length (65 kDa) MMP-2, but not C-terminal-truncated MMP-2 or other MMPs [222]. These studies suggest that TIMP-2 bound to the plasma membrane in a non-MT1-MMP-mediated manner also provides the mechanism for recruiting proMMP-2 to the cell surface. Thus, this system may be an alternative pathway that participates in proMMP-2 activation by MT1-MMP.

While it is generally considered that proMMP-2 is activated by cell surface expressed MT-MMP, normal human skin fibroblasts cultured in a type I collagen lattice activate proMMP-2 intracellularly and secrete active MMP-2 [238]. The activator is inhibited by EDTA and is enriched in the Golgi fraction. No significant changes were observed in mRNA and protein levels for MT1-MMP during lattice culture. Thus, it is not clear whether this intracellular activation of proMMP-2 is dependent on the activity of MT-MMPs. Membrane-anchoring of MT1-MMP through the transmembrane domain is essential for its ability to activate proMMP-2 [176], but soluble forms of MT1-MMP [240, 345], and MT2-MMP [173] were shown to activate proMMP-2 directly. MT1-MMP [345] and MT2-MMP [173] are inhibited by TIMP-2 and TIMP-3, but not by TIMP-1.

N-Terminal phenylalanine is required for the full enzymic activity of collagenases and stromelysin 1

The expression of full enzymic activity of at least some matrixins requires correct processing of the propeptide for MMP-1, MMP-3 and MMP-8. Treatment of proMMP-1 with trypsin, plasmin or APMA results in only about 10% of known collagenolytic activity [169, 211, 220, 253, 317]. Plasmin generates an initial 47-kDa species, which is then converted to a 42-kDa form of [Leu65]MMP-1 by autolysis [317] (Fig. 21(a)).

The subsequent reaction with MMP-3 (first isolated as procollagenase activator [334]) cleaves the Gln80–Phe81 bond and generates the fully active [Phe81]MMP-1 [317]. This specific bond is also cleaved by other matrixins (see above). Incubation of proMMP-1 with APMA initially generates a 42-kDa intermediate, which then undergoes autoproteolysis and results in [Val82]MMP-1 and [Leu83]MMP-1. Activation of proMMP-1 with rat chymase generates [Val82]MMP-1 [319]. These species of the enzyme possess a limited collagenolytic activity, at most about 40% of full activity.

When proMMP-3 is activated by trypsin, the Arg84–Thr85 bond, two residues from the C-terminal Phe83 in the catalytic domain, is cleaved [237]. The resulting [Thr85]MMP-3 exhibits only about 20% of the activity of [Phe83]MMP-3 [159]. MMP-3 with several extra residues attached at the N-terminal Phe83 also has significantly reduced activity [159]. In addition, these MMP-3 forms exhibited significantly different substrate specificity from that of the correctly processed [Phe83]MMP-3 [159].

Treatment of proMMP-8 with $HgCl_2$ results in [Met80]MMP-8, which only exhibits about 25% of the full collagenolytic activity [232]. Specific cleavage of the Gly78–Phe79 bond by MMP-3 [232] or MMP-10 [229] is also required for full collagenolytic activity of MMP-8. X-Ray crystallographic studies of [Phe79]MMP-8 [287] indicated that the correctly processed N-terminal Phe forms a salt bridge between its ammonium group and the side chain carboxylate of Asp232 in helix C. Without the Phe79, the N-terminal hexapeptide Met80-Leu-Thr-Pro-Gly-Asn85 is disordered [168]; it is considered, therefore, that the formation of the N-terminal salt bridge is important for the expression of the full enzymic activity. Nonetheless, it is not clearly understood how the formation of the salt bridge influences activity, since the geometry of the active site of the two forms of MMP-8 is essentially identical [287]. It may, however, influence the stabilization of the active site of the transition state, or the mobile N-terminal segment may interfere with substrate binding to the extended substrate binding groove of MMPs [287].

Processing at the hinge region

Many activated MMPs undergo further autoproteolysis. This may result in inactivation of the enzyme [160, 216, 255, 272] or specific processing at the hinge region [182, 228, 230, 271]. Human MMP-3 has been isolated as 45-kDa and 28-kDa species [273]. The 28-kDa species is the catalytic domain lacking the C-terminal hemopexin-like domain [271] but the two forms had no significant differences in proteolytic activities against protein substrates [273]. In the case of MMP-1, MMP-8 and MMP-13, loss of the C-terminal domain abolishes their activities to cleave triple helical regions of interstitial collagens [182, 228, 230, 250], although the catalytic domains retain their activity towards other protein or peptide substrates. The susceptible bonds are the Pro250–Ile251 bond in the hinge region of MMP-1 [182] and the Gly242–Leu243 and Pro247–Ile248 bonds [230] in MMP-8. Substitution of Ser for Ile251 of human MMP-1 results in a full-length enzyme resistant to autoproteolysis at the hinge region [267, 348]. Activation of proMMP-2 by plasma membrane also generates the C-terminal truncated 45-kDa form [222, 337]. This form is not readily inhibited by TIMP-2 that is bound to the cell surface, whereas activity of the 65-kDa full-length MMP-2 is regulated by TIMP-2 [222]. The 45-kDa MMP-2 has very low affinity to TIMP-2 ($K_i = 315$ nM) and barely binds to TIMP-1 [278]. The autolysis of the 45-kDa MMP-2 is largely prevented in the presence of collagens [222], suggesting that such species of MMP-2 may be effective in a pericellular environment.

Activation of progelatinase– TIMP complexes

ProMMP-2 binds to TIMP-2 [187, 203, 235, 311] and TIMP-4 [161] and proMMP-9 binds to TIMP-1 [344]. These complexes are formed by specific interaction of the C-terminal domain of the progelatinase and the C-terminal domain of the TIMP [199, 215, 254, 257, 266, 347]. Because the N-terminal inhibitory domains of TIMPs are free, the progelatinase–TIMP complexes can inhibit active MMPs [221, 235, 269]. These complexes therefore introduce further complexity in the regulation of MMP activity and progelatinase activation.

Interaction of proMMP-2 and TIMP-2 is stabilized by salt bridges formed between the C-terminal tail of TIMP-2 and of proMMP-2, and the approximate K_D value is estimated to be 50 pM [347]. The K_D value determined between the isolated hemopexin domain and TIMP-2 is estimated to be 61.6 nM by surface plasmon resonance [278]. The difference between these values may in part be due to the methods employed but also due to partial contribution of the N-terminal domains of proMMP-2 for the binding to TIMP-2 [201]. A TIMP-2 nicked at the Lys155–Asn156 bond in the C-terminal domain was found in the medium of human bladder carcinomas [247]. This form lacks the ability to bind to proMMP-2, although it retains inhibitory activity. Treatment of TIMP-2 with trypsin generates similar species of TIMP-2 [190]. Further studies [172, 315] suggest that the C-terminal interaction of proMMP-2 and TIMP-2 is critical for the cell surface action of proMMP-2. On the other hand, when the proMMP-2–TIMP-2 complex is treated with APMA, the activated proMMP-2 is rapidly inhibited by TIMP-2 before it loses its propeptide [221]. Similar events takes place on the cell surface when proMMP-2 is activated, but the inhibition by TIMP-2 in this case is slower, probably because it requires the dissociation of TIMP-2 from the membrane [222].

ProMMP-9 is activated by a number of proteinases, including MMPs (see above). When proMMP-9 exists as a complex of proMMP-9–TIMP-1, TIMP-1 readily inhibits other MMPs. A ternary complex, proMMP-9–TIMP-1–MMP-3, is found when the complex is reacted with MMP-3 [269]. Although this ternary complex partially dissociates into free proMMP-9 and the TIMP-1–MMP-3 complex, activation of proMMP-9 requires an excess of MMP-3 or other MMP relative to the complex. An alternative pathway for the activation of the proMMP-9–TIMP-1 complex is specific destruction of TIMP-1. Leukocyte elastase preferentially inactivates TIMP-1 by cleaving the Val69–Cys70 bond [263]. This proMMP-9 bound to TIMP-1 is readily activatable by MMP-3 [223]. Trypsin also inactivates TIMP-1 [274], but this enzyme activates proMMP-9 more readily than inactivating TIMP-1. Thus, in this case the activated MMP-9 was already inhibited by the cognate TIMP-1 [223]. Treatment of the proMMP-9–TIMP-1 complex with APMA activates proMMP-9 and the propeptide is removed by autolysis, but the activated MMP-9 is readily inhibited by TIMP-1 [269]. Under these conditions activity of MMP-9 cannot be detected.

Biological significance of stepwise activation mechanisms

Many matrixins are activated in a stepwise manner through generation of an intermediate form, and the final activation process is often catalysed by an active MMP. This suggests that MMP intermediates formed during activation may interact with endogenous inhibitors such as TIMPs and α-macroglobulins before they are fully activated to a mature form. Thus, the stepwise activation processes provide a more finely controlled mechanism for the degradative pathways of extracellular matrix components that are critical during development and morphogenesis. Indeed, many intermediates of matrixins do interact and form a complex with TIMPs. Examples are shown for MMP-1 [191], MMP-3 [159, 318, 339], MMP-9 [339] and MMP-13 [228]. However, the complex formed between the MMP-3 intermediate and TIMP-1 is not as tight as that of the fully activated MMP-3 and TIMP-1 [318], suggesting that a limited enzymic activity may be expressed during activation, even in the presence of endogenous inhibitors, especially in tissues that have abundant extracellular matrix components.

Protein substrates of the MMPs

Introduction

As emphasized in a recent review by Woessner [147] the natural substrates of MMPs acting *in vivo* remain largely unknown and unexplored. Although large compendia of substrates have been gathered, e.g. Chandler *et al.* [30], such lists comprise proteins that have been tested *in vitro* as potential substrates. This is no guarantee that the substrate in question is degraded *in vivo* by the enzyme tested, it merely shows that the potential exists. Furthermore, the number of known constituents of the extracellular matrix is now well over 150 and only a small fraction of these have been tested as substrates. To establish the natural substrates it is necessary to isolate degradation products from the tissue and show by sequence analysis that the bond cleavage matches that produced by the MMP in question. This has been done in the case of three substrates: interstitial collagens, aggrecan and link protein (see below). However, ambiguities still remain: often several different MMPs will digest the same protein at the same bond, and the fragments produced may undergo further digestion by other enzymes in the matrix. Another path to follow would be to show the absence of substrate degradation in a gene knockout experiment. In the section that follows, we will summarize known substrates, but with the three exceptions noted, there is no clear proof that any of these substrates is degraded *in vivo* by a particular MMP. A preliminary overview is provided by Table 7.

Interstitial collagens of types I, II and III

Properties of interstitial collagens as substrates

Collagen is an extracellular protein that is very difficult to digest with a wide range of proteolytic enzymes. The reason is that its tightly wound triple helical structure provides little opportunity for the proteinase to bring a single strand of collagen into its active site. Enzymes that can accomplish this have usually been named collagenases because of their unique ability to digest this protein. It was the search for such an enzyme in vertebrate tissues that led to the first discovery of a collagenase in higher organisms by Gross and Lapière in 1962 [1523].

A study of the cleavage specificity of MMP-1 for type I collagen [1495] indicates that one would expect as many as 31 cleavage sites along one collagen alpha-chain. The fact that only a single site is cleaved suggests that the surrounding sequence is of critical importance. It has been suggested that a relatively low content of imino acid residues in the immediate vicinity of the scissile bond may be conducive to partial unfolding of the helix in this region, permitting one strand to enter the active centre [1531]. This is also supported by the observation that high levels of trypsin could digest collagen at about the same point as later found for collagenase [1630]. Later studies [1591] showed that type III collagen is split at an Arg–Gly bond that is only five residues towards the C-end of the alpha chain from the Gly–Ile bond split by collagenase. Type III is also cleaved by neutrophil elastase, pronase and thermolysin when in soluble tropocollagen form, but not when in reconstituted

Table 7 Some common protein substrates for the matrix metalloproteinases

Enzyme:	MMP-1	MMP-2	MMP-3	MMP-7	MMP-8	MMP-9	MMP-10	MMP-12	MMP-13	MMP-14
Collagens:										
Type I	**[1521]**	**[1442]**	*[1447]*	[1713]	[1535]	*[1606]*		[1473]	**[1703]**	**[1623]**
Type II	**[1592]**				[1450]	*[1606]*			[1703]	**[1623]**
Type III	**[1592]**	[1719]	[1618]	*[1713]*	[1535]	*[1606]*	[1618]		[1703]	**[1623]**
Type IV	*[1697]*	[1603]	[1474]	[1595]	*[1527]*	[1606]	[1618]	[1473]	*[1703]*	
Type V	*[1697]*	[1603]	[1618]	*[1713]*	*[1527]*	[1606]	[1618]		*[1703]*	
Type VI	*[1629]*	*[1506]*	*[1629]*			*[1551]*			[1552]	
Type VII	[1663]	[1477]	[1657]							
Type VIII	[1650]									
Type IX	*[1510]*	*[1625]*	[1625]						[1552]	
Type X	[1660]	[1698]	[1718]	*[1673]*					[1552]	
Type XI	[1510]	[1675]	[1718]			[1533]			*[1552]*	
Type XIV	*[1670]*		*[1670]*	*[1670]*		[1670]			[1552]	
Gelatin I	[1702]	[1603]	[1474]	**[1713]**		[1643]	[1618]	[1473]	[1699]	[1541]
dnp-peptide	[1471]	[1680]	[1513]	[1713]		[1643]				
C1q	[1589]				[1496]	*[1613]*			[1489]	
Elastin	*[1664]*	[1626]	[1474]	[1641]		[1604]	[1604]	[1455]		
ECM proteins:										
Fibronectin	[1507]	[1477]	**[1709]**	[1713]	*[1714]*	*[1714]*	[1618]	[1473]	[1552]	[1623]
Vitronectin	[1542]	[1542]	[1542]	[1542]		[1542]		[1473]		[1623]
Laminin	[1632]	[1626]	[1474]	[1595]		[1597]		[1473]		[1623]
Entactin	**[1584]**	**[1584]**	**[1584]**	**[1584]**				**[1520]**		[1479]
Tenascin	[1540]	[1674]	[1540]	[1540]		*[1540]*				
SPARC		[1657]	[1657]	[1657]		**[1657]**			[1552]	
Aggrecan	**[1499]**	**[1503]**	**[1502]**	**[1503]**	**[1499]**	**[1503]**	**[1502]**	[1473]	**[1498]**	[1623]
Link protein	**[1616]**	**[1616]**	**[1615]**	**[1616]**		**[1616]**	**[1616]**			
Decorin	*[1539]*	[1539]	[1539]	[1539]		[1539]				
Myelin basic	[1472]	[1472]	[1472]	[1472]		[1517]		[1473]		
Regulatory:										
Autolytic act	[1602]	[1605]	[1543]	[1661]	[1557]	[1608]			[1553]	
α_2-Macroglob	[1491]	*[1581]*	[1491]		[1606]	[1597]		[1455]	[1614]	[1623]
Ovostatin	[1491]		[1491]		[1562]					
α1-PI	[1483]	[1581]	[1581]	**[1672]**	[1485]	**[1485]**		**[1454]**		[1623]
α1-AntiChym	[1483]	[1581]	[1581]			*[1484]*				
IL-1α	*[1544]*	*[1545]*	*[1545]*			*[1545]*				
IL-1β	[1544]	[1545]	[1545]			[1545]				
ProTNFalpha	[1515]	[1515]	[1515]	[1515]		[1515]		[1473]		[1479]
IGFBP-3	[1504]	[1504]	[1504]							
Substance P		**[1611]**	[1580]		[1486]	[1486]				
T kininogen		*[1652]*	[1652]			*[1652]*				
Casein	[1601]	*[1446]*	[1474]	[1641]		[1575]	[1655]		[1565]	
CM-Transferrin			[1525]	[1441]		[1597]				

[**bold**], substrate is digested and bond specificity has also been determined; [plain], substrate is digested; [*italic*], substrate is not digested.

fibrils [1462]. These results and studies with various MMPs (below) suggest that type III may be more flexible and susceptible in the cleavage region than types I and II.

Dissociated collagen molecules are more susceptible to hydrolysis than molecules in fibrils. Part of this is due to concealment of the substrate from the enzyme when only the surface is exposed, but part must be due to increased stabilization of the helix by interaction with neighbours. Furthermore, crosslinking of collagen molecules in native fibres further retards the rate of enzymic attack [1526].

Evidence that MMP-1 digests types I, II and III

Gross and Lapière, 1962 [1523] first showed digestion of reconstituted type I collagen fibrils by culture medium from resorbing tadpole tail fin. The collagen cleavage products were visualized by gel electrophoresis [1524, 1651]. Human skin fibroblasts were shown to produce a similar enzyme (MMP-1) acting on type I collagen [1490], furthermore, rabbit corneal MMP-1 degraded the trimer of $\alpha1(I)_3$ [1481]. Type II collagen is digested by human synovial MMP-1 [1715, 1716]. Using tadpole collagenase to digest chick type III collagen provided a useful tool to show the location of disulfide bridging in the C-terminal portion of the triple helix [1469].

In this early work, debate arose as to whether MMP-1 could cleave native polymeric collagen fibrils or only collagen that had been solubilized (often with the help of pepsin treatment). Thus, it was proposed [1564] that there could be no cleavage of human synovial collagen by synovial MMP-1 until the coating of proteoglycan material had first been removed by other proteinases. However, tadpole collagenase can digest polymeric collagen [1522]. When synovial MMP-1 is applied to insoluble tendon and soft connective tissue collagen, they are digested at rates 1/100 and 1/20 that of reconstituted fibrils [1715]. So it appears likely that proteoglycans, such as decorin, and minor collagen types, such as XII, that may coat collagen fibres, may slow down, but do not block,

the action of collagenase. The crosslinks of native collagen will further retard the rate relative to reconstituted fibrils [1526]. It is difficult to think of a situation in which MMP-1 would be acting alone *in vivo*, it is invariably accompanied by other MMPs which could facilitate collagen digestion.

Using a highly refined method of precipitating collagenase fragments of type I collagen into the so-called SLS (segment-long-spacing) form [1465, 1466] permitted assignment of the primary collagenase cleavage site at a position about 3/4 of the molecular length in from the N-terminus within a few residues. Subsequently, the specific bonds cleaved by human and frog collagenase in type I collagen were identified by sequence analysis [1521, 1531]; the $\alpha1(I)$ chain is cleaved at Gly775–Ile and the $\alpha2(I)$ chain at Gly775–Leu (the numbering was added later). MMP-1 cleaves type II at the corresponding Gly–Ile bond and type III at the Gly–Leu bond [1591, 1592]; these cleavages produce the so-called TC^A and TC^B fragments (TC = tropocollagen).

Collagen is one of the few substrates where it has been possible to demonstrate that the degradation products found *in vivo* have exposed amino acid termini that match the sequence specificity of the MMP that is postulated to digest it. This has been shown by the development of specific antibodies that recognize the neoepitopes exposed when the α-chains are cleaved [1459]. These epitopes for type II collagen are Gly-Pro-Hyp-Gly-Pro-Gln-Gly (C-end of TC^B) and Leu-Ala-Gly-Gln-Arg-Gly (N-end of TC^A); specific antibodies for these epitopes indicate damage to type II collagen in osteoarthritic human cartilage due to collagenase. Cartilage can produce four distinct collagenases (MMP-1, -8, -13 and -14) [1468, 1667], so their individual contributions to degradation cannot be estimated by this method; however, degradation in explant cultures was inhibited by Roche compound RS 102,481 which is most effective against MMP-13 [1459]. A second approach strongly implicating collagenases in collagen degradation is the production of transgenic mice containing type I collagen mutated at the collagenase cleavage site [1560, 1572].

It has long been recognized that MMP-1 is difficult to extract from tissues and almost all studies have been conducted using enzyme produced in tissue culture. Pardo *et al.* [1631] were among the first to demonstrate that collagenase was bound to collagen. Extraction from tissue requires stringent conditions such as 60 °C [1635] or 5 M urea [1711]. Collagen affinity columns were used to purify active MMP-1 [1456]. While there is some affinity of latent collagenase for collagen fibrils [1689] it was found [1701] that binding was negligible and that the active form binds tightly. Therefore, it is currently not clear why the latent enzyme is difficult to extract.

Kinetics of collagen digestion by MMP-1: substrate species and enzyme species

Rabbit MMP-1 digests type III collagen about five times faster than type II collagen [1592] and cow type I collagen about five times faster than type II [1467]. The tadpole enzyme produces similar results [1529]. Extensive studies have been done using human MMP-1 and various species of collagen of types I, II, III [1700]. Relative rates (k_{cat}/K_m) at 25°C are shown below (human and type I set equal to 100):

	Type		
	I	II	III
Human	100	0.75	604
Calf	64	2.5	
Guinea pig	37		38
Rat	32	6.1	-

A more detailed analysis of type III [1696] explores the difference between soluble and fibrillar type III of various species (k_{cat}/K_m, Cat = 100):

	Soluble III 25 °C	Fibrillar III 37 °C
Human	90	75
Dog	75	89
Cat	100	100
Guinea pig	2.8	6.7
Chick	0.86	1.5
Type I guinea pig	3.6	41

These large differences are attributed to the different stability of the triple helix in the different species; the susceptibility to collagenase being roughly parallel to the susceptibility to trypsin. The differences between soluble and fibrillar (627 versus 60.7, respectively, for k_{cat}/K_m in M^{-1}h^{-1} for the cat type III) are probably due to the fact that only surface collagen is available as a substrate for fibrillar collagen. Similar differences between soluble and fibrillar collagens have been noted for human MMP-1 acting on calf type I and III [1462] and tadpole collagenase acting on guinea pig type I [1458].

It is difficult to compare MMP-1 from different species because the purity, temperature of assay and type of collagen substrate are rarely constant between reports. Activities range from a low of 312 units (μg collagen digested/min/mg enzyme) [1715] to a maximum of 53 000 units for pig synovial MMP-1 [1687]. Bovine and human MMP-1 were compared on four substrates [1682] (some of the values are taken from [1700]). The k_{cat}/K_m for human MMP-1 on human type I is set to 100:

	Human MMP-1	Bovine MMP-1
Human type I	100	38
Human type II	0.75	2.1
Human type III	633	221
Bovine type I		40

This human enzyme has a specific activity on human type I of about 6100 μg/min/mg enzyme at 37 °C.

Digestion of collagen relatives gelatin, telopeptides, dnp-peptide and C1q

The action of MMP-1 on gelatin is not very striking, there being no action of human MMP-1 on gelatin I [1586]. The rates of human MMP-1 on gelatins of type I, II, III, IV, V [1702] and the effects of MMP-1 on gelatin type I [1715] and the kinetics of digestion [1548] have been determined. A band of MMP-1 can be visualized on gelatin zymograms [1684].

The action of rabbit MMP-2 on gelatins has been studied [1603, 1605]. Human MMP-2 digests gelatin better than type IV collagen [1477, 1626], while stromelysin digests type I gelatin [1525, 1627, 1628] as well as other types [1618]. Both human [1641] and rat matrilysin [1441, 1713] digest gelatins of type I, III, IV and V. Activity is largely against the $\alpha2(I)$ chain [1441]. Human gelatinase B acts on gelatins of type I, II, III, IV, V [1606, 1643, 1676]. Purified human MMP-9 digests type I gelatin at the rate of 15 000 μg/min at 37 °C [1597], whilst rat MMP-9 has been shown to digest gelatins I, II, V [1575, 1613, 1653]. Human MMP-12 attacks a large range of substrates including gelatin I [1473]. Rat MMP-13 acts on both the $\alpha1$ and $\alpha2$ chains of gelatin I, and the rate of gelatin digestion by human MMP-13 is 100.6 μg/min/nanomole enzyme compared to 106.7 for MMP-8 [1553]. Recently, MMP-14 [1541, 1623] and MMP-16 [1668] have been reported to digest type I gelatin.

MMP-1 is not a telopeptidase [1560], although collagenase 3 is. Purified rabbit stromelysin digested telopeptides of type I collagen [1513] and this was confirmed for human MMP-3 [1627, 1628]. MMP-3 cleaves the telopeptides of type II collagen [1585, 1717, 1718]. Mouse MMP-13 cleaves the N-telopeptide of type I collagen internal to the lysine-derived crosslink [64], permitting solubilization of fibrils [1565].

In 1977 Masui *et al.* [1582] developed a series of synthetic substrates incorporating the sequence of the collagen $\alpha1(I)$ chain that is cleaved by the interstitial collagenase of the frog. The most specific substrate was the octapeptide dinitrophenyl-Pro-Gln-Gly Ile-Ala-Gly-Gln-D-Arg; however, this peptide proved to be cleaved by a number of MMPs (and by proteinases of other classes). The action of tadpole collagenase, presumably MMP-1, but possibly MMP-18 [1534], and MMP-8 and MMP-9 from human polymorphonuclear leukocytes [1556, 1643] have been tested on this peptide. Early studies of human MMP-1 [1708], MMP-2 [1662], rabbit MMP-3 [1513] and rat MMP-7 [1713] all showed that this peptide was a suitable substrate. Other enzymes cleaving this peptide include human MMP-8 [1686], pig MMP-1, -2 and -3 [1679, 1680], human MMP-7 [1478] and mouse MMP-13 [1565]. These results indicate that cleavage of the collagen helix involves more than the immediate sequence that is cut; the hemopexin domain of MMP-1 develops a critical interaction with the collagen molecule, and the catalytic domain without the hemopexin cleaves this peptide but fails to cleave collagen [1601].

The complement component C1q is treated here because it possesses a short collagen-like triple-helical tail; human synovial MMP-1 can cut this protein [1589] at a potential cleavage site Gly Ile/Leu at position 39–40 in the three strands. Human MMP-8 can also cleave C1q [1496] but rat MMP-9 is ineffective [1613].

Collagenase 2 (MMP-8) action on interstitial collagens

The $\alpha1(I)$ trimer is cleaved by human MMP-8 [1612] as is type II collagen [1645], although later studies suggest this result may have been due to the slow rate. MMP-8 digests type I equally or better than type III [1527, 1535, 1606], and can digest soluble type II but not when it is in the form of native fibres [1450]. The cleavage rate is dependent on the extent of glycosylation [1636]. Bond specificity in type I collagen has been determined [1528].

Collagenase 3 (MMP-13) action on interstitial collagens

The recent discovery of collagenase 2 in mouse and rat suggests caution in interpreting all earlier rat/mouse collagenase results as due to MMP-13. Values of k_{cat}/K_m for the action of MMP-13 for seven species of type I, two of type II and three of type III all fell in the range of 5–16 $\mu\text{M}^{-1}\text{h}^{-1}$ [1703]. However, human MMP-1 had rates varying by almost two orders of magnitude with the same substrates. It was noticed [1559] that mice were not affected by a mutation in collagen type I at the collagenase cleavage site, due to the ability of MMP-13 to degrade collagen by means of its telopeptidase action. Rates of digestion of different collagens by human MMP-13 have been reported [1553, 1644]; type II collagen appears to be the best substrate. Human MMP-13 digests human type II more rapidly that do MMP-1 and MMP-8 [1459].

Action of MMP-2, -3 and -7 on interstitial collagens

Of collagens type I, II, III, IV, V and VI, only type VI is resistant to MMP-1, -2, and -3 [1629]. MMP-2 has no action on the interstitial collagens type I, II, or III [1570] for both human [1477] and rat MMP-2 [1653]. However, high levels of MMP-2 will slowly cleave type I, whereas MMP-9 does not [1442]. Binding of MMP-2 to type I collagen has also been demonstrated [1443, 1444]. MMP-2 does digest type III [1719] but not type II collagen [1464a], and stromelysin digests type III [1618] but not type I [1706]. Rat MMP-7 does not digest collagens of types I, III, IV, V [1713]; likewise, human MMP-7 fails to digest types I and III [1595].

Action of MMP-9 through -18 on interstitial collagens

Several groups report that human MMP-9 has no action on type I collagen [1558, 1639] nor on types II or III [1606]. The rat enzyme also has no action on I, II or III [1480, 1613, 1648]. However, the picture is not completely clear; it has been claimed that MMP-9 does cleave type III [1695] and that MMP-9 is able to digest type I at $30\,^{\circ}\mathrm{C}$ and type III at $37\,^{\circ}\mathrm{C}$ [1719].

Recombinant MMP-10 is reported [1618] to digest type III collagen. MMP-14 has been shown [1623] to digest collagen I (guinea pig), II (cow) and III (man); typical bond cleavages at Gly775–Ile/Leu were observed in $\alpha 1(I)$ and $\alpha 2(I)$, but there was also cleavage at Gly781–Ile in $\alpha 2(I)$.

MMP-18, also known as collagenase 4, cleaves type I collagen [1681].

Other (non-interstitial) collagen types

Type IV collagen

Type IV collagen is digested by human MMP-2 [1570]; there was action on both the pro$\alpha 1$(IV) and pro$\alpha 2$(IV) chains [1654]. Type IV collagen is digested to a greater extent than type V [1477], although the opposite has been reported [1626]. Mouse MMP-2 [1494] and rabbit MMP-2 [1605] also degrade type IV. There is evidence that native type IV collagen is not digested, but only the denatured ($37\,^{\circ}\mathrm{C}$) or pepsinized forms [1577, 1719]. The rate of digestion is also affected by the degree of glycosylation [1599].

MMP-3 digests type IV collagen [1457, 1513, 1604, 1618, 1627, 1706], and rabbit [1474, 1513] and rat [1688] stromelysin also digest type IV; but again only at higher temperatures where some denaturation may occur [1688]. Type IV collagen is digested by matrilysin of human [1595, 1604] and rat [1713] origin. MMP-1, and -8, the interstitial collagenases, do not digest type IV collagen [1527, 1700], and rat MMP-13 is inactive [1703], but human collagenase 3 does digest it [1552].

Purified human MMP-9 acts on type IV collagen and this has been reported by many groups [1596, 1597, 1604, 1606, 1695, 1707]. Type IV is degraded more slowly than type V [1596, 1597] and, as with MMP-2, there is evidence that type IV may be denatured prior to digestion [1577]. Conflicting reports have been presented for the action of rat MMP-9 on type IV; digestion is reported at $37\,^{\circ}\mathrm{C}$ but not at $30\,^{\circ}\mathrm{C}$ [1480], others also find no digestion [1648] or action at $30\,^{\circ}\mathrm{C}$ [1575].

Recombinant MMP-10 digests type IV [1618]. Stromelysin 3 has weak proteolytic activity; however, if the C-terminal 175 residues are removed from mouse MMP-11, it is then capable of digesting type IV. Human MMP-11 acted only after the mutation of Ala255 to Pro [1619]. Human macrophage elastase attacks a large range of substrates including type IV collagen [1473, 1520].

Type V collagen

Human [1603] and rabbit MMP-2 digest pepsin-soluble type V collagen. Human MMP-2 digests IV better than V [1477], although see [1626]. Stromelysin digests type V [1618] but the interstitial collagenases MMP-1 [1700], MMP-8 [1527] and rat MMP-13 [1703] do not.

Human MMP-9 digests collagen type V [1606, 1639, 1695], and type V is digested more rapidly than type IV [1596, 1597]. Type V is not digested when in its native state, but only when denatured [1577]. Similar debate revolves around rat MMP-9, which is said to act on native type V [1480, 1575] or only on the denatured collagen [1613, 1648, 1653]. In regard to these arguments, the observations that even the serine proteinases plasmin and thrombin can degrade type V at 35 °C, but not below 33 °C [1569], indicate that the non-helical portions of the type V molecule are readily exposed to enzyme attack once the molecule begins to unfold near body temperature.

Collagen of types VI to XX

In studies on the digestion by MMP-1, -2 and -3 of collagens I, II, III, IV, V and VI, none of these enzymes could digest **type VI** [1629]. This has been confirmed [1551] and extended to include MMP-9. The $\alpha 3$ chain of type VI is digested by MMP-2 [1609].

Type VII collagen is susceptible to human MMP-1 and -2 [1663]; MMP-2 digests type VII faster than type V [1477]. Stromelysin also digests type VII [1659]. **Type VIII** collagen is digested by human MMP-1 [1649, 1650]. **Type IX** collagen is not digested by MMP-1 [1510, 1511, 1625], whereas the situation with MMP-2 is not clear [1464a, 1625]. Stromelysin cleaves type IX [1492, 1717, 1718] as does human MMP-13 [1552].

Type X collagen is digested with MMP-1 [1475, 1660], but might require the synergistic action of cathepsin B [1673]. Digestion of type X could occur at 25 °C [1510] and cleavage occurs at two positions [1512]. MMP-2 can also attack type X [1475] making two cuts [1698]. Stromelysin and MMP-13 action has been shown [1552, 1717, 1718], whilst matrilysin is said not to act on type X [1552].

Type XI collagen was earlier known as the $1\alpha 2\alpha 3\alpha$ collagen of cartilage; it is digested by MMP-1 [1493] and MMP-2 [1603, 1675, 1720, 1721]. Only the first two of the three chains are cleaved [1568]. Stromelysin has telopeptidase action on type XI [1717, 1718], and type XI is cleaved by MMP-9 [1533, 1638, 1639], but not by MMP-13 [1552].

There is no action of MMP-1, MMP-3 or MMP-7 on **type XIV** collagen, but MMP-9 [1670] does digest it as does human MMP-13 [1552].

Elastin

Early reports indicated that MMP-2 and MMP-3 digested elastin [1474, 1604, 1605, 1626], but it has been reported more recently that while MMP-1 and MMP-3 do not digest elastin [1664], MMP-2 and MMP-9 do [1549, 1604]. Rat MMP-7 has only a weak effect on elastin [1441, 1713] but human MMP-7 is about 10 times more active [1641]. MMP-10 also digests elastin [1604].

Mouse macrophage digestion of elastin was traced to a metalloproteinase [1452, 1455, 1704], and extended in demonstrations that both mouse and human MMP-12 digest insoluble elastin [1665, 1666]. Human MMP-12 has been the subject of several other studies [1473, 1520]. The elastin bonds cleaved by mouse and human MMP-12 have been determined [1587].

Glycoproteins

Fibronectin/vitronectin

Early studies of fibronectin [1590] indicated that it inhibited collagenase action on type III collagen; however, the effect was attributed to binding of fibronectin to the substrate. More recently, MMP-1 was shown to digest fibronectin [1507]. Variable results are reported for MMP-2: negative [1570, 1719] and positive [1477, 1507, 1626]. A number of reports indicate that stromelysin 1 digests fibronectin [1507, 1513, 1525, 1600, 1618, 1627, 1628, 1706] and specific points of cleavage have been identified [1709].

Rat [1713] and human [1641] matrilysin both cleave fibronectin and many similar reports have appeared [1441, 1507, 1595, 1692]. MMP-8 does not cleave fibronectin [1714] and many reports agree that MMP-9 also does not act on this substrate [1480, 1613, 1714, 1719]. Recombinant human MMP-10 degrades fibronectin [1618] as does MMP-12 [1520] and MMP-13 [1552]. Both MMP-14 and -15 digest fibronectin [1479, 1623] and MMP-1, -2, -3, -7 and -9 digest vitronectin [1542, 1622]. Human MMP-12 [1473] and MMP-14 [1623] also digest vitronectin.

Laminin

A difficulty here is that we are gradually coming to recognize that there are a great many different laminins composed of three variable chains. It is not always possible to determine which type was used as substrate, so this may account for some of the conflicting reports. Thus, Collier *et al.* [1477] claim human MMP-2 does not attack laminin, whereas Okada *et al.* [1626] do find digestion; a similar opposing pair is found in [1719] and [1516]. Digestion by rabbit [1474, 1513] and human stromelysin 1 [1457, 1627, 1628, 1706] is clear. Matrilysin attacks laminin [1595], whereas results with MMP-9 are ambiguously positive [1597] or negative [1719]. Mouse MMP-11 with the C-terminal 175 residues removed can digest laminin, but the human enzyme acts only after mutation of Ala255 to Pro. Metalloelastase [1473, 1520] attacks laminin and MMP-14 digests laminin-1 [1623].

Entactin/nidogen, tenascin, SPARC

Entactin/nidogen is cleaved by MMP-1, -2, -3 and -7 [1519, 1584, 1671]. MMP-12 [1520] and MMP-14 and -15 [1479] act as well. Tenascin is digested by MMP-1 and MMP-3, but not by MMP-2 or MMP-9 [1540]. Siri *et al.* [1674] agree to the action of MMP-9 but claim that MMP-2, as well as MMP-3 and MMP-7, attack tenascin. MMP-13, MMP-14 and MMP-15 are also reported to digest tenascin [1479, 1552]. SPARC, also known as BM-40 and osteonectin, is digested by MMP-2, -3, -9 and -13 [1657] and the bonds that are cleaved have been identified; matrilysin may also be added to the list [1519]. Mouse MMP-13 cleaves mouse BM-40 in the C-helix [1583].

Miscellaneous glycoproteins

Myelin basic protein is digested by a number of MMPs (-1, -2, -3, -7, -9) [1472]. Several sites of cleavage by MMP-9 have been identified [1517, 1640]. This substrate is also cleaved by MMP-12. Galectin 3 is digested at the Ala62–Tyr bond by MMP-2 and MMP-9 [1621]. Integrin was not digested by MMP-2 [1488] but the β4 integrin was cleaved by MMP-7 [1693]. α2-HS glycoprotein is digested by MMP-2 and -7, but MMP-9 binds without cleaving [1620]. Fibulins 1 and 2 are digested by MMP-3 and -7 [1658]. Carbohydrate binding protein 30 (CAP 30) is cut by hamster kidney MMP-7 and -9 [1588]. The amelogenin of tooth enamel is processed by human MMP-20 (enamelysin) [1573].

Proteoglycans

Aggrecan

Aggrecan is a large proteoglycan with a core of 2300 amino acids and over 100 chains of chondroitin sulfate and keratan sulfate giving a total M_r of about 2 000 000. Near the N-terminus are two globular domains (G1 and G2) and the stretch of amino acids between these two points is particularly susceptible to proteolytic attack. In the earlier literature, names such as PPL (protein polysaccharide large) or cartilage proteoglycan were used. Early work focused on the effects of stromelysin 1 on aggrecan [1447, 1470, 1474, 1513, 1604, 1627, 1628, 1706, 1712]. In later work the site of cleavage was identified as VDIPEN~FFVGVG [1500, 1502, 1536, 1669]. Most of the MMPs tested appear to work on aggrecan and most cleave at this same site. Thus MMP-1 [1499, 1536], MMP-2 [1503, 1604, 1626], MMP-7 [1499, 1503, 1604, 1713], MMP-9 of dog [1476], rat [1480] and human [1503, 1604], MMP-10 [1502, 1604], MMP-12 [1473], MMP-13 [1498], MMP-14 [1479, 1623] and MMP-15 [1479] all have similar action on the core protein of aggrecan. Of course, action is not limited to this single bond, prolonged incubation of enzyme and substrate leads to extensive degradation. Multiple sites of cleavage by MMP-3 have been identified [1647]. MMP-1 and MMP-8 also cleave at several sites [1499].

Recently researchers in this area were given a severe jolt by the discovery that the cut ends of aggrecan found in human arthritic disease (in joint fluid and in cartilage extracts) arose from cleavage at the sequence NITE~GEAR [1500, 1538, 1574, 1656]. This cleavage was not produced by the known MMPs but the cleavage was sensitive to hydroxamate inhibitors of MMPs. To date, the only MMP found to cut at this site is human neutrophil collagenase (MMP-8) [1501]. However, this is not the major site of cleavage for MMP-8 [1445, 1499]. A search for 'aggrecanase', possibly a member of the ADAM family, is under-way, aided by the use of monoclonal antibodies specific for the cut ends of aggrecan produced by this enzyme [1537] and by MMPs [1497, 1537].

These results expose the difficulty in relating a specific MMP action to its normal *in vivo* matrix substrate. All results were consistent with MMPs, particularly MMP-3, having aggrecan as a natural substrate. However, examination of the cut ends produced by MMP-3 and the ends occurring naturally *in vivo* showed that a completely different enzyme (with very closely related properties) was involved. More recently, it has been possible to show that human cartilage aggrecan in arthritis does also contain ends produced by MMPs [1563], but there remain a number of contending MMPs with the same specificity that are candidates for producing this cleavage.

Link protein and others

In the case of **link protein**, which aids in the binding of the N-terminus of aggrecan to hyaluronic acid, there is a much clearer case for cleavage by MMPs [1470, 1571, 1615, 1617]. The protein is cleaved at Leu-25–Leu by MMP-2, -7 and -10, and at His16–Leu by MMP-9 [1616]. These cleavage sites can be shown in the products isolated from arthritic cartilage [1616].

MMP-1 has no action on **decorin** but MMP-2 cleaves at seven characterized sites [1539]. MMP-3, -7 and -9 do not cleave decorin [1539]. Glial hyaluronate-binding protein of 60 kDa is cut by MMP-1, MMP-2 and MMP-9 as a product arising from **versican** [1637]. Versican is also cleaved by MMP-7 [1550]. **Perlecan** is degraded by MMP-3 and MMP-13 [1705] as well as by MMP-14 and MMP-15 [1479]. MMP-12 releases chondroitin sulfate and heparan sulfate from basement membrane proteoglycans [1520].

Regulatory peptides and proteins

Interleukin 1α is not cleaved by MMP-2, MMP-3

or MMP-9, whereas interleukin 1β is [1545]. The cleavage site of MMP-2 is Glu25–Ile [1544]. MMP-1 cleaves IL-1β [1530, 1545]. **Tumour necrosis factor α** (TNFα) is cleaved from its precursor by TACE (TNFα converting enzyme) which is a member of the ADAM family [1464, 1598]; this enzyme is closely related to the MMPs but is not inhibited by TIMP. The MMPs release TNFα in a similar manner: MMP-3 [1514], MMP-3 [1463, 1515] and MMP-1, -2, -7 and -9 [1515] all liberate TNFα. Similar results have been obtained with MMP-12 [1473], MMP-14 and MMP-15 [1479].

IGBFP (**insulin-like growth factor binding proteins**) are cleaved by a variety of cellular proteinases including MMPs. The site of cleavage of IGBFP-3 has been shown for MMP-1, -2 and -3 [43, 1504, 1505]. MMP-2 cleaves IGBFP-5 [1685] and MMP-11 attacks IGFBP-1 [1578]. IGBFP-2 is degraded by MMP-1 [1642]. **Substance P** is digested by MMP-2 [1611], MMP-3 [1580], MMP-8 [1486] and MMP-9 [1449, 1486]. **T-Kininogen** resists the action of MMP-2, -3 or -9 [1652]. MMP-8 digests **angiotensin** I and II and **bradykinin**; MMP-9 digests all but bradykinin [1486]. The ectodomain of the **FGF** receptor 1 is cleaved by MMP-2 [1567] and the heparin-binding **EGF** is released by MMP-3 [1683].

Constituents of the blood plasma/serum

The **macroglobulin** inhibitors are large molecules (up to 760 000 Da) that are cleaved in a bait region, producing a conformational change that physically traps the proteinase, which may also become bound through interaction with a thiol ester. Further details are given under *macroglobulins* in the section on MMP inhibitors. In view of the odd specificity of collagenase for the Gly–Ile bond in the triple helix, it was somewhat surprising to find that MMP-1 was inhibited by α2-macroglobulin [1491, 1677]. MMP-1 also includes the egg macroglobulin, **ovostatin** [1491], and indeed it cleaves five different members of the macroglobulin family [1677, 1678]. Surprisingly, the cleavage of α2-macroglobulin by MMP-1 is much faster than the cleavage of collagen [1491, 1678]. MMP-3 is inhibited by both α2-macroglobulin and ovostatin [1491] as is MMP-8 [1482, 1562, 1606]. MMP-9 [1597], MMP-11 [1634], mouse MMP-12 [1455] and MMP-14 [1623] all react with α2-macroglobulin. The trapping of MMPs by α2-macroglobulin provides a useful tool to identify active enzymes by zymography, the trapped enzymes appearing at the top of nonreducing gels [93].

Both **α1-proteinase inhibitor** (α1PI) and **α1-antichymotrypsin** (α1AC) are digested by MMP-1 but this does not result in inhibition of the enzyme [1483, 1581]. A study of MMP-1, -3, -7 and -9 showed that MMP-7 cleaved α1PI most rapidly [1672]. α1PI, α1AC, **antithrombin III** are all cleaved by MMP-3 but not by MMP-2 [1581]. Winyard *et al.* [1710] also observed destruction of α1PI. Matrilysin rapidly cleaves α1PI at the Phe352–Leu bond [1722]. MMP-7 digests α1PI 30 times more effectively than MMP-9, 70-fold more than MMP-1 and 180-fold more than MMP-3 [1672]. The same cleavage at Phe352–Leu is reported for neutrophil MMP (this could be MMP-8 or MMP-9) [1485], both are known to act on α1PI [1554, 1575, 1691]. α1PI is one of the few substrates found to date for MMP-11 [1634]. Mouse MMP-12 digests α1PI, but not if it is complexed with a proteinase [1451, 1452], cleavage is at the Pro357–Met bond [1454]. Human MMP-12 [1473, 1520] and MMP-14 [1623] digest α1PI. The **plasmin C1-inhibitor** is cleaved at Ala439–Ile by MMP-8 [1555] and also at Pro357–Met [1484].

Rabbit MMP-1 cleaves serum **amyloid** A [1593]. MMP-3 also acts on amyloid A [1593]. MMP-2 releases β-amyloid protein from the precursor [1594]. The MMP-2 cleavage is in the β-peptide [1646] (but compare negative results in [1694]). This digestion by MMP-2 [1566] is not an α secretase action. MMP-9 cleaves amyloid β-peptide [1448]. MMP-3 digests the γ-Gly404–Ala link in **fibrin** [1461]. **Fibrinogen** is resistant to the action of MMP-9 [1613]; however, rat MMP-9 does attack it [1575]. **Plasminogen** is affected by MMP-3 action [1460], and MMP-7 and MMP-9 have been shown to digest specifically at the Pro466–Val bond and the Pro465–Pro bond, respectively [1633] to produce angiostatin. Rat MMP-12 also releases angiostatin [1487]. Mouse MMP-12 digests immunoglobulins IgG2a, IgG2b of mouse origin [1453]. Finally, rat MMP-7 activates **prourokinase** by cleavage at the Glu143–Leu bond [1579].

Proteins commonly used as assay substrates

Casein has been widely used as an MMP substrate. MMP-1 can digest casein, even after collagenolytic activity is lost by removal of the hemopexin domain [1601]. Rabbit MMP-3 was one of the first MMPs shown to digest casein [1474, 1513]; the human enzyme has also been studied [1525, 1627, 1628]. MMP-7 digests casein [1595], but MMP-9 has little effect [1653]. The rat MMP-9 is also said to have no effect [1613] but this is controverted [1575]. MMP-11 has only a weak effect on casein zymography, and then only after removal of the hemopexin domain [1607]. Similarly, removal of 175 residues from the C-end of mouse MMP-11 gave it the ability to digest casein [1619]; the human enzyme had activity after mutation of Ala255 to Pro. MMP-13 of mouse [1565] and MMP-16 [1668] have been reported to digest casein.

Reduced and carboxymethylated **transferrin** has been used as a substrate as introduced by Nagase for MMP-3 [1525, 1610, 1690]. It has been applied to the measurement of activity or zymography of rat MMP-7 [1441, 1547] and MMP-9 [1597]. Rat MMP-9 is said not to digest transferrin [1613]. **Azocoll** has been used to measure MMP-2 [1626] and rat matrilysin [1441, 1661].

Specificity requirements of the MMPs

Information about the specificity of the various MMPs is gathered in several ways: the site at which the MMPs cleave natural substrates of known sequence can be determined; their action on peptide substrates of defined sequence can be measured (usually be determining the specificity constant k_{cat}/K_m); synthetic inhibitors can be tested for binding affinity; and the structure of the binding site can be examined by X-ray crystallography and by NMR. In the first two cases, which will be the topic of this section, one is limited to combinations of the 20 known amino acids, whereas further subtleties of the binding pockets can be explored by the last two methods, which will be taken up in their appropriate sections.

In the case of the examination of natural substrates, it is often difficult to obtain kinetic data that would permit comparison of one substrate with another; as a consequence, there is a tendency to place more emphasis on the kinetic data from small peptides, typically blocked hexa- or octapeptides, even though these compounds are often digested much more slowly than are protein substrates. A recent review of MMP specificity has been provided by Nagase and Fields [92] and reviews of individual MMPs will be cited in the following subsections.

Collagenase 1 (MMP-1)

Various protein substrates cleaved by MMP-1 have been discussed in the section on MMP substrates. Here we will consider the specific bonds cleaved in such substrates (see Table 8). The peptide positions to either side of the scissile bond are designated P4 through P4′—it is likely that an octapeptide is about the length of peptide that can be accommodated by the active centre (see below).

It is apparent that no striking pattern emerges to form a strong consensus sequence. The most prominent feature is the appearance of Leu/Ile at P1′ in 27/37 cases and the rare appearance of any charged residue other than Glu. Moreover, it is difficult to compare these various substrates because, for the most part, there is no uniformity of assay conditions and no way to compare the action of the enzyme in a quantitative fashion. Therefore, some of these bonds such as that in α2-macroglobulin are cleaved at rates several orders of magnitude faster than collagen [1314, 1491] whereas collagen-like peptides may be cleaved at 1/100 the rate of collagen [1240].

Collagenase has attracted a great deal of attention with respect to the synthesis of synthetic substrates that would be suitable for purposes of routine assay; this would avoid the cumbersome assays involving collagen in fibrous or soluble form. Masui *et al.* in 1977 [1279] were the first to synthesize such substrates, based on the sequence about the scissile bond in collagen; they prepared a series of seven dinitrophenyl derivatives with six to eight amino acid residues (e.g. dnp-PLG$\sim$IAGR) in which the final arginine was in the D-configuration to protect it from carboxypeptidase action. It took only a short while to recognize that other enzymes could attack this same substrate, and in fact, all MMPs that have been tested appear to digest it. A much better peptide was prepared by Weingarten and Feder [1324, 1326] using a thioester peptide: Ac-Pro-Leu-Gly-S$\sim$Leu-Leu-Gly-OEt. This has a k_{cat}/K_m of $26\,350\,\mathrm{M}^{-1}\mathrm{s}^{-1}$, making it about 350 times better than the Masui compound shown above. A study of the P1′ position showed that Leu was slightly better than Ile and both were much better than Phe or Val [1325]. A series of 10 blocked hexapeptides was also studied [1310]. An initial study on a series of

Table 8 Protein sequences cleaved by collagenase 1 (MMP-1)

Substrate	Sequence								Ref.
	P4	P3	P2	P1	P1'	P2'	P3'	P4'	
Aggrecan, bovine	Ile–Pro–Glu–Asn Phe–Phe–Gly–Val345								[1242]
Aggrecan, bovine	Thr–Ser–Glu–Asp Leu–Val–Val–Gln445								[1242]
α1-Antichymotrypsin, human	Leu–Leu–Ser–Ala Leu–Val–Glu–Thr365[a]								[1235]
Collagen α1(I), calf, chick	Gly–Pro–Gln–Gly Ile–Ala–Gly–Gln779								[1250,1283]
Collagen α2(I), chick	Gly–Pro–Gln–Gly Ile–Leu–Gly–Ala779								[1238]
Collagen α1(II), human	Gly–Pro–Gln–Gly Leu–Ala–Gly–Gln779								[1283]
Collagen α1(III), human liver	Gly–Pro–Leu–Gly Ile–Ala–Gly–Ile779								[1283]
Collagen α1(III), human skin	Gly Leu–Ala–Gly–Leu779								[1283]
Collagen α1(III), calf	Gly–Pro–Leu–Gly Ile–Ala–Gly–Leu779								[1283]
Collagen α1(X), human	Gly–Pro–Ala–Gly Leu–Ser–Val–Leu148[a]								[1328]
Collagen α1(X), human	Gly–Pro–Ala–Gly Ile–Val–Thr–Lys476[a]								[1328]
IGFBP-3, human	Leu–Arg–Ala–Tyr Leu–Leu–Pro–Ala103								[1245]
α1-Inhibitor-3, rat (2J)	Glu–Ser–Leu–Ala Val–Glu–Ser–Glu670								[1313]
α1-Inhibitor-3, rat (27J)	Ser–Ala–Pro–Pro Val–Val–Ala–Val687								[1313]
Link protein, human	Arg–Ala–Ile–His Ile–Gln–Ala–Glu20								[1294]
α2-Macroglobulin, human	Gly–Pro–Glu–Gly Leu–Arg–Val–Gly683								[1313]
α2-Macroglobulin, rat	Ala–Ala–Tyr–His Leu–Val–Ser–Gln685								[1313]
α2-Macroglobulin, rat	Met–Asp–Ala–Phe Leu–Glu–Ser–Ser695								[1313]
α1-Macroglobulin, rat	Glu–Pro–Gln–Ala Leu–Ala–Met–Ser687								[1313]
α1-Macroglobulin, rat	Gln–Ala–Leu–Ala Met–Ser–Ala–Ile689								[1313]
MMP-1, human autocleavage	Pro–Val–Gln–Pro Ile–Gly–Pro–Gln254								[1231]
Ovostatin, chicken	Pro–Ser–Tyr–Phe Leu–Asn–Ala–Gly677								[1239]
Pregnancy zone protein, human	Tyr–Glu–Ala–Gly Leu–Gly–Val–Val689								[1313]
Pregnancy zone protein, human	Ala–Gly–Leu–Gly Val–Val–Glu–Arg691								[1313]
Pregnancy zone protein, human	Ala–Gly–Leu–Gly Ile–Ser–Ser–Thr761								[1313]
ProMMP-1, human	Pro–Val–Gln–Pro Met–Ser–Ile–Pro273								[1231]
ProMMP-1, human	Asp–Ala–Glu–Thr Leu–Lys–Val–Met68								[1318]
ProMMP-1, human	Thr–Leu–Lys–Val Met–Lys–Gln–Pro71								[1318]
ProMMP-1, human	Val–Ala–Gln–Phe Val–Leu–Thr–Glu85								[1318]
ProMMP-1, human	Ala–Gln–Phe–Val Leu–Thr–Glu–Gly86								[1318]
ProMMP-2, human	Glu–Ser–Cys–Asn Leu–Phe–Val–Leu41								[1232]
ProTNFα, human	Leu–Ala–Gln–Ala Val–Arg–Ser–Ser??								[1247]
α1-Protease inhibitor, human	Gly–Ala–Met–Phe Leu–Glu–Ala–Ile356								[1235, 1278]
α1-Protease inhibitor, human	Glu–Ala–Ile–Pro Met–Ser–Ile–Pro361								[1235]

[a]Number includes signal peptide.

16 octapeptides based on the collagen scissile bond [1241] was later extended to 60 peptides [1291]. This series showed that the enzyme recognizes seven positions from P3 to P4' and additional residues at either end did not improve cleavage. Some of the best residues were found to be:

P3	P2	P1	P1'	P2'	P3'	P4'
Pro	Met	Ala	Leu	Trp	Ala	Thr
	Leu	Pro	Met	Phe	Ser	Pro

Many more will be found in the paper [1291]. It is noteworthy that the glycine residue at P1 in collagen is not among the best substituents at this position.

Another good peptide was prepared [1268] taking advantage of the combination of a fluorescent group (Mca) and a quenching group (Dpa) in the same peptide, so that fluorescence develops only upon cleavage. This peptide Mca-Pro-Leu-Gly~Leu-Dpa-Ala-Arg-NH$_2$ has a k_{cat}/K_m of

14 800 $\mathrm{M}^{-1}\mathrm{s}^{-1}$ which makes it about half as effective as the Weingarten peptide. As with many of the peptides, it is cleaved by a wide variety of MMPs, so it cannot be used to assay crude extracts. The Glaxo group [1226, 1281] explored the concept of substituting amino acids with other groups of similar size; a total of 88 groups was tested at each of the four positions from P2 to P2'. A good peptide was Dnp-Pro-Cha-Abu~Smc-His-Ala-D-Arg, where Cha is cyclohexylalanine, Abu is α-aminobutyric and Smc is *S*-methylcysteine. Other synthetic substrates for MMP-1 have been proposed [1225, 1228, 1289, 1323] including Ac-Pro-Leu-Ala-S~Nva-Trp-NH$_2$ [1317]. This last has k_{cat}/K_m of 123 000 $\mathrm{M}^{-1}\mathrm{s}^{-1}$ at 25 °C, making it the best peptide reported to date. A peptide based on the scissile bond of α_2-macroglobulin (a much better substrate than collagen) was very poorly cleaved [1323]. A summary of some 90 peptide substrates based on the collagen sequence may be found in [92].

Gelatinase A (MMP-2)

Protein substrates for this MMP are tabulated in Table 9. In spite of the name 'gelatinase' this enzyme is able to attack native type I collagen molecules [1220] as well as elastin and many other proteins. There is perhaps a clearer specificity apparent in this table: Gly-Pro-Ala-Gly~Leu-Xaa-Gly-Xaa, although this is heavily influenced by the large number of collagen and gelatin substrates. The enzyme also appears to be more tolerant of charged residues throughout the binding site than is MMP-1.

The assay of gelatinase A with synthetic substrates was initiated by Seltzer *et al.* [1308]. The collagenase peptides [1279] were tested and found to work also for MMP-2, e.g. dnp-PQG~IAGQR and dnp-PLG~IAGR were cleaved. A further 15 peptides were synthesized [1310] and it was shown that at least six residues were required for optimal cleavage. The best peptide in this group was the blocked hexapeptide: Ac-PLG~LAG-OEt. In general, the same peptides were cleaved by MMP-1 and at similar rates. A further ten peptides produced the best to date: Ac-PLG~VRG-OEt with a k_{cat}/K_m of 14 176 $\mathrm{M}^{-1}\mathrm{s}^{-1}$ at 37 °C. A return to the concept of using dnp as a fluorescent group led to dnp-PLG~LWA $_D$R with a $k_{cat}/K_m = 124\,000\,\mathrm{M}^{-1}\mathrm{s}^{-1}$ [1315, 1316]. The thiopeptide used for MMP-1 [1326] is also suitable for MMP-2 [1336]. Of a series of fluorogenic peptides developed for the assay of various MMPs, the best compound for MMP-2 was dnp-PLG~MWSR which has $k_{cat}/K_m = 8400\,\mathrm{M}^{-1}\mathrm{s}^{-1}$ at 23 °C [1292]. A further 50 octapeptides were compared using MMP-2, -7 and -9 [1293]. The best for MMP-2 was GPQG~IFGQ with $k_{cat}/K_m = 694\,\mathrm{M}^{-1}\mathrm{s}^{-1}$ at 30 °C. Systematic variation at each position gave the following first and second choices at each position:

P4	P3	P2	P1		P1'	P2'	P3'	P4'
Gly	Pro	Leu	Gly		Met	Leu	Ser	His
			Tyr	Ala		Ile	Phe	Ala	Gln

A series of peptides containing collagen-like sequences were cleaved, but when these peptides were cyclized they were neither substrates nor inhibitors [1273]. Another collagen-like peptide Pro-Gln-Gly-Ile-Mel-Gly is cleaved by gelatinase A (Mel = melphalan, a group with antitumour action) [1320]. Probably the best synthetic substrate found to date is that developed by Knight *et al.* [1269]: Mca-Pro-Leu-Gly~Leu-Dpa-Ala-Arg-NH$_2$. This is basically the peptide of Stack and Gray [1315] in which Trp has been replaced by Dpa (N^3-dnp-L-2,3-diaminopropionic acid). The value of $k_{cat}/K_m = 629\,000\,\mathrm{M}^{-1}\mathrm{s}^{-1}$ at 37 °C; this is about 40 times higher than the value for MMP-1 on the same peptide. An improvement on the thiol peptide was introduced in 1994 [1317], namely, Ac-Pro-Leu-Ala~^{S}Nva-Trp-NH$_2$ (norvaline in thiol linkage), which has $k_{cat}/K_m = 167\,000\,\mathrm{M}^{-1}\mathrm{s}^{-1}$ at 25 °C. Another excellent peptide for MMP-2 is Dabcyl-Gaba-Pro-Gln-Gly~Leu-Glu (EDANS)-Ala-Lys-NH$_2$ [1225]. This has $k_{cat}/K_m = 619\,000\,\mathrm{M}^{-1}\mathrm{s}^{-1}$ at 37 °C, about the same as that of the Knight peptide. The most recent series of peptides to be studied extends the

Substrate	Sequence								Ref.
	P4	P3	P2	P1	P1′	P2′	P3′	P4′	
Aggrecan, bovine	Ile–Pro–Glu–Asn				Phe–Phe–Gly–Val345				[1244]
β-Amyoid protein, human	His–His–Gln–Lys				Leu–Val–Phe–Phe20				[1285, 1305]
Collagen α1(I), calf	Gly–Pro–Gln–Gly				Ile–Ala–Gly–Gln779				[1220]
Collagen α2(I), calf	Gly–Pro–Gln–Gly				Leu–Leu–Gly–Ala779				[1220]
Collagen α1(X), human	Gly–Pro–Ala–Gly				Leu–Ser–Val–Leu148[a]				[1328]
Collagen α1(X), human	Gly–Pro–Ala–Gly				Ile–Val–Thr–Lys476[a]				[1328]
Decorin, human	Asp–Ala–Ala–Ser				Leu–Lys–Gly–Leu214				[1252]
FGF-Receptor 1, human	Arg–Pro–Ala–Val				Met–Thr–Ser–Pro372				[1272]
Galectin 3, human	Pro–Pro–Gly–Ala				Tyr–His–Gly–Ala66				[1298]
Gelatin α1(I), guinea pig	Gly–Ala–Hyp–Gly				Leu–Glx–Gly–His551				[1308]
Gelatin α1(I), rat	Gly–Pro–Gln–Gly				Val–Arg–Gly–Glu194				[1309]
Gelatin α1(I), rat	Gly–Pro–Ala–Gly				Val–Gln–Gly–Pro281				[1309]
Gelatin α1(I), rat	Gly–Pro–Ser–Gly				Leu–Hyp–Gly–Pro305				[1309]
Gelatin α1(I), rat	Gly–Pro–Ala–Gly				Glu–Arg–Gly–Ser335				[1309]
Gelatin α1(I), rat	Gly–Ala–Lys–Gly				Leu–Thr–Gly–Ser365				[1309]
Gelatin α1(I), rat	Gly–Pro–Ala–Gly				Gln–Asp–Gly–Pro386				[1309]
Gelatin α1(I), rat	Gly–Pro–Ala–Gly				Phe–Ala–Gly–Pro638				[1309]
Gelatin α1(I), rat	Gly–Pro–Ile–Gly				Asn–Val–Gly–Ala680				[1309]
Gelatin α1(I), rat	Gly–Pro–Hyl–Gly				Ser–Arg–Gly–Ala689				[1309]
IGFBP-3, human	Leu–Arg–Ala–Tyr				Leu–Leu–Pro–Ala103				[1245]
Interleukin 1β, human	Gly–Pro–Tyr–Glu				Leu–Lys–Ala–Leu29				[1255]
Laminin-5, γ2, human	Thr–Ala–Ala–Ala				Leu–Thr–Ser–Cys590				[1248]
Link protein, human	Arg–Ala–Ile–His				Ile–Gln–Ala–Glu20				[1294]
Link protein, human	Gly–Pro–His–Leu				Leu–Val–Glu–Ala29				[1294]
α2-Macroglobulin, human	Gly–Pro–Glu–Gly				Leu–Arg–Val–Gly683				[1222]
α2-Macroglobulin, human	Gly–His–Ala–Arg				Leu–Val–His–Val700				[1222]
Pregnancy zone protein, human	Gln–Leu–Gly–Thr				Tyr–Asn–Val–Ile69				[1222]
ProMMP-1, human	Asp–Val–Ala–Gln				Phe–Val–Leu–Tyr84				[1232]
ProMMP-2, human	Asp–Val–Ala–Asn				Tyr–Asn–Phe–Phe84				[1232]
ProMMP-13, human	Asp–Val–Gly–Glu				Tyr–Asn–Val–? 88				[1267]
ProTNFα, human	Leu–Ala–Gln–Ala				Val–Arg–Ser–Ser80				[1247]
SPARC/BM-40, human	Thr–Val–Ala–Glu				Val–Thr–Glu–Val24				[1306]
SPARC/BM-40, human	Gly–Ala–Asn–Pro				Val–Gln–Val–Glu34				[1306]
SPARC/BM-40, human	His–Pro–Val–Glu				Leu–Leu–Ala–Arg201				[1306]
Substance P, human	Lys–Pro–Gln–Gln				Phe–Phe–Gly–Leu10				[1290]

[a]Number includes signal peptide.

peptide out to P6 and P6′, noting a twofold effect of substituents at P5 and P5′ [1334].

Stromelysin 1 (MMP-3)

Protein substrates of MMP-3 with known cleavage sites are presented in Table 10. Charged residues are accommodated well in P4, Ala and Gly predo-minate in P2 as do Leu and Ile in P1′. A wide variety of residues are found in the remaining positions. The ability to cleave Glu–Xaa bonds was thought to be pertinent to the cleavage of aggrecan in arthritis. However, the critical bond in aggrecan is not cleaved by stromelysin, leading to the postulate that there is a specific 'aggrecanase' [1243].

Studies of MMP-3 action on synthetic peptides

Substrate	Sequence								Ref.
	P4	P3	P2	P1	P1′	P2′	P3′	P4′	
Aggrecan, bovine	Ile–Pro–Glu–Asn				Phe–Phe–Gly–Val345				[1244]
α1-Antichymotrypsin, human	Leu–Leu–Ser–Ala				Leu–Val–Glu–Thr365[a]				[1278]
Antithrombin III	Ile–Ala–Gly–Arg				Ser–Leu–Asn–Pro389				[1278]
Collagen α1(II) N-telopeptide	Ala–Gly–Gly–Ala				Gln–Met–Gly–Val119				[1333]
Collagen α1(II) N-telopep-cow	Gln–Met–Gly–Val				Met–Gln–Gly–Pro123				[1333]
Collagen α1(IX) NC 2, bovine	Met–Ala–Ala–Ser				Leu–Lys–Arg–Pro601				[1333]
Collagen α2(IX) NC 2, bovine					Ala–Lys–Arg–Glu				[1333]
Collagen α3(IX) NC 2, bovine					Leu–Arg–Lys–Pro				[1333]
Collagen α1(XI) N–telopeptide	Gln–Ala–Gln–Ala				Ile–Leu–Gln–Gln				[1333]
Decorin, human	Asp–Ala–Ala–Ser				Leu–Lys–Gly–Leu214				[1252]
Fibrinogen, human gamma chain	His–Leu–Gly–Gly				Ala–Lys–Gln–Ala408				[1227]
Fibronectin, human	Pro–Phe–Ser–Pro				Leu–Val–Ala–Thr693				[1330]
Fibulin-2, mouse	Pro–Pro–Ser–Ser				Ile–Arg–Val–Xaa213				[1307]
Fibulin-2, mouse	Pro–Pro–Ala–Pro				Val–Gln–Ala–Lys237				[1307]
Fibulin-2, mouse	Arg–Val–Ser–Glu				Met–Glu–Met–Ala547				[1307]
IGFBP-3, human	Leu–Arg–Ala–Tyr				Leu–Leu–Pro–Ala103				[1245]
IGFBP-3, human	Ala–Pro–Gly–Asn				Ala–Ser–Glu–Ser113				[1245]
IGFBP-3, human	Phe–Ser–Ser–Glu				Ser–Lys–Arg–Glu180				[1245]
Insulin B chain, bovine	Leu–Val–Glu–Ala				Leu–Tyr–Leu–Val18				[1223]
Insulin B chain, bovine	Glu–Ala–Leu–Tyr				Leu–Val–Cys–Gly20				[1223]
Link protein, human	Arg–Ala–Ile–His				Ile–Gln–Ala–Glu20				[1294]
α₂-Macroglobulin, human	Gly–Pro–Glu–Gly				Leu–Arg–Val–Gly683				[1239]
α₂-Macroglobulin, human	Arg–Val–Gly–Phe				Tyr–Glu–Ser–Asp689				[1239]
MMP-8, human autocleavage	Ala–Ile–Tyr–Gly				Leu–Ser–Ser–Asn246				[1262]
MMP-8, human autocleavage	Ser–Ser–Asn–Pro				Ile–Gln–Pro–Thr251				[1262]
Nidogen, mouse	Phe–Leu–Ala–Asp				Leu–Asp–Thr–Thr86				[1280]
Nidogen, mouse	Gly–Leu–Gly–Asn				Val–Tyr–Tyr–Arg95				[1280]
Nidogen, mouse	Val–Val–Phe–Ser				Tyr–Asn–Thr–Gly355				[1280]
Nidogen, mouse	Val–Ala–Pro–Pro				Ile–His–Gln–Gly900				[1280]
Nidogen, mouse	Gly–Arg–Ala–Ser				Leu–His–Gly–Gly980				[1280]
Nidogen, mouse	Glu–Pro–Thr–Thr				Ile–Ile–Arg–Gln988				[1280]
Nidogen, mouse	Asn–Pro–Arg–Gly				Ile–Val–Thr–Asp1042				[1280]
Nidogen, mouse	Lys–Thr–Asn–Ser				Val–Ile–Ala–Met1146				[1280]
Ovostatin, chicken	Leu–Asn–Ala–Gly				Phe–Thr–Ala–Ser681				[1239]
ProMMP-1, human	Asp–Val–Ala–Gln				Phe–Val–Leu–Thr84				[1318]
ProMMP-3, human	Asp–Thr–Leu–Glu				Val–Met–Arg–Lys72				[1288]
ProMMP-3, human	Asp–Val–Gly–His				Phe–Arg–Thr–Phe86				[1288]
ProMMP-7, human	Asp–Val–Ala–Glu				Tyr–Ser–Leu–Phe82				[1253]
ProMMP-8, human	Asp–Ser–Gly–Gly				Phe–Met–Leu–Thr82				[1266]
ProMMP-9, human	Arg–Val–Ala–Glu				Met–Arg–Gly–Glu44				[1299]
ProMMP-9, human	Asp–Leu–Gly–Arg				Phe–Gln–Thr–Phe91				[1299]
ProMMP-13, human	Ser–Phe–Phe–Gly				Leu–Glu–Val–Thr61				[1260]
ProTNFα, human	Leu–Ala–Gln–Ala				Val–Arg–Ser–Ser80				[1247]
α1-Proteinase inhibitor, human	Glu–Ala–Ile–Pro				Met–Ser–Ile–Pro361				[1278]
SPARC/BM-40, human	Thr–Val–Ala–Glu				Val–Thr–Glu–Val24				[1306]
SPARC/BM-40, human	His–Pro–Val–Glu				Leu–Leu–Ala–Arg201				[1306]
Substance P	Lys–Pro–Gln–Gln				Phe–Phe–Gly–Leu10				[1319]

[a]Number includes signal peptide.

are relatively recent, although it was noted [1246] that rabbit stromelysin cleaved the dnp-PLG~IAGR-NH$_2$ peptide [1279]. Human MMP-3 was tested on 17 peptides, including the dnp peptide and found not to act. However, a series of 16 analogues of substance P (last line, Table 10) showed good cleavage of 11 [1319]. This study was extended [1251, 1295] and the analogue Ac-Arg-Pro-Ala-Pro-Gln-Gln~Phe-Phe-Gly-Leu-Nle-NH$_2$ was shown to have a $k_{cat}/K_m = 17\,000$ M^{-1} s^{-1} at 25 °C. The pH optimum is biphasic with a maximum at 6.0 and a shoulder at 8.5. This helps to explain why early studies of aggrecan digestion indicated an acid pH optimum for stromelysin [1223]. The thiopeptolide MMP-1 substrate [1326] is cleaved by recombinant stromelysin with $k_{cat}/K_m = 7800$ M^{-1} s^{-1} at 22 °C [1335]. A fluorogenic substrate dnp-PYA~YWMR was developed with $k_{cat}/K_m = 2400$ M^{-1} s^{-1} at 23 °C [1292]. The Mca peptide of Knight *et al.* [1269] mentioned previously is a relatively poor substrate for MMP-3, with $k_{cat}/K_m = 23\,000$ M^{-1} s^{-1} at 37 °C compared to 629 000 for MMP-2. Similarly, the Bz-PLA~LWNH(CH$_2$)$_4$N(CH$_3$)$_2$ peptide is cleaved at 1/3 the rate of MMP-1 [1228]. A thioester peptide [1317] is cleaved at 1/7 the rate that MMP-2 cleaves. A more successful peptide was developed by Nagase *et al.* [1289]: Mca-RPKPVE~ZWRK(dnp)-NH$_2$ (where Z is Norvaline). This has $k_{cat}/K_m = 218\,000$ M^{-1} s^{-1} at 37 °C, which is 20 times higher than the value for MMP-9, whilst MMP-1 and -2 have no action. The peptide of Beekman *et al.* [1225] noted above is cleaved at only 1/15 the rate found with MMP-2. Thus, synthetic substrates are beginning to be developed that can discriminate among the various MMPs. However, there are none that can be used directly and unambiguously with crude tissue extracts or culture medium.

A two-pronged approach has been taken to studying the P1′ specificity of MMP-3: mutations in the binding site and modification of the substrate residue. The peptide dnp-RPLA~XWRS is best when X is Leu ($k_{cat}/K_m = 96\,000$ M^{-1} s^{-1} at 30 °C), this decreases to 43 000 M^{-1} s^{-1} with Phe, 17 000 with Tyr and 4400 with Trp substitutions [1327]. The most recent approach has been to use phage display methods to examine millions of hexapeptide sequences as substrates for stromelysin [1312]. On this basis the best peptide was synthesized: Ac-PFE~LRA-NH$_2$ that has $k_{cat}/K_m = 126\,000$ M^{-1} s^{-1} at 37 °C. Replacement of Glu with Ala reduces this by 50%.

Matrilysin (MMP-7)

The cleavage points in various protein sequences are presented in Table 11. The results are not too different from MMP-1: Pro is frequently found at P3, Gly at P1 but also Glu (not found in MMP-1), and Leu/Ile at P1′. The other positions do not seem to be very selective. It is surprising that the literature values for entactin and nidogen are different except for one match because this is the same protein (entactin/nidogen) from mouse in both cases.

The literature on synthetic substrates for matrilysin is not extensive. The enzyme is able to digest the octapeptide substrate for collagenase [1279] dnp-PLG~IAGR according to Woessner and Taplin [1332]. The Stack and Gray substrate [1315] dnp-PLG~LWAR-NH$_2$ is cleaved with $k_{cat}/K_m = 13\,000$ M^{-1} s^{-1} at 37 °C [1234]. The so-called Knight peptide Mca-PLG~LDpaAR-NH$_2$ [1269] is cleaved with $k_{cat}/K_m = 169\,000$ M^{-1} s^{-1} or about 14 times better than the Stack peptide. In an examination of about 50 octapeptides [1293] comparing MMP-7 to MMP-2 and -9, the best was GPQA~IAGQ which had $k_{cat}/K_m = 580$ M^{-1} s^{-1} at 30 °C. Examination of various subsites indicated that P3 = Pro, P2 = Leu/Met, P1 = Ala, P1′ = Leu, P2′ = Trp and P3′ = Met are good choices. Study of variations in the peptide dnp-RPLA~XWRS indicates that X = Leu is best, with $k_{cat}/K_m = 190\,000$ M^{-1} s^{-1} at 30 °C [1327]. This is reduced sharply to 4000 or less by substituting Tyr, Phe or Trp. Another reasonable peptide substrate is Dns-PLA~LWAR with $k_{cat}/K_m = 172\,000$ M^{-1} s^{-1} at 37 °C [1230]. Again, there has been a study of millions of hexapeptides derived from a phage display library [1312]. The best of the blocked peptides

Table 11 Protein sequences cleaved by matrilysin (MMP-7)

Substrate	Sequence								Ref.
	P4	P3	P2	P1	P1'	P2'	P3'	P4'	
Aggrecan, bovine	Ile	Pro	Glu	Asn	Phe	Phe	Gly	Val345	[1244]
Aggrecan, bovine	Thr	Ser	Glu	Asp	Leu	Val	Val	Gln445	[1244]
Decorin, human	His	Leu	Arg	Glu	Leu	His	Leu	Asp247	[1252]
Decorin, human				N	Asp	Glu	Ala	Ser–Gly5	[1252]
Entactin, mouse	Leu	Ile	Gly	Glu	Leu	Ser	Phe	Tyr38	[1311]
Entactin, mouse	Ser	His	Leu	Gly	Leu	Glu	Asp	Val279	[1311]
Entactin, mouse	Ser	Pro	Asp	Ala	Leu	Gln	Asn	Pro641	[1311]
Entactin, mouse	Arg	Ala	Gln	Cys	Ile	Tyr	Met	Gly751	[1311]
Fibulin-2, mouse	Leu	Leu	Pro	Arg	Phe	Arg	Ala	Glu382	[1307]
Fibulin-2, mouse	Glu	Ala	Leu	Ser	Leu	Gly	Thr	Glu557	[1307]
Gelatin α2(I), rat	Gly	Pro	Ser	Gly	Ile	Thr	Gly	Pro716	[1219]
Gelatin α2(I), rat	Gly	Pro	Gln	Gly	Leu	Leu	Gly	Ala779	[1219]
Gelatin α2(I), rat	Gly	Pro	Leu	Gly	Ile	Ala	Gly	Pro812	[1219]
Insulin B chain, bovine	Leu	Val	Glu	Ala	Leu	Tyr	Leu	Val18	[1219]
Insulin B chain, bovine	Glu	Ala	Leu	Tyr	Leu	Val	Cys	Gly20	[1219]
Link protein, human	Gly	Pro	His	Leu	Leu	Val	Glu	Ala29	[1294]
Nidogen, mouse	Ser	Ile	Glu	Asn	Leu	Ala	Lys	Ser224	[1280]
Nidogen, mouse	Ser	His	Leu	Gly	Leu	Glu	Asp	Val279	[1280]
Nidogen, mouse	Gly	Arg	Ala	Ser	Leu	His	Gly	Gly980	[1280]
Nidogen, mouse	Ser	Pro	Glu	Gly	Ile	Ala	Leu	Asp999	[1280]
Nidogen, mouse	Glu	Val	Ala	Lys	Met	Asp	Gly	Ser1023	[1280]
Nidogen, mouse	Leu	Pro	Asn	Gly	Leu	Thr	Phe	Asp1087	[1280]
Plasminogen, human	Ala	Pro	Pro	Pro	Val	Val	Leu	Leu470	[1300]
ProMMP-2 activation	Asp	Val	Ala	Asn	Tyr	Asn	Phe	Phe84	[1233]
ProMMP-7, autocleavage	Asp	Val	Ala	Glu	Tyr	Ser	Leu	Phe81	[1234]
α1-Proteinase inhibitor, human	Glu	Ala	Ile	Pro	Met	Ser	Ile	Pro361	[1337]
α1-Proteinase inhibitor, oxid.	Gly	Ala	Met	Phe	Leu	Glu	Ala	Ile356	[1337]
ProTNFα, human	Leu	Ala	Gln	Ala	Val	Arg	Ser	Ser80	[1247]
Prourokinase, human	Pro	Pro	Glu	Glu	Leu	Lys	Phe	Gln147	[1277]
SPARC/BM-40, human	Thr	Val	Ala	Glu	Val	Thr	Glu	Val24	[1306]
SPARC/BM-40, human	Glu	Val	Gly	Glu	Phe	Asp	Asp	Gly41	[1306]
SPARC/BM-40, human	His	Pro	Val	Glu	Leu	Leu	Ala	Arg201	[1306]

based on these results was Ac-PLE~LRA-NH$_2$ with $k_{cat}/K_m = 177\,000 \,\mathrm{M}^{-1}\,\mathrm{s}^{-1}$.

Neutrophil collagenase (MMP-8)

There are few data on the cleavage of protein substrates by collagenase 2, these are presented in Table 12. This is insufficient to draw much of a conclusion about specificity.

The study of synthetic substrates in this case exceeds the study of protein cleavage. In an examination of a few octapeptide substrates [1275] and a fluorogenic substrate [1292], the peptide dnp-PLA~LPAR had $k_{cat}/K_m = 190\,000 \,\mathrm{M}^{-1}\,\mathrm{s}^{-1}$ at 23 °C. Examination of a series of some 50 octapeptides [1291] indicated that GPQG~IWGQ was one of the best with k_{cat}/K_m of 11 000 $\mathrm{M}^{-1}\,\mathrm{s}^{-1}$. Systematic variation indicated that Pro was best in P3, Leu in P2, Glu or Ala at P1, Tyr at P1', Trp at P2' and Ser at P3'. Kinetic studies on substance P indicates that the cleavage (Table 12) at Gln–Phe occurs with $k_{cat}/K_m = 88\,000 \,\mathrm{M}^{-1}\,\mathrm{s}^{-1}$ at 37 °C [1237].

Substrate	Sequence								Ref.
	P4	P3	P2	P1	P1′	P2′	P3′	P4′	
Aggrecan, bovine	Ile–Pro–Glu–Asn Phe–Phe–Gly–Val345								[1242]
Aggrecan, bovine	Thr–Ser–Glu–Asp Leu–Val–Val–Gln445								[1242]
Aggrecan, bovine	Thr–Glu–Gly–Glu Ala–Arg–Gly–Ser377								[1243]
C-1 inhibitor, human plasma	Ala–Ala–Ser–Ala Ile–Ser–Val–Ala443								[1265]
ProMMP-8, human	Phe–Gly–Leu–Asn Val–Thr–Gly–Lys57								[1266]
ProMMP-8, human	Glu–Thr–Leu–Asp Met–Met–Lys–Lys68								[1266]
ProMMP-8, human	Ser–Gly–Gly–Phe Met–Leu–Thr–Pro83								[1266]
ProMMP-8, human	Gly–Gly–Phe–Met Leu–Thr–Pro–Gly84								[1266]
α1-Protease inhibitor, human	Gly–Ala–Met–Phe Leu–Glu–Ala–Ile356								[1263]
α1-Protease inhibitor, human	Glu–Ala–Ile–Pro Met–Ser–Ile–Pro361								[1263]
Substance P, human	Lys–Pro–Gln–Gln Phe–Phe–Gly–Leu10								[1237, 1322]

Gelatinase B (MMP-9)

The action of this enzyme has been characterized on a number of protein substrates (Table 13). Again, there is not a strong pattern of specificity. P3 is almost always occupied by Pro or Ala and P1′ is frequently rich in Leu and Ile, a number of Arg residues appear at P2′ and Gly at P3′.

A gelatinase from neutrophils was shown to hydrolyse the Masui [1279] peptide dnp-PQG~IAGQR [1304] and this result was repeated later with purified enzyme [1270]. The studies previously cited [1293] for MMP-2 and MMP-7 also compared MMP-9 action on some 50 octapeptides. GPLG~IAGQ was a good substrate for MMP-9 with $k_{cat}/K_m = 280\,\mathrm{M}^{-1}\mathrm{s}^{-1}$ at 30 °C. Systematic variation at different positions indicated Pro at P3 and Leu at P2 were best, Gly at P1 and Met at P1′ were good, and Leu at P2′, Ser at P3′ and His at P4′ were optimal. It must be emphasized that not all 20 amino acids were tried at each position and a peptide with all the optimal components was not synthesized. Such a detailed substitution was made in the case of dnp-Pro-Cha-Gly~Cys(Me)-His-Ala-Lys-(Nma)-NH$_2$ where Cha is β-cyclohexyla-lanine and Nma is N-methylanthranilate [1226]. This had $k_{cat}/K_m = 69\,400\,\mathrm{M}^{-1}\mathrm{s}^{-1}$ for MMP-9 compared to 3350 for MMP-1 at 23 °C. Then 352 peptides were synthesized based on the region P2–P2′ and substituting 88 groups at each position;

this included all of the amino acids plus a great many additional chemical groups [1281]. The best substrate for MMP-9 proved to be: dnp-Pro-Cha-Abu~Smc-His-Ala-DArgNH$_2$ where Abu is α-aminobutyric and Smc is S-methylcysteine. This peptide is derived directly from the original Stack and Gray peptide [1315] and is stated to be 5.4 times better with respect to k_{cat}/K_m; unfortunately, no absolute values are available, but it is presumably close to that reported in [1226] as $69\,400\,\mathrm{M}^{-1}\mathrm{s}^{-1}$ (above). Nagase *et al.* [1289] prepared three fluorogenic substrates, of which the best for MMP-9 was Mca-RPKPYA~NvaMK(dnp)NH$_2$ which had $k_{cat}/K_m = 55\,300\,\mathrm{M}^{-1}\mathrm{s}^{-1}$ at 37 °C. This was indistinguishable from the values for MMP-2 and MMP-3, so there is little discriminatory power. Xia *et al.* [1334] examined a dodecapeptide sequence of collagen and found that residues at P5 and P5′ have an influence, Pro at P5 was six times better than hydroxyproline. Finally, mention may be made again of Dabcyl-Gaba-Pro-Gln-Gly~Leu-Glu(EDANS)-Ala-Lys-NH$_2$ [1225] which has a k_{cat}/K_m of $206\,000\,\mathrm{M}^{-1}\mathrm{s}^{-1}$ at 37 °C, or about 1/3 the value found for MMP-2.

Stromelysin 2 (MMP-10)

Studies of protein specificity of this enzyme are few in number (see Table 14). It is too early to say

Table 13 Protein sequences cleaved by gelatinase B (MMP-9)

Substrate	Sequence								Ref.
	P4	P3	P2	P1	P1′	P2′	P3′	P4′	
Aggrecan, bovine	Ile	Pro	Glu	Asn	Phe	Phe	Gly	Val345	[1244]
C-1 inhibitor, human plasma	Ala	Ala	Ser	Ala	Ile	Ser	Val	Ala443	[1265]
C-1 inhibitor, human plasma	Ser	Ala–Ile	Ser	Val–Ala	Arg	Thr445			[1265]
Collagen α1(V), human	Gly	Pro	Pro	Gly	Val	Val	Gly	Pro443	[1296]
Collagen α2(V), human	Gly	Pro	Pro	Gly	Leu	Arg	Gly	Glu449	[1296]
Collagen α1(XI), human	Gly	Pro	Gly	Gly	Val	Val	Gly	Pro443	[1296]
Galectin 3, human	Pro	Pro	Gly	Ala	Tyr	His	Gly	Ala66	[1298]
Link protein, human	Arg	Ala	Ile	His	Ile	Gln	Ala	Glu20	[1294]
α2-Macroglobulin, human	Gly	Pro	Glu	Gly	Leu	Arg	Val	Gly683	[1222]
α2-Macroglobulin, human	Gly	His	Ala	Arg	Leu	Val	His	Val700	[1222]
Myelin basic protein, human	Val	His	Phe	Phe	Lys	Asn	Ile	Val94	[1303]
Myelin basic protein, human	Arg	Gly	Leu	Ser	Leu	Ser	Arg	Phe114	[1303]
Myelin basic protein, human	Leu	Ser	Arg	Phe	Ser	Trp	Gly	Ala118	[1303]
Myelin basic protein, human	Arg	Ala	Ser	Asp	Tyr	Lys	Ser	Ala137	[1303]
Plasminogen, human	Val	Ala	Pro	Pro	Pro	Val	Val	Leu469	[1300]
Pregnancy zone protein, human	Ala	Phe	Cys	Leu	Ser	Glu	Asp	Ala757	[1222]
ProMMP-9, human autocleavage	Phe	Pro	Gly	Asp	Leu	Arg	Thr	Asn19	[1321]
ProMMP-9, human autocleavage	Arg	Val	Ala	Glu	Met	Arg	Gly	Glu44	[1321]
ProMMP-9, human autocleavage	Gly	Pro	Ala	Leu	Leu	Leu	Leu	Gln56	[1321]
ProMMP-9, human autocleavage	Pro	Ala	Leu	Leu	Leu	Leu	Gln	Lys57	[1321]
ProMMP-9, human autocleavage	Thr	Leu	Lys	Ala	Met	Arg	Thr	Pro78	[1321]
ProMMP-9, human autocleavage	Ser	Arg	Ser	Glu	Leu	Asn	Gln	Val670	[1321]
ProMMP-9, human autocleavage	Asp	Ala	Ile	Ala	Glu	Ile	Gly	Asn510	[1321]
ProMMP-9, human autocleavage	Val	Pro	Glu	Ala	Leu	Met	Tyr	Pro402	[1321]
α1-Protease inhibitor, human	Gly	Ala	Met	Phe	Leu	Glu	Ala	Ile356	[1236]
α1-Protease inhibitor, human	Glu	Ala	Ile	Pro	Met	Ser	Ile	Pro361	[1236]
ProTNFα, human	Leu	Ala	Gln	Ala	Val	Arg	Ser	Ser??	[1247]
SPARC/BM-40, human	Gly	Ala	Asn	Pro	Val	Gln	Val	Glu34	[1306]
SPARC/BM-40, human	Asn	Pro	Val	Gln	Val	Glu	Val	Gly36	[1306]
SPARC/BM-40, human	His	Pro	Val	Glu	Leu	Leu	Ala	Arg201	[1306]
Substance P, human	Lys	Pro	Gln	Gln	Phe	Phe	Gly	Leu10	[1237]

much about bond specificity for MMP-10. It hydrolyses Mca-RPKPVE~ZWRK(dnp)-NH$_2$ (X = norvaline) with $k_{cat}/K_m = 170\,000\ \mathrm{M}^{-1}\mathrm{s}^{-1}$ at 37 °C [86].

Stromelysin 3 (MMP-11)

A table is not warranted for this enzyme. For a long time its action on proteins was thought to be very weak. Then it was found [1301] that the serpin α1-proteinase inhibitor was cleaved at the Ala-Ala-Gly-Ala~Met-Phe-Leu-Glu354 bond and α2-macroglobulin at the Arg-Val-Gly-Phe~Tyr-Glu-Ser-Asp688 bond. The insulin-like growth factor binding protein 1 is also cleaved [1276] at Lys-Ala-Leu-His~Val-Thr-Asn-Ile144. Recently [1287] MMP-11 was tested on 13 synthetic peptides of which the best was Dns-Pro-Leu-Ala~Cys(OMeBzl)-Trp-Ala-Arg-NH$_2$; $k_{cat}/K_m = 16\,700\ \mathrm{M}^{-1}\mathrm{s}^{-1}$, which is about 100 times less than the value for MMP-14.

Metalloelastase

Macrophage elastase digested the bovine B chain of insulin [1257]. Cleavage was seen at

Substrate	Sequence								Ref.
	P4	P3	P2	P1	P1′	P2′	P3′	P4′	
Link protein, human	Arg–Ala–Ile–His				Ile–Gln–Ala–Glu20				[1294]
Link protein, human	Gly–Pro–His–Leu				Leu–Val–Glu–Ala29				[1294]
ProMMP-1, human	Asp–Val–Ala–Gln				Phe–Val–Leu–Tyr84				[1331]
ProMMP-8, human	Asp–Ser–Gly–Gly				Phe–Met–Leu–Tyr82				[1261]
ProMMP-10, human	Asp–Val–Gly–His				Phe–Ser–Ser–Phe85				[264]
ProMMP-10, human	Gly–Ile–Gln–Ser				Leu–Tyr–Gly–Pro247				[264]

two sites: Leu-Val-Glu-Ala~Leu-Tyr~Leu-Val-CysO$_3$H-Gly20. Mouse macrophage elastase digested α1-proteinase inhibitor at Glu-Ala-Ile-Pro~Met-Ser-Ile-Pro357 [1224]. Elastin digestion indicates that cleavage most commonly involves Leu/Ile in the P1′ position [1282]. A synthetic peptide developed for pancreatic elastase appears to work with MMP-12 from human monocytes: 6-(N-CBZ-Ala-Ala-Ala-amido)quinoline, but this has not been verified with isolated enzyme [1221]. Human MMP-12 was tested on a series of nine peptides of the type dnp-Arg-Pro-Leu-Ala~Leu-Trp-Arg-Ser-NH$_2$ with variations only at P1′ [1249]. It was found that Leu was best at P1′ and gave $k_{cat}/K_m = 18\,000\,\mathrm{M}^{-1}\,\mathrm{s}^{-1}$. Interestingly, this appears to be the only MMP that can tolerate Arg at P1′. In the same study, α1-proteinase inhibitor

was found to be cleaved at Arg-Pro-Phe-Glu~Val-Lys-Asp-Thr203 and Gly-Ala-Met-Phe~Leu-Glu-Ala-Ile356.

Collagenase 3 (MMP-13)

This enzyme has not been extensively studied with respect to cleavage specificity. Although the rat enzyme has been known since the beginning of work in the field and was thought to have similar properties to MMP-1. A study of its action on gelatin [1329] showed that the main residues at P1′ were Leu > Ile, Phe, Ala, Val. Cleavage sites in proteins produced by human MMP-13 are tabulated (Table 15). The human enzyme attacks the Knight peptide [1269] and digests itself in activa-

Table 15 Protein sequences cleaved by collagenase 3 (MMP-13)

Substrate	Sequence								Ref.
	P4	P3	P2	P1	P1′	P2′	P3′	P4′	
α_1-Antichymotrypsin	Leu–Leu–Ser–Ala				Leu–Val–Glu–Thr366				[1260]
Collagen type II	Gly–Pro–Gln–Gly				Leu–Ala–Gly–Gln779				[1284]
Collagen type II	Gly–Leu–Ala–Gly				Gln–Arg–Gly–Ile782				[1284]
ProMMP-9, human	Arg–Val–Ala–Glu				Met–Arg–Gly–Glu44				[1264]
ProMMP-9, human	Asp–Leu–Gly–Arg				Phe–Gln–Thr–Phe91				[1264]
ProMMP-13, human, autocleavage	Asp–Val–Gly–Glu				Tyr–Asn–Val–Phe88				[1259]
ProMMP-13, human, autocleavage	Gly–Pro–Ser–Gly				Leu–Leu–Ala–His169				[1259]
ProMMP-13, human, autocleavage	Gly–Ile–Gln–Ser				Leu–tyr–Gly–Pro249				[1259]
SPARC/BM-40, human	Gly–Ala–Asn–Pro				Val–Gln–Val–Glu34				[1306]
SPARC/BM-40, human	Asn–Pro–Val–Gln				Val–Glu–Val–Gly36				[1306]
SPARC/BM-40, human	His–Pro–Val–Glu				Leu–Leu–Ala–Arg201				[1306]

Substrate	Sequence								Ref.
	P4	P3	P2	P1	P1′	P2′	P3′	P4′	
ProMMP-2, human	Glu	Ser	Cys	Asn	Leu	Phe	Val	Leu66	[1258]
ProMMP-13, human	Asn	Leu	Ala	Gly	Ile	Leu	Lys	Glu39	[1267]
ProMMP-13, human	Asp	Val	Gly	Glu	Tyr	Asn	Val	Phe88	[1267]

tion at Glu Tyr84 [1260]. The synthetic octapeptide for collagenase [1279] is also cleaved [1271].

Membrane-type matrix metalloproteinase 1 (MT1-MMP)

Again, due to its relatively recent discovery, there is little specificity information concerning this proteinase. Information about the activation of other MMPs is found in Table 16. A series of Dns-PLA~XWAR-NH$_2$ peptides containing unusual groups in P1′ = (X) was tested [1287], the best of these, Cys(OMeBzl) gave $k_{cat}/K_m = 1\,590\,000$ M^{-1} s^{-1}, which was 38 times higher than the value for the best natural amino acid, Leu. A similar series of dnp-Arg-Pro-Leu-Ala~Xaa-Trp-Arg-Ser-NH$_2$ peptides was substituted at P1′ with natural amino acids; here Leu or Tyr was best, with Trp slightly behind [1254].

Remaining MMPs

Membrane type matrix metalloproteinase 2 (MT2-MMP, MMP-15) cleaves the Knight peptide [1269] with $k_{cat}/K_m = 84\,000$ M^{-1} s^{-1} at 25 °C [1229] which is 70% of the rate with MMP-14 and 13% of the rate with MMP-2. MMP-19 also cleaves this peptide, the $k_{cat}/K_m = 1320$ M^{-1} s^{-1} [1302]. The peptide Mca-PLA~NvaDpaAR-NH$_2$ by contrast has a value of 19 600 M^{-1} s^{-1}. Human enamelysin (MMP-20) cleaves these same two substrates with $k_{cat}/K_m = 45\,500$ M^{-1} s^{-1} and 58 800 M^{-1} s^{-1}, respectively [1274]. The sea urchin hatching enzyme, envelysin, has been tested on a large series of natural peptides [1297], the pattern is similar to that for most mammalian MMPs, namely, the presence of Leu, Ile, Phe or Tyr at P1′. The bacterial fragilysin cleaves itself at YVIC~LREN17 and FTAS~LKSN79 and actin at GILT~LKYP130 and VMVG~MGNK50 [1286].

Inhibition of the MMPs

In addition to knowledge gained by studies of enzyme specificity, a good deal can also be learned about the active centre by the use of inhibitors. First, we will discuss inhibitors that act by chelation or replacing zinc at the active centre. This will be followed by information on inhibitors that act by non-specific means, such as affecting the substrate, and inhibitors that are modelled on the pro-peptides of MMPs. Inhibition by proteins in the macroglobulin class provides some information about specificity (much of it tabulated in the section on MMP specificity). Finally, the design of synthetic peptide derivatives or peptide mimetics will be considered. The inhibition by TIMPs will be treated in a separate section.

Common chelators

The earliest recognition that collagenase was a metalloproteinase came when tadpole collagenase was found to be inhibited by using 2 mM EDTA or 5 mM cysteine (the -SH group can coordinate to the zinc atom) [961]. It was soon shown that human skin collagenase was blocked by the same inhibitors [872], and synovial collagenase [879] and rat, goat and human bone collagenase [885] by EDTA. This led to the early use of these two compounds applied directly to the eye to treat alkali burns of the cornea [844]. It was such work on rabbit cornea that led to the conclusion that collagenase was a zinc enzyme [832]; a variety of chelators including 1,10-phenanthroline (Fig. 22(a)), dithizone and 8-OH-quinoline-5-sulfonate were tested, the latter of which proved more powerful than the magnesium/calcium chelator EDTA. Acetyl-cysteine (Fig. 22(b)) and D-penicil-lamine (Fig. 22(c)) were also studied. Similarly, mouse macrophage elastase or MMP-12 was inhibited by dithiothreitol (DTT), EDTA and 1,10-phenanthroline [825] as well as by high zinc concentrations; whereas Zn·EDTA reversed the effects of EDTA. Phosphoramidon, a typical inhibitor of metalloproteinases, had very little effect on MMP-12 or MMP-1. This compound, and Zincov, fail to inhibit stromelysin [821], but EDTA, phenanthroline and cysteine do inhibit stromelysin [863]. Zincov (Fig. 22(d)) at high levels (0.5 mM) inhibits matrilysin, but phosphoramidon (25 μM) does not [1031]. An antitumour agent related to EDTA, namely razoxane, blocks the action of MMP-2 on type IV collagen [926]. Captopril, an -SH containing inhibitor of angiotensin converting enzyme, blocks MMP-2 and MMP-9 action in zymograms when used at high level (20–40 mg/ml) and this is reversed by added zinc ion [967, 1003].

After many years there was a return to D-penicillamine [972] which was shown to inhibit MMP-2 and MMP-9 with $IC_{50} = 1$–2 mg/ml, and in zymography 0.5 mM inhibits MMP-9 [971]. The bisphosphate clodronate inhibits MMP-1 [1021] and MMP-8 [1020] with $IC_{50} = 150$ μM. A chelator from plants, betulinic acid, inhibits both MMP-1 and MMP-3 with $K_i = 1.3$ μM [1011]; this compound may mimic a hydroxamate (see below). The mechanism of chelator action has been worked out in some detail using 1,10-phenanthroline and the related 4-(2-pyridylazo)resorcinol [1008]. The first step is rapid and involves chelation of the zinc in the active centre; the second step is slow and involves the removal of zinc from the enzyme and joining with a second molecule of chelator. This same mechanism also applies to MMP-2, -3, -8 and -9.

Closely related to chelation is the replacement of the zinc atom at the active centre. A series of lanthanide salts were tested on MMP-1, -2 and -3,

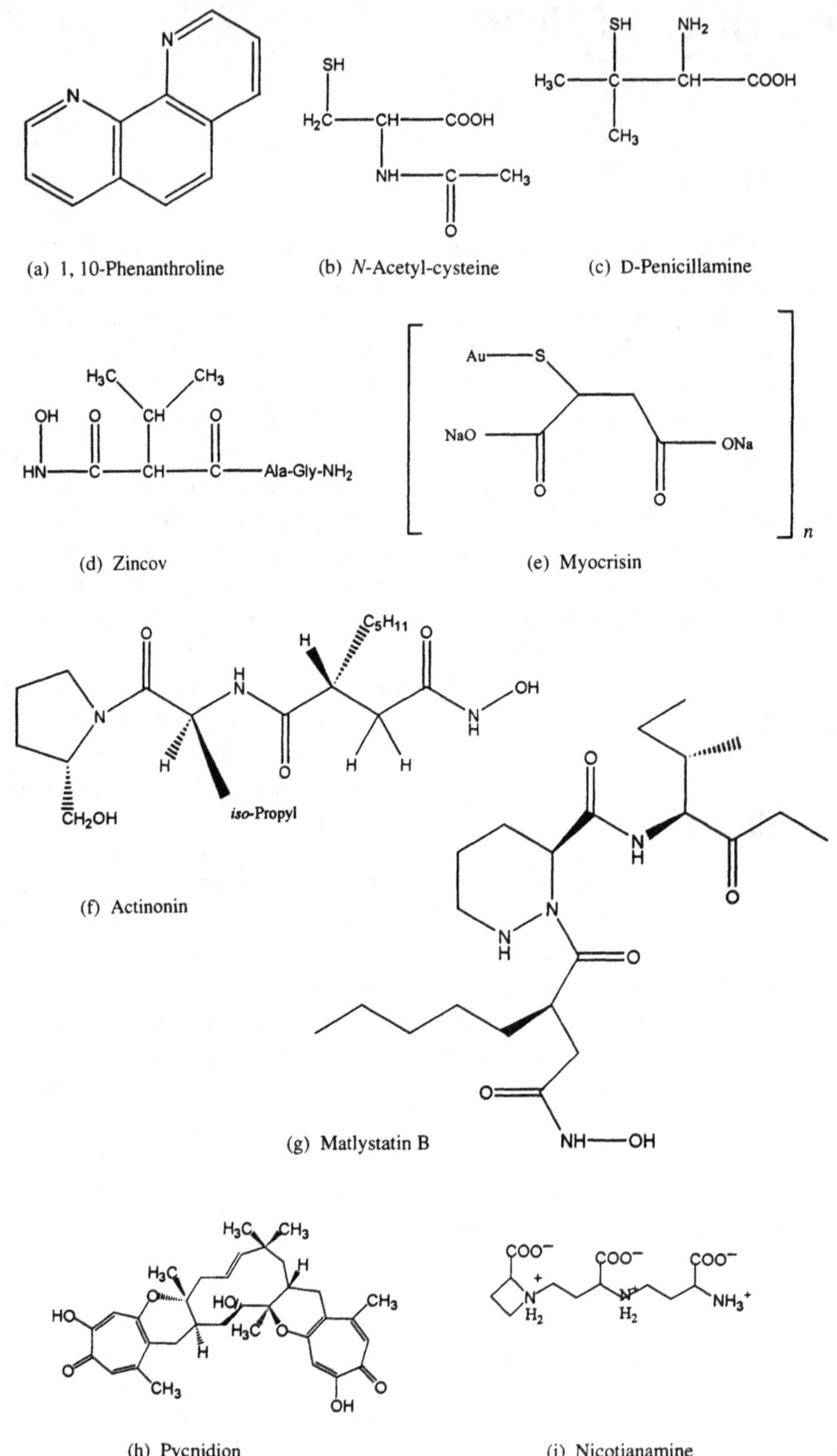

Figure 22 (a)–(i). Caption overleaf

(j) Gelastatin B

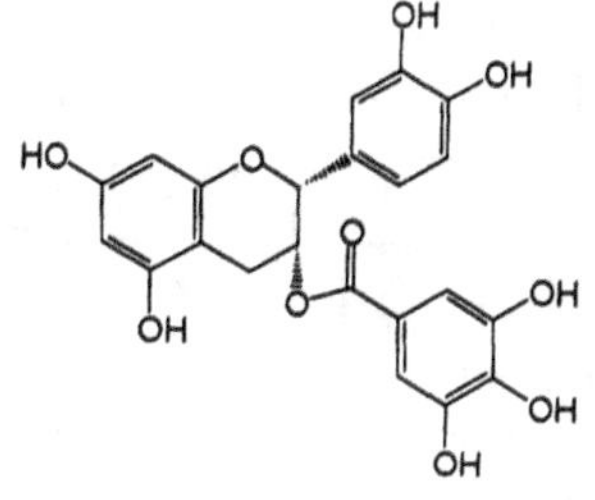

(k) (–)-Epicatechin gallate

(l) Eriochrome Black T

(m) Minocycline

Figure 22 (j)–(m)
Chemical structures of natural inhibitors of the MMPs.

and it was found that all were inhibited [878]. Samarium^{3+} was best for inhibiting human synovial MMP-2 and MMP-3, whereas lanthanum^{3+} was best for MMP-1. Erbium and lutecium were relatively poor. This line of work has not been followed further. However, a great deal of work has focused on gold compounds because of their importance in treating rheumatoid arthritis. Aurothiomaleate (250 μg/ml) inhibited MMP-1 almost completely, comparable to 2 mM DTT [1034]. One must be alert for other interpretations of results *in vivo* because this compound inhibits monocyte production of MMP-1 [978]. A series of six gold compounds in use, or under test, for clinical application were tested against human neutrophil collagenase, MMP-8 [944]; the best was myocrisin (Fig. 22(e)) with an IC$_{50}$ = 3.5 nM when used in the assay of latent MMP-8 together with an activator. When MMP-8 was already active the IC$_{50}$ increased to 130 μM. Further studies [945] indicate that gold compounds work best when the gold atom is released, e.g. gold is released from myocrisin if *p*-chloromercuribenzoate binds to the thiol groups. This gold atom is postulated to bind to a second heavy-metal binding site distinct from the active site. Zinc will reverse the gold inhibi-

tion, but a high zinc concentration, > 100 μM, is itself inhibitory. Cadmium and copper can also bind at this second site and inhibit the enzyme.

Antibiotics

A number of natural antibiotics have been used as inhibitors of MMPs; in most cases these seem to act by chelation. Erythromycin blocks MMP-1 in blister fluid of epidermolysis bullosa patients [986] and MMP-8 in rat kidney [941]; neither enzyme was purified. A powerful inhibitor of MMP-1 was found in *Actinomycetes*; actinonin (Fig. 22(f)) has K_i = 1.4 μM and its structure contains a natural hydroxamate that is a very effective chelator of zinc [880]. Synthesis of analogues of actinonin showed that the portion fitting in enzyme pocket S1′ of MMP-1 is important; separating the hydroxamate spatially from the P1′ group led to a 100-fold decrease in inhibition [934] (Table 17, Pat. 4918105). Anthracycline antibiotics, such as daunorubicin, are less effective: K_i = 92 μM [927]. The actinomycete *Actinomadura* is the source of a series of powerful hydroxamate compounds named matlystatin (Fig. 22(g)) [977]; the mixture inhibits

MMP-2 and MMP-9 with $IC_{50} = 0.3$ and $0.56\,\mu M$, respectively [1018]. The structure was determined [913] and this led to the synthesis of the best compound in the form of 10 isomers [1014]. The best of these isomers inhibited MMP-9, MMP-2 and MMP-3 with $K_i = 0.52$, 0.61 and $0.12\,\mu M$, respectively. Further modification at the P1′ position with a nonyl group led to $IC_{50} = 1.2$ nM on MMP-9 and 38 nM on MMP-2 [1016]. A very closely related compound from *Streptomyces*, YL-01869P, has an $IC_{50} = 1.6\,\mu M$ [1015]; this was later renamed YM-24074 and the structure was elucidated [996] as a hydroxamate compound. *Chaunopycnis alba* produces pyridoxatin, a cyclic hydroxamate with an $IC_{50} = 15\,\mu M$ on MMP-2 [932]. Another *Streptomyces* inhibitor is BE16627B or *N*-(*N*-hydroxy-2-isobutylsuccinamoyl)-Ser-Val which inhibits MMP-2, MMP-3 and MMP-9 with $IC_{50} = 0.6$–$0.8\,\mu M$ [966]. Rifamycin and relatives from *Streptomyces* inhibit MMP-1 with IC_{50} about $15\,\mu M$ [868].

A different type of inhibitor was obtained from the soil fungus *Phoma*, a bistropolone named pycnidion (Fig. 22(h)). This inhibits MMP-3 with $IC_{50} = 31\,\mu M$; the crystal structure is known [912]. β-Lactam antibiotics have been suggested as MMP inhibitors [995], but no specific MMP has been demonstrated to be blocked so this remains questionable [888]. A chelator originally purified from tobacco leaves, nicotianamine (Fig. 22(i)), has recently been isolated from *Streptomyces*; this inhibits MMP-2 and MMP-9 with $IC_{50} = 0.23$ and $1.0\,\mu M$, respectively [1013]. The related mugineic acid from barley root has $IC_{50} = 0.2$ and $0.41\,\mu M$ [1013]. Both inhibitors are reversed by adding zinc ions. Finally, mention may be made of two closely related inhibitors from the fungus *Westerdykella* named gelastatin A and B (Fig. 22(j)); the names may be misleading—both inhibit gelatinase A [933]. The mixture of inhibitors has $IC_{50} = 0.63\,\mu M$ for MMP-2.

Inhibitors not acting by chelation

It has been shown [962] that the digestion of procollagen with pepsin leads to the production of peptide fragments from the C- and N-terminal propeptides. There are four fractions among these that inhibit tadpole collagenase. This was suggested as a feedback regulation to reduce degradation at times when collagen synthesis was high. Studies of (+)catechin show that it reduces collagen digestion by MMP-1 [931]; this was attributed to coating of the substrate to reduce accessibility by the enzyme. Oxidation of the catechin flavenoid to polymers may make it more inhibitory [1026]. A separation of tea catechins into six types showed that (-)epicatechin gallate (Fig. 22(k)) was best, blocking MMP-1 action at $100\,\mu g/ml$ [943]. A further clue to the possibility that these compounds act by binding to the substrate is the demonstration that various collagen-digesting enzymes, including those of the serine and cysteine classes, are also inhibited [943]. This same compound at the same level was shown to inhibit digestion of type IV collagen by MMP-2 and MMP-9 [997]; however, it was also shown that the enzymes could interact with catechin affinity columns. A related compound GG6–10 galloylglucose inhibited MMP-2 and MMP-9 in zymography to the extent of 50% at a level of $100\,\mu g/ml$ [820]. In a similar vein, the addition of fibronectin to collagen type III at a level of $200\,\mu g/ml$ reduced digestion by MMP-1 by 76% [953]; there was no effect on digestion of collagen I or II.

The dye Eriochrome black T (Fig. 22(l)) was shown to inhibit MMP-1 but not MMP-2 [1000]; $IC_{50} = 45\,\mu M$ for the action of MMP-1 on fibrillar collagen and $K_i = 8\,\mu M$ for the action on soluble collagen. It has no effect on matrilysin [1031]. This compound was further studied later [959]; it is more effective ($10\,\mu M$ range) than the related compound suramin. Suramin has been suggested to inhibit cartilage stromelysin [815]. These compounds are sulfated. More highly sulfated compounds have been used to treat arthritis. For example, glycosaminoglycan polysulfate was used by Altman *et al.* [817] to treat a dog model of osteoarthritis; the levels of MMP-3 were reduced and a direct inhibitory effect on the enzyme was suggested. An interesting idea has been put forward recently by Zhou [1039]. They started with a monoclonal antibody that inhibited MMP-9 and isolated the variable fragment of a single chain immunoglobulin. This was then cloned and expressed as the VH-linker-Vκ protein and shown to be comparable to the pepsin fragment F(ab')$_2$ in inhibiting MMP-9.

Propeptides of MMPs as inhibitors

The propeptides contain the conserved sequence PRCGV/NPD in which the cysteine is coordinated to the zinc atom at the active site. This has led to the concept that peptides from the larger 80-residue propeptides, containing the critical cysteine, would inhibit active MMPs. Such inhibition should be much more powerful than that of cysteine alone because the neighbouring residues are presumably adapted to bind to specific sites near the active centre. Stetler-Stevenson *et al.* [1010] were the first to pursue this idea; they prepared the peptide TMRKPRCGNPDVAN of 14 residues matching the MMP-2 propeptide; this inhibited active MMP-2 at a level of 10 μM. Usefulness in a tumour invasiveness culture assay was soon shown [952] implicating MMP-2 in invasion. A similar peptide was based on the sequence of MMP-3, MRKPRCGVPDVG, which was whittled down to Ac-RCGVPD-NH$_2$ and Ac-RCGVP-NH$_2$; both MMP-1 and MMP-3 were inhibited by these short peptides [884]. The best compound was Ac-*iso*Cys-Gly-Tyr(2,6-dichloro-benzyl)-Pro-NH$_2$, which had an IC$_{50}$ = 3 μM for MMP-3. Further efforts to produce still smaller synthetic peptides led to the synthesis of Ac-RCGV-NH$_2$ which had IC$_{50}$ = 39 μM for MMP-3 and Ac-CVG-NH$_2$ with IC$_{50}$ = 62 μM [911].

A different approach was to prepare mutants in this region of the rat MMP-3 propeptide [982]. Examining mutations of MHKPRCGVPDV it was found that certain changes had no effect, enzymes expressed with these mutants remained latent until activated. Other mutants became partially active during expression, particularly those in the two prolines. However, the RC pair was most critical—changes here produced enzyme that was already degraded before it could be isolated. Even the Arg to Lys mutant was unstable.

Tetracyclines

The tetracycline group of antibiotics is known to have chelating properties and this may account for their ability to inhibit MMPs. Golub *et al.* in 1983 [890] were the first to note that such compounds, used in treating periodontal disease, could inhibit collagenase from gingival tissues and fluid—both rat and human. Minocycline (Fig. 22(m)) given pharmacologically reduced the activity of collagenase in gingiva and *in vitro* it inhibited MMP-8 about 90% at a dose of 20 μg/ml (40 μM). While there are clear inhibitory effects *in vitro*, the effects referred to as inhibition of collagenase *in vivo* are more difficult to interpret. It seems likely in retrospect that the 'inhibition' *in vivo* was actually a down-regulation of the enzyme protein in the tissue. If one removes the tissue or fluid and assays it with collagen substrate, there will be a considerable dilution of the inhibitor, so the detection of reduced collagenase is most likely due to reduced amount of enzyme, not to inhibition during the assay. This does not detract from the importance, still growing today, of the use of tetracyclines in the treatment of diseases, but it suggests the mechanism *in vivo* may involve factors other than, or in addition to, direct inhibition of the MMP activity. A case in point is the observation that after treatment of patients with tetracyclines has stopped, collagenase activity remains low for at least 7 days [892], although the antibiotic levels should have declined sharply in this period.

A comparison of various tetracyclines on rat

MMP-8 indicated about 90% inhibition by minocycline, doxycycline or tetracycline all at a concentration of 16 µg/ml. An unidentified collagenase activity on membranes of melanoma cells was inhibited by 20 µg/ml minocycline. The effect of tetracyclines does not depend on bactericidal action; a modified tetracycline lacking antibacterial action was found to be just as effective as minocycline in inhibiting synovial collagenase or rat cartilage collagenase [891]. Human synovial collagenase was inhibited 70% by 20 µg/ml minocycline [905]. A more detailed study was performed by Burns *et al.* [846] using purified rabbit corneal MMP-1 and the dnp–octapeptide substrate [947]. The IC_{50} for doxycycline was 15 µM, for minocycline 190 µM and for tetracycline 350 µM. Tetracycline was shown to inhibit human MMP-2 100% at the level of 50 µM. Direct assay of collagenase recovered from mouth washings in periodontal disease showed tetracycline to have an $IC_{50} = 260$ µM while that for doxycycline was 18 µM [950]. Doxycyline appears to act quite differently on human MMP-8 compared to MMP-1; 100% inhibition required 100 µM for MMP-8 but 600 µM for MMP-1 [1004]. A subsequent measurement with finer resolution gave corresponding values of 30 µM and 280 µM, respectively [1012]; or a ninefold difference.

Pig epithelial cells in culture produce MMP-1; this is inhibited by doxycycline at 0.5 mM and by a modified tetracycline without antibacterial activity at 1 mM [969]. Collagenase 2 (MMP-8) is inhibited 55% by doxycycline at 30 µM [1001, 1002]; however, the proenzyme is destabilized by doxycycline, making it susceptible to fragmentation with trypsin or APMA. A test of 23 different tetracyclines and derivatives on MMP-9 showed that minocycline inhibited 50% at 0.1 mM and that the best compound was 4-epichlorotetracycline which inhibited 85% at 60 µM [981]. The difficulty in comparing results between laboratories, assay methods, animal sources, etc. is exposed by the report on MMP-13 [939] in which doxycycline is required at quite high levels ($IC_{50} = 0.5$ mM) and a subsequent report, involving some of the same authors, which gives the value as 40 µM [906].

The *in vivo* effects of tetracyclines are difficult to assess: while enzyme inhibition is likely under some treatment regimens, it is also likely that enzyme synthesis is depressed. Thus, doxycycline causes skin keratinocytes to down-regulate the mRNA for MMP-2 [1022]. Treatment of cultures with 10 µM doxycycline significantly reduces mRNA by 6 h, and with 100 µM the level is decreased by 40%. Effects of reducing the enzyme may be more important than inhibiting the enzyme directly. Tetracycline down-regulates the mRNA for MMP-3 and this is shown not to be by way of the AP-1 site [925]. Similarly, 50 µM doxycycline inhibits degradation of type II collagen by MMP-8 *in vitro*, but *in vivo* it is shown to down-regulate both mRNA and protein [910]. Finally, it may be mentioned that chemical modifications to various tetracyclines show that the important portion of the molecule in inhibiting MMPs by chelation is the carbonyl oxygen at C11 and the –OH at C12 [1005]. A series of 10 patents dealing with the use of tetracyclines in inhibiting MMPs may be found in Table 17 (below).

Macroglobulins

α_2-Macroglobulin

Early studies [871, 914] indicated that human serum contains anti-collagenase (MMP-1) activity. The inhibitory fraction was identified as serum α-globulins, and the responsible inhibitors were thought to be α_2-macroglubulin and α_1-proteinase inhibitor (α_1-antitrypsin) [870, 992]; however, it was shown that α_1-proteinase inhibitor failed to inhibit vertebrate collagenases [827, 948]. α_2-Macroglobulin (α_2M), on the other hand, was shown to inhibit tadpole collagenase [812], mouse bone collagenase (MMP-13) [991], rabbit and human corneal collagenase [830, 831] and rabbit synovial collagenase [1028]. It was reported [814] that the majority of collagenase activity in rheumatoid synovial fluids was found as a complex with α_2M, suggesting that α_2M is a major regulator of collagenase activity in the joint *in vivo*. The collagenase-α_2M complex was shown to be dissociated by treating with 3 M NaSCN [813]. The major inhibitor of collagenase in human serum was shown to be α_2M [837, 1035].

Human α_2M is a plasma glycoprotein with a molecular mass of 725 kDa, consisting of four identical subunits of about 180 kDa that are linked in pairs by disulfide bonds and assembled noncovalently [7, 121]. It binds to and inhibits most endopeptidases regardless of their substrate specificity. The binding of a proteinase and α_2M is initiated by proteolysis of the 'bait' region located near the middle of the subunit, which in turn triggers large conformational changes in α_2M and entraps the proteinase [826]. The active site of the entrapped proteinase is unblocked but its action on large protein substrates is sterically hindered [826]. For example, the α_2M–collagenase complex can hydrolyse a synthetic substrate, but it cannot cleave triple helical collagens. As a result of conformational changes the C-terminal domains (residues 1314–1451) [1007, 1023] become exposed for interaction with the cellular receptor for α_2M–proteinase complexes, and the complex is ultimately endocytosed [121]. The conformational changes of α_2M upon proteolysis were demonstrated by electron microscopy [839, 1019]. The crystal structure of the methylamine-treated α_2M (similar structural conformation to that induced by proteinase binding) was resolved at 10 Å [818]. Each subunit of α_2M contains a characteristic activatable intrachain β-cysteinyl-γ-glutamyl thiolester bond. Hydrolysis of the bait region causes the activation and cleavage of the thiolester bond, resulting in the covalent attachment of the proteinase through formation of an ε-lysyl-γ-glutimyl bond with an ε-amino group of Lys in the proteinase [7]. The thiolester bond is found in all members of the α_2-macroglobulin family from various species except chicken ovostatin (ovomacroglobulin) [964, 968]. However, this covalent attachment of the proteinase is not the mechanism of enzyme inhibition [993].

Although an early work indicated that the interaction of α_2M and rabbit and human collagenases (MMP-1) occurred very slowly [1028], kinetic studies with human MMP-1 provided the k_2/K_i value of 2.8×10^6 M^{-1} s^{-1} [873]. This indicates that α_2M is about 150-fold better as a substrate for MMP-1 than human type I collagen ($k_{cat}/K_m = 1.8 \times 10^4$ M^{-1} s^{-1}) [1027]. MMP-3 reacts with α_2M more slowly ($k_2/K_i = 5.6 \times 10^4$ M^{-1} s^{-1}) than MMP-1 [873]. The rapid binding properties with MMP-1 suggest that α_2M is the major regulator of collagenolysis, especially in the fluid phase. This concept is supported by the competition experiment in which equimolar amounts of α_2M and TIMP-1 were reacted with MMP-1: all MMP-1 bound to α_2M [854]. Other MMPs shown to interact with α_2M include MMP-8 [928], human MMP-2 [965], rabbit MMP-3 [850], rat MMP-7 [1030], human MMP-9 [957], rat MMP-9 [1029], human MMP-11 [984], human MMP-12 [816] and mouse MMP-12 [824]. Human MMP-14 was recently reported to cleave α_2M [979].

Related proteins

Rabbits, rodents and guinea pigs contain **α_1-macroglobulin** (α_1M), with a slightly faster electrophoretic migration than α_2M. α_1M and α_2M are homologous and share a similar quaternary structure. α_2M in rodents is an acute phase protein and its synthesis is significantly enhanced in response to tissue injury or inflammation [897], while rat α_1M only increased by a factor of 2 or less. α_1M from rabbit [829], guinea pig [915] and rat [1006] have been reported to react with MMP-1.

In the plasma of healthy nonpregnant women **pregnancy zone protein** (PZP) is a trace protein (10–30 µg/ml; men, < 10 µg/ml) [883], but during pregnancy the level of PZP increases rapidly in the first trimester and it may reach levels of approximately 10 mg/ml in the third trimester. PZP is a close homologue of α_2M, but it is a dimeric molecule of 360 kDa consisting of two 180-kDa subunits that are linked by disulfide bonds [994]. Each dimer has the proteinase-binding units, and upon reaction with a proteinase two dimers loosely form a tetramer. Human MMP-1 cleaves the bait region of PZP [1006]. Human MMP-2 and MMP-9 also bind to PZP [819].

Rat α_1-inhibitor-3 (α_1I3) is a monomeric inhibitor of 180 kDa belonging to the α-macroglobulin family [874] and is a negative acute phase protein [887]. It has also been called murinoglobulin [990]. During the acute phase response to tissue injury or inflammation, the plasma levels of a set of proteins (acute phase proteins) increase significantly. While α_2M in rats is a typical acute phase protein, in response to tissue injury the serum level of α_1I3 falls rapidly. In normal rats, the serum concentration of α_1I3 is 6–10 mg/ml, but it decreases to the level of 1–2 mg/ml in response to tissue injury [940]. Human MMP-1 cleaves α_1I3, indicating that it is an inhibitor of MMPs [1006]. Rat α_1I3 inhibits rat MMP-7 and MMP-13 [1040].

Ovomacroglobulins (ovostatins) are members of the α-macroglobulin family but they are found exclusively in avian and reptilian egg whites. All share structural homology with α_2M. Chicken ovomacroglobulin was first characterized as a weak proteinase inhibitor [921]; it was shown to be a better inhibitor for metalloproteinases such as MMP-1 and thermolysin, but a poor inhibitor for serine and cysteine proteinases [964]. Chicken ovomacroglobulin is the exception among the members of the family, for it lacks an internal thiolester bond [964, 968]. Nonetheless, it reacts with metalloproteinases in a very similar manner to α_2M [963]. Kinetic studies of chicken ovomacroglobulin and human MMP-1 indicated a k_2/K_i value of $1.8 \times 10^3 \, \text{M}^{-1}\text{s}^{-1}$, considerably slower than with human α_2M [873].

Synthetic inhibitors

The development of MMP inhibitors has had commercial pharmaceutical implications since the beginning; therefore, a table of patents (Table 17) has been included to show the progression of the field. Some of the compounds found there have never been published in the regular scientific literature. There are additional compounds to be found in European and Japanese patents, but only those from the US are readily accessed by the internet without payment of fees. It is possible to learn a great deal about the binding pocket specificity at the active centre of MMPs by the use of synthetic inhibitors. That is because one can introduce a variety of groups not found in natural amino acids. The majority of inhibitors prepared to date are based on the chelation of zinc by the use of thiol, carboxyl, phosphorus and hydroxamate functions. The earliest synthetic inhibitor was developed in 1978 by Yankeelov *et al.* [1038] based on the inhibition of clostridial collagenase by histidine and mammalian collagenase by cysteine: mercaptoimidazolylpropionic acid (Fig. 23(a)). This had an IC_{50} for MMP-1 acting on collagen of 3.8 mM compared to 8.7 mM for cysteine (certainly a very modest beginning). Other early inhibitors included mercaptomethyl compounds and ureides of substituted naphthoic acids (see Table 17, Pat. 4235885, 4275076). The idea of adding a thiol group to a cleavage product of collagenase was more successful: replacing Ile with HSCH(CH$_2$CH(CH$_3$)$_2$)CO- in Ile-Ala-Gly-Gln-D-Arg-NH$_2$ [947] provided an inhibitor with $IC_{50} = 10 \, \mu\text{M}$ (Fig. 23(b)) [904]. Placing the -SH at the opposite end of the molecule also provided an effective inhibitor of collagenase with $IC_{50} = 9 \, \mu\text{M}$ (Fig. 23(c)) [861]. Other SH compounds are found in Patents 4371465 and 4374765. The fruitful idea of introducing a carboxyalkyl moiety at the amino end of Leu-Tyr-amide to direct a chelating carboxyl group toward the zinc in place of the scissile bond resulted in a collagenase inhibitor with $IC_{50} = 1.7 \, \mu\text{M}$ [949]. Compound SC40827 (Fig. 23(d)) (Pat. 4568666) based on this principle was used to inhibit bone resorption in culture [866] and ovulation by perfused rat ovaries [842]—in both cases probably due to inhibition of MMP-13, thought at the time to be MMP-1.

A series of peptide derivatives (Table 17, Pat. 4771037, 4771038) produced compounds effective in inhibiting both collagenase and stromelysin [869]. These fell into three classes and the best inhibitors were: thiol, Ac-Trp-Leu-SH with $IC_{50} = 1.6 \, \mu\text{M}$ on MMP-1; carboxyalkyl, COOH-CH(iBu)-Leu-Phe amide with $IC_{50} = 1.9 \, \mu\text{M}$ on

Patent No.	Inventor (N)[a]	M/D/YR	Assignee	Type of compound
4216164	Bernstein, S.	08/05/80	Amer. Cyanamide	Ureides of naphtholsulfonic acids
4216165	Bernstein, S.	08/05/80	Amer. Cyanamide	Aminobenzamidonaphthalene sulfonates
4216166	Bernstein, S.	08/05/80	Amer. Cyanamide	Nitrobenzamidonaphthalenesulfonic acids
4235885	Sundeen, J. E. (1)	11/25/79	E. R. Squibb & Sons	Mercapto compounds
4263293	Sundeen, J. E. (1)	04/21/81	E. R. Squibb & Sons	Heterocyclic amides
4275076	Conrow, R. B. (1)	06/23/81	Amer. Cyanamide	Ureides of naphtholsulfonic acids
4276284	Brown, S. I.	06/30/81	–	EDTA, acetyl cysteine
4297275	Sundeen, J. E. (1)	10/27/81	E. R. Squibb & Sons	Mercapto compounds
4297372	Bernstein, S.	10/27/81	Amer. Cyanamide	Ureides of naphtholsulfonic acids
4327111	Sundeen, J. E. (1)	04/27/82	E. R. Squibb & Sons	N-Mercaptoacyl-propionamides
4361574	Grant, N. H. (1)	11/30/82	Am. Home Products	Phenylthizaolobenzimidazoles
4367233	Clark, D. E. (1)	01/04/83	Am. Home Products	Phenylthizaolobenzimidazoles
4371465	McGregor, W. H.	02/01/83	Am. Home Products	Cysteine peptides
4371466	McGregor, W. H.	02/01/83	Am. Home Products	Cysteine peptides
4374765	McGregor, W. H.	02/22/83	Am. Home Products	R–GCGEE–NH$_2$ peptides
4382081	Sundeen, J. E. (1)	05/03/83	E. R. Squibb & Sons	R–S–CH$_2$–CHR^1–...
4391824	Siuta, G. J. (1)	07/05/83	Amer. Cyanamide	Ureylenebis(hydroxynaphthalenesulfonate)
4424354	Sundeen, J. E. (1)	01/03/84	E. R. Squibb & Sons	Mercapto compounds
4457936	Drager, E. (1)	07/03/84	Hoechst AG	Hydroxyphenylthiazole carboxylate
4511504	McCullagh, K. (2)	04/16/85	G. D. Searle	Carboxyalkyl compounds
4568666	McCullagh, K. (2)	02/04/86	G. D. Searle	Carboxyalkyl compounds
4595700	Donald, D. K. (3)	06/17/86	G. D. Searle	Thiol compounds
4599204	Poletto, J. F. (1)	07/08/86	Amer. Cyanamide	Oxalylamides of naphthalene sulfonic acid
H000083	Poletto, J. F. (1)	07/01/86	Amer. Cyanamide	Oxalylamides of naphthalene sulfonic acid
4599361	Dickens, J. P. (3)	07/08/86	G. D. Searle	Hydroxamates
4599966	Donald, D. K. (3)	07/21/87	G. D. Searle	Thiol compounds
4609667	Clark, D. E. (1)	09/02/86	Amer. Home Products	Phenacetyl-L-cysteines
4687841	Spilburg, C. A. (1)	08/18/87	Monsanto	Peptide hydroxamates
4704383	McNamara, T. F. (2)	11/03/87	Res. Found. SUNY	Non-antibacterial tetracyclines
4720486	Spilburg, C. A. (1)	01/19/88	Monsanto	Peptide hydroxamates
4735945	Sakamoto, S. (1)	04/05/88	Vipont Labs	Sanguinarine
4743587	Dickens, J. P. (3)	05/10/88	G. D. Searle	Hydroxamates
4771037	Roberts, R. A. (1)	09/13/88	ICI Americas	N-Carboxylalkyl compounds
4771038	Wolanin, D. J. (1)	09/13/88	ICI Americas	Peptide hydroxamates
4866082	Hayward, M. (1)	09/12/89	Am. Home Products	Cysteine thiol derivatives
4885283	Broadhurst, M. J. (4)	12/05/89	Hoffmann–La Roche	Phosphinic acid derivatives
4918105	Cartwright, T. (3)	04/17/90	SA Lab. R. Bellon	Actinonin antibiotic
4925833	McNamara, T. F. (2)	05/15/90	Res. Found SUNY	Tetracyclines
4935404	Hunter, D. J. (2)	06/19/90	Beecham Group	Phosphonates
4935411	McNamara, T. F. (2)	06/19/90	Res. Found. SUNY	Non-antibacterial tetracycline
4935412	McNamara, T. F. (2)	06/19/90	Res. Found SUNY	Non-antibacterial tetracyclines
4937243	Markwell, R. E. (2)	06/26/90	Beecham Group	Thiol carboxylic acid derivatives
4996358	Handa, B. K. (2)	02/26/91	Hoffmann–La Roche	Amino acid hydroxamates
5006651	Broadhurst, M. J. (4)	04/09/91	Hoffmann–La Roche	Phosphinic acid derivatives
5010097	Markwell, R. E. (3)	04/23/91	Beecham Pharm.	Thiol derivatives
5100874	Odake, S. (4)	03/31/92	Fuji Yakuhin KKK	Tetrapeptide hydroxamates
5109000	Markwell, R. E. (2)	04/28/92	Beecham Group	Thiol carboxylic acids
5114953	Galardy, R. E. (2)	05/19/92	Univ. Florida	Hydroxamates
5114960	Lawton, G. (2)	05/19/92	Hoffmann–La Roche	Isoxazole derivatives
5124322	Hughes, I.	06/23/92	Beecham Group	Dipeptide with cleavable disulfide

Continued

Patent No.	Inventor (N)[a]	M/D/YR	Assignee	Type of compound
5183900	Galardy, R. E. (1)	02/02/93	–	Hydroxamate derivatives
5189178	Galardy, R. E. (1)	02/23/93	–	Hydroxamate derivatives
5212163	Ward, R. W. (2)	05/18/93	Beecham Group	Phosphorus compounds
5239078	Galardy R. E. (2)	08/24/93	Glycomed	Hydroxamate
5240958	Campion, C. (3)	08/31/93	Brit. Biotech.	Hydroxamate
5256657	Singh, J. (4)	10/26/93	Sterling Winthrop	Succinimide derivatives
5258371	Golub, L. M. (9)	05/29/93	Kureha Co.	Non-antibacterial tetracycline
5260059	Acott, T. S. (2)	11/09/93	State of Oregon	Chelators, EDTA, TIMP
5268384	Galardy, R. E.	12/07/93	–	Hydroxamates
5270326	Galardy, R. E. (2)	12/14/93	Univ. of Florida	Hydroxamates
5270447	Liotta, L. A. (2)	12/14/93	USA–DHHS	PRCGNPD MMP propeptides
5280106	Liotta, L. A. (2)	01/18/94	–	Peptides around RKPRC of MMP-2
5300501	Porter, J. R. (3)	04/05/94	Celltech	Hydroxamate
5300674	Crimmin, M. J. (2)	04/05/94	Brit. Biotech.	Hydroxamate
5304549	Broadhurst, M. J. (3)	04/19/94	Hoffmann–La Roche	Phosphinic derivatives of amino acids
5304604	Davidson, A. H. (2)	04/19/94	Brit. Biotech.	Hydroxamate
5308839	Golub, L. M. (3)	05/03/94	Res. Found. SUNY	Non-antibacterial tetracycline
5310759	Bockman, R. S.	05/10/94	–	PGE blocks MMP-1 gene expression
5310763	Campion, C. (3)	05/10/94	British Biotech.	Hydroxamate
5318964	Broadhurst, M. J. (3)	06/07/94	Hoffmann–La Roche	Hydroxamate
5321017	Golub, L. M. (3)	06/14/94	Res. Found. SUNY	Non-antibacterial tetracycline
5326760	McElroy, A. B. (4)	07/05/94	Glaxo	Aminobutanoic acid cpds
5387610	Gray, R. D. (2)	02/07/95	Res. Corp. Technol.	Hydroxamate
5391486	Okuyama, A. (6)	02/21/95	Banyu	BE16627 from *Streptomyces*
5403952	Hagmann, W. (2)	04/04/95	Merck	Cyclic hydroxamates
5412145	Crimmin, M. J. (2)	05/02/95	British Biotech.	P2′-modified hydroxamates
5447929	Broadhurst, M. J. (3)	09/05/95	Hoffmann–La Roche	Hydroxamate
5453438	Campion, C. (3)	09/26/95	British Biotech.	Hydroxamate
5455258	MacPherson, L. J. (1)	10/03/95	Ciba–Geigy	Arylsulfonamide substituted hydroxamate
5455262	Schwartz, M. A. (1)	10/03/95	Florida State Univ.	Mercaptosulfide
5459135	Golub, L. M. (3)	10/17/95	Res. Found. SUNY	Non-antibacterial tetracycline
5470834	Schwartz, M. A. (1)	11/28/95	Florida State Univ.	Sulfoximines and sulfodiimine cpds
5473100	Isomura, Y. (4)	12/05/95	Yamanouchi Pharm.	Hydroxamate
5506242	MacPherson, L. J. (2)	04/09/96	Ciba–Geigy	Arylsulfonamide hydroxamate
5514677	Davidson, A. M. (2)	05/07/96	British Biotech.	Hydroxamate
5514716	Gowravaram, M. R. (6)	05/07/96	Sterling Winthrop	Hydroxamates and carboxylates
5525629	Crimmin, M. J. (2)	06/11/96	British Biotech.	Hydroxamate
5530128	Porter, J. R. (3)	06/25/96	Celltech	N-Sulfonamino dipeptide hydroxamate
5530161	Campion, C. (3)	06/25/96	British Biotech.	Hydroxamate
5532265	Gijbels, K. (1)	07/02/96	Stanford Univ.	Tripeptide hydroxamate
5550216	Miyazaki, K.	08/27/96	Oriental Yeast	Beta-amyloid peptide blocks MMP-2
5552162	Lee, R.	09/03/96	Arch Develop. Corp.	Stimulate MMP-1
5552419	MacPherson, L. J. (1)	09/03/96	Ciba–Geigy	Arylsulfonamide hydroxamate
5569665	Porter, J. R (3)	10/29/96	Celltech	Peptide hydroxamates
5580570	Robertson, S. M. (1)	12/03/96	Alcon Labs.	Hydroxamate treatment
5585356	Liotta, L. A. (2)	12/17/96	–	Propeptides from MMP-2
5594006	Sakamoto, M. (5)	01/14/97	Otsuka Pharmaceut.	Carbostyril derivatives
5595885	Stetler-Stevenson W. G. (2)	01/21/97	USA–DHHS	TIMP-2 peptides
5602156	Kohn, E. C. (1)	02/11/97	USA–DHHS	Inhibit MMP expression
5612215	Draper, K. G. (4)	03/18/97	Ribozyme Pharm.	Ribozymes targeting stromelysin RNA

Continued

Patent No.	Inventor (N)[a]	M/D/YR	Assignee	Type of compound
5614625	Broadhurst, M. J. (2)	03/25/97	Hoffmann–La Roche	Tricyclic hydroxamate
5616605	Gray, R. D. (2)	04/01/97	Res. Corpt. Tech.	Peptide inhibitors
5618844	Gowravaram, M. R. (6)	04/08/97	Sanofi, SA	Hydroxamate and carboxylic acid
5618925	Dupont, E. (2)	04/08/97	Labs. Aeterna	Shark cartilage extracts
5622863	Mayerl, F. (2)	04/22/97	Bristol–Myers Squibb	Tropolone cpds
5626865	Harris, D. H. (2)	05/06/97	Advanced Corneal Sys	Inhibitors to treat cornea
5627206	Hupe, D. (2)	05/06/97	Warner–Lambert	Tricyclic MMP-2 inhibitor
5629343	Hagmann, W. (1)	05/13/97	Merck	Mercaptoacyl peptides
5639746	Yelm, K. E.	06/17/97	Procter & Gamble	Hydroxamates
5643752	Hawkins, P. R. (1)	07/01/97	Incyte Pharm.	TIMP-4 DNA
5643908	Sugimura, Y. (3)	07/01/97	Sankyo Co.	Hydroxamates
5643964	Dickens, J. P. (2)	07/01/97	British Biotech	Amino acid derivatives
5646167	MacPherson, L. J. (1)	07/08/97	Ciba–Geigy	Arylsulfonamide hydroxamates
5646316	Jacobson, A. R. (2)	07/08/97	OsteoArthritis Sci.	Bile acid–hydroxamate cpds
5652227	Teronen, O. P. (2)	07/29/97	–	Bisphosphonate chelators
5652262	Crimmin, M. J. (2)	07/29/97	British Biotech.	Hydroxamate
5659061	Glazier, A.	08/19/97	Drug Innov. Design	Phosphoramide mustards
5665753	Frazee, J. S. (2)	09/09/97	SmithKline Beecham	Imidazole hydroxamates
5665764	Hupe, D. (4)	09/09/97	Warner–Lambert	Tricyclic cpds
5665777	Fesik, S. W. (7)	09/09/97	Abbott Labs.	Biphenyl hydroxamates
5668122	Fife, R. S. (1)	09/16/97	–	Tetracyclines to treat cancer
5672583	Chapman, K. (5)	09/30/97	Merck	Carboxypeptidyl derivatives
5672598	De, B. (3)	09/30/97	Procter & Gamble	Lactam hydroxamates
5672615	MacPherson, L. J. (1)	09/30/97	Novartis	Arylsulfonamide hydroxamate
5677282	Oleksyszyn, J. (1)	10/14/97	Proscript	Amino acid amide thiadiazole
5679700	Caldwell, C. G. (4)	10/21/97	Merck	Phosphinic acid peptides
5684152	Ponpipom, M. M. (1)	11/04/97	Merck	Carboxyalkyl peptides
5686422	Gray, R. D. (5)	11/11/97	Res. Corp. Tech.	Thiol peptides
5691381	Jacobson, I. C. (2)	11/25/97	DuPont Merck	Hydroxamate and carbocyclic acids
5691382	Crimmin, M. J. (2)	11/25/97	British Biotech.	Hydroxamate for TNFα and MMP
5696082	Crimmin, M. J. (2)	12/09/97	British Biotech.	Hydroxamates
5696147	Galardy, R. E.	12/09/97	Glycomed	Hydroxamates
5698575	Watanabe, K. (2)	12/16/97	Kureha Chem. Ind.	Chromone derivatives
5698671	Stetler–Stevenson, W. G. (2)	12/16/97	USA–DHHS	TIMP peptides
5698690	Broadhurst, M. J. (2)	12/16/97	Hoffmann–La Roche	Hydroxamates
5698706	Baxter, A. D. (2)	12/16/97	Chiroscience	Heterocyclic amides with SH
5700838	Dickens, J. P. (2)	12/23/97	British Biotech.	Hydroxamates
5703092	Xue, C-B. (3)	12/30/97	DuPont Merck	Hydroxamates for TNFα and MMP
5710167	Broadhurst, M. J. (2)	01/20/98	Hoffmann–La Roche	Tricyclic hydroxamates
5712300	Jacobsen, E. J.	01/27/98	Pharmacia & Upjohn	Hydroxamate derivatives
5714465	Langley, K. E. (2)	02/03/98	Amgen	194 amino acids of TIMP
5714491	Morphy, J. R. (1)	02/03/98	Celltech	Peptide derivatives
5716953	Norcini, G. (3)	02/10/98	Zambon Group	Phosphinic acid derivatives
5731293	Watanabe, K. (3)	03/24/98	Kureha Chem. Ind.	Esculetin derivatives
5731295	Draper, K. G. (4)	03/24/98	Ribozyme Pharm.	Ribozymes to reduce MMP-3
5731441	Broadhurst, M. J. (2)	03/24/98	Hoffmann–La Roche	Tricyclic hydroxamates
5739123	Norcini, G. (3)	04/14/98	Zambon Group	Phosphinic acid derivatives
5744442	Richards, C. D. (3)	04/28/98	Bristol–Myers	Cytokine to decrease MMPs
5747514	Beckett, R. P. (3)	05/05/98	British Biotech	Hydroxamate and carboxylic acids
5753653	Bender, S. L. (1)	05/19/98	Agouron Pharm.	Arylsulfonamide hydroxamate

Continued

Patent No.	Inventor (N)[a]	M/D/YR	Assignee	Type of compound
5756545	O'Brien, P. M. (1)	05/26/98	Warner Lambert	Biphenylsulfonamide hydroxamates
5760285	Norcini, G. (2)	06/02/98	Zambon Group	Phosphonyldipeptides
5763621	Beckett, R. P. (2)	06/09/98	British Biotech	Hydroxamates
5770588	McNamara, T. F. (3)	06/23/98	Res. Found SUNY	Non-antibacterial tetracyclines
5770624	Parker, D. T.	06/23/98	Novartis	Arylsulfamide hydroxamates
5773428	Castelhano, A. L. (7)	06/30/98	Syntex	Carboxyalkyl compounds
5773438	Levy, D. E. (7)	06/30/98	Glycomed	Hydroxamates

[a]Number of co-inventors.

MMP-3; and hydroxamate, $HONHCO\text{-}CH_2C(n\text{-}pentyl)CO\text{-}Val\text{-}AlaNH_2$ with $IC_{50} = 450$ nM on collagenase and 34 nM on MMP-3. This was among the earliest indications that one compound was likely to inhibit more than one MMP. Representatives of each group were tested on chondrocyte stromelysin, with the hydroxamate clearly ahead [852]. The same compounds also inhibited proteoglycan release when applied to organ culture of rabbit cartilage [851]. Further thiol compounds ($IC_{50} = 1$–$4\,\mu M$) [903] and Z-Pro-Leu-Gly-NHOH with $IC_{50} = 40\,\mu M$ [956] were developed. Z-Pro-Ala-3-aminooxypropyl-Leu-Ala-Gly-OEt had $IC_{50} = 60\,\mu M$ for collagenase [1024]. For a review on the study of inhibitors up to 1987 see [61]. The major classes, except for phosphorus compounds, had been established by then, based largely on earlier studies of metalloproteinases such as thermolysin, and most attention had been devoted to collagenase. Compounds were just edging into the IC_{50} nanomolar range.

The first compounds developed to inhibit neutrophil collagenase (MMP-8) were phosphonamidates such as $Cbz\text{-}Gly^P\text{-}Leu\text{-}Ala\text{-}Gly$ with $K_i = 14\,\mu M$ [955]. A phthaloyl-Gly^P-Ile-Trp-benzylamide had $K_i = 25$ nM on MMP-1 [929]. A systematic study of phosphoramidates, on the other hand, did not produce a compound with K_i better than $1.5\,\mu M$ [930]. A hydroxamate based on Leu-Tyr, SC44463, (Fig. 23(e)) (Pat. 4743587) was found to inhibit gelatinase A with $IC_{50} = 2\,\mu g/ml$ and MMP-1 at $10\,\mu g/ml$ [987]. This same compound was shown to inhibit ovulation in perfused rat ovaries and to have an $IC_{50} = 1.2$ nM and 10 nM

against MMP-13 and MMP-7, respectively [847]. Placing the NHOH function at the C-terminus of a series of 17 tripeptides did not give striking results, the best peptide Z-Pro-Leu-Ala-NHOH had $IC_{50} = 1\,\mu M$ for collagenase 1 [975]. Extension of this concept to tetrapeptides [976] (Pat. 5100874) did not produce further improvement [974]. The peptide Leu-Phe-Ala-NH_2 having CH_2SH in place of NH_2 on Leu has an $IC_{50} = 10$ nM for MMP-1 acting an a synthetic dnp substrate but only $3\,\mu M$ acting on collagen; nonetheless, the compound was effective in treating corneal burns of rabbits [846]. A systematic study of such compounds showed Tyr(O-Met) to be optimal at P1′ and Nal-Ala-NH_2 at P2′ and P3′, with K_i as low as 14 nM against MMP-1 [864]. Lelièvre *et al.* [935] developed compounds ending in hydroxyaminocarbonyl- which appear to make a bidentate compound with zinc. With Leu at P1′, Tyr at P2′ and benzylamide at P3′, a $K_i = 4$ nM against MMP-1 was obtained [935]; this compound prevented TIMP binding to the enzyme. A brief review of progress by industrial groups through 1989 can be found in [916].

The study of MMP inhibitors followed a similar path during the early 1990s. The world was still awaiting the first three-dimensional structure of an MMP on which to base a rational design. In the interim, work continued on the various groups of compounds. Cysteine derivatives were added to the C-terminus of inhibitors [848], mercaptocarboxylic acid derivatives of dipeptides were developed with $IC_{50} = 1.8$ nM [833] (Fig. 23(f)) (Pat. 5124322), and a series of mercaptan

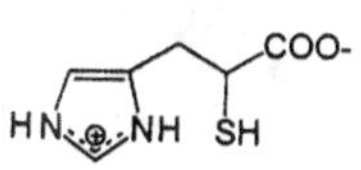

(a) Mercaptoimidazolylpropionate

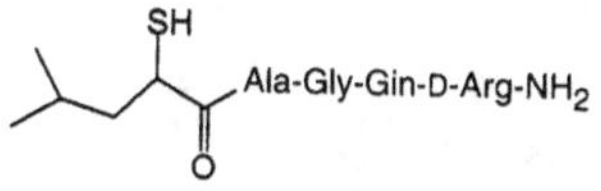

(b) Tadpole collagenase inhibitor

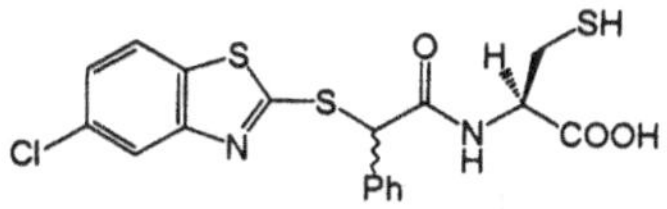

(c) Wyeth 45,368

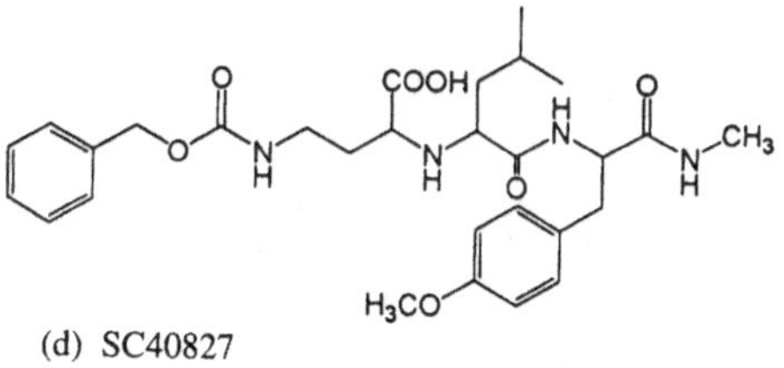

(d) SC40827

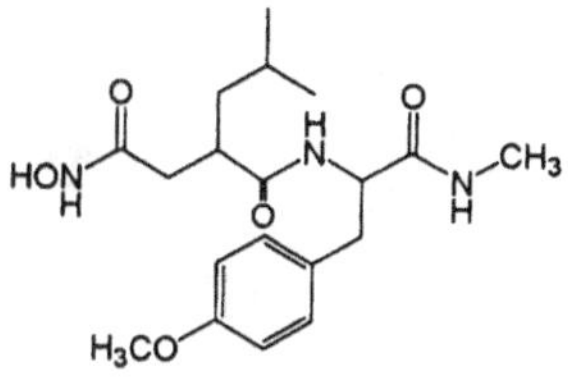

(e) SC44463

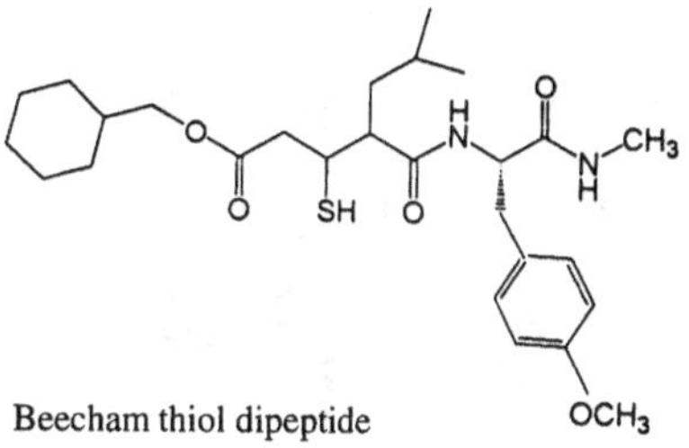

(f) Beecham thiol dipeptide

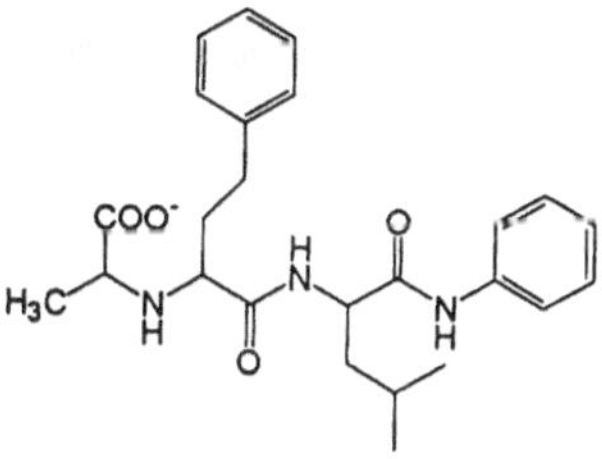

(g) Merck carboxyalkyl cpd.

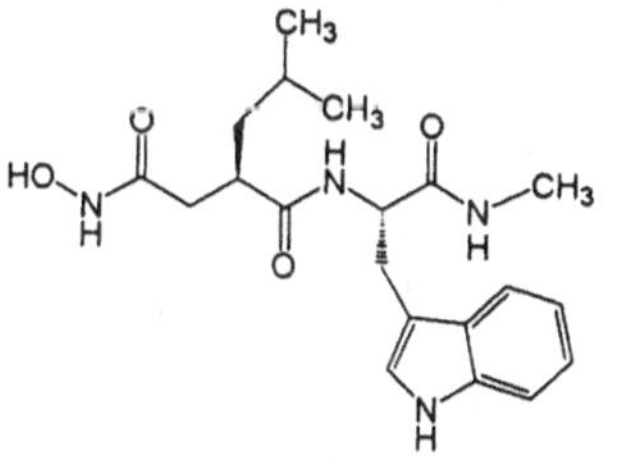

(h) Galardin, GM-60001

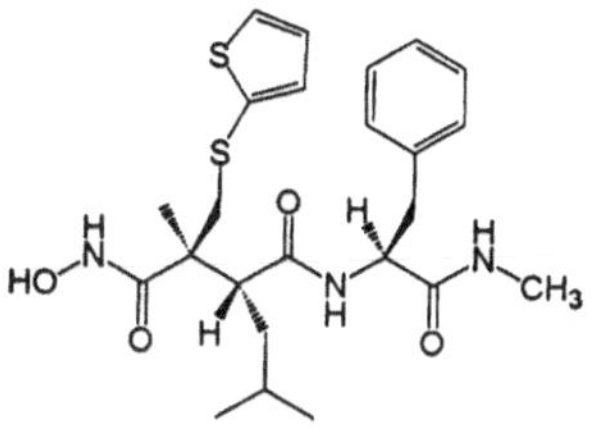

(i) BB-94, Batimastat

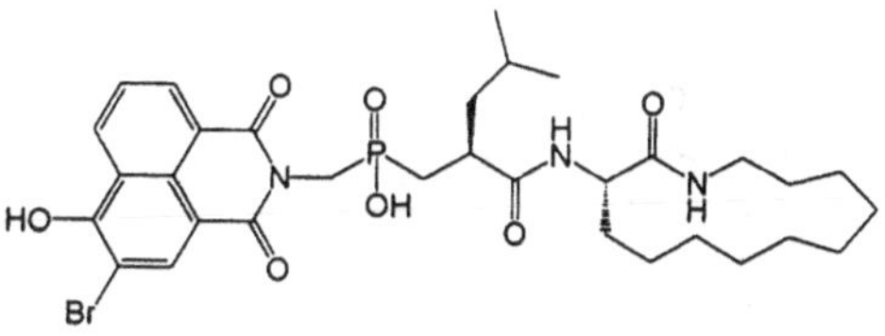

(j) Ro 31-7467

Figure 23 (a)–(j) Caption opposite

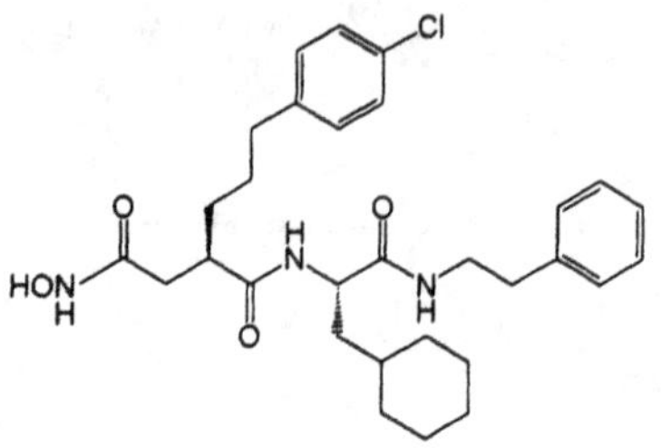

(k) Celltech hydroxamate

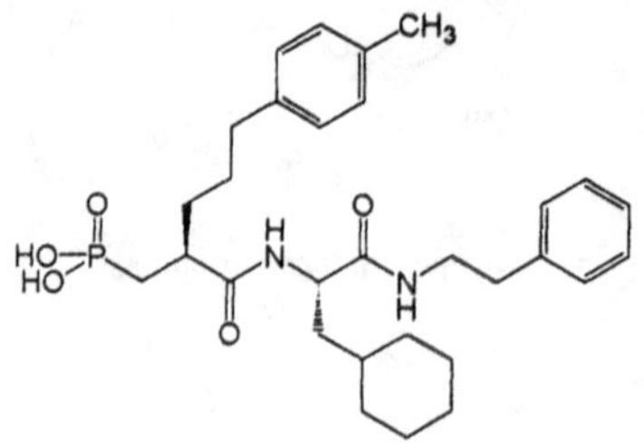

(l) Celltech phosphonic acid

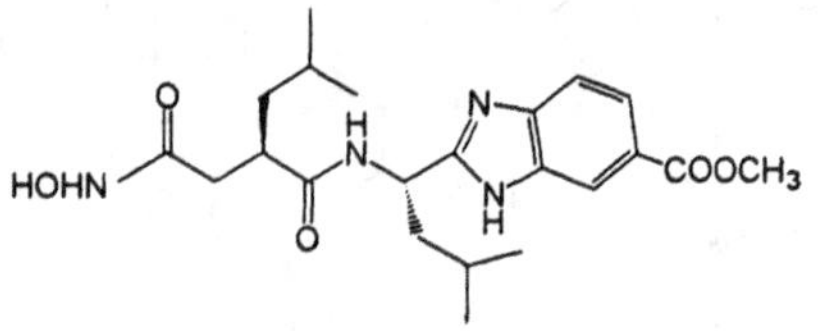

(m) Roche benzimidazole hydroxamate

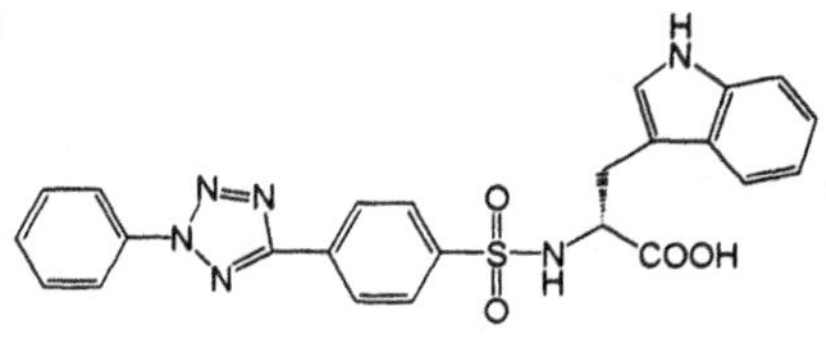

(n) Shionogi *N*-sulfonamino acid compound

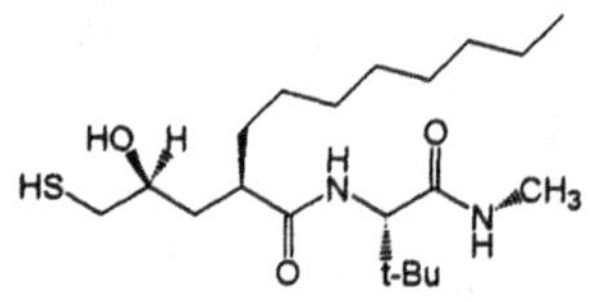

(o) Wyeth succinyl mercaptoalcohol

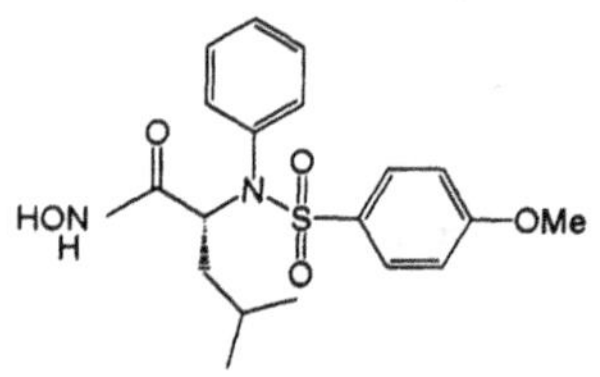

(p) Novartis sulfonamide hydroxamate

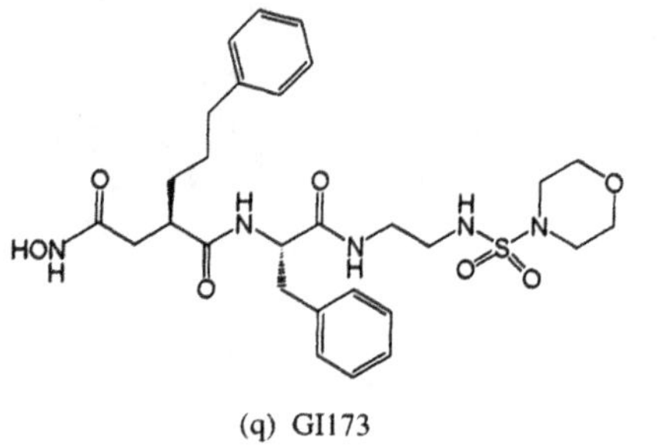

(q) GI173

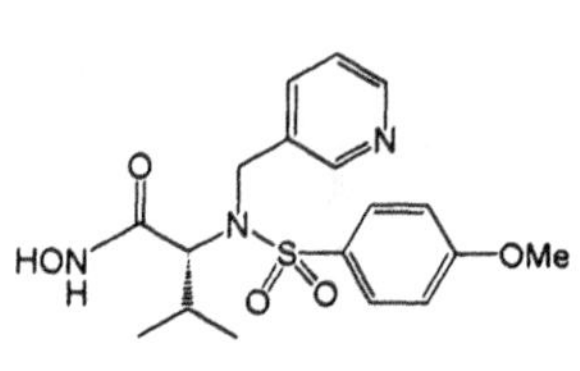

(r) CGS 27023A

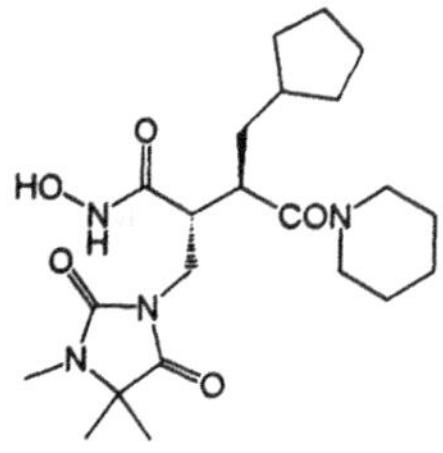

(s) Ro 32-3555

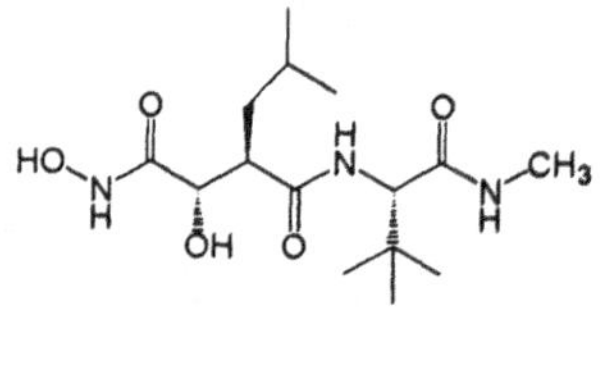

(t) BB-2516, Marimastat

Figure 23 (k)–(t)
Chemical structures of synthetic inhibitors of MMPs prepared by commercial sources.

compounds were developed that gave $IC_{50} = 42$ nM for MMP-8 and 56 nM for MMP-1 [999]. *N*-Carboxyalkyl compounds were explored by Stack *et al.* [1009] and used for affinity purification of MMP-1, -2 and -3. Much more effective compounds in this group were prepared by Glaxo [843] (Pat. 5326760) and by Merck [856], and were used by the latter in an effort to find selective inhibitors of various MMPs; thus a derivative of Gly-Leu produced a $K_i = 6.5$ nM for rabbit MMP-3 while the corresponding values for MMP-2 and MMP-9 were 410 and 760, respectively. The same compound showed only small discrimination among human MMP-1, -2, -3 and -9. The mechanism of one such compound (Fig. 23(g)) on MMP-3 was worked out [922], interestingly this enzyme shows a maximum inhibition (and activity) at pH 5.5. Hydroxamates continued to improve [970] (Pat. 5318964). The compound GM-6001 or galardin [907] (Fig. 23(h)) (Pat. 5239078) was one of the first to have K_i in the picomolar range (400 pM); the compound, HONH-CO-CH$_2$-CH(*i*-BU)CO-L-Trp-NHMe, was tested with MMP-1 acting on a peptide substrate. A hydroxamate with $IC_{50} = 1\,\mu$M against MMP-9 was developed [946]. An early hydroxamate, BB-94 or batimastat (Fig. 23(i)) (Pat. 5240958), was applied *in vivo* to block human tumour growth in nude mice [1025]; this compound inhibited MMP-1, -2, -3, -7 and -9 with IC_{50} in the range 3–20 nM.

Phosphinic acids were developed for selectivity against MMP-1 relative to MMP-2 and MMP-3 [970]; the best compound, Ro31–7476 (Fig. 23(j)) (Pat. 5304549), was 12 times better against MMP-1. Phosphinic and phosphonic compounds were also developed at Merck and SmithKline Beecham [835, 898, 920] (Pat. 5212163). There was a developing recognition that effective pharmaceutic agents would have to discriminate against one or more of the MMPs, rather than just blocking all of them without selectivity. This problem was approached in a systematic fashion by the Celltech group [958, 985] (Pat. 5300501) who found a hydroxamate (Fig. 23(k)) that inhibited MMP-2 with $K_i = 30$ pM; MMP-3, 7.3 nM; and MMP-1, 2440 nM. The best carboxylic acid derivative gave $K_i = 0.9$ nM for MMP-2, 320 nM for MMP-3 and 20 000 nM for MMP-1, and the best phosphonic acid derivative (Fig. 23(l)) gave 2.5 for MMP-2, 277 for MMP-3 and 17 000 for MMP-1. The patent literature to the end of this period has been reviewed [15].

With respect to the past four years we will consider the elucidation of crystal structures of MMPs, the use of combinatorial libraries, improvements to existing inhibitors, development of new inhibitors, the emphasis shifting to clinical applications requiring development of peptidomimetics, oral availability, clinical trials, and specificity limitation to MMPs. Recent developments in the field have been reviewed [12–14, 20, 53, 77, 101, 151]. [20] gives structures of some 60 compounds. Many of these reviews cover parts of the European and Japanese patent literature not covered in Table 17. An NMR structure for stromelysin was developed in 1993 [895, 896] by use of an *N*-carboxyalkyl inhibitor [856]. This revealed a deep hydrophobic pocket at S1′, and a shallow S2′ and an open S3′. This was followed not long after with the first crystal structure, that of human MMP-8 or neutrophil collagenase, inhibited with a tripeptide hydroxamate [956], which showed a similar pattern of binding pockets [838]. In both cases, the inhibitors did not reveal much about binding to the amino side of the scissile bond. However, the availability of structures promoted the use of SAR (specificity-activity relationship), studies of docking of model inhibitors, and verification of binding modes by crystallographic studies of inhibitor–enzyme complexes. Grams *et al.* [901] explored the binding of batimastat to MMP-8 and did the docking of eight related compounds. This information, together with the earlier work on Pro-Leu-Gly-NHOH provided information about the enzyme pocket from S3 through S3′. Other inhibitors of the hydroxamate and thiol group were examined by Grams *et al.* [902]. One compound, HONHCOCH(*i*Bu)CO-Ala-Gly-NH$_2$, showed an odd mode of binding in which the P2′ and P3′ groups bent around and inserted into the S1′ pocket. This observation was extended to produce a 500-fold improvement in malonic acid derivatives of hydroxamates [841, 900]. The MMP-8 structure was also used to develop a new cysteine compound with $IC_{50} = 180$ nM [960]. A comparison of BB-1909 binding to MMP-8 was made by docking to structures of MMP-1, -2, -3 and -7; this gave a reasonable explanation of the various IC_{50} values with these different enzymes in relation to differences in their binding pockets [834].

NMR was used to determine the conformation of inhibitory peptides bound to MMP-3 [893, 894]. An interesting approach to inhibitor design has been advanced by Hajduk *et al.* [909, 980] in finding two compounds that ligate to different parts of the MMP-3 active centre as determined by NMR, then ligating the two to provide 15 nM inhibitors. More recently, X-ray crystallography of MMP-3 has been used to evaluate carboxylate inhibitors [859] (Pat. 5691381), and a computational method to design MMP-1 inhibitors based on X-ray and NMR data, a so-called pharmacophoric model, has been developed [889]. A review of inhibitor design based on X-ray crystallography is found in [867]. Matrilysin (MMP-7) has received less attention than the other MMPs from 1 through 8, but the first crystal structure [845] involved the use of three inhibitors of very similar structure, differing only in the zinc-chelating moiety: hydroxamate, carboxylate and sulfodiimine. The K_i value for these three compounds was 0.03, 0.85 and 4.0 μM, respectively, which correlates with the strength of zinc binding.

A recent development is the introduction of combinatorial methods for preparing MMP inhibitors. The solid phase synthesis of dipeptide cysteine compounds as MMP-1 inhibitors was a start in this direction [882]. The combination of structure-based ligand design with combinatorial chemistry has produced novel MMP inhibitors [988]. The Merck group [876] have produced combinatorial libraries of *N*-carboxyalkyl tripeptides as MMP-3 inhibitors; the best compounds to date have $IC_{50} = 400$ nM. At the same time, the same group [875] used classical syntheses to prepare P1' biphenylethyl carboxyalkyl dipeptides that inhibit MMP-3 with $K_i = 2$ nM. So combinatorial methods have yet to fulfil their promise. A quick one-dimensional method of NMR has been developed for rapid screening of such libraries for binding to MMP-3 [908]. Another interesting approach has been to look for combinatorial tetrapeptides that would inhibit gelatinases—in this case there is no chelating group, only peptides that bind well but are not cleaved. A compound effective at 1 mM is His-ε-aminocaproic acid-βAla-His [881].

Continuing work on the hydroxamate group of inhibitors has included the introduction of a lactam ring at P2'–P3' [836, 853]; such compounds have $K_i = 25$ nM for MMP-1. Introduction of a heteroatom at P1' gives compounds with $K_i = 7$ nM against MMP-3 [899], whereas piperazic acid analogues at P3' give an IC_{50} of only 0.6 μM [973]. Introduction of benzamidazole in place of the amide bond between P2'–P3' (Fig. 23(m)) gives $IC_{50} = 1$–5 nM for MMP-1, -3 and -7 [858]. However, sulfonamide or urea in this position is much less satisfactory [865]. A quaternary hydroxy group at P1 produced inhibitors with K_i in the subnanomolar level for MMP-9 [923]. A large series of dipeptide mimics produced a hydroxamate with $K_i = 0.18$ nM for MMP-8, 0.57 nM for MMP-9 and 0.39 nM for MMP-2 [937] (Pat. 5773438). These recent modifications have not generally increased the ability of the compounds to inhibit MMPs relative to earlier compounds, rather, their purpose has been to achieve stability, particularly in *in vivo* situations where peptides would be rapidly cleaved by other proteinases. There is also a growing recognition that MMP inhibitors are frequently effective on other metalloproteinases, particularly the cell-membrane tumour necrosis factor α (TNFα) convertase activity [886, 951] which belongs to a different metalloproteinase family. The cleavage of low affinity IgE receptors on the cell surface is also inhibited by hydroxamates related to BB-94 [822, 823].

Carboxyalkyl compounds have received almost as much attention as hydroxamates. A series of homophenylalanine derivatives at P1' provided a compound distinguishing between MMPs [989] (Pat. 5684152); the $IC_{50} = 3$ nM for MMP-2, 18 nM for MMP-3 and 5900 nM for MMP-1. Extending the alkyl group at P1′ gave $K_i = 0.29\,\mu M$ for MMP-3 [877]; however, other extended groups at this position enhance the fit to the S1′ pocket of MMP-2. [954]. Modified dipeptides gave $K_i = 20$ nM for MMP-1 and 5 nM for MMP-9 [857]. Introduction of a toluenesulfonamide at P1 gave a compound with $IC_{50} = 50$ nM for MMP-3 [1036]. N-Sulfonamide derivatives of a compound with the COOH function at the C-end gave an inhibitor (Fig. 23(n)) with good inhibition of gelatinase A and B ($IC_{50} = 19$ and 32 nM, respectively) but poor inhibition ($>1\,\mu M$) for MMP-1, -3 or -7 [1017]. Selective inhibition of MMP-8 is provided by macrocyclic derivatives [860].

There has been one recent paper on phosphinic acids [849] (Pat. 5679700) and on thiol compounds [919]. A new class of compounds, malonyl and succinyl mercaptoketones and mercaptoalcohols, can be added to this latter group [936], these are active in the low nanomolar range on MMP-1, MMP-3 and MMP-9 (Fig. 23(o)). Attention has remained focused on the MMPs from 1 through 9; however, recently sulfonamide hydroxamate inhibitors have been developed for macrophage elastase (MMP-12) [924] (Fig. 23(p)) (Pat. 5770624) with $IC_{50} = 6$ nM. Hydroxamate compounds have been developed for membrane-type matrix metalloproteinase 1 (MMP-14); these are selective for MMP-14 relative to MMP-1, but are equally potent in inhibiting MMP-9 [1037].

With respect to drugs that can be taken orally, work has continued on hydroxamates [918], mercaptoacyl inhibitors [828] and carboxyalkyl dipeptides [855] to develop orally active compounds. In early studies of batimastat administered intraperitoneally to inhibit human ovarian carcinoma growth in nude mice [864a], it was possible to administer a diastereomer that had low activity on MMPs to demonstrate that the positive effects on tumour growth were due to inhibition of MMPs and not to general chelating powers of the inhibitor. It is not possible to review all such studies in biological systems. However, mention may be made of selective inhibitors of gelatinases and collagenase to dissect their individual roles in bone resorption in cultured mouse calvariae [917]; both types are needed to get complete inhibition. Various hydroxamates produced by Glaxo and Celltech [862] were used to inhibit growth and metastasis of various tumour types in mice. The compound, GI173 (Fig. 23(q)), had $IC_{50} = 50$ pM for MMP-2, 3 nM for MMP-13, 8 nM for MMP-3 and 37 nM for MMP-1. A non-peptide sulfonamide-based hydroxamate of Novartis, CGS 27023A (Fig. 23(r)), blocked cartilage degradation in a rabbit model when given orally [942]. This was the most effective of 52 compounds tested. Ro 32–3555, Trochate^{TM} of Roche (Fig. 23(s)) [938] selectively inhibits the three collagenases MMP-1, -8 and -13 relative to MMP-3, -2 and -9. It was effective in oral treatment of a rat arthritis model. Marimastat, BB-2516 (Fig. 23(t)), also gave positive results in animal models [840]. Batimastat has now been in clinical trials for some time. In phase I trials of cancer patients, the drug was given intraperitoneally to nine patients with various types of malignancy [1032]. There was some stabilization of the cancer, but the trial was discontinued due to abdominal pain and the availability of oral MMP inhibitors. A phase II trial on patients with ascites showed promising early results [983]. A review of six inhibitors from six companies currently in clinical trial is presented in [24].

Biological roles of the MMPs

Normal physiological processes

The contributions of MMPs to biological processes are difficult to assess. For the most part, our knowledge stems from correlations—that is, enzyme levels are high in a given situation and therefore are postulated to play an important role. The role cannot be proven by use of inhibitors, such as the natural TIMPs or the synthetic hydroxamates, because these do not, in general, inhibit only a single MMP. Moreover, we do not yet know any pathologies directly attributable to a mutation in one of the MMPs. It is not possible to survey the thousands of papers dealing with the biological roles of MMPs, so attention will be given to review articles in various areas of interest and these will be limited to the past three years only.

In an overview of the matrixin family, one of us [147] has presented a list of normal functions of MMPs grouped under the headings of reproduction, development and maintenance. A good deal of attention has been given to female reproductive processes: implantation, ovulation, mammary development, uterine involution, endometrial changes in the menstrual cycle, etc. A general overview of MMPs in reproduction is provided in [57]. Ovarian function [75], ovulation [128], mammary gland development and involution [137] and menstruation [69, 110] receive individual attention. With regard to development, there are reviews on implantation of the blastocyst [5], placental function [65], fetal membranes and their rupture [25, 98], enamel formation [120], kidney morphogenesis [124] and frog embryogenesis [119]. Topics under tissue maintenance include bone [99] and mineralization [48], normal skin function [62] as well as its ageing [138] and

wound healing [97], endocrinology [109] and nuclear receptors [112]. Other reviews discuss the importance of cell surface binding of MMPs and the consequent regulation of proteolysis [9, 31, 136].

Pathological processes involving MMPs

Areas of involvement of MMPs in pathological processes include tissue destruction, fibrotic diseases and weakening of the matrix [147]. The major topic of interest in this field is the invasion and metastasis of cancer. General overviews of MMPs and cancer within the past three years include [11, 29, 33, 39, 56, 102, 122]. The use of MMP inhibitors in treating cancers [24] and the role of angiogenesis [76, 78] may be included here. More specialized reviews discussing tumours restricted to certain parts of the body include pancreatic carcinomas [22, 23], head and neck carcinomas [21], gliomas [46, 134], brain tumours [107], gastric cancer [113] and melanomas [79]. One review is given over entirely to MMP-11 and cancer [10].

Arthritis is probably second in importance after cancer. Recent reviews include [20, 26, 27, 35]. Closely related to this is the topic of loosening of hip joint replacements due to MMP action [125, 126]. Oral pathology in general [106] and periodontal disease specifically are reviewed in [54, 108]. Under fibrotic diseases one may include atherosclerosis (and restenosis) which are reviewed in [67, 133, 148]. Infarction and heart disease have been reviewed [32, 131, 132]. Liver fibrosis [6] and renal fibrosis [40] also fall under the fibrotic diseases. Endometriosis is reviewed in [96].

Pathologies in which the ECM is weakened include heart failure [58] and aortic aneurysms [105, 127]. There are a number of reviews concerned with the central nervous system [47, 50, 149] and multiple sclerosis [30]. Haematological disorders have been reviewed by Guedez *et al.* [52].

Role of individual MMPs established by transgenic experiments

Under this heading one can place experiments that provide more direct evidence of the role of MMPs than do the method of correlation or the use of nonspecific inhibitors likely to affect several MMPs at once. Even these methods are not foolproof, they often affect other parts of the genome and, in the case of the MMPs, often lead to an up-regulation or down-regulation of unaffected MMPs and TIMPs as the organism attempts to restore a balance between synthesis and degradation. A good review of knockout results with MMPs is found in [116]. In this section, enzymes will be considered in numerical order rather than in the order of development of this branch of the field.

MMP-1

Human MMP-1 was introduced as a transgene under the control of the haptoglobin promoter into the mouse (which does not have its own MMP-1) [353a]. This led to expression of MMP-1 in the lungs with disruption of the alveolar walls and coalescence of alveolar spaces. In other words, a model was produced that very closely mimicked human pulmonary emphysema. A similar construct led to a strain of mice that overexpressed collagenase in the suprabasal level of the epidermis [354]. This produced skin disorders: acanthosis, hyperkeratosis and epithelial hyperplasia. Treating the skin with chemical agents to induce and pro-mote cancer resulted in a greatly increased incidence of skin cancer relative to control mice.

On the other hand, introduction of antisense RNA for MMP-1 into a human melanoma cell line resulted in cells that looked and grew as normally [355]. The cells produced normal amounts of gelatinase A; however, they had greatly inhibited ability to invade collagen I, IV or Matrigel. Introduction of a mutation in the $\alpha 1(I)$ chain of collagen at the Gly–Ile bond where collagenase normally cleaves, produced mice that could grow to adulthood, but that showed fibrosis of the skin similar to scleroderma and fibrotic changes in postpartum uterus [363]. While it was originally thought that such experiments showed the effect of an inability of MMP-1 to function, it turned out that mice lack MMP-1 and that they depend largely on MMP-13 to break down collagen. In these animals, the breakdown proved to be due largely to action of MMP-13 on N-telopeptides of collagen molecules which led to fibre dissolution [361].

MMP-2

Antisense RNA to MMP-2 was introduced into human glomerular mesangial cells, and also ribozymes acting on MMP-2 were introduced. In both cases, there was loss of inflammatory phenotype and departure from the cell cycle. Exogenous MMP-2 added to the medium restored the cells to their original phenotype [375]. A knockout mouse was produced which was shown to have a decreased response to B16-BL6 melanoma cells or Lewis lung carcinoma cells implanted intradermally or injected i.v. [360]. This indicates a role of MMP-2 in angiogenesis and tumour progression. Further work [359] showed that gelatinase A is not an α- or β-secretase with respect to β-amyloid processing.

MMP-3

Early experiments concerned the transgenic over-expression of the active form of stromelysin 1 in mice. These animals showed precocious development of mammary alveoli and in lactation there was loss of basement membrane, decreased alveolar size and little casein secretion [374]. There is a clear role of MMP-3 in remodelling the extracellular matrix during branching morphogenesis [381]. Mammary tumours in such animals showed apoptosis and fewer tumours developed [379]. A review of these studies is found in [377].

Rat pheochromocytoma cells transfected with antisense had less ability to penetrate Matrigel with their neurites [370]. Adenocarcinoma cells of mice lost their ability to invade Matrigel when transfected with antisense to MMP-3 [364]. A specific ribozyme that inhibits MMP-3 expression was injected into rabbit joints affected by IL-1 induced arthritis; this reduced MMP-3 mRNA in the synovial tissue.

An MMP-3 knockout mouse has been developed [368]. When collagen was injected to produce arthritis of the paws there was no change in loss of cartilage or in cleavage of aggrecan at the Asn341–Phe bond, indicating MMP-3 was not central to these processes. In the oestrous cycle and postpartum involution of the rat, MMP-3 is increased; in the knockout mouse there appears to be compensation by up-regulation of stromelysin 2 and matrilysin [372]. Evidence is found that the activation of MMP-2 is impaired in these knockouts [362]. Knockout of the Ets2 transcription factor DNA-binding domain leads to death of the embryos by 8.5 days; fibroblast growth factor fails to induce MMP-3 or MMP-13 [382].

MMP-7

Antisense transfection of a human colon cancer cell line producing MMP-7 reduced tumourigenicity, whereas transfection with active MMP-7 in another line not producing MMP-7 increased tumourigenicity [380]. Similar experiments with over- or under-expression of MMP-7 in other colon cancer cell lines led to increased and decreased invasiveness through Matrigel [383]. It was found [356] that antisense to MMP-7 in human colon cancer cells inhibited metastasis to the liver of nude mice and reduced invasive potential [367].

Knockout mice deficient in MMP-7 were followed through the oestrous cycle and postpartum involution of the uterus. There was a considerable up-regulation of MMP-3 and MMP-10 as a compensatory mechanism [372]. Tumourigenesis was suppressed in these animals [378]. Mice that overexpressed human MMP-7 under control of the mouse mammary tumour promoter were found to have normal-appearing mammary glands, but these produced casein in virgin animals [371], the testes showed disintegration of interstitial tissue around the seminiferous tubules.

Other MMPs

Introduction of a hammerhead ribozyme that inhibited MMP-9 expression blocked metastasis in a rat sarcoma model [358], and antisense RNA reduced the ability of malignant human lung cells to invade Matrigel [366]. Gelatinase B knockout mice show abnormal growth plate vascularization and ossification; a delay in growth led to eightfold lengthening of the growth plate [376]. Antisense RNA for MMP-11 was introduced into 3T3 cells and reduced their tumourigenicity. On the other hand, MCF7 cells, not normally expressing MMP-11, were transfected to produce the enzyme; they developed more tumours in a shorter time [369]. A knockout for MMP-11 resulted in a phenotype that was resistant to tumour induction by dimethylbenzanthracene [365]. A knockout mouse of macrophage elastase has been developed [373], macrophages from these mice have reduced ability to digest the extracellular matrix and cannot penetrate Matrigel. Inhalation of cigarette smoke by these mice did not lead to emphysema, whereas control mice developed the condition [357]. Various tumour cell lines were transfected with MMP-14; higher levels of enzyme led to higher activation of gelatinase A and higher invasion of Matrigel by the cells [384]. Partial activation of MMP-2 led to higher effects than low levels or complete activation.

In summary, no MMP knockout has been lethal

[116]. This is attributed to the redundancy of the MMPs such that other members of the family increase in expression to make up the deficit in the missing member. On the other hand, suppression of expression of a number of the MMPs reduces the ability of cells to invade Matrigel membrane, suggesting that several MMPs participate and that the process goes poorly if all are not present. It will probably be necessary to perturb the MMPs two or three at a time to see more striking effects.

Functions of the TIMPs

Inhibition of MMP activity

It is generally believed by those in the field that inhibition of proteolytic activity is the major function of the TIMPs. The growth effects are not widely recognized and have not received nearly the attention given to enzyme inhibition. The first point to be considered is whether all TIMPs inhibit all MMPs, and whether this is quantitatively the same for all cases. A brief overview is provided by Table 18, which shows all cases of inhibition that have been tested to date. This table is qualitative and only one case is found where inhibition was not observed. In all cases studied to date, the inhibition is stoichiometric, one molecule of TIMP inhibits one molecule of active MMP, as might be expected from the mechanism revealed by X-ray crystallography [428].

In the earliest days, TIMP-1 was observed to inhibit MMP-1, -2, -3 [479] and -8 [491]. Purified TIMP-1 was shown to inhibit MMP-1 with 1:1 stoichiometry [495]. The name TIMP was not coined until 1981 [401] when rabbit TIMP was shown to inhibit the four MMPs above plus MMP-9. The inhibition of neutrophil MMP-8 and -9 went slowly and it was difficult to get to 100% inhibition [463]. Highly purified MMP-3 reacted well with TIMP [423]. Human TIMP-1 was extracted directly from cartilage and showed typical properties: sensitivity to trypsin digestion and to reduction and alkylation, stability to heat and low pH, ability to bind to concanavalin A [413]; this preparation inhibited rat MMP-7 and MMP-13. Human MMP-7 is also inhibited [409, 475]. An early purification of rabbit MMP-2 led to demonstration of 1:1 binding to TIMP [461]. Albin *et al.* [387] showed that human alveolar macrophages produced both elastase (MMP-12) and TIMP-1 and these could interact.

At the time TIMP-MMP inhibition was being investigated, properties of TIMP were also under scrutiny. Con A binding is due to glycosylation of TIMP-1 [484, 487]. The complex of TIMP to MMP is not due either to cleavage of a bond in TIMP nor to covalent bond formation, as is common for serine proteinase inhibitors. The complex can be dissociated by acid pH [451] or EDTA at low pH [460]; active TIMP is recovered, but the enzyme activity is usually denatured. Dissociation is also accomplished by passing through a monoclonal anti-MMP-1 antibody column [504]. A nonprotein factor of low M_r (ESAF, endothelial cell stimulating angiogenesis factor) has been reported to free active MMP-1 and TIMP-1 from their complex [456], but the exact nature of the factor has never been established. If a hydroxamate inhibitor is first bound to the enzyme, this occludes the active centre and TIMP cannot bind [452]. TIMP-1 efficacy has been compared to that of α_2-macroglobulin on MMP-1 using mixtures of the two inhibitors [402]: MMP-1 will bind preferentially to α_2-macroglobulin, but if it is first bound to TIMP, it will not exchange. The assignment of the six disulfide bonds in TIMP-1 has been achieved [500]; this clarified the great heat stability of the molecule and its sensitivity to reduction.

In recent times there have been further discoveries of MMPs, generally accompanied by testing their response to TIMP. Stromelysin 2 and 3 (MMP-10 and -11) are blocked [464, 501]. MMP-14 is normally on the cell surface; its ability to attack matrix underlying the cell or to activate MMP-2 on the cell surface is not blocked by TIMP-1 [411, 440]. MMP-16, in a form lacking the transmembrane domain, is blocked by TIMP-1 but only 1/10 as well as by TIMP-2 [454]. Envelysin from sea urchins is blocked by TIMP-1 [472] as is the soybean metalloproteinase [455].

Table 18 Inhibition of matrix metalloproteinases by TIMPs

MMP No.	TIMP-1	TIMP-2	TIMP-3	TIMP-4
MMP-1	X	X	X	X
MMP-2	X	X	X	X
MMP-3	X	X	X	X
MMP-7	X	X	X	X
MMP-8	X	X	–	–
MMP-9	X	X	X	X
MMP-10	X	X	–	–
MMP-11	X	–	–	–
MMP-12	X	–	–	–
MMP-13	X	X	X	–
MMP-14	No	X	X	–
MMP-15	–	X	X	–
MMP-16	X	X	–	–
MMP-17	–	–	–	–
MMP-18	–	–	–	–
MMP-19	–	X	–	–
MMP-20	–	–	–	–
XMMP	–	–	–	–
Envelysin	X	–	–	–
Soybean	X	–	–	–

X, inhibition has been observed; –, not studied.

The existence of a second inhibitor (TIMP-2) was clearly established about 1989. Early studies showed that it inhibited MMP-1, -2, -3 and -7 [385, 412, 489]. This inhibitor also blocks MMP-8 [446], MMP-9 [414], MMP-10 [501], MMP-13 [445], MMP-14 [507], MMP-15 [400], MMP-16 [454] and MMP-19 [474].

TIMP-3 was shown in 1995 to block MMP-1, -2, -3 and -9 [390]; its effect on MMP-9 appears to be stronger than that of TIMP-1 or -2 [417]. It blocks MMP-13 [445], MMP-14 [497] and MMP-15 [400]. It is recognized that MMPs can act as sheddases, e.g. as TNFα converting enzymes [45], and that TIMPs will block such action. However, there is a recent observation that TIMP-3, but not TIMP-1 or -2, can block the shedding of the IL-6 receptor [429]; whether this action is due to inhibiting an MMP or whether TIMP-3 is able to block an enzyme from the astacin family remains to be seen. TIMP-4 action on MMPs has been studied by only one group [453], it inhibits MMP-1, -2, -3, -7 and -9.

In considering the inhibition of MMPs by TIMPs, it should not be overlooked that most of the MMPs undergo an autolytic cleavage step during their activation by mercurials or other proteinases. One would expect the TIMPs to be able to block this, often final, step in activation because the zinc atom has become exposed. Just to mention a few cases: TIMP-2 can inhibit the activation step of MMP-1, -2, -3 and -7 [489]. Oddly, TIMP-1 does not seem to block the autolytic cleavage of MMP-8 at the Asn53–Val54 link but does block the final cleavage at Phe79–Met80 [397]. Possibly, the propeptide is still of sufficient size to exclude the inhibitor at the first step. Further information on this role of TIMPs may be found in the section on proenzyme activation and in the review of Nagase *et al.* [94].

What domains of the enzyme and inhibitor are involved in TIMP inhibition?

It was noted that collagenase could undergo fragmentation producing the catalytic domain and the hemopexin domain [405]; the catalytic domain was first reported not to be inhibited by TIMP-1. Later refinement of these studies, however, showed that the catalytic domain of MMP-1 is inhibited by TIMP-1, but there is a considerable stabilizing effect of the hemopexin domain on this binding [396]. Collagenase 2 catalytic domain can still bind TIMP-1 and TIMP-2 [446]. Within the catalytic domain of MMP-1, it is necessary to have the zinc binding site for TIMP-1 attachment, deletion of this binding region eliminates TIMP attachment [490]. MMP-1 and -3 with deleted hemopexin domains were compared to MMP-7 which naturally lacks the domain, all had a relatively low rate of MMP–TIMP association, but MMP-7 was poorest in terms of ability to form complexes stable to SDS [502]. MMP-9 may lose its hemopexin domain in some leukaemia cells, it is then more poorly inhibited by TIMP-1, which might contribute to invasiveness of the cells [476]. On the other hand, the fibronectin domain of gelatinase (MMP-2) can be deleted without effect on binding of TIMP-1 or -2 [462].

With respect to the domains of TIMP-1, the N-terminal portion (residues 1–126) possess the MMP inhibitor activity; it binds and inhibits active MMP-1, -2, -3, -7 and -9 [459]. The inhibitory power is weaker than wild-type TIMP on MMP-2 and about equal on MMP-3 and -9. With MMP-10 an SDS-stable complex is still produced, a property unique to that enzyme (when studied under the conditions used in this particular laboratory, cf. [502] immediately above). Similar results have been found for TIMP-2: the N-terminal domain possesses the inhibitory activity [415] and this binding has been characterized by NMR [499]. The C-terminal domain does not bind tightly to MMPs [447]. With regard to the role of the C-domain in MMP inhibition, particularly of gelatinase A [139], the enhancement of binding by this region is attributed to holding down the C-end of TIMP on the hemopexin domain in such a way that the N-end is properly aligned with the active site of the enzyme. This is supported by observations of conformational change of MMP-2 when TIMP-1 binds, a shift not seen when hydroxamate inhibitors bind [408]. Of course, the C-terminal portion

of certain TIMPs is involved in binding to proenzymes of MMP-2 and -9, as discussed below. TIMP-3 is distinguished from the other TIMPs by being firmly anchored to the extracellular matrix. This binding is not seen in C-truncated TIMP-3 [450] and binding is greatly reduced if the C-terminal from TIMP-2 is substituted in chimeric form. Also, TIMP-3 lacking its C-domain is no longer able to inhibit MMP-9 although it retains some action on MMP-2 [469].

Quantifying the interactions of TIMPs with MMPs

This is a difficult problem; different animal species are used as sources of MMPs and TIMPs, different assay methods are employed and different levels of enzyme are examined. This makes comparisons of results between laboratories difficult. Moreover, the tight binding of TIMP leads to dissociation constants that are at the edge of possible measurement, so that many groups just report the on-rate of association of enzyme and inhibitor. Some also report an IC_{50} value which is of limited value because of the 1:1 stoichiometry of the complex.

The first measurements [494] showed a value of $K_i < 10^{-9}$ M for MMP-1–TIMP-1, and for rabbit MMP-1 and TIMP-1, a value of $K_D = 1.4 \times 10^{-10}$ M [403]. A similar value was obtained for MMP-3: 3.8×10^{-10} M [37]. A more qualitative approach [492] involved the preparation of complexes of TIMP-1 or TIMP-2 with various MMPs, which were passed through a molecular sieve column to see if there was any dissociation. MMP-1, -2, -3 and -9 remained firmly complexed, but MMP-7 dissociated from TIMPs. The kinetics of TIMP-1–MMP-1 binding has been studied in detail [488]. Inhibition proceeds in two steps: a rapid reversible complex forms with $K_D = 8$ nM followed by slow formation of a tight complex. The overall $K_D = 0.1$ nM. This shifts to 2.7 nM if the hemopexin domain of MMP-1 is absent. A similar shift of K_i from <0.2 nM to 1.2 nM has been reported [458]. In the case of MMP-3, removal of the hemopexin domain caused a change in K_i from 8.3×10^{-10} M to 6.0×10^{-9} M [393]. The shift with MMP-2 was from $K_i = 2$ pM to $K_i = 63$ pM [468], and for MMP-8 from 2.8×10^{-9} M to 18×10^{-9} M [441]. These changes emphasize what was said above about the hemopexin domain assisting in the binding of TIMP to enzyme.

A comparison of TIMP-1 with TIMP-2 has been carried out on MMP-2 and MMP-9 [432]. The IC_{50} for reaction with active MMP-2 was 0.4 and 0.5 nM for TIMP-1 and -2, respectively. However, with the 45-kDa fragment of MMP-2 (lacking hemopexin) the value for TIMP-1 was 20 times higher than that for TIMP-2; with MMP-9 the ratio was about 6. In comparison, MMP-7 gave a $K_i = 370$ pM for TIMP-1 versus 66 pM for TIMP-2 [467]. TIMP-1, -2 and -3 have been compared on MMP-13; the k_{on} rate ($M^{-1} s^{-1} \times 10^{-6}$) was 7.8, 1.4 and 10.2 for the three forms, respectively [444]; these rates diminished by a factor of 17, 2 and 28 when the hemopexin domain was removed. The IC_{50} values for TIMP-4 acting on several MMPs are 19, 3, 45, 8 and 83 nM for MMP-1, -2, -3, -7 and -9, respectively [453].

The removal of the C-terminal domain of TIMP-1 had little effect on its inhibition of MMP-3 ($k_{on} = 1\,900\,000\,M^{-1}\,s^{-1}$ decreased to $1\,117\,000\,M^{-1}\,s^{-1}$) [85]. However, with MMP-9 the decrease was from $11\,100\,000\,M^{-1}\,s^{-1}$ to $80\,000\,M^{-1}\,s^{-1}$, while matrilysin showed no change. When TIMP-2 was truncated k_{on} went from $300\,000\,M^{-1}\,s^{-1}$ to $5000\,M^{-1}\,s^{-1}$ for MMP-3 and showed no change with MMP-9 [466]. As discussed below, TIMP-2 forms a complex with latent MMP-2 through its hemopexin domain. If one adds TIMP-2 to active MMP-2 this interaction still occurs and increases k_{on}. Thus, $k_{on} = 3.4 \times 10^{-6}\,M^{-1}\,s^{-1}$ for TIMP-1 inhibition and 29.4 for TIMP-2 [498]. This value falls to $4.0 \times 10^{-6}\,M^{-1}\,s^{-1}$ if the hemopexin domain of MMP-2 is removed and to 0.157 if the C-domain of TIMP-2 is removed. Interestingly, just the last nine amino acids of the C-end of TIMP-2 are important in determining k_{on}, and their removal has the same effect as deleting the hemopexin domain [498]. Many of these issues concerning domains in MMP-2 and -9 and TIMP-1 and -2

have been explored recently using plasmon resonance methods; an internally consistent set of values of K_D and K_i appears to result but the values are much higher than those reported above as obtained by other methods. This has been attributed to the need to immobilize the enzyme for the measurements [433].

In sum, there are very few exceptions to the rule that all TIMPs inhibit all MMPs; however, there are wide variations in the efficacy of the different TIMPs with respect to each MMP. The importance of these differences in the regulation of MMP activity in various tissues and disease processes remains to be explored.

Binding of TIMPs at sites other than the active centre of MMPs

We have already discussed the effect of interaction of the C-domain of TIMP with the C-domain of MMPs as a means to orient the inhibitory end of TIMP toward the active centre, thereby increasing the rate of enzyme–inhibitor association. However, in addition to this general phenomenon, there are special cases of interaction involving TIMP-2 and the proenzyme form of MMP-2, TIMP-1 and the zymogen of MMP-1, and TIMP-2 and MMP-14 in relation to cell surface activation of MMP-2. These three cases are all discussed in the section on enzyme activation and will be reviewed only briefly here.

The binding of TIMP-2 to proMMP-2 was first reported in 1989 [425, 483]. Examination of a mixture of the two proteins showed that 79% consisted of the proMMP-2–TIMP-2 complex and 21% was a tetramer of two molecules of each [443]. This binding to proenzyme did not compromise the ability of TIMP to inactivate a molecule of active MMP [448], indicating that the amino end was not involved in the binding. It is not possible to activate proMMP-2 fully when it is complexed to TIMP-2, e.g. APMA leads to only 5–10% of the activity seen with inhibitor-free enzyme [448,

505]. This is attributed to interaction of the amino end of TIMP-2 with the active centre. The K_D for this interaction of TIMP-2 and active MMP-2 is 0.2×10^{-12} M [433]. The complex appears to protect MMP-2 from autodegradation [442]. Binding of TIMP-2 to the hemopexin domain is not dependent on the active centre [410] but does depend on the C-terminal nine residues of TIMP-2 [498].

Similarly the binding of TIMP-1 to proMMP-9 was reported also in 1989 [465, 496]. Gelatin is able to interfere with this complex formation [465]. Formation of the complex prevents dimerization of proMMP-9, complexation with MMP-1 and activation by MMP-3 [426]; these effects are due to binding of TIMP-1 to the hemopexin domain [486]. The complex can also inhibit other MMPs, such as MMP-3 lacking its hemopexin domain [449]. The interaction with proMMP-9 leaves the TIMP-1 molecule susceptible to attack by neutrophil elastase, which can be used to remove the TIMP and permit activation by MMP-3 [435]. Further details of the complex are found in the section on activation.

Finally, proMMP-2 and TIMP-2 appear to interact with the cell surface, thus anchoring the proenzyme for activation by the membrane type metalloproteinase, MMP-14 [406, 418]. It is likely that there is a three-way complex involving the two MMPs and TIMP, with TIMP linking proMMP-2 to active MMP-14 [434, 478, 485]. It is suggested that a TIMP-2–MMP-14 complex acts as a cell receptor for proMMP-2 [399]. It has recently been found that TIMP-3 can also participate in this process much the same as TIMP-2 [497].

Effects of TIMPs on cell growth

Although this topic is not directly relevant to the role of TIMPs in regulating the activity of MMPs, brief mention will be made. It is necessary to distinguish direct growth effects of TIMPs on cells as contrasted with effects of protection of the extracellular matrix which might directly influence cell growth [41]. Thus, bFGF signalling to the cell involves interaction with heparan sulfate proteoglycans; protecting these from degradation might enhance cell growth. Nonetheless, there appear to be many genuine examples of TIMP effects directly on cells.

There has been considerable interest in erythroid potentiating factor produced by T-lymphocytes [427]. Numerous studies culminated in purification and sequencing of the factor [424]. A few months later, TIMP was successfully sequenced and, surprisingly, turned out to be the same protein [416]. Those studying growth effects had not noticed proteinase inhibition and vice versa. In fact, there was some reluctance in the MMP field to accept the growth-promoting properties of TIMP. However studies soon showed effects of TIMP-1 on growth of keratinocytes [395] and on a wide range of bovine and human cell types [431]. Classical cell receptors were found [391, 394]. A fibroblast elongation factor proved to be TIMP-1 [386], and a factor that stimulated steroid hormone synthesis in rat testis proved to be a complex of TIMP-1 with procathepsin L [398]. TIMP-1 stimulated rabbit epithelial cell proliferation [477] and acted as an autocrine growth factor in human scleroderma fibroblasts [439].

TIMP-2, although first believed not to promote growth [431], was soon shown to possess erythroid potentiating activity [482]. However, a wide range of human, cow and mouse cell lines that are stimulated by TIMP-2 have been found [430]. SV40-transformed fibroblasts produced a factor stimulating growth of normal fibroblasts; this proved to be TIMP-2 [470]. In the case of four human cell lines, the growth effect of TIMP-2 was dependent on insulin in the medium; without insulin the effect was inhibitory [471].

The growth effects were shown in various ways to be independent of proteinase inhibition. Thus, a mutation of TIMP-1 that was no longer inhibitory still promoted growth [404]. Removal of the C-terminal domain also did not destroy growth activity. The complexed forms TIMP-1–MMP-9 and TIMP-2–MMP-2 failed to show growth effects [430], and destruction of TIMPs by reduction and alkylation did not eliminate growth effects. The growth signalling of both TIMP-1 and TIMP-2 involves protein phosphorylation [503], which in turn depends on a cAMP-dependent protein kinase [407]. Not all cells respond to all TIMPs; thus, fibrosarcoma cells HT-1080 are stimulated by TIMP-2 but not by TIMP-1 [407]. Effects can even be negative: TIMP-2 inhibits the growth of human microvascular endothelial cells, but TIMP-1 has no effect and the synthetic inhibitor of MMPs, BB-94, also has no effect [457]. Finally, TIMP-3 overexpression has been found to induce DNA synthesis in rat aortic smooth muscle cells, but this leads on to apoptosis. Direct addition of TIMP-3 has the same effect, but the MMP inhibitor Ro-31–9370 and TIMP-1 and -2 had no effect [392]. It is possible [480] that apoptosis is induced due to protection of cell surface TNFα receptors from MMP degradation; but this is not consistent with the failure of other TIMPs and inhibitors to produce the effect.

Implication of TIMPs in biological processes

There is not space to fully develop this topic; TIMPs are important in malignancy, regulation of cell morphology and organ morphogenesis, inhibition of angiogenesis, steroidogenesis and tissue remodelling of all sorts (see recent review [49]). In establishing the biological role of TIMPs, it is important to see if there are any effects observed when the gene is deleted or transfected to give overexpression. The first efforts in this direction were experiments in which mouse 3T3 cells were transfected to express antisense RNA to TIMP-1 [438]. This reduced cell secretion of TIMP and led to development of oncogenicity—the cells became tumourigenic and metastatic. There is evidence that the effect was due in part to the cells producing matrix components that favoured oncogenicity [437]. On the other hand, if mouse melanoma B16-F10 cells were transfected to overproduce TIMP-1, this suppressed tumourigenic and metastatic properties of the cells [436]. A

different approach was to disrupt the gene in an embryonic cell line (TIMP-1 is on the X chromosome); the result was increased invasiveness of the cells in basement membrane test systems [389]. Addition of exogenous TIMP-1 reversed this effect.

Transgenic mice have been developed that overexpress TIMP-1 or stromelysin. If stromelysin is overexpressed, the fetal alveolar epithelial cells undergo unscheduled apoptosis. If this strain of mice is crossbred with the overexpressers of TIMP-1 the effect is abolished. The apoptosis is attributed to overdigestion of entactin in the matrix around the cells [388]. If lymphoma cells are injected in the skin of mice, they develop tumours and metastasize to the liver. If mice are overexpressers of TIMP-1, the tumours are decreased and there is less metastasis. However, if they are overexpressers, there is a doubling of metastatic foci and shortened survival time. In general, then, the results of TIMP expression appear to be related to their ability to block MMP activity. However, there are other effects, as well. Thus, when TIMP-1 is lacking in mice, their ovaries produce less TIMP-2 and TIMP-3 during normal cycling [473]. In rat breast carcinoma cell lines, overexpression of TIMP-1 leads to increased growth of tumours accompanied by up-regulation of VEGF [506]. If different types of tumour cells are made deficient in TIMP-1 and then injected in mice, the results may be either enhanced or diminished tumourigenesis dependent on cell type. However, if the cells are injected in TIMP-1 deficient mice, the results are unaffected, indicating that the tumour cell genotype and not the host genotype is the major determinant [481].

TIMP-3 is the only TIMP (or MMP) this far associated with a disease, namely, Sorsby's fundus dystrophy. This is due to point mutations in TIMP-3 [493] including Ser156Cys [422], Tyr168Cys [421] and Ser181Cys [420]. TIMP-3 is found to deposit within Bruch's membrane [419] in the form of dimers, presumably linked by S–S bridges [450]. These dimers have not lost their ability to inhibit MMPs, but they retain the property of TIMP-3 of binding to the extracellular matrix.

Bibliography

Books

1. **Birkedal-Hansen, H., Z. Werb, H. G. Welgus, and H. E. Van Wart.** 1992. *Matrix metalloproteinases and inhibitors.* 504 pp., Gustav Fischer, Stuttgart.
2. **Parks, W. C. and R. P. Mecham.** 1998. *Matrix metalloproteinases.* 362 pp., Academic Press, San Diego.
3. **Barrett, A. J., N. D. Rawlings, and J. F. Woessner.** 1998. *Handbook of proteolytic enzymes.* 1666 pp., Academic Press, London.

Reviews

4. **Alexander, C. M. and Z. Werb.** 1989. Proteinases and extracellular matrix remodeling. *Curr. Opin. Cell. Biol.* **1**: 974–982.
5. **Aplin, J. D.** 1996. The cell biology of human implantation. *Placenta* **17**: 269–275.
6. **Arthur, M. J.** 1997. Matrix degradation in liver: a role in injury and repair. *Hepatology* **26**: 1069–1071.
7. **Barrett, A. J.** 1981. α_2-Macroglobulin. *Methods Enzymol.* **80**: 737–754.
8. **Barrett, A. J.** (ed.) 1995. Methods in Enzymology. Vol. 248. Proteolytic Enzymes: Aspartic and Metallo Peptidases. Academic Press, San Diego.
9. **Basbaum, C. B. and Z. Werb.** 1996. Focalized proteolysis: spatial and temporal regulation of extracellular matrix degradation at the cell surface. *Curr. Opin. Cell. Biol.* **8**: 731–738.
10. **Basset, P., J. P. Bellocq, O. Lefebvre, A. Noël, M. P. Chenard, C. Wolf, P. Anglard, and M. C. Rio.** 1997. Stromelysin-3: a paradigm for stroma-derived factors implicated in carcinoma progression. *Crit. Rev. Oncol. Hematol.* **26**: 43–53.
11. **Basset, P., A. Okada, M. P. Chenard, R. Kannan, I. Stoll, P. Anglard, J. P. Bellocq, and M. C. Rio.** 1997. Matrix metalloproteinases as stromal effectors of human carcinoma progression: therapeutic implications. *Matrix Biol.* **15**: 535–541.
12. **Beckett, R. P.** 1996. Recent advances in the field of matrix metalloproteinase inhibitors. *Expert Opin. Therap. Pat.* **6**: 1305–1315.
13. **Beckett, R. P., A. H. Davidson, A. H. Drummond, P. Huxley, and M. Whittaker.** 1996. Recent advances in matrix metalloproteinase inhibitor research. *Drug Discovery Today* **1**: 16–26.
14. **Beckett, R. P. and M. Whittaker.** 1998. Matrix metalloproteinase inhibitors 1998. *Expert Opin. Therap. Pat.* **8**: 259–282.
15. **Beeley, N. R. A., P. R. J. Ansell, and A. J. P. Docherty.** 1994. Inhibitors of matrix metalloproteinases (MMP's). *Curr. Opin. Ther. Pat.* 4: 7–16.
16. **Birkedal-Hansen, H., C. M. Cobb, R. E. Taylor, and H. M. Fullmer.** 1975. Activation of latent bovine gingival collagenase. *Arch. Oral Biol.* **20**: 681–685.
17. **Birkedal-Hansen, H., W. G. Moore, M. K. Bodden, L. J. Windsor, B. Birkedal-Hansen, A. DeCarlo, and J. A. Engler.** 1993. Matrix metalloproteinases: a review. *Crit. Rev. Oral Biol. Med.* **4**: 197–250.
18. **Bode, W.** 1995. A helping hand for collagenases: the haemopexin-like domain. *Structure* **3**: 527–530.
19. **Bode, W., F. X. Gomis-Rüth, and W. Stöcker.** 1993. Astacins, serralysins, snake venom and matrix metalloproteinases exhibit identical zinc-binding environments (HEXXHXXGXXH and Met-turn) and topologies and should be grouped into a common family, the 'metzincins'. *FEBS Lett.* **331**: 134–140.
20. **Bottomley, K. M., W. H. Johnson, and D. S. Walter.** 1998. Matrix metalloproteinase inhibitors in arthritis. *J. Enz. Inhib.* **13**: 79–101.
21. **Boyd, D.** 1996. Invasion and metastasis. *Cancer Metastasis Rev.* **15**: 77–89.
22. **Bramhall, S. R.** 1997. The matrix metalloproteinases and their inhibitors in pancreatic cancer. From molecular science to a clinical application. *Int. J. Pancreatol.* **21**: 1–12.
23. **Bramhall, S. R., J. P. Neoptolemos, G. W. H. Stamp, and N. R. Lemoine.** 1997. Imbalance of expression of matrix metalloproteinases (MMPs) and tissue inhibitors of the matrix metalloproteinases (TIMPs) in human pancreatic carcinoma. *J. Pathol.* **182**: 347–355.
24. **Brown, P. D.** 1997. Matrix metalloproteinase inhibitors in the treatment of cancer. *Med. Oncol.* **14**: 1–10.
25. **Bryant-Greenwood, G. D.** 1998. The extracellular matrix of the human fetal membranes: structure and function. *Placenta* **19**: 1–11.
26. **Cawston, T.** 1998. Matrix metalloproteinases and TIMPs—properties and implications for the rheumatic diseases. *Mol. Med. Today* **4**: 130–137.
27. **Cawston, T. E.** 1996. Metalloproteinase inhibitors and the prevention of connective tissue breakdown. *Pharmacol. Ther.* **70**: 163–182.
28. **Cawston, T. E.** 1998. Interstitial collagenase. In *Handbook of Proteolytic Enzymes.* A. J. Barrett, N. D. Rawlings, and J. F. Woessner (eds), pp. 1155–1162. Academic Press, London.
29. **Chambers, A. F. and L. M. Matrisian.** 1997. Changing views of the role of matrix metalloproteinases in metastasis. *J. Natl. Cancer Inst.* **89**: 1260–1270.

30. **Chandler, S., K. M. Miller, J. M. Clements, J. Lury, D. Corkill, D. C. Anthony, S. E. Adams, and A. J. Gearing.** 1997. Matrix metalloproteinases, tumor necrosis factor and multiple sclerosis: an overview. *J. Neuroimmunol.* **72**: 155–161.

31. **Chen, W. T.** 1996. Proteases associated with invadopodia, and their role in degradation of extracellular matrix. *Enzyme Protein* **49**: 59–71.

32. **Cleutjens, J. P.** 1996. The role of matrix metalloproteinases in heart disease. *Cardiovasc. Res.* **32**: 816–821.

33. **Cockett, M. I., G. Murphy, M. L. Birch, J. P. O'Connell, T. Crabbe, A. T. Millican, I. R. Hart, and A. J. Docherty.** 1998. Matrix metalloproteinases and metastatic cancer. *Biochem. Soc. Symp.* **63**: 295–313.

34. **Collier, I. E. and G. I. Goldberg.** 1998. Gelatinase B. In *Handbook of Proteolytic Enzymes.* A. J. Barrett, N. D. Rawlings, and J. F. Woessner (eds), pp. 1305–1210. Academic Press, London.

35. **Cunnane, G., K. M. Hummel, U. Muller-Ladner, R. E. Gay, and S. Gay.** 1998. Mechanism of joint destruction in rheumatoid arthritis. *Arch. Immunol. Ther. Exp. (Warsz.)* **46**: 1–7.

36. **Dioszegi, M., P. Cannon, and H. E. Van Wart.** 1995. Vertebrate collagenases. *Methods Enzymol.* **248**: 413–431.

37. **Docherty, A. J. P. and G. Murphy.** 1990. The tissue metalloproteinase family and the inhibitor TIMP: a study using cDNAs and recombinant proteins. *Ann. Rheum. Dis.* **49**: 469–479.

38. **Douglas, D. A., Y. E. Shi, and Q. X. A. Sang.** 1997. Computational sequence analysis of the tissue inhibitor of metalloproteinase family. *J. Protein Chem.* **16**: 237–255.

39. **Duffy, M. J. and K. McCarthy.** 1998. Matrix metalloproteinases in cancer—prognostic markers and targets for therapy. *Int. J. Oncol.* **12**: 1343–1348.

40. **Eddy, A. A.** 1996. Molecular insights into renal interstitial fibrosis. *J. Am. Soc. Nephrol.* **7**: 2495–2508.

41. **Edwards, D. R., P. P. Beaudry, T. D. Laing, V. Kowal, K. J. Leco, P. A. Leco, and M. S. Lim.** 1996. The roles of tissue inhibitors of metalloproteinases in tissue remodelling and cell growth. *Int. J. Obes. Relat. Metab. Disord.* **20**: S9-S15

42. **Emonard, H. and J.-A. Grimaud.** 1990. Matrix metalloproteinases. A review. *Cell. Mol. Biol.* **36**: 131–153.

43. **Fowlkes, J. L., K. M. Thrailkill, D. M. Serra, K. Suzuki, and H. Nagase.** 1995. Matrix metalloproteinases as insulin-like growth factor binding protein-degrading proteinases. *Prog. Growth Factor Res.* **6**: 255–263.

44. **Gache, C., T. Lepage, C. Ghiglione, F. Emily-Fenouil, and G. Lhomond.** 1998. Envelysin. In *Handbook of Proteolytic Enzymes.* A. J. Barrett, N. D. Rawlings, and J. F. Woessner (eds), pp. 1195–1199. Academic Press, London.

45. **Gearing, A. J., P. Beckett, M. Christodoulou, M. Churchill, J. M. Clements, M. Crimmin, A. H. Davidson, A. H. Drummond, W. A. Galloway, R. Gilbert,** *et al.* 1995. Matrix metalloproteinases and processing of pro-TNF-α. *J. Leukoc. Biol.* **57**: 774–777.

46. **Giese, A. and M. Westphal.** 1996. Glioma invasion in the central nervous system. *Neurosurgery* **39**: 235–250.

47. **Giraudon, P., S. Buart, A. Bernard, N. Thomasset, and M. F. Belin.** 1996. Extracellular matrix-remodeling metalloproteinases and infection of the central nervous system with retrovirus human T-lymphotropic virus type I (HTLV-I). *Prog. Neurobiol.* **49**: 169–184.

48. **Golub, E. E.** 1996. Enzymes in mineralizing systems: state of the art. *Connect. Tissue Res.* **35**: 183–188.

49. **Gomez, D. E., D. F. Alonso, H. Yoshiji, and U. P. Thorgeirsson.** 1997. Tissue inhibitors of metalloproteinases—structure, regulation and biological functions. *Eur. J. Cell. Biol.* **74**: 111–122.

50. **Gottschall, P. E. and S. Deb.** 1996. Regulation of matrix metalloproteinase expressions in astrocytes, microglia and neurons. *Neuroimmunomodulation* **3**: 69–75.

51. **Greenwald, R. A. and L. M. Golub.** 1994. *Inhibition of matrix metalloproteinases: therapeutic potential.* Vol. **732**, 507 pp. New York Academy of Sciences, New York.

52. **Guedez, L., M. S. Lim, and W. G. Stetler-Stevenson.** 1996. The role of metalloproteinases and their inhibitors in hematological disorders. *Crit. Rev. Oncogen.* **7**: 205–225.

53. **Hagmann, W. K., M. W. Lark, and J. W. Becker.** 1996. Inhibition of matrix metalloproteinases. *Ann. Rep. Med. Chem* **31**: 231–240.

54. **Havemose-Poulsen, A. and P. Holmstrup.** 1997. Factors affecting IL-1-mediated collagen metabolism by fibroblasts and the pathogenesis of periodontal disease: a review of the literature. *Crit. Rev. Oral Biol. Med.* **8**: 217–236.

55. **Hayakawa, T.** 1994. Tissue inhibitors of metalloproteinases and their cell growth-promoting activity. *Cell. Struct. Funct.* **19**: 109–114.

56. **Heino, J.** 1996. Biology of tumor cell invasion: interplay of cell adhesion and matrix degradation. *Int. J. Cancer* **65**: 717–722.

57. **Hulboy, D. L., L. A. Rudolph, and L. M. Matrisian.** 1997. Matrix metalloproteinases as mediators of reproductive function. *Mol. Hum. Reprod.* **3**: 27–45.

58. **Janicki, J. S., S. C. Tyagi, S. E. Campbell, H. K. Reddy, and J. R. Henegar.** 1995. Progressive ventricular dilatation in heart failure: The role of myocardial collagenase. *Dev. Cardiovasc. Med.* **169**: 261–273.

59. **Jeffrey, J. J.** 1998. Collagenase 3. In *Handbook of Proteolytic Enzymes.* A. J. Barrett, N. D. Rawlings, and J.

F. Woessner (eds), pp. 1167–1170. Academic Press, London.

60. **Jeffrey, J. J.** 1998. Interstitial collagenases. In *Matrix Metalloproteinases*. W. C. Parks and R. P. Mecham (eds), pp. 15–42. Academic Press, San Diego.

61. **Johnson, W. H., N. A. Roberts, and N. Borkakoti.** 1987. Collagenase inhibitors: their design and potential therapeutic use. *J. Enz. Inhib.* **2**: 1–22.

62. **Kahari, V. M. and U. Saarialho-Kere.** 1997. Matrix metalloproteinases in skin. *Exp. Dermatol.* **6**: 199–213.

63. **Knäuper, V. and G. Murphy.** 1998. Membrane-type matrix metalloproteinases and cell surface-associated activation cascades for matrix metalloproteinases. In *Matrix Metalloproteinases*. W. C. Parks and R. P. Mecham (eds), pp. 199–218. Academic Press, San Diego.

64. **Krane, S. M.** 1995. Is collagenase (matrix metalloproteinase-1) necessary for bone and other connective tissue remodeling? *Clin. Orthop.* **313**: 47–53.

65. **Lala, P. K. and G. S. Hamilton.** 1996. Growth factors, proteases and protease inhibitors in the maternal-fetal dialogue. *Placenta* **17**: 545–555.

66. **Lazarus, G. S., J. R. Daniels, J. Lian, and M. C. Burleigh.** 1972. Role of granulocyte collagenase in collagen degradation. *Am. J. Pathol.* **68**: 565–578.

67. **Libby, P., U. Schoenbeck, F. Mach, A. P. Selwyn, and P. Ganz.** 1998. Current concepts in cardiovascular pathology: the role of LDL cholesterol in plaque rupture and stabilization. *Am. J. Med.* **104**: 14S–18S.

68. **Maeda, H., T. Okamoto, and T. Akaike.** 1998. Human matrix metalloprotease activation by insults of bacterial infection involving proteases and free radicals. *Biol. Chem.* **379**: 193–200.

69. **Marbaix, E., I. Kokorine, J. Donnez, P. J. Courtoy, and Y. Eeckhout.** 1995. [Collagenase and related matrix metalloproteinases: a prominent role in the initiation of menstruation?] [French]. *M. S.-Med. Sci.* **11**: 1261–1269.

70. **Massova, I., L. P. Kotra, R. Fridman, and S. Mobashery.** 1998. Matrix metalloproteinases structures, evolution, and diversification. *FASEB J.* **12**: 1075–1095.

71. **Matrisian, L. M.** 1990. Metalloproteinases and their inhibitors in matrix remodeling. *Trends Genet.* **6**: 121–125.

72. **Matrisian, L. M.** 1992. The matrix-degrading metalloproteinases. *BioEssays* **14**: 455–463.

73. **Matrisian, L. M.** 1998. Stromelysin 2. In *Handbook of Proteolytic Enzymes*. A. J. Barrett, N. D. Rawlings, and J. F. Woessner (eds), pp. 1178–1180. Academic Press, London.

74. **McGeehan, G. and J. S. Graham.** 1998. Soybean metalloproteinase 1. In *Handbook of Proteolytic Enzymes*. A. J. Barrett, N. D. Rawlings, and J. F. Woessner (eds), pp. 1190–1192. Academic Press, London.

75. **McIntush, E. W. and M. F. Smith.** 1998. Matrix metalloproteinases and tissue inhibitors of metalloproteinases in ovarian function. *Revs. Reprod.* **3**: 23–30.

76. **Mignatti, P. and D. B. Rifkin.** 1996. Plasminogen activators and matrix metalloproteinases in angiogenesis. *Enzyme Protein* **49**: 117–137.

77. **Morphy, J. R., T. A. Millican, and J. R. Porter.** 1995. Matrix metalloproteinase inhibitors: current status. *Curr. Med. Chem.* **2**: 743–762.

78. **Moses, M. A.** 1997. The regulation of neovascularization by matrix metalloproteinases and their inhibitors. *Stem Cells* **15**: 180–189.

79. **Mueller, B. M.** 1996. Different roles for plasminogen activators and metalloproteinases in melanoma metastasis. *Curr. Top. Microbiol. Immunol.* **213**: 65–80.

80. **Murphy, G.** 1998. Gelatinase A. In *Handbook of Proteolytic Enzymes*. A. J. Barrett, N. D. Rawlings, and J. F. Woessner (eds), pp. 199–1205. editors. Academic Press, London.

81. **Murphy, G. and T. Crabbe.** 1995. Gelatinases A and B. *Methods Enzymol.* **248**: 470–484.

82. **Murphy, G. and V. Knäuper.** 1997. Relating matrix metalloproteinase structure to function: why the 'hemopexin' domain? *Matrix Biol.* **15**: 511–518.

83. **Murphy, G. and J. J. Reynolds.** 1993. Extracellular matrix degradation. In *Connective Tissue and Its Heritable Disorders: Molecular, Genetic, and Medical Aspects*. P. M. Royce and B. Steinmann (eds), pp. 287–316. Wiley-Liss, Inc., New York.

84. **Murphy, G. and F. Willenbrock.** 1995. Tissue inhibitors of matrix metalloendopeptidases. *Methods Enzymol.* **248**: 496–510.

85. **Murphy, G., F. Willenbrock, T. Crabbe, M. O'Shea, R. Ward, S. Atkinson, J. O'Connell, and A. Docherty.** 1994. Regulation of matrix metalloproteinase activity. *Ann. N. Y. Acad. Sci.* **732**: 31–41.

86. **Nagase, H.** 1995. Human stromelysins 1 and 2. *Methods Enzymol.* **248**: 449–470.

87. **Nagase, H.** 1996. Matrix Metalloproteinases. In *Zinc Metalloproteases in Health and Disease*. N. M. Hooper (ed), pp. 153–204. Taylor & Francis, London.

88. **Nagase, H.** 1997. Activation mechanisms of matrix metalloproteinases. *Biol. Chem.* **378**: 151–160.

89. **Nagase, H.** 1998. Stromelysin 1. In *Handbook of Proteolytic Enzymes*. A. J. Barrett, N. D. Rawlings, and J. F. Woessner (eds), pp. 1172–1178. Academic Press, London.

90. **Nagase, H.** 1998. Stromelysins 1 and 2. In *Matrix Metalloproteinases*. W. C. Parks and R. P. Mecham (eds), pp. 43–84. Academic Press, San Diego.

91. **Nagase, H., A. J. Barrett, and J. F. Woessner, Jr.** 1992. Nomenclature and glossary of the matrix metalloproteinases. *Matrix Suppl.* **1**: 421–424.

92. **Nagase, H. and G. B. Fields.** 1996. Human matrix

metalloproteinase specificity studies using collagen sequence-based synthetic peptides. *Biopolymers* **40**: 399–416.

93. **Nagase, H., Y. Itoh, and S. Binner.** 1994. Interaction of α_2-macroglobulin with matrix metalloproteinases and its use for identification of their active forms. *Ann. N. Y. Acad. Sci.* **732**: 294–302.

94. **Nagase, H., K. Suzuki, Y. Itoh, C. C. Kan, M. R. Gehring, W. Huang, and K. Brew.** 1996. Involvement of tissue inhibitors of metalloproteinases (TIMPS) during matrix metalloproteinase activation. *Adv. Exp. Med. Biol.* **389**: 23–31.

95. **Obiso, R. J., Jr. and T. D. Wilkins.** 1998. Fragilysin. In *Handbook of Proteolytic Enzymes.* A. J. Barrett, N. D. Rawlings, and J. F. Woessner (eds), pp. 1211–1213. Academic Press, London.

96. **Osteen, K. G., K. L. Bruner, and K. L. Sharpe-Timms.** 1996. Steroid and growth factor regulation of matrix metalloproteinase expression and endometriosis. *Semin. Reprod. Endocrinol.* **14**: 247–255.

97. **Parks, W. C., B. D. Sudbeck, G. R. Doyle, and U. K. Saariahlo-Kere.** 1998. Matrix metalloproteinases in tissue repair. In *Matrix Metalloproteinases.* W. C. Parks and R. P. Mecham (eds), pp. 263–297. Academic Press, San Diego. 263–297.

98. **Parry, S. and J. F. Strauss, III.** 1998. Mechanisms of disease—premature rupture of the fetal membranes. *New Engl. J. Med.* **338**: 663–670.

99. **Partridge, N. C., H. W. Walling, S. R. Bloch, T. H. Omura, P. T. Chan, A. T. Pearman, and W. Y. Chou.** 1996. The regulation and regulatory role of collagenase in bone. *Crit. Rev. Eukaryot. Gene Expr.* **6**: 15–27.

100. **Pei, D. and S. J. Weiss.** 1998. Stromelysin 3. In *Handbook of Proteolytic Enzymes.* A. J. Barrett, N. D. Rawlings, and J. F. Woessner (eds), pp. 1187–1190. Academic Press, London.

101. **Porter, J. R., T. A. Millican, and J. R. Morphy.** 1995. Recent developments in matrix metalloproteinase inhibitors. *Expert Opin. Therap. Pat.* **5**: 1287–1296.

102. **Powell, W. C. and L. M. Matrisian.** 1996. Complex roles of matrix metalloproteinases in tumor progression. *Curr. Top. Microbiol. Immunol.* **213**: 1–21.

103. **Rawlings, N. D. and A. J. Barrett.** 1998. Introduction: family M10 of interstitial collagenase (clan MB). In *Handbook of Proteolytic Enzymes.* A. J. Barrett, N. D. Rawlings, and J. F. Woessner (eds), pp. 1144–1147. Academic Press, London.

104. **Rawlings, N. D. and A. J. Barrett.** 1995. Evolutionary families of metallopeptidases. *Methods Enzymol.* **248**: 183–228.

105. **Reilly, J. M.** 1996. Plasminogen activators in abdominal aortic aneurysmal disease. *Ann. N. Y. Acad. Sci.* **800**: 151–156.

106. **Reynolds, J. J. and M. C. Meikle.** 1997. The functional balance of metalloproteinases and inhibitors in tissue degradation: relevance to oral pathologies. *J. R. Coll. Surg. Edinb.* **42**: 154–160.

107. **Rooprai, H. K. and D. McCormick.** 1997. Proteases and their inhibitors in human brain tumours—a review. *Anticanc. Res.* **17**: 4151–4162.

108. **Ryan, M. E., S. Ramamurthy, and L. M. Golub.** 1996. Matrix metalloproteinases and their inhibition in periodontal treatment. *Curr. Opin. Periodontol.* **3**: 85–96.

109. **Salamonsen, L. A.** 1996. Matrix metalloproteinases and their tissue inhibitors in endocrinology. *Trends Endocrinol. Metab.* **7**: 28–34.

110. **Salamonsen, L. A. and D. E. Woolley.** 1996. Matrix metalloproteinases in normal menstruation. *Hum. Reprod.* **11**: 124–133.

111. **Sang, Q. A. and D. A. Douglas.** 1996. Computational sequence analysis of matrix metalloproteinases. *J. Protein Chem.* **15**: 137–160.

112. **Schroen, D. J. and C. E. Brinckerhoff.** 1996. Nuclear hormone receptors inhibit matrix metalloproteinase (MMP) gene expression through diverse mechanisms. *Gene Expr.* **6**: 197–207.

113. **Schwartz, G. K.** 1996. Invasion and metastases in gastric cancer: *in vitro* and *in vivo* models with clinical correlations. *Semin. Oncol.* **23**: 316–324.

114. **Seiki, M.** 1998. Membrane-type matrix metalloproteinase 1. In *Handbook of Proteolytic Enzymes.* A. J. Barrett, N. D. Rawlings, and J. F. Woessner (eds), pp. 1192–1195. Academic Press, London.

115. **Senior, R. M. and S. D. Shapiro.** 1998. Macrophage elastase. In *Handbook of Proteolytic Enzymes.* A. J. Barrett, N. D. Rawlings, and J. F. Woessner (eds), pp. 1180–1183. Academic Press, London.

116. **Shapiro, S. D.** 1997. Mighty mice: transgenic technology 'knocks out' questions of matrix metalloproteinase function. *Matrix Biol.* **15**: 527–533.

117. **Shapiro, S. D. and R. M. Senior.** 1998. Macrophage elastase (MMP-12). In *Matrix Metalloproteinases.* W. C. Parks and R. P. Mecham (eds), pp. 185–197. Academic Press, San Diego.

118. **Shi, Y.-B. and Q.-X. A. Sang.** 1998. Collagenase 4. In *Handbook of Proteolytic Enzymes.* A. J. Barrett, N. D. Rawlings, and J. F. Woessner (eds), pp. 1170–1172. Academic Press, London.

119. **Shi, Y. B. and A. Ishizuya-Oka.** 1996. Biphasic intestinal development in amphibians: embryogenesis and remodeling during metamorphosis. *Curr. Top. Dev. Biol.* **32**: 205–235.

120. **Smith, C. E.** 1998. Cellular and chemical events during enamel maturation. *Crit. Rev. Oral Biol. Med.* **9**: 128–161.

121. **Sottrup-Jensen, L.** 1989. α-Macroglobulins: structure, shape, and mechanism of proteinase complex formation. *J. Biol. Chem.* **264**: 11539–11542.

122. **Stetler-Stevenson, W. G., R. Hewitt, and M.**

Corcoran. 1996. Matrix metalloproteinases and tumor invasion: from correlation and causality to the clinic. *Semin. Cancer Biol.* **7**: 147 154.

123. **Stöcker, W. and W. Bode.** 1995. Structural features of a superfamily of zinc-endopeptidases: the metzincins. *Curr. Opin. Struct. Biol.* **5**: 383 390.

124. **Stuart, R. O., E. J. Barros, E. Ribeiro, and S. K. Nigam.** 1995. Epithelial tubulogenesis through branching morphogenesis: relevance to collecting system development. *J. Am. Soc. Nephrol.* **6**: 1151–1159.

125. **Takagi, M.** 1996. Neutral proteinases and their inhibitors in the loosening of total hip prostheses. *Acta Orthop. Scand. Suppl.* **271**: 3 29.

126. **Takagi, M., S. Santavirta, H. Ida, M. Ishii, J. Mandelin, and Y. T. Konttinen.** 1998. Matrix metalloproteinases and tissue inhibitors of metalloproteinases in loose artificial hip joints. *Clin. Orthop. Rel. Res.* **352**: 35 45.

127. **Thompson, R. W. and W. C. Parks.** 1996. Role of matrix metalloproteinases in abdominal aortic aneurysms. *Ann. N. Y. Acad. Sci.* **800**: 157 174.

128. **Tsafriri, A.** 1995. Ovulation as a tissue remodelling process. Proteolysis and cumulus expansion. *Adv. Exp. Med. Biol.* **377**: 121 140.

129. **Tschesche, H.** 1995. Human neutrophil collagenase. *Methods Enzymol.* **248**: 431–449.

130. **Tschesche, H. and M. Pieper.** 1998. Neutrophil collagenase. In *Handbook of Proteolytic Enzymes.* A. J. Barrett, N. D. Rawlings, and J. F. Woessner (eds), pp. 1162 1167. Academic Press, London.

131. **Tyagi, S. C.** 1997. Proteinases and myocardial extracellular matrix turnover. *Mol. Cell. Biochem.* **168**: 1 12.

132. **Tyagi, S. C., V. S. Bheemanathini, D. Mandi, H. K. Reddy, and D. J. Voelker.** 1996. Role of extracellular matrix metalloproteinases in cardiac remodelling. *Heart Failure Rev.* **1**: 73–80.

133. **Tyagi, S. C., L. Meyer, S. Kumar, R. A. Schmaltz, H. K. Reddy, and D. J. Voelker.** 1996. Induction of tissue inhibitor of metalloproteinase and its mitogenic response to endothelial cells in human atherosclerotic and restenotic lesions. *Can. J. Cardiol.* **12**: 353 362.

134. **Uhm, J. H., N. P. Dooley, J.-G. Villemure, and V. W. Yong.** 1997. Mechanisms of glioma invasion: role of matrix-metalloproteinases. *Can. J. Neurol. Sci.* **24**: 3 15.

135. **Vu, T. H. and Z. Werb.** 1998. Gelatinase B: structure, regulation and function. *In Matrix Metalloproteinases.* W. C. Parks and R. P. Mecham (eds), pp. 115 148. Academic Press, San Diego.

136. **Werb, Z.** 1997. ECM and cell surface proteolysis: regulating cellular ecology. *Cell* **91**: 439 442.

137. **Werb, Z., C. J. Sympson, C. M. Alexander, N. Thomasset, L. R. Lund, A. MacAuley, J. Ashkenas, and M. J. Bissell.** 1996. Extracellular matrix remodeling and the regulation of epithelial-stromal interactions during differentiation and involution. *Kidney Int. Suppl.* **54**: S68-S74.

138. **West, M. D.** 1994. The cellular and molecular biology of skin aging. *Arch. Dermatol.* **130**: 87–95.

139. **Willenbrock, F. and G. Murphy.** 1994. Structure-function relationships in the tissue inhibitors of metalloproteinases. *Am. J. Respir. Crit. Care Med.* **150**: S165-S170

140. **Wilson, C. L. and L. M. Matrisian.** 1996. Matrilysin: an epithelial matrix metalloproteinase with potentially novel functions. *Int. J. Biochem. Cell. Biol.* **28**: 123–136.

141. **Wilson, C. L. and L. M. Matrisian.** 1998. Matrilysin. In *Matrix Metalloproteinases.* W. C. Parks and R. P. Mecham (eds), pp. 149 184. Academic Press, San Diego. 149 184.

142. **Woessner, J. F.** 1998. Matrilysin. In *Handbook of Proteolytic Enzymes.* A. J. Barrett, N. D. Rawlings, and J. F. Woessner (eds), pp. 1183–1187. Academic Press, London.

143. **Woessner, J. F., Jr.** 1968. Biological mechanisms of collagen resorption. In *Treatise on Collagen.* B. S. Gould (ed), Vol. 2B, pp. 253–330. Academic Press, New York.

144. **Woessner, J. F., Jr.** 1991. Matrix metalloproteinases and their inhibitors in connective tissue remodeling. *FASEB J.* **5**: 2145 2154.

145. **Woessner, J. F., Jr.** 1994. The family of matrix metalloproteinases. *Ann. N. Y. Acad. Sci.* **732**: 11–21.

146. **Woessner, J. F., Jr.** 1995. Matrilysin. *Methods Enzymol.* **248**: 485–495.

147. **Woessner, J. F., Jr.** 1998. The matrix metalloproteinase family. In *Matrix Metalloproteinases.* W. C. Parks and R. P. Mecham (eds), pp. 1 14. Academic Press, San Diego.

148. **Ye, S., S. Humphries, and A. Henney.** 1998. Matrix metalloproteinases—implication in vascular matrix remodelling during atherogenesis. *Clin. Sci.* **94**: 103–110.

149. **Yong, V. W., C. A. Krekoski, P. A. Forsyth, R. Bell, and D. R. Edwards.** 1998. Matrix metalloproteinases and diseases of the CNS. *Trends Neurosci.* **21**: 75–80.

150. **Yu, A. E., A. N. Murphy, and W. G. Stetler-Stevenson.** 1998. 72-kDa gelatinase (gelatinase A): structure, activation, regulation and substrate specificity. In *Matrix Metalloproteinases.* W. C. Parks and R. P. Mecham (eds), pp. 85 113. Academic Press, San Diego.

151. **Zask, A., J. I. Levin, L. M. Killar, and J. S. Skotnicki.** 1996. Inhibition of matrix metalloproteinases: structure based design. *Curr. Pharm. Des.* **2**: 624–661.

Activation of MMP zymogens

152. **Abe, S. and Y. Nagai.** 1972. Evidence for the presence of a latent form of collagenase in human rheumatoid synovial fluid. *J. Biochem. (Tokyo)* **71**: 919–922.

153. **Abramson, S. R., G. E. Conner, H. Nagase, I.**

Neuhaus, and J. F. Woessner, Jr. 1995. Characterization of rat uterine matrilysin and its cDNA. Relationship to human pump-1 and activation of procollagenases. *J. Biol. Chem.* **270**: 16016–16022.

154. **Atkinson, S. J., T. Crabbe, S. Cowell, R. V. Ward, M. J. Butler, H. Sato, M. Seiki, J. J. Reynolds, and G. Murphy.** 1995. Intermolecular autolytic cleavage can contribute to the activation of progelatinase A by cell membranes. *J. Biol. Chem.* **270**: 30479–30485.

155. **Azzam, H. S. and E. W. Thompson.** 1992. Collagen-induced activation of the M_r 72,000 type IV collagenase in normal and malignant human fibroblastoid cells. *Cancer Res.* **52**: 4540–4544.

156. **Baramova, E. N., K. Bajou, A. Remacle, C. L'Hoir, H. W. Krell, U. H. Weidle, A. Noël, and J. M. Foidart.** 1997. Involvement of PA/plasmin system in the processing of pro-MMP-9 and in the second step of pro-MMP-2 activation. *FEBS Lett.* **405**: 157–162.

157. **Bauer, E. A., G. P. Stricklin, J. J. Jeffrey, and A. Z. Eisen.** 1975. Collagenase production by human skin fibroblasts. *Biochem. Biophys. Res. Commun.* **64**: 232–240.

158. **Becker, J. W., A. I. Marcy, L. L. Rokosz, M. G. Axel, J. J. Burbaum, P. M. Fitzgerald, P. M. Cameron, C. K. Esser, W. K. Hagmann, J. D. Hermes,** *et al.* 1995. Stromelysin-1: three-dimensional structure of the inhibited catalytic domain and of the C-truncated proenzyme. *Protein Sci.* **4**: 1966–1976.

159. **Benbow, U., G. Buttice, H. Nagase, and M. Kurkinen.** 1996. Characterization of the 46-kDa intermediates of matrix metalloproteinase 3 (stromelysin 1) obtained by site-directed mutation of phenylalanine 83. *J. Biol. Chem.* **271**: 10715–10722.

160. **Bergmann, U., A. Tuuttila, W. G. Stetler-Stevenson, and K. Tryggvason.** 1995. Autolytic activation of recombinant human 72 kilodalton type IV collagenase. *Biochemistry* **34**: 2819–2825.

161. **Bigg, H. F., Y. E. Shi, Y. L. E. Liu, B. Steffensen, and C. M. Overall.** 1997. Specific, high affinity binding of tissue inhibitor of metalloproteinases-4 (TIMP4) to the COOH-terminal hemopexin-like domain of human gelatinase A—TIMP-4 binds progelatinase A and the COOH-terminal domain in a similar manner to TIMP-2. *J. Biol. Chem.* **272**: 15496–15500.

162. **Birkedal-Hansen, H., C. M. Cobb, R. E. Taylor, and H. M. Fullmer.** 1974. Bovine gingival collagenase: demonstration and initial characterization. *J. Oral Pathol.* **3**: 232–238.

163. **Birkedal-Hansen, H., C. M. Cobb, R. E. Taylor, and H. M. Fullmer.** 1976. Activation of fibroblast procollagenase by mast cell proteases. *Biochim. Biophys. Acta* **438**: 273–286.

164. **Birkedal-Hansen, H., C. M. Cobb, R. E. Taylor, and H. M. Fullmer.** 1976. Synthesis and release of procollagenase by cultured fibroblasts. *J. Biol. Chem.* **251**: 3162–3168.

165. **Birkedal-Hansen, H. and R. E. Taylor.** 1982. Detergent-activation of latent collagenase and resolution of its component molecules. *Biochem. Biophys. Res. Commun.* **107**: 1173–1178.

166. **Bläser, J., V. Knäuper, A. Osthues, H. Reinke, and H. Tschesche.** 1991. Mercurial activation of human polymorphonuclear leucocyte procollagenase. *Eur. J. Biochem.* **202**: 1223–1230.

167. **Bläser, J., S. Triebel, H. Reinke, and H. Tschesche.** 1992. Formation of a covalent Hg-Cys-bond during mercurial activation of PMNL procollagenase gives evidence of a cysteine-switch mechanism. *FEBS Lett.* **313**: 59–61.

168. **Bode, W., P. Reinemer, R. Huber, T. Kleine, S. Schnierer, and H. Tschesche.** 1994. The X-ray crystal structure of the catalytic domain of human neutrophil collagenase inhibited by a substrate analogue reveals the essentials for catalysis and specificity. *EMBO J.* **13**: 1263–1269.

169. **Brinckerhoff, C. E., K. Suzuki, T. I. Mitchell, F. Oram, C. I. Coon, R. D. Palmiter, and H. Nagase.** 1990. Rabbit procollagenase synthesized and secreted by a high-yield mammalian expression vector requires stromelysin (matrix metalloproteinase-3) for maximal activation. *J. Biol. Chem.* **265**: 22262–22269.

170. **Brown, P. D., A. T. Levy, I. M. K. Margulies, L. A. Liotta, and W. G. Stetler-Stevenson.** 1990. Independent expression and cellular processing of M_r 72,000 type IV collagenase and interstitial collagenase in human tumorigenic cell lines. *Cancer Res.* **50**: 6184–6191.

171. **Burleigh, M. C., Z. Werb, and J. J. Reynolds.** 1977. Evidence that species specificity and rate of collagen degradation are properties of collagen, not collagenase. *Biochim. Biophys. Acta* **494**: 198–208.

172. **Butler, G. S., M. J. Butler, S. J. Atkinson, H. Will, T. Tamura, S. S. van Westrum, T. Crabbe, J. Clements, M. P. d'Ortho, and G. Murphy.** 1998. The TIMP2 membrane type 1 metalloproteinase 'receptor' regulates the concentration and efficient activation of progelatinase A—a kinetic study. *J. Biol. Chem.* **273**: 871–880.

173. **Butler, G. S., H. Will, S. J. Atkinson, and G. Murphy.** 1997. Membrane-type-2 matrix metalloproteinase can initiate the processing of progelatinase A and is regulated by the tissue inhibitors of metalloproteinases. *Eur. J. Biochem.* **244**: 653–657.

174. **Cameron, P. M., A. I. Marcy, L. L. Rokosz, and J. D. Hermes.** 1995. Use of an active-site inhibitor of stromelysin to elucidate the mechanism of prostromelysin activation. *Bioorg. Chem.* **23**: 415–426.

175. **Cao, J., A. Rehemtulla, W. Bahou, and S. Zucker.** 1996. Membrane type matrix metalloproteinase 1

activates pro-gelatinase A without furin cleavage of the N-terminal domain. *J. Biol. Chem.* **271**: 30174–30180.

176. **Cao, J., H. Sato, T. Takino, and M. Seiki.** 1995. The C-terminal region of membrane type matrix metalloproteinase is a functional transmembrane domain required for pro-gelatinase A activation. *J. Biol. Chem.* **270**: 801–805.

177. **Capodici, C., G. Muthukumaran, M. A. Amoruso, and R. A. Berg.** 1989. Activation of neutrophil collagenase by cathepsin G. *Inflammation* **13**: 245–258.

178. **Carmeliet, P., L. Moons, R. Lijnen, M. Baes, V. Lemaître, P. Tipping, A. Drew, Y. Eeckhout, S. Shapiro, F. Lupu, and D. Collen.** 1997. Urokinase-generated plasmin activates matrix metalloproteinases during aneurysm formation. *Nature Genet.* **17**: 439–444.

179. **Cawston, T. E., E. Mercer, and J. A. Tyler.** 1981. The activation of latent pig synovial collagenase. *Biochim. Biophys. Acta* **657**: 73–83.

180. **Cawston, T. E., G. Murphy, E. Mercer, W. A. Galloway, B. L. Hazleman, and J. J. Reynolds.** 1983. The interaction of purified rabbit bone collagenase with purified rabbit bone metalloproteinase inhibitor. *Biochem. J.* **211**: 313–318.

181. **Chen, L. C., M. E. Noelken, and H. Nagase.** 1993. Disruption of the cysteine-75 and zinc ion coordination is not sufficient to activate the precursor of human matrix metalloproteinase 3 (stromelysin 1). *Biochemistry* **32**: 10289–10295.

182. **Clark, I. M. and T. E. Cawston.** 1989. Fragments of human fibroblast collagenase. Purification and characterization. *Biochem. J.* **263**: 201–206.

183. **Cowell, S., V. Knäuper, M. L. Stewart, M.-P. d'Ortho, H. Stanton, R. M. Hembry, C. López-Otín, J. J. Reynolds, and G. Murphy.** 1998. Induction of matrix metalloproteinase activation cascades based on membrane-type 1 matrix metalloproteinase—associated activation of gelatinase A, gelatinase B and collagenase 3. *Biochem. J.* **331**: 453–458.

184. **Crabbe, T., C. Ioannou, and A. J. P. Docherty.** 1993. Human progelatinase A can be activated by autolysis at a rate that is concentration-dependent and enhanced by heparin bound to the C-terminal domain. *Eur. J. Biochem.* **218**: 431–438.

185. **Crabbe, T., J. P. O'Connell, B. J. Smith, and A. J. P. Docherty.** 1994. Reciprocated matrix metalloproteinase activation: a process performed by interstitial collagenase and progelatinase A. *Biochemistry* **33**: 14419–14425.

186. **Crabbe, T., F. Willenbrock, D. Eaton, P. Hynds, A. F. Carne, G. Murphy, and A. J. P. Docherty.** 1992. Biochemical characterization of matrilysin. Activation conforms to the stepwise mechanisms proposed for other matrix metalloproteinases. *Biochemistry* **31**: 8500–8507.

187. **Curry, V. A., I. M. Clark, H. Bigg, and T. E. Cawston.** 1992. Large inhibitor of metalloproteinases (LIMP) contains tissue inhibitor of metalloproteinases (TIMP)-2 bound to 72,000-M_r progelatinase. *Biochem. J.* **285**: 143–147.

188. **Davis, G. E.** 1991. Identification of an abundant latent 94-kDa gelatin-degrading metalloprotease in human saliva which is activated by acid exposure: implications for a role in digestion of collagenous proteins. *Arch. Biochem. Biophys.* **286**: 551–554.

189. **DeCarlo, A. A., Jr., L. J. Windsor, M. K. Bodden, G. J. Harber, B. Birkedal-Hansen, and H. Birkedal-Hansen.** 1997. Activation and novel processing of matrix metalloproteinases by a thiol-proteinase from the oral anaerobe *Porphyromonas gingivalis*. *J. Dent. Res.* **76**: 1260–1270.

190. **DeClerck, Y. A., T. D. Yean, Y. Lee, J. M. Tomich, and K. E. Langley.** 1993. Characterization of the functional domain of tissue inhibitor of metalloproteinases-2 (TIMP-2). *Biochem. J.* **289**: 65–69.

191. **DeClerck, Y. A., T. D. Yean, H. S. Lu, J. Ting, and K. E. Langley.** 1991. Inhibition of autoproteolytic activation of interstitial procollagenase by recombinant metalloproteinase inhibitor MI/TIMP-2. *J. Biol. Chem.* **266**: 3893–3899.

192. **Desrivières, S., H. Lu, N. Peyri, C. Soria, Y. Legrand, and S. Menashi.** 1993. Activation of the 92 kDa type IV collagenase by tissue kallikrein. *J. Cell. Physiol.* **157**: 587–593.

193. **Eeckhout, Y. and G. Vaes.** 1977. Further studies on the activation of procollagenase, the latent precursor of bone collagenase. Effects of lysosomal cathepsin B, plasmin and kallikrein, and spontaneous activation. *Biochem. J.* **166**: 21–31.

194. **Fang, K. C., W. W. Raymond, J. L. Blount, and G. H. Caughey.** 1997. Dog mast cell α-chymase activates progelatinase B by cleaving the Phe[88]-Gln[89] and Phe[91]-Glu[92] bonds of the catalytic domain. *J. Biol. Chem.* **272**: 25628–25635.

195. **Fang, K. C., W. W. Raymond, S. C. Lazarus, and G. H. Caughey.** 1996. Dog mastocytoma cells secrete a 92-kD gelatinase activated extracellularly by mast cell chymase. *J. Clin. Invest.* **97**: 1589–1596.

196. **Ferry, G., M. Lonchampt, L. Pennel, G. de Nanteuil, E. Canet, and G. C. Tucker.** 1997. Activation of MMP-9 by neutrophil elastase in an in vivo model of acute lung injury. *FEBS Lett.* **402**: 111–115.

197. **Fini, M. E., M. J. Karmilowicz, P. L. Ruby, A. M. Beeman, K. A. Borges, and C. E. Brinckerhoff.** 1987. Cloning of a complementary DNA for rabbit proactivator. A metalloproteinase that activates synovial cell collagenase, shares homology with stromelysin and transin, and is coordinately regulated with collagenase. *Arthritis Rheum.* **30**: 1254–1264.

198. **Freimark, B. D., W. S. Feeser, and S. A. Rosenfeld.** 1994. Multiple sites of the propeptide region of human

stromelysin-1 are required for maintaining a latent form of the enzyme. *J. Biol. Chem.* **269**: 26982–26987.

199. **Fridman, R., T. R. Fuerst, R. E. Bird, M. Hoyhtya, M. Oelkuct, S. Kraus, D. Komarek, L. A. Liotta, M. L. Berman, and W. G. Stetler-Stevenson.** 1992. Domain structure of human 72-kDa gelatinase/type IV collagenase. Characterization of proteolytic activity and identification of the tissue inhibitor of metalloproteinase-2 (TIMP-2) binding regions. *J. Biol. Chem.* **267**: 15398–15405.

200. **Fridman, R., M. Toth, D. Pena, and S. Mobashery.** 1995. Activation of progelatinase B (MMP-9) by gelatinase A (MMP-2). *Cancer Res.* **55**: 2548–2555.

201. **Fujimoto, N., R. V. Ward, T. Shinya, K. Iwata, K. Yamashita, and T. Hayakawa.** 1996. Interaction between tissue inhibitor of metalloproteinases-2 and progelatinase A: immunoreactivity analyses. *Biochem. J.* **313**: 827–833.

202. **Galazka, G., L. J. Windsor, H. Birkedal-Hansen, and J. A. Engler.** 1996. APMA (4-aminophenylmercuric acetate) activation of stromelysin-1 involves protein interactions in addition to those with cysteine-75 in the propeptide. *Biochemistry* **35**: 11221–11227.

203. **Goldberg, G. I., B. L. Marmer, G. A. Grant, A. Z. Eisen, S. Wilhelm, and C. S. He.** 1989. Human 72-kilodalton type IV collagenase forms a complex with a tissue inhibitor of metalloproteases designated TIMP-2. *Proc. Natl. Acad. Sci. USA* **86**: 8207–8211.

204. **Goldberg, G. I., A. Strongin, I. E. Collier, L. T. Genrich, and B. L. Marmer.** 1992. Interaction of 92-kDa type IV collagenase with the tissue inhibitor of metalloproteinases prevents dimerization, complex formation with interstitial collagenase, and activation of the proenzyme with stromelysin. *J. Biol. Chem.* **267**: 4583–4591.

205. **Grant, G. A., A. Z. Eisen, B. L. Marmer, W. T. Roswit, and G. I. Goldberg.** 1987. The activation of human skin fibroblast procollagenase. Sequence identification of the major conversion products. *J. Biol. Chem.* **262**: 5886–5889.

206. **Gruber, B. L., M. J. Marchese, K. Suzuki, L. B. Schwartz, Y. Okada, H. Nagase, and N. S. Ramamurthy.** 1989. Synovial procollagenase activation by human mast cell tryptase dependence upon matrix metalloproteinase 3 activation. *J. Clin. Invest.* **84**: 1657–1662.

207. **Gunja-Smith, Z. and J. F. Woessner, Jr.** 1993. Activation of cartilage stromelysin-1 at acid pH and its relation to enzyme pH optimum and osteoarthritis. *Agents Actions* **40**: 228–231.

208. **Haas, T. L., S. J. Davis, and J. A. Madri.** 1998. Three-dimensional type I collagen lattices induce coordinate expression of matrix metalloproteinases MT1-MMP and MMP-2 in microvascular endothelial cells. *J. Biol. Chem.* **273**: 3604–3610.

209. **Harper, E., K. J. Bloch, and J. Gross.** 1971. The zymogen of tadpole collagenase. *Biochemistry* **10**: 3035–3041.

210. **Harper, E. and J. Gross.** 1972. Collagenase, procollagenase and activator relationships in tadpole tissue cultures. *Biochem. Biophys. Res. Commun.* **48**: 1147–1152.

211. **He, C. S., S. M. Wilhelm, A. P. Pentland, B. L. Marmer, G. A. Grant, A. Z. Eisen, and G. I. Goldberg.** 1989. Tissue cooperation in a proteolytic cascade activating human interstitial collagenase. *Proc. Natl. Acad. Sci. USA* **86**: 2632–2636.

212. **Hibbs, M. S., K. A. Hasty, J. M. Seyer, A. H. Kang, and C. L. Mainardi.** 1985. Biochemical and immunological characterization of the secreted forms of human neutrophil gelatinase. *J. Biol. Chem.* **260**: 2493–2500.

213. **Holz, R. C., S. P. Salowe, C. K. Smith, G. C. Cuca, and L. Que, Jr.** 1992. EXAFS evidence for a 'cysteine switch' in the activation of prostromelysin. *J. Am. Chem. Soc.* **114**: 9611–9614.

214. **Hook, R. M., C. W. Hook, and S. I. Brown.** 1973. Fibroblast collagenase partial purification and characterization. *Invest. Ophthalmol.* **12**: 771–776.

215. **Howard, E. W. and M. J. Banda.** 1991. Binding of tissue inhibitor of metalloproteinases 2 to two distinct sites on human 72-kDa gelatinase. Identification of a stabilization site. *J. Biol. Chem.* **266**: 17972–17977.

216. **Howard, E. W., E. C. Bullen, and M. J. Banda.** 1991. Regulation of the autoactivation of human 72-kDa progelatinase by tissue inhibitor of metalloproteinases-2. *J. Biol. Chem.* **266**: 13064–13069.

217. **Imai, K., Y. Yokohama, I. Nakanishi, E. Ohuchi, Y. Fujii, N. Nakai, and Y. Okada.** 1995. Matrix metalloproteinase 7 (matrilysin) from human rectal carcinoma cells. Activation of the precursor, interaction with other matrix metalloproteinases and enzymic properties. *J. Biol. Chem.* **270**: 6691–6697.

218. **Ishibashi, M., A. Ito, K. Sakyo, and Y. Mori.** 1987. Procollagenase activator produced by rabbit uterine cervical fibroblasts. *Biochem. J.* **241**: 527–534.

219. **Ishihara, T., A. Hayasaka, T. Yamaguchi, F. Kondo, and H. Saisho.** 1998. Immunohistochemical study of transforming growth factor beta-1, matrix metalloproteinase-2,9, tissue inhibitors of metalloproteinase-1,2, and basement membrane components at pancreatic ducts in chronic pancreatitis. *Pancreas* **17**: 412–418.

220. **Ito, A. and H. Nagase.** 1988. Evidence that human rheumatoid synovial matrix metalloproteinase 3 is an endogenous activator of procollagenase. *Arch. Biochem. Biophys.* **267**: 211–216.

221. **Itoh, Y., S. Binner, and H. Nagase.** 1995. Steps involved in activation of the complex of pro-matrix metalloproteinase 2 (progelatinase A) and tissue inhibitor of metalloproteinases (TIMP)-2 by 4-aminophenylmercuric acetate. *Biochem. J.* **308**: 645–651.

222. **Itoh, Y., A. Ito, K. Iwata, K. Tanzawa, Y. Mori, and H. Nagase.** 1998. Plasma membrane-bound tissue inhibitor of metalloproteinases (TIMP)-2 specifically inhibits matrix metalloproteinase 2 (gelatinase A) activated on the cell surface. *J. Biol. Chem.* **273**: 24360–24367.

223. **Itoh, Y. and H. Nagase.** 1995. Preferential inactivation of tissue inhibitor of metalloproteinases-1 that is bound to the precursor of matrix metalloproteinase 9 (progelatinase B) by human neutrophil elastase. *J. Biol. Chem.* **270**: 16518–16521.

224. **Keski-Oja, J., J. Lohi, A. Tuuttila, K. Tryggvason, and T. Vartio.** 1992. Proteolytic processing of the 72,000-Da type IV collagenase by urokinase plasminogen activator. *Exp. Cell Res.* **202**: 471–476.

225. **Kinoshita, T., H. Sato, A. Okada, E. Ohuchi, K. Imai, Y. Okada, and M. Seiki.** 1998. TIMP-2 promotes activation of progelatinase A by membrane-type 1 matrix metalloproteinase immobilized on agarose beads. *J. Biol. Chem.* **273**: 16098–16103.

226. **Kleiner, D. E. and W. G. Stetler-Stevenson.** 1994. Quantitative zymography: detection of picogram quantities of gelatinases. *Anal. Biochem.* **218**: 325–329.

227. **Knäuper, V., S. Kramer, H. Reinke, and H. Tschesche.** 1990. Characterization and activation of procollagenase from human polymorphonuclear leucocytes. N-terminal sequence determination of the proenzyme and various proteolytically activated forms. *Eur. J. Biochem.* **189**: 295–300.

228. **Knäuper, V., C. López-Otín, B. Smith, G. Knight, and G. Murphy.** 1996. Biochemical characterization of human collagenase-3. *J. Biol. Chem.* **271**: 1544–1550.

229. **Knäuper, V., G. Murphy, and H. Tschesche.** 1996. Activation of human neutrophil procollagenase by stromelysin 2. *Eur. J. Biochem.* 235: 187–191.

230. **Knäuper, V., A. Osthues, Y. A. DeClerck, K. E. Langley, J. Bläser, and H. Tschesche.** 1993. Fragmentation of human polymorphonuclear-leucocyte collagenase. *Biochem. J.* 291: 847–854.

231. **Knäuper, V., B. Smith, C. López-Otín, and G. Murphy.** 1997. Activation of progelatinase B (proMMP-9) by active collagenase-3 (MMP-13). *Eur. J. Biochem.* **248**: 369–373.

232. **Knäuper, V., S. M. Wilhelm, P. K. Seperack, Y. A. DeClerck, K. E. Langley, A. Osthues, and H. Tschesche.** 1993. Direct activation of human neutrophil procollagenase by recombinant stromelysin. *Biochem. J.* **295**: 581–586.

233. **Knäuper, V., H. Will, C. López-Otín, B. Smith, S. J. Atkinson, H. Stanton, R. M. Hembry, and G. Murphy.** 1996. Cellular mechanisms for human procollagenase-3 (MMP-13) activation. Evidence that MT1-MMP (MMP-14) and gelatinase A (MMP-2) are able to generate active enzyme. *J. Biol. Chem.* **271**: 17124–17131.

234. **Koklitis, P. A., G. Murphy, C. Sutton, and S. Angal.** 1991. Purification of recombinant human prostromelysin. Studies on heat activation to give high-M_r and low-M_r active forms, and a comparison of recombinant with natural stromelysin activities. *Biochem. J.* **276**: 217–221.

235. **Kolkenbrock, H., D. Orgel, A. Hecker-Kia, W. Noack, and N. Ulbrich.** 1991. The complex between a tissue inhibitor of metalloproteinases (TIMP-2) and 72-kDa progelatinase is a metalloproteinase inhibitor. *Eur. J. Biochem.* **198**: 775–781.

236. **Kruze, D. and E. Wojtecka.** 1972. Activation of leucocyte collagenase proenzyme by rheumatoid synovial fluid. *Biochim. Biophys. Acta* **285**: 436–446.

237. **Lark, M. W., L. A. Walakovits, T. K. Shah, J. Vanmiddlesworth, P. M. Cameron, and T. Y. Lin.** 1990. Production and purification of prostromelysin and procollagenase from IL-1 beta-stimulated human gingival fibroblasts. *Connect. Tiss. Res.* **25**: 49–65.

238. **Lee, A. Y., K. T. Akers, M. Collier, L. Li, A. Z. Eisen, and J. L. Seltzer.** 1997. Intracellular activation of gelatinase A (72-kDa type IV collagenase) by normal fibroblasts. *Proc. Natl. Acad. Sci. USA* **94**: 4424–4429.

239. **Lees, M., D. J. Taylor, and D. E. Woolley.** 1994. Mast cell proteinases activate precursor forms of collagenase and stromelysin, but not of gelatinases A and B. *Eur. J. Biochem.* **223**: 171–177.

239a. **Lehti, K., J. Lohi, H. Valtanen, and J. Keski-Oja.** 1998. Proteolytic processing of membrane-type-1 matrix metalloproteinase is associated with gelatinase A activation at the cell surface. *Biochem. J.* **334**: 345–353.

240. **Lichte, A., H. Kolkenbrock, and H. Tschesche.** 1996. The recombinant catalytic domain of membrane-type matrix metalloproteinase-1 (MT1-MMP) induces activation of progelatinase A and progelatinase A complexed with TIMP-2. *FEBS Lett.* **397**: 277–282.

241. **Lim, Y. T., Y. Sugiura, W. E. Laug, B. Sun, A. Garcia, and Y. A. DeClerck.** 1996. Independent regulation of matrix metalloproteinases and plasminogen activators in human fibrosarcoma cells. *J. Cell. Physiol.* 167: 333–340.

242. **Macartney, H. W. and H. Tschesche.** 1980. Latent collagenase from human polymorphonuclear leucocytes and activation to collagenase by removal of a inhibitor. *FEBS Lett.* 119: 327–332.

243. **Mazzieri, R., L. Masiero, L. Zanetta, S. Monea, M. Onisto, S. Garbisa, and P. Mignatti.** 1997. Control of type IV collagenase activity by components of the

urokinase-plasmin system: a regulatory mechanism with cell-bound reactants. *EMBO J.* **16**: 2319–2332.

244. **McLaughlin, B., T. Cawston, and J. B. Weiss.** 1991. Activation of the matrix metalloproteinase inhibitor complex by a low molecular weight angiogenic factor. *Biochim. Biophys. Acta* **1073**: 295–298.

245. **McLaughlin, B. and J. B. Weiss.** 1996. Endothelial-cell-stimulating angiogenesis factor (ESAF) activates progelatinase A (72 kDa type IV collagenase), prostromelysin 1 and procollagenase and reactivates their complexes with tissue inhibitors of metalloproteinases: a role for ESAF in non-inflammatory angiogenesis. *Biochem. J.* **317**: 739–745.

246. **Michaelis, J., M. C. Vissers, and C. C. Winterbourn.** 1992. Different effects of hypochlorous acid on human neutrophil metalloproteinases: activation of collagenase and inactivation of collagenase and gelatinase. *Arch. Biochem. Biophys.* **292**: 555–562.

247. **Miyazaki, K., K. Funahashi, Y. Numata, N. Koshikawa, K. Akaogi, Y. Kikkawa, H. Yasumitsu, and M. Umeda.** 1993. Purification and characterization of a two-chain form of tissue inhibitor of metalloproteinases (TIMP) type 2 and a low molecular weight TIMP-like protein. *J. Biol. Chem.* **268**: 14387–14393.

248. **Montgomery, A. M., Y. A. De Clerck, K. E. Langley, R. A. Reisfeld, and B. M. Mueller.** 1993. Melanoma-mediated dissolution of extracellular matrix: contribution of urokinase-dependent and metalloproteinase-dependent proteolytic pathways. *Cancer Res.* **53**: 693–700.

249. **Morodomi, T., Y. Ogata, Y. Sasaguri, M. Morimatsu, and H. Nagase.** 1992. Purification and characterization of matrix metalloproteinase 9 from U937 monocytic leukaemia and HT1080 fibrosarcoma cells. *Biochem. J.* **285**: 603–611.

250. **Murphy, G., J. A. Allan, F. Willenbrock, M. I. Cockett, J. P. O'Connell, and A. J. P. Docherty.** 1992. The role of the C-terminal domain in collagenase and stromelysin specificity. *J. Biol. Chem.* **267**: 9612–9618.

251. **Murphy, G., S. Atkinson, R. Ward, J. Gavrilovic, and J. J. Reynolds.** 1992. The role of plasminogen activators in the regulation of connective tissue metalloproteinases. *Ann. N. Y. Acad. Sci.* **667**: 1–12.

252. **Murphy, G., U. Bretz, M. Baggiolini, and J. J. Reynolds.** 1980. The latent collagenase and gelatinase of human polymorphonuclear neutrophil leucocytes. *Biochem. J.* **192**: 517–525.

253. **Murphy, G., M. I. Cockett, P. E. Stephens, B. J. Smith, and A. J. P. Docherty.** 1987. Stromelysin is an activator of procollagenase. A study with natural and recombinant enzymes. *Biochem. J.* **248**: 265–268.

254. **Murphy, G., A. Houbrechts, M. I. Cockett, R. A. Williamson, M. O'Shea, and A. J. P. Docherty.** 1991. The N-terminal domain of tissue inhibitor of metalloproteinases retains metalloproteinase inhibitory activity. *Biochemistry* **30**: 8097–8102.

255. **Murphy, G., C. G. McAlpine, C. T. Poll, and J. J. Reynolds.** 1985. Purification and characterization of a bone metalloproteinase that degrades gelatin and types IV and V collagen. *Biochim. Biophys. Acta* **831**: 49–58.

256. **Murphy, G., J. P. Segain, M. O'Shea, M. Cockett, C. Ioannou, O. Lefebvre, P. Chambon, and P. Basset.** 1993. The 28-kDa N-terminal domain of mouse stromelysin-3 has the general properties of a weak metalloproteinase. *J. Biol. Chem.* **268**: 15435–15441.

257. **Murphy, G., F. Willenbrock, R. V. Ward, M. I. Cockett, D. Eaton, and A. J. P. Docherty.** 1992. The C-terminal domain of 72 kDa gelatinase A is not required for catalysis, but is essential for membrane activation and modulates interactions with tissue inhibitors of metalloproteinases. *Biochem. J.* **283**: 637–641.

258. **Nagase, H., C. E. Brinckerhoff, C. A. Vater, and E. D. Harris, Jr.** 1983. Biosynthesis and secretion of procollagenase by rabbit synovial fibroblasts. Inhibition of procollagenase secretion by monensin and evidence for glycosylation of procollagenase. *Biochem. J.* **214**: 281–288.

259. **Nagase, H., T. E. Cawston, M. de Silva, and A. J. Barrett.** 1982. Identification of plasma kallikrein as an activator of latent collagenase in rheumatoid synovial fluid. *Biochim. Biophys. Acta* **702**: 133–142.

260. **Nagase, H., J. J. Enghild, K. Suzuki, and G. Salvesen.** 1990. Stepwise activation mechanisms of the precursor of matrix metalloproteinase 3 (stromelysin) by proteinases and (4-aminophenyl)mercuric acetate. *Biochemistry* **29**: 5783–5789.

261. **Nagase, H., R. C. Jackson, C. E. Brinckerhoff, C. A. Vater, and E. D. Harris, Jr.** 1981. A precursor form of latent collagenase produced in a cell-free system with mRNA from rabbit synovial cells. *J. Biol. Chem.* **256**: 11951–11954.

262. **Nagase, H., Y. Ogata, K. Suzuki, J. J. Enghild, and G. Salvesen.** 1991. Substrate specificities and activation mechanisms of matrix metalloproteinases. *Biochem. Soc. Trans.* **19**: 715–718.

263. **Nagase, H., K. Suzuki, T. E. Cawston, and K. Brew.** 1997. Involvement of a region near valine-69 of tissue inhibitor of metalloproteinases (TIMP)-1 in the interaction with matrix metalloproteinase 3 (stromelysin 1). *Biochem. J.* **325**: 163–167.

264. **Nakamura, H., Y. Fujii, E. Ohuchi, E. Yamamoto, and Y. Okada.** 1998. Activation of the precursor of human stromelysin 2 and its interactions with other matrix metalloproteinases. *Eur. J. Biochem.* **253**: 67–75.

265. **Nicholson, R., G. Murphy, and R. Breathnach.** 1989. Human and rat malignant-tumor-associated mRNAs encode stromelysin-like metalloproteinases. *Biochemistry* **28**: 5195–5203.

266. **O'Connell, J. P., F. Willenbrock, A. J. P. Docherty, D. Eaton, and G. Murphy.** 1994. Analysis of the role of the COOH-terminal domain in the activation, proteolytic activity, and tissue inhibitor of metalloproteinase interactions of gelatinase B. *J. Biol. Chem.* **269**: 14967–14973.

267. **O'Hare, M. C., V. A. Curry, R. E. Mitchell, and T. E. Cawston.** 1995. Stabilisation of purified human collagenase by site-directed mutagenesis. *Biochem. Biophys. Res. Commun.* **216**: 329–337.

268. **Ogata, Y., J. J. Enghild, and H. Nagase.** 1992. Matrix metalloproteinase 3 (stromelysin) activates the precursor for the human matrix metalloproteinase 9. *J. Biol. Chem.* **267**: 3581–3584.

269. **Ogata, Y., Y. Itoh, and H. Nagase.** 1995. Steps involved in activation of the pro-matrix metalloproteinase 9 (progelatinase B)-tissue inhibitor of metalloproteinases-1 complex by 4-aminophenylmercuric acetate and proteinases. *J. Biol. Chem.* **270**: 18506–18511.

270. **Okada, Y., Y. Gonoji, K. Naka, K. Tomita, I. Nakanishi, K. Iwata, K. Yamashita, and T. Hayakawa.** 1992. Matrix metalloproteinase 9 (92-kDa gelatinase/type IV collagenase) from HT 1080 human fibrosarcoma cells. Purification and activation of the precursor and enzymic properties. *J. Biol. Chem.* **267**: 21712–21719.

271. **Okada, Y., E. D. Harris, Jr., and H. Nagase.** 1988. The precursor of a metalloendopeptidase from human rheumatoid synovial fibroblasts. Purification and mechanisms of activation by endopeptidases and 4-aminophenylmercuric acetate. *Biochem. J.* **254**: 731–741.

272. **Okada, Y., T. Morodomi, J. J. Enghild, K. Suzuki, A. Yasui, I. Nakanishi, G. Salvesen, and H. Nagase.** 1990. Matrix metalloproteinase 2 from human rheumatoid synovial fibroblasts. Purification and activation of the precursor and enzymic properties. *Eur. J. Biochem.* **194**: 721–730.

273. **Okada, Y., H. Nagase, and E. D. Harris, Jr.** 1986. A metalloproteinase from human rheumatoid synovial fibroblasts that digests connective tissue matrix components. Purification and characterization. *J. Biol. Chem.* **261**: 14245–14255.

274. **Okada, Y., S. Watanabe, I. Nakanishi, J. Kishi, T. Hayakawa, W. Watorek, J. Travis, and H. Nagase.** 1988. Inactivation of tissue inhibitor of metalloproteinases by neutrophil elastase and other serine proteinases. *FEBS Lett.* **229**: 157–160.

275. **Okamoto, T., T. Akaike, T. Nagano, S. Miyajima, M. Suga, M. Ando, K. Ichimori, and H. Maeda.** 1997. Activation of human neutrophil procollagenase by nitrogen dioxide and peroxynitrite—a novel mechanism for procollagenase activation involving nitric oxide. *Arch. Biochem. Biophys.* **342**: 261–274.

276. **Okamoto, T., T. Akaike, M. Suga, S. Tanase, H. Horie, S. Miyajima, M. Ando, Y. Ichinose, and H. Maeda.** 1997. Activation of human matrix metalloproteinases by various bacterial proteinases. *J. Biol. Chem.* **272**: 6059–6066.

277. **Okumura, Y., H. Sato, M. Seiki, and H. Kido.** 1997. Proteolytic activation of the precursor of membrane type 1 matrix metalloproteinase by human plasmin. A possible cell surface activator. *FEBS Lett.* **402**: 181–184.

278. **Olson, M. W., D. C. Gervasi, S. Mobashery, and R. Fridman.** 1997. Kinetic analysis of the binding of human matrix metalloproteinase-2 and -9 to tissue inhibitor of metalloproteinase (TIMP-1 and TIMP-2). *J. Biol. Chem.* **272**: 29975–29983.

279. **Oronsky, A. L., R. J. Perper, and H. C. Schroder.** 1973. Phagocytic release and activation of human leukocyte procollagenase. *Nature* **246**: 417–419.

280. **Overall, C. M. and J. Sodek.** 1990. Concanavalin A produces a matrix-degradative phenotype in human fibroblasts. Induction and endogenous activation of collagenase, 72-kDa gelatinase, and Pump-1 is accompanied by the suppression of the tissue inhibitor of matrix metalloproteinases. *J. Biol. Chem.* **265**: 21141–21151.

281. **Owens, M. W., S. A. Milligan, D. Jourdheuil, and M. B. Grisham.** 1997. Effects of reactive metabolites of oxygen and nitrogen on gelatinase A activity. *Am. J. Physiol. -Lung Cell. Mol. Physiol.* **273**: L445-L450

282. **Park, A. J., L. M. Matrisian, A. F. Kells, R. Pearson, Z. Y. Yuan, and M. Navre.** 1991. Mutational analysis of the transin (rat stromelysin) autoinhibitor region demonstrates a role for residues surrounding the 'cysteine switch'. *J. Biol. Chem.* **266**: 1584–1590.

283. **Pei, D. and S. J. Weiss.** 1995. Furin-dependent intracellular activation of the human stromelysin-3 zymogen. *Nature* **375**: 244–247.

284. **Pei, D. and S. J. Weiss.** 1996. Transmembrane-deletion mutants of the membrane-type matrix metalloproteinase-1 process progelatinase A and express intrinsic matrix-degrading activity. *J. Biol. Chem.* **271**: 9135–9140.

285. **Peppin, G. J. and S. J. Weiss.** 1986. Activation of the endogenous metalloproteinase, gelatinase, by triggered human neutrophils. *Proc. Natl. Acad. Sci. USA* **83**: 4322–4326.

286. **Puente, X. S., A. M. Pendás, E. Llano, G. Velasco, and C. López-Otín.** 1996. Molecular cloning of a novel membrane-type matrix metalloproteinase from a human breast carcinoma. *Cancer Res.* **56**: 944–949.

287. **Reinemer, P., F. Grams, R. Huber, T. Kleine, S. Schnierer, M. Piper, H. Tschesche, and W. Bode.** 1994. Structural implications for the role of the N terminus in the 'superactivation' of collagenases. A crystallographic study. *FEBS Lett.* **338**: 227–233.

288. **Saari, H., K. Suomalainen, O. Lindy, Y. T. Konttinen, and T. Sorsa.** 1990. Activation of latent human neutrophil collagenase by reactive oxygen species and serine proteases. *Biochem. Biophys. Res. Commun.* **171**: 979–987.

289. **Saarinen, J., N. Kalkkinen, H. G. Welgus, and P. T. Kovanen.** 1994. Activation of human interstitial procollagenase through direct cleavage of the Leu[83]-Thr[84] bond by mast cell chymase. *J. Biol. Chem.* **269**: 18134–18140.

290. **Sakamoto, S., M. Sakamoto, P. Goldhaber, and M. J. Glimcher.** 1978. Mouse bone collagenase. Purification of the enzyme by heparin-substituted Sepharose 4B affinity chromatography and preparation of specific antibody to the enzyme. *Arch. Biochem. Biophys.* **188**: 438–449.

291. **Sakamoto, S., M. Sakamoto, A. Matsumoto, M. Nagayama, and M. J. Glimcher.** 1981. Chick bone collagenase inhibitor and latency of collagenase. *Biochem. Biophys. Res. Commun.* **103**: 339–346.

292. **Salowe, S. P., A. I. Marcy, G. C. Cuca, C. K. Smith, I. E. Kopka, W. K. Hagmann, and J. D. Hermes.** 1992. Characterization of zinc-binding sites in human stromelysin-1: stoichiometry of the catalytic domain and identification of a cysteine ligand in the proenzyme. *Biochemistry* **31**: 4535–4540.

293. **Sang, Q. A., M. K. Bodden, and L. J. Windsor.** 1996. Activation of human progelatinase A by collagenase and matrilysin: activation of procollagenase by matrilysin. *J. Protein Chem.* **15**: 243–253.

294. **Sang, Q. X., H. Birkedal-Hansen, and H. E. Van Wart.** 1995. Proteolytic and non-proteolytic activation of human neutrophil progelatinase B. *Biochim. Biophys. Acta* **1251**: 99–108.

295. **Sang, Q. X., E. W. Thompson, D. Grant, W. G. Stetler-Stevenson, and S. W. Byers.** 1991. Soluble laminin and arginine-glycine-aspartic acid containing peptides differentially regulate type IV collagenase messenger RNA, activation, and localization in testicular cell culture. *Biol. Reprod.* **45**: 387–394.

296. **Santavicca, M., A. Noël, H. Angliker, I. Stoll, J. P. Segain, P. Anglard, M. Chretien, N. Seidah, and P. Basset.** 1996. Characterization of structural determinants and molecular mechanisms involved in pro-stromelysin-3 activation by 4-aminophenylmercuric acetate and furin-type convertases. *Biochem. J.* **315**: 953–958.

297. **Sato, H., T. Kinoshita, T. Takino, K. Nakayama, and M. Seiki.** 1996. Activation of a recombinant membrane type 1-matrix metalloproteinase (MT1-MMP) by furin and its interaction with tissue inhibitor of metalloproteinases (TIMP)-2. *FEBS Lett.* **393**: 101–104.

298. **Sato, H., T. Takino, Y. Okada, J. Cao, A. Shinagawa, E. Yamamoto, and M. Seiki.** 1994. A matrix metalloproteinase expressed on the surface of invasive tumour cells. *Nature* **370**: 61–65.

299. **Saus, J., S. Quinones, Y. Otani, H. Nagase, E. D. Harris, Jr., and M. Kurkinen.** 1988. The complete primary structure of human matrix metalloproteinase-3. Identity with stromelysin. *J. Biol. Chem.* **263**: 6742–6745.

300. **Schor, A. M., S. L. Schor, J. B. Weiss, R. A. Brown, S. Kumar, and P. Phillips.** 1980. Stimulation by a low-molecular-weight angiogenic factor of capillary endothelial cells in culture. *Br. J. Cancer* **41**: 790–797.

301. **Sellers, A., E. Cartwright, G. Murphy, and J. J. Reynolds.** 1977. Evidence that latent collagenases are enzyme-inhibitor complexes. *Biochem. J.* **163**: 303–307.

302. **Seltzer, J. L., A. Y. Lee, K. T. Akers, B. Sudbeck, E. A. Southon, E. A. Wayner, and A. Z. Eisen.** 1994. Activation of 72-kDa type IV collagenase/gelatinase by normal fibroblasts in collagen lattices is mediated by integrin receptors but is not related to lattice contraction. *Exp. Cell Res.* **213**: 365–374.

303. **Shapiro, S. D., C. J. Fliszar, T. J. Broekelmann, R. P. Mecham, R. M. Senior, and H. G. Welgus.** 1995. Activation of the 92-kDa gelatinase by stromelysin and 4-aminophenylmercuric acetate. Differential processing and stabilization of the carboxyl-terminal domain by tissue inhibitor of metalloproteinases (TIMP). *J. Biol. Chem.* **270**: 6351–6356.

304. **Shapiro, S. D., G. L. Griffin, D. J. Gilbert, N. A. Jenkins, N. G. Copeland, H. G. Welgus, R. M. Senior, and T. J. Ley.** 1992. Molecular cloning, chromosomal localization, and bacterial expression of a murine macrophage metalloelastase. *J. Biol. Chem.* **267**: 4664–4671.

305. **Shapiro, S. D., D. K. Kobayashi, and T. J. Ley.** 1993. Cloning and characterization of a unique elastolytic metalloproteinase produced by human alveolar macrophages. *J. Biol. Chem.* **268**: 23824–23829.

306. **Shinkai, H. and Y. Nagai.** 1977. A latent collagenase from embryonic human skin explants. *J. Biochem. (Tokyo)* **81**: 1261–1268.

307. **Simpson, J. W. and A. C. Taylor.** 1974. Regulation of gingival collagenase: a possible role for a mast cell factor. *Proc. Soc. Exp. Biol. Med.* **145**: 42–47.

308. **Sopata, I. and A. M. Dancewicz.** 1974. Presence of a gelatin-specific proteinase and its latent form in human leucocytes. *Biochim. Biophys. Acta* **370**: 510–523.

309. **Sopata, I. and J. Wize.** 1979. A latent gelatin specific proteinase of human leucocytes and its activation. *Biochim. Biophys. Acta* **571**: 305–312.

309a. **Sorsa, T., T. Salo, T., E. Koivunen, Y. T. Konttinen, U. Bergmann, A. Tuuttila, E. Niemi, O. Teronen, P. Heikkila, H. Tschesche** *et al.* 1997. Activation of type IV procollagenases by human tumor-associated trypsin-2. *J. Biol. Chem.* **272**: 21067–21074.

310. **Springman, E. B., E. L. Angleton, H. Birkedal-Hansen, and H. E. Van Wart.** 1990. Multiple modes of activation of latent human fibroblast collagenase:

evidence for the role of a Cys[73] active-site zinc complex in latency and a 'cysteine switch' mechanism for activation. *Proc. Natl. Acad. Sci. USA* **87**: 364–368.

311. **Stetler-Stevenson, W. G., H. C. Krutzsch, and L. A. Liotta.** 1989. Tissue inhibitor of metalloproteinase (TIMP-2). A new member of the metalloproteinase inhibitor family. *J. Biol. Chem.* **264**: 17374–17378.

312. **Stetler-Stevenson, W. G., H. C. Krutzsch, M. P. Wacher, I. M. Margulies, and L. A. Liotta.** 1989. The activation of human type IV collagenase proenzyme. Sequence identification of the major conversion product following organomercurial activation. *J. Biol. Chem.* **264**: 1353–1356.

313. **Stricklin, G. P., E. A. Bauer, J. J. Jeffrey, and A. Z. Eisen.** 1977. Human skin collagenase: isolation of precursor and active forms from both fibroblast and organ cultures. *Biochemistry* **16**: 1607–1615.

314. **Stricklin, G. P., J. J. Jeffrey, W. T. Roswit, and A. Z. Eisen.** 1983. Human skin fibroblast procollagenase: mechanisms of activation by organomercurials and trypsin. *Biochemistry* **22**: 61–68.

315. **Strongin, A. Y., I. Collier, G. Bannikov, B. L. Marmer, G. A. Grant, and G. I. Goldberg.** 1995. Mechanism of cell surface activation of 72-kDa type IV collagenase. Isolation of the activated form of the membrane metalloprotease. *J. Biol. Chem.* **270**: 5331–5338.

316. **Strongin, A. Y., B. L. Marmer, G. A. Grant, and G. I. Goldberg.** 1993. Plasma membrane-dependent activation of the 72-kDa type IV collagenase is prevented by complex formation with TIMP-2. *J. Biol. Chem.* **268**: 14033–14039.

317. **Suzuki, K., J. J. Enghild, T. Morodomi, G. Salvesen, and H. Nagase.** 1990. Mechanisms of activation of tissue procollagenase by matrix metalloproteinase 3 (stromelysin). *Biochemistry* **29**: 10261–10270.

318. **Suzuki, K., C. C. Kan, W. Hung, M. R. Gehring, K. Brew, and H. Nagase.** 1998. Expression of human pro-matrix metalloproteinase 3 that lacks the N-terminal 34 residues in *Escherichia coli*—autoactivation and interaction with tissue inhibitor of metalloproteinase 1 (TIMP-1). *Biol. Chem.* **379**: 185–191.

319. **Suzuki, K., M. Lees, G. F. Newlands, H. Nagase, and D. E. Woolley.** 1995. Activation of precursors for matrix metalloproteinases 1 (interstitial collagenase) and 3 (stromelysin) by rat mast-cell proteinases I and II. *Biochem. J.* **305**: 301–306.

320. **Suzuki, K., H. Nagase, A. Ito, J. J. Enghild, and G. Salvesen.** 1990. The role of matrix metalloproteinase 3 in the stepwise activation of human rheumatoid synovial procollagenase. *Biol. Chem. Hoppe Seyler* **371**: 305–310.

321. **Takino, T., H. Sato, A. Shinagawa, and M. Seiki.** 1995. Identification of the second membrane-type matrix metalloproteinase (MT-MMP-2) gene from a human placenta cDNA library. MT-MMPs form a unique membrane-type subclass in the MMP family. *J. Biol. Chem.* **270**: 23013–23020.

322. **Treadwell, B. V., J. Neidel, M. Pavia, C. A. Towle, M. E. Trice, and H. J. Mankin.** 1986. Purification and characterization of collagenase activator protein synthesized by articular cartilage. *Arch. Biochem. Biophys.* **251**: 715–723.

323. **Triebel, S., J. Bläser, H. Reinke, V. Knäuper, and H. Tschesche.** 1992. Mercurial activation of human PMN leucocyte type IV procollagenase (gelatinase). *FEBS Lett.* **298**: 280–284.

324. **Tschesche, H., V. Knäuper, S. Kramer, J. Michaelis, R. Oberhoff, and H. Reinke.** 1992. Latent collagenase and gelatinase from human neutrophils and their activation. *Matrix Suppl.* **1**: 245–255.

325. **Tschesche, H., J. Michaelis, U. Kohnert, J. Fedrowitz, and R. Oberhoff.** 1989. Tissue kallikrein effectively activates latent matrix degrading metalloenzymes. In *Kinins 5, Part A—Adv. Exp. Med. Biol 247.* K. Abe, H. Moriya, and S. Fujii (eds), pp. 545–548. Plenum, New York.

326. **Tyree, B., J. L. Seltzer, J. Halme, J. J. Jeffrey, and A. Z. Eisen.** 1981. The stoichiometric activation of human skin fibroblast pro-collagenase by factors present in human skin and rat uterus. *Arch. Biochem. Biophys.* **208**: 440–443.

327. **Unemori, E. N., M. J. Bair, E. A. Bauer, and E. P. Amento.** 1991. Stromelysin expression regulates collagenase activation in human fibroblasts. Dissociable control of two metalloproteinases by interferon-γ. *J. Biol. Chem.* **266**: 23477–23482.

328. **Vaes, G.** 1972. Multiple steps in the activation of the inactive precursor of bone collagenase by trypsin. *FEBS Lett.* **28**: 198–200.

329. **Vaes, G.** 1972. The release of collagenase as an inactive proenzyme by bone explants in culture. *Biochem. J.* **126**: 275–289.

330. **Valle, K.-J. and E. A. Bauer.** 1979. Biosynthesis of collagenase by human skin fibroblasts in monolayer culture. *J. Biol. Chem.* **254**: 10115–10122.

331. **Vallee, B. L. and D. S. Auld.** 1990. Zinc coordination, function, and structure of zinc enzymes and other proteins. *Biochemistry* **29**: 5647–5659.

332. **Van Wart, H. E. and H. Birkedal-Hansen.** 1990. The cysteine switch: a principle of regulation of metalloproteinase activity with potential applicability to the entire matrix metalloproteinase gene family. *Proc. Natl. Acad. Sci. USA* **87**: 5578–5582.

333. **Vater, C. A., C. L. Mainardi, and E. D. Harris, Jr.** 1978. Activation in vitro of rheumatoid synovial collagenase from cell cultures. *J. Clin. Invest.* **62**: 987–992.

334. **Vater, C. A., H. Nagase, and E. D. Harris, Jr.** 1983. Purification of an endogenous activator of

procollagenase from rabbit synovial fibroblast culture medium. *J. Biol. Chem.* **258**: 9374–9382.

335. **Vater, C. A., H. Nagase, and E. D. Harris, Jr.** 1986. Proactivator-dependent activation of procollagenase induced by treatment with EGTA. *Biochem. J.* **237**: 853–858.

336. **Vissers, M. C. and C. C. Winterbourn.** 1988. Activation of human neutrophil gelatinase by endogenous serine proteinases. *Biochem. J.* **249**: 327–331.

337. **Ward, R. V., S. J. Atkinson, J. J. Reynolds, and G. Murphy.** 1994. Cell surface-mediated activation of progelatinase A: demonstration of the involvement of the C-terminal domain of progelatinase A in cell surface binding and activation of progelatinase A by primary fibroblasts. *Biochem. J.* **304**: 263–269.

338. **Ward, R. V., S. J. Atkinson, P. M. Slocombe, A. J. P. Docherty, J. J. Reynolds, and G. Murphy.** 1991. Tissue inhibitor of metalloproteinases-2 inhibits the activation of 72 kDa progelatinase by fibroblast membranes. *Biochim. Biophys. Acta* **1079**: 242–246.

339. **Ward, R. V., R. M. Hembry, J. J. Reynolds, and G. Murphy.** 1991. The purification of tissue inhibitor of metalloproteinases-2 from its 72 kDa progelatinase complex. Demonstration of the biochemical similarities of tissue inhibitor of metalloproteinases-2 and tissue inhibitor of metalloproteinases-1. *Biochem. J.* **278**: 179–187.

340. **Weeks, J. G.** 1974. Collagenase in the postpartum rat uterus. Extraction characterization and mechanism of inhibition by estradiol. *Diss. Abstr. Int. B.* **35**: 107.

341. **Weiss, S. J., G. Peppin, X. Ortiz, C. Ragsdale, and S. T. Test.** 1985. Oxidative autoactivation of latent collagenase by human neutrophils. *Science* **227**: 747–749.

342. **Werb, Z., C. L. Mainardi, C. A. Vater, and E. D. Harris, Jr.** 1977. Endogenous activiation of latent collagenase by rheumatoid synovial cells. Evidence for a role of plasminogen activator. *N. Engl. J. Med.* **296**: 1017–1023.

343. **Wilhelm, S. M., I. E. Collier, A. Kronberger, A. Z. Eisen, B. L. Marmer, G. A. Grant, E. A. Bauer, and G. I. Goldberg.** 1987. Human skin fibroblast stromelysin: structure, glycosylation, substrate specificity, and differential expression in normal and tumorigenic cells. *Proc. Natl. Acad. Sci. USA* **84**: 6725–6729.

344. **Wilhelm, S. M., I. E. Collier, B. L. Marmer, A. Z. Eisen, G. A. Grant, and G. I. Goldberg.** 1989. SV40-transformed human lung fibroblasts secrete a 92-kDa type IV collagenase which is identical to that secreted by normal human macrophages. *J. Biol. Chem.* **264**: 17213–17221.

345. **Will, H., S. J. Atkinson, G. S. Butler, B. Smith, and G. Murphy.** 1996. The soluble catalytic domain of membrane type 1 matrix metalloproteinase cleaves the propeptide of progelatinase A and initiates autoproteolytic activation. Regulation by TIMP-2 and TIMP-3. *J. Biol. Chem.* **271**: 17119–17123.

346. **Will, H. and B. Hinzmann.** 1995. cDNA sequence and mRNA tissue distribution of a novel human matrix metalloproteinase with a potential transmembrane segment. *Eur. J. Biochem.* **231**: 602–608.

347. **Willenbrock, F., T. Crabbe, P. M. Slocombe, C. W. Sutton, A. J. P. Docherty, M. I. Cockett, M. O'Shea, K. Brocklehurst, I. R. Phillips, and G. Murphy.** 1993. The activity of the tissue inhibitors of metalloproteinases is regulated by C-terminal domain interactions: a kinetic analysis of the inhibition of gelatinase A. *Biochemistry* **32**: 4330–4337.

348. **Williams, D. H. and E. J. Murray.** 1994. Specific amino acid substitutions in human collagenase cause decreased autoproteolysis and reveal a requirement for a second zinc atom for catalytic activity. *FEBS Lett.* **354**: 267–270.

349. **Windsor, L. J., H. Birkedal-Hansen, B. Birkedal-Hansen, and J. A. Engler.** 1991. An internal cysteine plays a role in the maintenance of the latency of human fibroblast collagenase. *Biochemistry* **30**: 641–647.

350. **Windsor, L. J., H. Grenett, B. Birkedal-Hansen, M. K. Bodden, J. A. Engler, and H. Birkedal-Hansen.** 1993. Cell type-specific regulation of SL-1 and SL-2 genes. Induction of the SL-2 gene but not the SL-1 gene by human keratinocytes in response to cytokines and phorbolesters. *J. Biol. Chem.* **268**: 17341–17347.

351. **Woessner, J. F., Jr.** 1977. A latent form of collagenase in the involuting rat uterus and its activation by a serine proteinase. *Biochem. J.* **161**: 535–542.

352. **Yang, M., K. Hayashi, M. Hayashi, J. T. Fujii, and M. Kurkinen.** 1996. Cloning and developmental expression of a membrane-type matrix metalloproteinase from chicken. *J. Biol. Chem.* **271**: 25548–25554.

353. **Yang, M. Z., M. T. Murray, and M. Kurkinen.** 1997. A novel matrix metalloproteinase gene (XMMP) encoding vitronectin-like motifs is transiently expressed in Xenopus laevis early embryo development. *J. Biol. Chem.* **272**: 13527–13533.

See also review articles: 66, 68, 88.

Biological activity of MMPs

353a. **D'Armiento, J., S. S. Dalal, Y. Okada, R. A. Berg, and K. Chada.** (1992) Collagenase expression in the lungs of transgenic mice causes pulmonary emphysema. *Cell* **71**: 955–961.

354. **D'Armiento, J., T. DiColandrea, S. S. Dalal, Y. Okada, M.-T. Huang, A. H. Conney, and K. Chada.** 1995. Collagenase expression in transgenic mouse skin causes hyperkeratosis and acanthosis and increases

susceptibility to tumorigenesis. *Mol. Cell. Biol.* **15**: 5732–5739.

355. **Durko, M., R. Navab, H. R. Shibata, and P. Brodt.** 1997. Suppression of basement membrane type IV collagen degradation and cell invasion in human melanoma cells expressing an antisense RNA for MMP-1. *Biochim. Biophys. Acta—Mol. Cell Res.* **1356**: 271–280.

356. **Hasegawa, S., N. Koshikawa, N. Momiyama, K. Moriyama, Y. Ichikawa, T. Ishikawa, M. Mitsuhashi, H. Shimada, and K. Miyazaki.** 1998. Matrilysin-specific antisense oligonucleotide inhibits liver metastasis of human colon cancer cells in a nude mouse model. *Int. J. Canc.* **76**: 812–816.

357. **Hautamaki, R. D., D. K. Kobayashi, R. M. Senior, and S. D. Shapiro.** 1997. Requirement for macrophage elastase for cigarette smoke-induced emphysema in mice. *Science* **277**: 2002–2004.

358. **Hua, J. and R. J. Muschel.** 1996. Inhibition of matrix metalloproteinase 9 expression by a ribozyme blocks metastasis in a rat sarcoma model system. *Cancer Res.* **56**: 5279–5284.

359. **Itoh, T., T. Ikeda, H. Gomi, S. Nakao, T. Suzuki, and S. Itohara.** 1997. Unaltered secretion of beta-amyloid precursor protein in gelatinase A (matrix metalloproteinase 2)-deficient mice. *J. Biol. Chem.* **272**: 22389–22392.

360. **Itoh, T., M. Tanioka, H. Yoshida, T. Yoshioka, H. Nishimoto, and S. Itohara.** 1998. Reduced angiogenesis and tumor progression in gelatinase A-deficient mice. *Cancer Res.* **58**: 1048–1051.

361. **Lemaître, V., A. Jungbluth, and Y. Eeckhout.** 1997. The recombinant catalytic domain of mouse collagenase-3 depolymerizes type I collagen by cleaving its aminotelopeptides. *Biochem. Biophys. Res. Commun.* **230**: 202–205.

362. **Lijnen, H. R., J. Silence, B. Vanhoef, and D. Collen.** 1998. Stromelysin-1 (MMP-3)-independent gelatinase expression and activation in mice. *Blood* **91**: 2045–2053.

363. **Liu, X., H. Wu, M. Byrne, J. Jeffrey, S. Krane, and R. Jaenisch.** 1995. A targeted mutation at the known collagenase cleavage site in mouse type I collagen impairs tissue remodeling. *J. Cell Biol.* **130**: 227–237.

364. **Lochter, A., A. Srebrow, C. J. Sympson, N. Terracio, Z. Werb, and M. J. Bissell.** 1997. Misregulation of stromelysin-1 expression in mouse mammary tumor cells accompanies acquisition of stromelysin-1-dependent invasive properties. *J. Biol. Chem.* **272**: 5007–5015.

365. **Masson, R., O. Lefebvre, A. Noël, M. El Fahime, M. P. Chenard, C. Wendling, F. Kebers, M. LeMeur, A. Dierich, J.-M. Foidart, P. Basset, and M.-C. Rio.** 1998. In vivo evidence that the stromelysin-3 metalloproteinase contributes in a paracrine manner to epithelial cell malignancy. *J. Cell Biol.* **140**: 1535–1541.

366. **Melong, R. K. and P. S. Miller.** 1996. Inhibition of human collagenase activity by antisense oligonucleoside methylphosphonates. *Antisense Nucleic Acid Drug Dev.* **6**: 273–280.

367. **Momiyama, N., N. Koshikawa, T. Ishikawa, Y. Ichikawa, S. Hasegawa, Y. Nagashima, M. Mitsuhashi, K. Miyazaki, and H. Shimada.** 1998. Inhibitory effect of matrilysin antisense oligonucleotides on human colon cancer cell invasion in vitro. *Mol. Carcinog.* **22**: 57–63.

368. **Mudgett, J. S., N. I. Hutchinson, N. A. Chartrain, A. J. Forsyth, J. McDonnell, I. I. Singer, E. K. Bayne, J. Flanagan, D. Kawka, C. F. Shen, K., et al.** 1998. Susceptibility of stromelysin 1-deficient mice to collagen-induced arthritis and cartilage destruction. *Arth. Rheum.* **41**: 110–121.

369. **Noël, A. C., O. Lefebvre, E. Maquoi, L. VanHoorde, M. P. Chenard, M. Mareel, J. M. Foidart, P. Basset, and M. C. Rio.** 1996. Stromelysin-3 expression promotes tumor take in nude mice. *J. Clin. Invest.* **97**: 1924–1930.

370. **Nordstrom, L. A., J. Lochner, W. Yeung, and G. Ciment.** 1995. The metalloproteinase stromelysin-1 (transin) mediates PC12 cell growth cone invasiveness through basal laminae. *Mol. Cell. Neurosci.* **6**: 56–68.

371. **Rudolph-Owen, L. A., P. Cannon, and L. M. Matrisian.** 1998. Overexpression of the matrix metalloproteinase matrilysin results in premature mammary gland differentiation and male infertility. *Mol. Biol. Cell.* **9**: 421–435.

372. **Rudolph-Owen, L. A., D. L. Hulboy, C. L. Wilson, J. Mudgett, and L. M. Matrisian.** 1997. Coordinate expression of matrix metalloproteinase family members in the uterus of normal, matrilysin-deficient, and stromelysin-1-deficient mice. *Endocrinology* **138**: 4902–4911.

373. **Shipley, J. M., R. L. Wesselschmidt, D. K. Kobayashi, T. J. Ley, and S. D. Shapiro.** 1996. Metalloelastase is required for macrophage-mediated proteolysis and matrix invasion in mice. *Proc. Natl. Acad. Sci. USA* **93**: 3942–3946.

374. **Sympson, C. J., R. S. Talhouk, C. M. Alexander, J. R. Chin, S. M. Clift, M. J. Bissell, and Z. Werb.** 1994. Targeted expression of stromelysin-1 in mammary gland provides evidence for a role of proteinases in branching morphogenesis and the requirement for an intact basement membrane for tissue-specific gene expression. *J. Cell Biol.* **125**: 681–693.

375. **Turck, J., A. S. Pollock, L. K. Lee, H. P. Marti, and D. H. Lovett.** 1996. Matrix metalloproteinase 2 (gelatinase A) regulates glomerular mesangial cell proliferation and differentiation. *J. Biol. Chem.* **271**: 15074–15083.

376. **Vu, T. H., J. M. Shipley, G. Bergers, J. E. Berger, J. A. Helms, D. Hanahan, S. D. Shapiro, R. M. Senior,**

and Z. Werb. 1998. MMP-9/gelatinase B is a key regulator of growth plate angiogenesis and apoptosis of hypertrophic chondrocytes. *Cell* **93**: 411–422.

377. **Werb, Z., C. J. Sympson, C. M. Alexander, N. Thomasset, L. R. Lund, A. MacAuley, J. Ashkenas, and M. J. Bissell.** 1996. Extracellular matrix remodeling and the regulation of epithelial-stromal interactions during differentiation and involution. *Kidney Int. Suppl.* **54**: S68–74.

378. **Wilson, C. L., K. J. Heppner, P. A. Labosky, B. L. Hogan, and L. M. Matrisian.** 1997. Intestinal tumorigenesis is suppressed in mice lacking the metalloproteinase matrilysin. *Proc. Natl. Acad. Sci. USA* **94**: 1402–1407.

379. **Witty, J. P., T. Lempka, R. J. Coffey, Jr., and L. M. Matrisian.** 1995. Decreased tumor formation in 7,12-dimethylbenzanthracene-treated stromelysin-1 transgenic mice is associated with alterations in mammary epithelial cell apoptosis. *Cancer Res.* **55**: 1401–1406.

380. **Witty, J. P., S. McDonnell, K. J. Newell, P. Cannon, M. Navre, R. J. Tressler, and L. M. Matrisian.** 1994. Modulation of matrilysin levels in colon carcinoma cell lines affects tumorigenicity in vivo. *Cancer Res.* **54**: 4805–4812.

381. **Witty, J. P., J. H. Wright, and L. M. Matrisian.** 1995. Matrix metalloproteinases are expressed during ductal and alveolar mammary morphogenesis, and misregulation of stromelysin-1 in transgenic mice induces unscheduled alveolar development. *Mol. Biol. Cell.* **6**: 1287–1303.

382. **Yamamoto, H., M. L. Flannery, S. Kupriyanov, J. Pearce, S. R. McKercher, G. W. Henkel, R. A. Maki, Z. Werb, R. G. Oshima, Y. Akiyama, et al.** 1998. Defective trophoblast function in mice with a targeted mutation of Ets2. *Genes Devel.* **12**: 1315–1326.

383. **Yamamoto, H., F. Itoh, Y. Hinoda, and K. Imai.** 1995. Suppression of matrilysin inhibits colon cancer cell invasion in vitro. *Int. J. Cancer* **61**: 218–222.

384. **Yu, M., E. T. Bowden, J. Sitlani, H. Sato, M. Seiki, S. C. Mueller, and E. W. Thompson.** 1997. Tyrosine phosphorylation mediates ConA-induced membrane type 1-matrix metalloproteinase expression and matrix metalloproteinase-2 activation in MDA-MB-231 human breast carcinoma cells. *Cancer Res.* **57**: 5028–5032.

See also review articles: 5, 6, 9, 10, 11, 20, 21, 22, 23, 24, 25, 26, 27, 29, 30, 31, 32, 33, 35, 39, 40, 46, 47, 48, 50, 54, 56, 57, 58, 62, 65, 67, 69, 75, 76, 78, 79, 96, 97, 98, 99, 102, 105, 106, 107, 108, 109, 110, 112, 113, 116, 119, 120, 124, 125, 126, 127, 128, 131, 132, 133, 134, 136, 137, 138, 147, 148, 149.

Biological role of TIMPs

385. **Adachi, Y. and H. Nohara.** 1989. Separation and partial characterization of three forms of collagenase inhibitor from bovine gingiva. *Arch. Oral Biol.* **34**: 431–436.

386. **Agrez, M. V., C. J. Meldrum, A. T. Sim, R. H. Aebersold, I. M. Clark, T. E. Cawston, and G. F. Burns.** 1995. A fibroblast elongation factor purified from colon carcinoma cells shares sequence identity with TIMP-1. *Biochem. Biophys. Res. Commun.* **206**: 590–600.

387. **Albin, R. J., R. M. Senior, H. G. Welgus, N. L. Connolly, and E. J. Campbell.** 1987. Human alveolar macrophages secrete an inhibitor of metalloproteinase elastase. *Am. Rev. Respir. Dis.* **135**: 1281–1285.

388. **Alexander, C. M., E. W. Howard, M. J. Bissell, and Z. Werb.** 1996. Rescue of mammary epithelial cell apoptosis and entactin degradation by a tissue inhibitor of metalloproteinases-1 transgene. *J. Cell Biol.* **135**: 1669–1677.

389. **Alexander, C. M. and Z. Werb.** 1992. Targeted disruption of the tissue inhibitor of metalloproteinases gene increases the invasive behavior of primitive mesenchymal cells derived from embryonic stem cells in vitro. *J. Cell Biol.* **118**: 727–739.

390. **Apte, S. S., B. R. Olsen, and G. Murphy.** 1995. The gene structure of tissue inhibitor of metalloproteinases (TIMP)-3 and its inhibitory activities define the distinct TIMP gene family. *J. Biol. Chem.* **270**: 14313–14318.

391. **Avalos, B. R., S. E. Kaufman, M. Tomonaga, R. E. Williams, D. W. Golde, and J. C. Gasson.** 1988. K562 cells produce and respond to human erythroid-potentiating activity. *Blood* **71**: 1720–1725.

392. **Baker, A. H., A. B. Zaltsman, S. J. George, and A. C. Newby.** 1998. Divergent effects of tissue inhibitor of metalloproteinase-1, -2, or -3 overexpression on rat vascular smooth muscle cell invasion, proliferation, and death in vitro—TIMP-3 promotes apoptosis. *J. Clin. Invest.* **101**: 1478–1487.

393. **Baragi, V. M., C. J. Fliszar, M. C. Conroy, Q. Z. Ye, J. M. Shipley, and H. G. Welgus.** 1994. Contribution of the C-terminal domain of metalloproteinases to binding by tissue inhibitor of metalloproteinases. C-terminal truncated stromelysin and matrilysin exhibit equally compromised binding affinities as compared to full-length stromelysin. *J. Biol. Chem.* **269**: 12692–12697.

394. **Bertaux, B. and W. Hornebeck.** 1993. [Tissue inhibitors of matrix metalloproteinases TIMP 1–2. Structures and functions]. [French]. *C. R. Seances Soc. Biol. Fil.* **187**: 192–200.

395. **Bertaux, B., W. Hornebeck, A. Z. Eisen, and L. Dubertret.** 1991. Growth stimulation of human keratinocytes by tissue inhibitor of metalloproteinases. *J. Invest. Dermatol.* **97**: 679–685.

396. **Bigg, H. F., I. M. Clark, and T. E. Cawston.** 1994. Fragments of human fibroblast collagenase: interaction with metalloproteinase inhibitors and substrates. *Biochim. Biophys. Acta* **1208**: 157–165.

397. **Bläser, J., V. Knäuper, A. Osthues, H. Reinke, and H. Tschesche.** 1991. Mercurial activation of human polymorphonuclear leucocyte procollagenase. *Eur. J. Biochem.* **202**: 1223–1230.

398. **Boujrad, N., S. O. Ogwuegbu, M. Garnier, C. H. Lee, B. M. Martin, and V. Papadopoulos.** 1995. Identification of a stimulator of steroid hormone synthesis isolated from testis. *Science* **268**: 1609–1612.

399. **Butler, G. S., M. J. Butler, S. J. Atkinson, H. Will, T. Tamura, S. S. van Westrum, T. Crabbe, J. Clements, M. P. d'Ortho, and G. Murphy.** 1998. The TIMP2 membrane type 1 metalloproteinase receptor regulates the concentration and efficient activation of progelatinase A—a kinetic study. *J. Biol. Chem.* **273**: 871–880.

400. **Butler, G. S., H. Will, S. J. Atkinson, and G. Murphy.** 1997. Membrane-type-2 matrix metalloproteinase can initiate the processing of progelatinase A and is regulated by the tissue inhibitors of metalloproteinases. *Eur. J. Biochem.* **244**: 653–657.

401. **Cawston, T. E., W. A. Galloway, E. Mercer, G. Murphy, and J. J. Reynolds.** 1981. Purification of rabbit bone inhibitor of collagenase. *Biochem. J.* **195**: 159–165.

402. **Cawston, T. E. and E. Mercer.** 1986. Preferential binding of collagenase to α_2-macroglobulin in the presence of the tissue inhibitor of metalloproteinases. *FEBS Lett.* **209**: 9–12.

403. **Cawston, T. E., G. Murphy, E. Mercer, W. A. Galloway, B. L. Hazleman, and J. J. Reynolds.** 1983. The interaction of purified rabbit bone collagenase with purified rabbit bone metalloproteinase inhibitor. *Biochem. J.* **211**: 313–318.

404. **Chesler, L., D. W. Golde, N. Bersch, and M. D. Johnson.** 1995. Metalloproteinase inhibition and erythroid potentiation are independent activities of tissue inhibitor of metalloproteinases-1. *Blood* **86**: 4506–4515.

405. **Clark, I. M. and T. E. Cawston.** 1989. Fragments of human fibroblast collagenase. Purification and characterization. *Biochem. J.* **263**: 201–206.

406. **Corcoran, M. L., M. R. Emmert-Buck, J. L. McClanahan, M. Pelina-Parker, and W. G. Stetler-Stevenson.** 1996. TIMP-2 mediates cell surface binding of MMP-2. *Adv. Exp. Med. Biol.* **389**: 295–304.

407. **Corcoran, M. L. and W. G. Stetler-Stevenson.** 1995. Tissue inhibitor of metalloproteinase-2 stimulates fibroblast proliferation via a cAMP-dependent mechanism. *J. Biol. Chem.* **270**: 13453–13459.

408. **Crabbe, T., S. M. Kelly, and N. C. Price.** 1996. An analysis of the conformational changes that accompany the activation and inhibition of gelatinase A. *FEBS Lett.* **380**: 53–57.

409. **Crabbe, T., F. Willenbrock, D. Eaton, P. Hynds, A. F. Carne, G. Murphy, and A. J. P. Docherty.** 1992. Biochemical characterization of matrilysin. Activation conforms to the stepwise mechanisms proposed for other matrix metalloproteinases. *Biochemistry* **31**: 8500–8507.

410. **Crabbe, T., S. Zucker, M. I. Cockett, F. Willenbrock, S. Tickle, J. P. O'Connell, J. M. Scothern, G. Murphy, and A. J. P. Docherty.** 1994. Mutation of the active site glutamic acid of human gelatinase A: effects on latency, catalysis, and the binding of tissue inhibitor of metalloproteinases-1. *Biochemistry* **33**: 6684–6690.

411. **d'Ortho, M. P., H. Stanton, M. Butler, S. J. Atkinson, G. Murphy, and R. M. Hembry.** 1998. MT1-MMP on the cell surface causes focal degradation of gelatin films. *FEBS Lett.* **421**: 159–164.

412. **De Clerck, Y. A., T.-D. Yean, B. J. Ratzkin, H. S. Lu, and K. E. Langley.** 1989. Purification and characterization of two related but distinct metalloproteinase inhibitors secreted by bovine aortic endothelial cells. *J. Biol. Chem.* **264**: 17445–17453.

413. **Dean, D. D. and J. F. Woessner, Jr.** 1984. Extracts of human articular cartilage contain an inhibitor of tissue metalloproteinases. *Biochem. J.* **218**: 277–280.

414. **DeClerck, Y. A., T. D. Yean, D. Chan, H. Shimada, and K. E. Langley.** 1991. Inhibition of tumor invasion of smooth muscle cell layers by recombinant human metalloproteinase inhibitor. *Cancer Res.* **51**: 2151–2157.

415. **DeClerck, Y. A., T. D. Yean, Y. Lee, J. M. Tomich, and K. E. Langley.** 1993. Characterization of the functional domain of tissue inhibitor of metalloproteinases-2 (TIMP-2). *Biochem. J.* **289**: 65–69.

416. **Docherty, A. J. P., A. Lyons, B. J. Smith, E. M. Wright, P. E. Stephens, T. J. R. Harris, G. Murphy, and J. J. Reynolds.** 1985. Sequence of human tissue inhibitor of metalloproteinases and its identity to erythroid-potentiating activity. *Nature* **318**: 66–69.

417. **Edwards, D. R., P. P. Beaudry, T. D. Laing, V. Kowal, K. J. Leco, P. A. Leco, and M. S. Lim.** 1996. The roles of tissue inhibitors of metalloproteinases in tissue remodelling and cell growth. *Int. J. Obes. Relat. Metab. Disord.* **20**: S9–S15

418. **Emmert-Buck, M. R., H. P. Emonard, M. L. Corcoran, H. C. Krutzsch, J. M. Foidart, and W. G. Stetler-Stevenson.** 1995. Cell surface binding of TIMP-2 and pro-MMP-2/TIMP-2 complex. *FEBS Lett.* **364**: 28–32.

419. **Fariss, R. N., S. S. Apte, B. R. Olsen, K. Iwata, and A. H. Milam.** 1997. Tissue inhibitor of metalloproteinases-3 is a component of Bruch's membrane of the eye. *Am. J. Pathol.* **150**: 323–328.

420. **Felbor, U., C. Benkwitz, M. L. Kein, K. Greenberg, C. Y. Gregory, and B. H. F. Weber.** 1997. Sorsby fundus dystrophy — reevaluation of variable expressivity in patients carrying a TIMP3 founder mutation. *Arch. Ophthalmol.* **115**: 1569–1571.

421. **Felbor, U., H. Stöhr, T. Amann, U. Schönherr, E. Apfelstedt-Sylla, and B. H. Weber.** 1996. A second

independent Tyr168Cys mutation in the tissue inhibitor of metalloproteinases-3 (TIMP3) in Sorsby's fundus dystrophy. *J. Med. Genet.* **33**: 233–236.

422. **Felbor, U., H. Stöhr, T. Amann, U. Schönherr, and B. H. Weber.** 1995. A novel Ser156Cys mutation in the tissue inhibitor of metalloproteinases-3 (TIMP3) in Sorsby's fundus dystrophy with unusual clinical features. *Hum. Mol. Genet.* **4**: 2415–2416.

423. **Galloway, W. A., G. Murphy, J. D. Sandy, J. Gavrilovic, T. E. Cawston, and J. J. Reynolds.** 1983. Purification and characterization of a rabbit bone metalloproteinase that degrades proteoglycan and other connective-tissue components. *Biochem. J.* **209**: 741–752.

424. **Gasson, J. C., D. W. Golde, S. E. Kaufman, C. A. Westbrook, R. M. Hewick, R. J. Kaufman, G. G. Wong, P. A. Temple, A. C. Leary, E. L. Brown,** *et al.* 1985. Molecular characterization and expression of the gene encoding human erythroid-potentiating activity. *Nature* **315**: 768–771.

425. **Goldberg, G. I., B. L. Marmer, G. A. Grant, A. Z. Eisen, S. Wilhelm, and C. S. He.** 1989. Human 72-kilodalton type IV collagenase forms a complex with a tissue inhibitor of metalloproteases designated TIMP-2. *Proc. Natl. Acad. Sci. USA* **86**: 8207–8211.

426. **Goldberg, G. I., A. Strongin, I. E. Collier, L. T. Genrich, and B. L. Marmer.** 1992. Interaction of 92-kDa type IV collagenase with the tissue inhibitor of metalloproteinases prevents dimerization, complex formation with interstitial collagenase, and activation of the proenzyme with stromelysin. *J. Biol. Chem.* **267**: 4583–4591.

427. **Golde, D. W., N. Bersch, S. G. Quan, and A. J. Lusis.** 1980. Production of erythroid-potentiating activity by a human T-lymphoblast cell line. *Proc. Natl. Acad. Sci. USA* **77**: 593–596.

428. **Gomis-Rüth, F. X., K. Maskos, M. Betz, A. Bergner, R. Huber, K. Suzuki, N. Yoshida, H. Nagase, K. Brew, G. P. Bourenkov, H. Bartunik, and W. Bode.** 1997. Mechanism of inhibition of the human matrix metalloproteinase stromelysin-1 by TIMP-1. *Nature* **389**: 77–81.

429. **Hargreaves, P. G., F. F. Wang, J. Antcliff, G. Murphy, J. Lawry, R. G. G. Russell, and P. I. Croucher.** 1998. Human myeloma cells shed the interleukin-6 receptor—inhibition by tissue inhibitor of metalloproteinase-3 and a hydroxamate-based metalloproteinase inhibitor. *Br. J. Haematol.* **101**: 694–702.

430. **Hayakawa, T., K. Yamashita, E. Ohuchi, and A. Shinagawa.** 1994. Cell growth-promoting activity of tissue inhibitor of metalloproteinases-2 (TIMP-2). *J. Cell Sci.* **107**: 2373–2379.

431. **Hayakawa, T., K. Yamashita, K. Tanzawa, E. Uchijima, and K. Iwata.** 1992. Growth-promoting activity of tissue inhibitor of metalloproteinases-1 (TIMP-1) for a wide range of cells. A possible new growth factor in serum. *FEBS Lett.* **298**: 29–32.

432. **Howard, E. W., E. C. Bullen, and M. J. Banda.** 1991. Preferential inhibition of 72- and 92-kDa gelatinases by tissue inhibitor of metalloproteinases-2. *J. Biol. Chem.* **266**: 13070–13075.

433. **Hutton, M., F. Willenbrock, K. Brocklehurst, and G. Murphy.** 1998. Kinetic analysis of the mechanism of interaction of full-length TIMP-2 and gelatinase A—evidence for the existence of a low-affinity intermediate. *Biochemistry* **37**: 10094–10098.

434. **Imai, K., E. Ohuchi, T. Aoki, H. Nomura, Y. Fujii, H. Sato, M. Seiki, and Y. Okada.** 1996. Membrane-type matrix metalloproteinase 1 is a gelatinolytic enzyme and is secreted in a complex with tissue inhibitor of metalloproteinases 2. *Cancer Res.* **56**: 2707–2710.

435. **Itoh, Y. and H. Nagase.** 1995. Preferential inactivation of tissue inhibitor of metalloproteinases-1 that is bound to the precursor of matrix metalloproteinase 9 (progelatinase B) by human neutrophil elastase. *J. Biol. Chem.* **270**: 16518–16521.

436. **Khokha, R.** 1994. Suppression of the tumorigenic and metastatic abilities of murine B16-F10 melanoma cells in vivo by the overexpression of the tissue inhibitor of the metalloproteinases-1. *J. Natl. Cancer Inst.* **86**: 299–304.

437. **Khokha, R., P. Waterhouse, P. Lala, M. Zimmer, and D. T. Denhardt.** 1991. Increased proteinase expression during tumor progression of cell lines down-modulated for TIMP levels: a new transformation paradigm? *J. Cancer Res. Clin. Oncol.* **117**: 333–338.

438. **Khokha, R., P. Waterhouse, S. Yagel, P. K. Lala, C. M. Overall, G. Norton, and D. T. Denhardt.** 1989. Antisense RNA-induced reduction in murine TIMP levels confers oncogenicity on Swiss 3T3 cells. *Science* **243**: 947–950.

439. **Kikuchi, K., T. Kadono, M. Furue, and K. Tamaki.** 1997. Tissue inhibitor of metalloproteinase 1 (TIMP-1) may be an autocrine growth factor in scleroderma fibroblasts. *J. Invest. Dermatol.* **108**: 281–284.

440. **Kinoshita, T., H. Sato, T. Takino, M. Itoh, T. Akizawa, and M. Seiki.** 1996. Processing of a precursor of 72-kilodalton type IV collagenase/gelatinase A by a recombinant membrane-type 1 matrix metalloproteinase. *Cancer Res.* **56**: 2535–2538.

441. **Kleine, T., S. Bartsch, J. Bläser, S. Schnierer, S. Triebel, M. Valentin, T. Gote, and H. Tschesche.** 1993. Preparation of active recombinant TIMP-1 from *Escherichia coli* inclusion bodies and complex formation with the recombinant catalytic domain of PMNL-collagenase. *Biochemistry* **32**: 14125–14131.

442. **Kleiner, D. E., Jr., A. Tuuttila, K. Tryggvason, and W. G. Stetler-Stevenson.** 1993. Stability analysis of latent and active 72-kDa type IV collagenase: the role of

tissue inhibitor of metalloproteinases-2 (TIMP-2). *Biochemistry* **32**: 1583–1592.

443. **Kleiner, D. E., Jr., E. J. Unsworth, H. C. Krutzsch, and W. G. Stetler-Stevenson.** 1992. Higher-order complex formation between the 72-kilodalton type IV collagenase and tissue inhibitor of metalloproteinases-2. *Biochemistry* **31**: 1665–1672.

444. **Knäuper, V., S. Cowell, B. Smith, C. López-Otín, M. O'Shea, H. Morris, L. Zardi, and G. Murphy.** 1997. The role of the C-terminal domain of human collagenase-3 (MMP-13) in the activation of procollagenase-3, substrate specificity, and tissue inhibitor of metalloproteinase interaction. *J. Biol. Chem.* **272**: 7608–7616.

445. **Knäuper, V., C. López-Otín, B. Smith, G. Knight, and G. Murphy.** 1996. Biochemical characterization of human collagenase-3. *J. Biol. Chem.* **271**: 1544–1550.

446. **Knäuper, V., A. Osthues, Y. A. DeClerck, K. E. Langley, J. Bläser, and H. Tschesche.** 1993. Fragmentation of human polymorphonuclear-leucocyte collagenase. *Biochem. J.* **291**: 847–854.

447. **Ko, Y. C., K. E. Langley, E. A. Mendiaz, V. P. Parker, S. M. Taylor, and Y. A. DeClerck.** 1997. The C-terminal domain of tissue inhibitor of metalloproteinases-2 is required for cell binding but not for antimetalloproteinase activity. *Biochem. Biophys. Res. Commun.* **236**: 100–105.

448. **Kolkenbrock, H., D. Orgel, A. Hecker-Kia, W. Noack, and N. Ulbrich.** 1991. The complex between a tissue inhibitor of metalloproteinases (TIMP-2) and 72-kDa progelatinase is a metalloproteinase inhibitor. *Eur. J. Biochem.* **198**: 775–781.

449. **Kolkenbrock, H., D. Orgel, A. Hecker-Kia, J. Zimmermann, and N. Ulbrich.** 1995. Generation and activity of the ternary gelatinase B/TIMP-1/LMW-stromelysin-1 complex. *Biol. Chem. Hoppe Seyler* **376**: 495–500.

450. **Langton, K. P., M. D. Barker, and N. McKie.** 1998. Localization of the functional domains of human tissue inhibitor of metalloproteinases-3 and the effects of a Sorsby's fundus dystrophy mutation. *J. Biol. Chem.* **273**: 16778–16781.

451. **Lefebvre, V. and G. Vaes.** 1989. The enzymatic evaluation of procollagenase and collagenase inhibitors in crude biological media. *Biochim. Biophys. Acta* **992**: 355–361.

452. **Lelièvre, Y., R. Bouboutou, J. Boiziau, D. Faucher, D. Achard, and T. Cartwright.** 1990. Low molecular weight, sequence based, collagenase inhibitors selectively block the interaction between collagenase and TIMP (tissue inhibitor of metalloproteinases). *Matrix* **10**: 292–299.

453. **Liu, Y. L. E., M. S. Wang, J. Greene, J. Su, S. Ullrich, H. Li, S. J. Sheng, P. Alexander, Q. X. A. Sang, and Y. E. Shi.** 1997. Preparation and characterization of recombinant tissue inhibitor of metalloproteinase 4 (TIMP-4). *J. Biol. Chem.* **272**: 20479–20483.

454. **Matsumoto, S. I., M. Katoh, S. Saito, T. Watanabe, and Y. Masuho.** 1997. Identification of soluble type of membrane-type matrix metalloproteinase-3 formed by alternatively spliced mRNA. *Biochim. Biophys. Acta—Gene Struct. Express.* **1354**: 159–170.

455. **McGeehan, G., W. Burkhart, R. Anderegg, J. D. Becherer, J. W. Gillikin, and J. S. Graham.** 1992. Sequencing and characterization of the soybean leaf metalloproteinase. Structural and functional similarity to the matrix metalloproteinase family. *Plant Physiol.* **99**: 1179–1183.

456. **McLaughlin, B., T. Cawston, and J. B. Weiss.** 1991. Activation of the matrix metalloproteinase inhibitor complex by a low molecular weight angiogenic factor. *Biochim. Biophys. Acta* **1073**: 295–298.

457. **Murphy, A. N., E. J. Unsworth, and W. G. Stetler-Stevenson.** 1993. Tissue inhibitor of metalloproteinases-2 inhibits bFGF-induced human microvascular endothelial cell proliferation. *J. Cell. Physiol.* **157**: 351–358.

458. **Murphy, G., J. A. Allan, F. Willenbrock, M. I. Cockett, J. P. O'Connell, and A. J. P. Docherty.** 1992. The role of the C-terminal domain in collagenase and stromelysin specificity. *J. Biol. Chem.* **267**: 9612–9618.

459. **Murphy, G., A. Houbrechts, M. I. Cockett, R. A. Williamson, M. O'Shea, and A. J. P. Docherty.** 1991. The N-terminal domain of tissue inhibitor of metalloproteinases retains metalloproteinase inhibitory activity. *Biochemistry* **30**: 8097–8102.

460. **Murphy, G., P. Koklitis, and A. F. Carne.** 1989. Dissociation of tissue inhibitor of metalloproteinases (TIMP) from enzyme complexes yields fully active inhibitor. *Biochem. J.* **261**: 1031–1034.

461. **Murphy, G., C. G. McAlpine, C. T. Poll, and J. J. Reynolds.** 1985. Purification and characterization of a bone metalloproteinase that degrades gelatin and types IV and V collagen. *Biochim. Biophys. Acta* **831**: 49–58.

462. **Murphy, G., Q. Nguyen, M. I. Cockett, S. J. Atkinson, J. A. Allan, C. G. Knight, F. Willenbrock, and A. J. P. Docherty.** 1994. Assessment of the role of the fibronectin-like domain of gelatinase A by analysis of a deletion mutant. *J. Biol. Chem.* **269**: 6632–6636.

463. **Murphy, G., J. J. Reynolds, U. Bretz, and M. Baggiolini.** 1982. Partial purification of collagenase and gelatinase from human polymorphonuclear leucocytes. Analysis of their actions on soluble and insoluble collagens. *Biochem. J.* **203**: 209–221.

464. **Murphy, G., J. P. Segain, M. O'Shea, M. Cockett, C. Ioannou, O. Lefebvre, P. Chambon, and P. Basset.** 1993. The 28-kDa N-terminal domain of mouse

stromelysin-3 has the general properties of a weak metalloproteinase. *J. Biol. Chem.* **268**: 15435–15441.

465. **Murphy, G., R. Ward, R. M. Hembry, J. J. Reynolds, K. Kuhn, and K. Tryggvason.** 1989. Characterization of gelatinase from pig polymorphonuclear leucocytes. A metalloproteinase resembling tumour type IV collagenase. *Biochem. J.* **258**: 463–472.

466. **Murphy, G., F. Willenbrock, T. Crabbe, M. O'Shea, R. Ward, S. Atkinson, J. O'Connell, and A. Docherty.** 1994. Regulation of matrix metalloproteinase activity. *Ann. N. Y. Acad. Sci.* **732**: 31–41.

467. **Murphy, G., F. Willenbrock, R. Ward, M. O'Shea, and A. Docherty.** 1993. Studies on the structure and function of the matrix metalloproteinases and their inhibitors. In *Proteolysis and Protein Turnover.* J. S. Bond and A. J. Barrett (eds), pp. 25–32. Portland Press, London.

468. **Murphy, G., F. Willenbrock, R. V. Ward, M. I. Cockett, D. Eaton, and A. J. P. Docherty.** 1992. The C-terminal domain of 72 kDa gelatinase A is not required for catalysis, but is essential for membrane activation and modulates interactions with tissue inhibitors of metalloproteinases. *Biochem. J.* **283**: 637–641.

469. **Negro, A., M. Onisto, L. Grassato, C. Caenazzo, and S. Garbisa.** 1997. Recombinant human TIMP-3 from *Escherichia coli*—synthesis, refolding, physico-chemical and functional insights. *Protein Eng.* **10**: 593–599.

470. **Nemeth, J. A. and C. L. Goolsby.** 1993. TIMP-2, a growth-stimulatory protein from SV40-transformed human fibroblasts. *Exp. Cell Res.* **207**: 376–382.

471. **Nemeth, J. A., A. Rafe, M. Steiner, and C. L. Goolsby.** 1996. TIMP-2 growth-stimulatory activity: a concentration- and cell type-specific response in the presence of insulin. *Exp. Cell Res.* **224**: 110–115.

472. **Nomura, K., T. Shimizu, H. Kinoh, Y. Sendai, M. Inomata, and N. Suzuki.** 1997. Sea urchin hatching enzyme (envelysin)—cDNA cloning and deprivation of protein substrate specificity by autolytic degradation. *Biochemistry* **36**: 7225–7238.

473. **Nothnick, W. B., P. Soloway, and T. E. Curry.** 1997. Assessment of the role of tissue inhibitor of metalloproteinase-1 (TIMP-1) during the periovulatory period in female mice lacking a functional TIMP-1 gene. *Biol. Reprod.* **56**: 1181–1188.

474. **Pendás, A. M., V. Knäuper, X. S. Puente, E. Llano, M. G. Mattei, S. Apte, G. Murphy, and C. López-Otín.** 1997. Identification and characterization of a novel human matrix metalloproteinase with unique structural characteristics, chromosomal location, and tissue distribution. *J. Biol. Chem.* **272**: 4281–4286.

475. **Quantin, B., G. Murphy, and R. Breathnach.** 1989. Pump-1 cDNA codes for a protein with characteristics similar to those of classical collagenase family members. *Biochemistry* **28**: 5327–5334.

476. **Ries, C., F. Lottspeich, K. H. Dittmann, and P. E. Petrides.** 1996. HL-60 leukemia cells produce an autocatalytically truncated form of matrix metalloproteinase-9 with impaired sensitivity to inhibition by tissue inhibitors of metalloproteinases. *Leukemia* **10**: 1520–1526.

477. **Saika, S., Y. Kawashima, Y. Okada, S. Tanaka, O. Yamanaka, Y. Ohnishi, and A. Ooshima.** 1998. Recombinant TIMP-1 and -2 enhance the proliferation of rabbit corneal epithelial cells in vitro and the spreading of rabbit corneal epithelium in situ. *Curr. Eye Res.* **17**: 47–52.

478. **Sato, H., T. Kinoshita, T. Takino, K. Nakayama, and M. Seiki.** 1996. Activation of a recombinant membrane type 1-matrix metalloproteinase (MT1-MMP) by furin and its interaction with tissue inhibitor of metalloproteinases (TIMP)-2. *FEBS Lett.* **393**: 101–104.

479. **Sellers, A., G. Murphy, M. C. Meikle, and J. J. Reynolds.** 1979. Rabbit bone collagenase inhibitor blocks the activity of other neutral metalloproteinases. *Biochem. Biophys. Res. Commun.* **87**: 581–587.

480. **Smith, M. R., H. F. Kung, S. K. Durum, N. H. Colburn, and Y. Sun.** 1997. TIMP-3 induces cell death by stabilizing TNF-alpha receptors on the surface of human colon carcinoma cells. *Cytokine* **9**: 770–780.

481. **Soloway, P. D., C. M. Alexander, Z. Werb, and R. Jaenisch.** 1996. Targeted mutagenesis of *Timp-1* reveals that lung tumor invasion is influenced by *Timp-1* genotype of the tumor but not by that of the host. *Oncogene* **13**: 2307–2314.

482. **Stetler-Stevenson, W. G., N. Bersch, and D. W. Golde.** 1992. Tissue inhibitor of metalloproteinase-2 (TIMP-2) has erythroid-potentiating activity. *FEBS Lett.* **296**: 231–234.

483. **Stetler-Stevenson, W. G., H. C. Krutzsch, and L. A. Liotta.** 1989. Tissue inhibitor of metalloproteinase (TIMP-2). A new member of the metalloproteinase inhibitor family. *J. Biol. Chem.* **264**: 17374–17378.

484. **Stricklin, G. P.** 1986. Human fibroblast tissue inhibitor of metalloproteinases: glycosylation and function. *Coll. Relat. Res.* **6**: 219–228.

485. **Strongin, A. Y., I. Collier, G. Bannikov, B. L. Marmer, G. A. Grant, and G. I. Goldberg.** 1995. Mechanism of cell surface activation of 72-kDa type IV collagenase. Isolation of the activated form of the membrane metalloprotease. *J. Biol. Chem.* **270**: 5331–5338.

486. **Strongin, A. Y., I. E. Collier, P. A. Krasnov, L. T. Genrich, B. L. Marmer, and G. I. Goldberg.** 1993. Human 92 kDa type IV collagenase: functional analysis of fibronectin and carboxyl-end domains. *Kidney Int.* **43**: 158–162.

487. **Sutton, C. W., J. A. O'Neill, and J. S. Cottrell.** 1994. Site-specific characterization of glycoprotein

carbohydrates by exoglycosidase digestion and laser desorption mass spectrometry. *Anal. Biochem.* **218**: 34–46.

488. **Taylor, K. B., L. J. Windsor, N. C. M. Caterina, M. K. Bodden, and J. A. Engler.** 1996. The mechanism of inhibition of collagenase by TIMP-1. *J. Biol. Chem.* **271**: 23938–23945.

489. **Umenishi, F., M. Umeda, and K. Miyazaki.** 1991. Efficient purification of TIMP-2 from culture medium conditioned by human hepatoma cell line, and its inhibitory effects on metalloproteinases and *in vitro* tumor invasion. *J. Biochem. (Tokyo)* **110**: 189–195.

490. **Vallon, R., R. Müller, D. Moosmayer, E. Gerlach, and P. Angel.** 1997. The catalytic domain of activated collagenase I (MMP-1) is absolutely required for interaction with its specific inhibitor, tissue inhibitor of metalloproteinases-1 (TIMP-1). *Eur. J. Biochem.* **244**: 81–88.

491. **Vater, C. A., C. L. Mainardi, and E. D. Harris, Jr.** 1979. Inhibitor of human collagenase from cultures of human tendon. *J. Biol. Chem.* **254**: 3045–3053.

492. **Ward, R. V., R. M. Hembry, J. J. Reynolds, and G. Murphy.** 1991. The purification of tissue inhibitor of metalloproteinases-2 from its 72 kDa progelatinase complex. Demonstration of the biochemical similarities of tissue inhibitor of metalloproteinases-2 and tissue inhibitor of metalloproteinases-1. *Biochem. J.* **278**: 179–187.

493. **Weber, B. H. F., G. Vogt, R. C. Pruett, H. Stöhr, and U. Felbor.** 1994. Mutations in the tissue inhibitor of metalloproteinases-3 (TIMP3) in patients with Sorsby's fundus dystrophy. *Nature Genet.* **8**: 352–356.

494. **Welgus, H. G., J. J. Jeffrey, A. Z. Eisen, W. T. Roswit, and G. P. Stricklin.** 1985. Human skin fibroblast collagenase: interaction with substrate and inhibitor. *Coll. Relat. Res.* **5**: 167–179.

495. **Welgus, H. G., G. P. Stricklin, A. Z. Eisen, E. A. Bauer, R. V. Cooney, and J. J. Jeffrey.** 1979. A specific inhibitor of vertebrate collagenase produced by human skin fibroblasts. *J. Biol. Chem.* **254**: 1938–1943.

496. **Wilhelm, S. M., I. E. Collier, B. L. Marmer, A. Z. Eisen, G. A. Grant, and G. I. Goldberg.** 1989. SV40-transformed human lung fibroblasts secrete a 92-kDa type IV collagenase which is identical to that secreted by normal human macrophages. *J. Biol. Chem.* **264**: 17213–17221.

497. **Will, H., S. J. Atkinson, G. S. Butler, B. Smith, and G. Murphy.** 1996. The soluble catalytic domain of membrane type 1 matrix metalloproteinase cleaves the propeptide of progelatinase A and initiates autoproteolytic activation. Regulation by TIMP-2 and TIMP-3. *J. Biol. Chem.* **271**: 17119–17123.

498. **Willenbrock, F., T. Crabbe, P. M. Slocombe, C. W. Sutton, A. J. P. Docherty, M. I. Cockett, M. O'Shea, K. Brocklehurst, I. R. Phillips, and G. Murphy.** 1993. The activity of the tissue inhibitors of metalloproteinases is regulated by C-terminal domain interactions: a kinetic analysis of the inhibition of gelatinase A. *Biochemistry* **32**: 4330–4337.

499. **Williamson, R. A., M. D. Carr, T. A. Frenkiel, J. Feeney, and R. B. Freedman.** 1997. Mapping the binding site for matrix metalloproteinase on the N-terminal domain of the tissue inhibitor of metalloproteinases-2 by NMR chemical shift perturbation. *Biochemistry* **36**: 13882–13889.

500. **Williamson, R. A., F. A. O. Marston, S. Angal, P. Koklitis, M. Panico, H. R. Morris, A. F. Carne, B. J. Smith, T. J. R. Harris, and R. B. Freedman.** 1990. Disulphide bond assignment in human tissue inhibitor of metalloproteinases (TIMP). *Biochem. J.* **268**: 267–274.

501. **Windsor, L. J., H. Grenett, B. Birkedal-Hansen, M. K. Bodden, J. A. Engler, and H. Birkedal-Hansen.** 1993. Cell type-specific regulation of SL-1 and SL-2 genes. Induction of the SL-2 gene but not the SL-1 gene by human keratinocytes in response to cytokines and phorbolesters. *J. Biol. Chem.* **268**: 17341–17347.

502. **Windsor, L. J., D. L. Steele, S. B. LeBlanc, and K. B. Taylor.** 1997. Catalytic domain comparisons of human fibroblast-type collagenase, stromelysin-1, and matrilysin. *Biochim. Biophys. Acta* **1334**: 261–272.

503. **Yamashita, K., M. Suzuki, H. Iwata, T. Koike, M. Hamaguchi, A. Shinagawa, T. Noguchi, and T. Hayakawa.** 1996. Tyrosine phosphorylation is crucial for growth signaling by tissue inhibitors of metalloproteinases (TIMP-1 and TIMP-2). *FEBS Lett.* **396**: 103–107.

504. **Yamashita, K., J. Zhang, L. Zou, H. Hayakawa, T. Noguchi, I. Kondo, O. Narita, N. Fujimoto, K. Iwata, and T. Hayakawa.** 1992. Dissociation of collagenase-tissue inhibitor of metalloproteinases-1 (TIMP-1) complex–its application for the independent measurements of TIMP-1 and collagenase activity in crude culture media and body fluids. *Matrix* **12**: 481–487.

505. **Yasumitsu, H., K. Miyazaki, F. Umenishi, N. Koshikawa, and M. Umeda.** 1992. Comparison of extracellular matrix-degrading activities between 64-kDa and 90-kDa gelatinases purified in inhibitor-free forms from human schwannoma cells. *J. Biochem. (Tokyo)* **111**: 74–80.

506. **Yoshiji, H., S. R. Harris, E. Raso, D. E. Gomez, C. K. Lindsay, M. Shibuya, C. C. Sinha, and U. P. Thorgeirsson.** 1998. Mammary carcinoma cells over-expressing tissue inhibitor of metalloproteinases-1 show enhanced vascular endothelial growth factor expression. *Int. J. Canc.* **75**: 81–87.

507. **Zucker, S., M. Drews, C. Conner, H. D. Foda, Y. A. De Clerck, K. E. Langley, W. F. Bahou, A. J. P. Docherty, and J. Cao.** 1998. Tissue inhibitor of

metalloproteinase-2 (TIMP-2) binds to the catalytic domain of the cell surface receptor, membrane type 1 matrix metalloproteinase 1 (MT1-MMP). *J. Biol. Chem.* **273**: 1216–1222.

See also review articles: 37, 41, 45, 85, 94, 139.

Domains of MMPs

508. **Allan, J. A., A. J. P. Docherty, P. J. Barker, N. S. Huskisson, J. J. Reynolds, and G. Murphy.** 1995. Binding of gelatinases A and B to type-I collagen and other matrix components. *Biochem. J.* **309**: 299–306.

509. **Allan, J. A., R. M. Hembry, S. Angal, J. J. Reynolds, and G. Murphy.** 1991. Binding of latent and high M_r active forms of stromelysin to collagen is mediated by the C-terminal domain. *J. Cell Sci.* **99**: 789–795.

510. **Azzo, W. and J. F. Woessner, Jr.** 1986. Purification and characterization of an acid metalloproteinase from human articular cartilage. *J. Biol. Chem.* **261**: 5434–5441.

511. **Bányai, L. and L. Patthy.** 1991. Evidence for the involvement of type II domains in collagen binding by 72 kDa type IV procollagenase. *FEBS Lett.* **282**: 23–25.

512. **Bányai, L., H. Tordai, and L. Patthty.** 1996. Structure and domain-domain interactions of the gelatin binding site of human 72-kilodalton type IV collagenase (gelatinase A, matrix metalloproteinase 2). *J. Biol. Chem.* **271**: 12003–12008.

513. **Bányai, L., H. Tordai, and L. Patthy.** 1994. The gelatin-binding site of human 72 kDa type IV collagenase (gelatinase A). *Biochem. J.* **298**: 403–407.

514. **Baragi, V. M., C. J. Fliszar, M. C. Conroy, Q. Z. Ye, J. M. Shipley, and H. G. Welgus.** 1994. Contribution of the C-terminal domain of metalloproteinases to binding by tissue inhibitor of metalloproteinases. C-terminal truncated stromelysin and matrilysin exhibit equally compromised binding affinities as compared to full-length stromelysin. *J. Biol. Chem.* **269**: 12692–12697.

515. **Becker, J. W., A. I. Marcy, L. L. Rokosz, M. G. Axel, J. J. Burbaum, P. M. Fitzgerald, P. M. Cameron, C. K. Esser, W. K. Hagmann, J. D. Hermes, *et al.* 1995. Stromelysin-1: three-dimensional structure of the inhibited catalytic domain and of the C-truncated proenzyme. *Protein Sci.* **4**: 1966–1976.

516. **Bode, W., P. Reinemer, R. Huber, T. Kleine, S. Schnierer, and H. Tschesche.** 1994. The X-ray crystal structure of the catalytic domain of human neutrophil collagenase inhibited by a substrate analogue reveals the essentials for catalysis and specificity. *EMBO J.* **13**: 1263–1269.

517. **Borkakoti, N., F. K. Winkler, D. H. Williams, A. D'Arcy, M. J. Broadhurst, P. A. Brown, W. H. Johnson, and E. J. Murray.** 1994. Structure of the catalytic domain of human fibroblast collagenase complexed with an inhibitor. *Nature Struct. Biol.* **1**: 106–110.

518. **Brooks, P. C., S. Strömblad, L. C. Sanders, T. L. von Schalscha, R. T. Aimes, W. G. Stetler-Stevenson, J. P. Quigley, and D. A. Cheresh.** 1996. Localization of matrix metalloproteinase MMP-2 to the surface of invasive cells by interaction with integrin $\alpha_v\beta_3$. *Cell* **85**: 683–693.

519. **Brownell, J., W. Earley, E. Kunec, B. A. Morgan, B. Olyslager, R. C. Wahl, and D. R. Houck.** 1994. Comparison of native matrix metalloproteinases and their recombinant catalytic domains using a novel radiometric assay. *Arch. Biochem. Biophys.* **314**: 120–125.

520. **Browner, M. F., W. W. Smith, and A. L. Castelhano.** 1995. Matrilysin-inhibitor complexes: common themes among metalloproteases. *Biochemistry* **34**: 6602–6610.

521. **Bu, C. H. and T. Pourmotabbed.** 1996. Mechanism of Ca^{2+}-dependent activity of human neutrophil gelatinase B. *J. Biol. Chem.* **271**: 14308–14315.

522. **Cao, J., A. Rehemtulla, W. Bahou, and S. Zucker.** 1996. Membrane type matrix metalloproteinase 1 activates pro-gelatinase A without furin cleavage of the N-terminal domain. *J. Biol. Chem.* **271**: 30174–30180.

523. **Cao, J., H. Sato, T. Takino, and M. Seiki.** 1995. The C-terminal region of membrane type matrix metalloproteinase is a functional transmembrane domain required for pro-gelatinase A activation. *J. Biol. Chem.* **270**: 801–805.

524. **Cha, J. and D. S. Auld.** 1997. Site-directed mutagenesis of the active site glutamate in human matrilysin— investigation of its role in catalysis. *Biochemistry* **36**: 16019–16024.

525. **Clark, I. M. and T. E. Cawston.** 1989. Fragments of human fibroblast collagenase. Purification and characterization. *Biochem. J.* **263**: 201–206.

526. **Collier, I. E., P. A. Krasnov, A. Y. Strongin, H. Birkedal-Hansen, and G. I. Goldberg.** 1992. Alanine scanning mutagenesis and functional analysis of the fibronectin-like collagen-binding domain from human 92-kDa type IV collagenase. *J. Biol. Chem.* **267**: 6776–6781.

527. **Crabbe, T., C. Ioannou, and A. J. P. Docherty.** 1993. Human progelatinase A can be activated by autolysis at a rate that is concentration-dependent and enhanced by heparin bound to the C-terminal domain. *Eur. J. Biochem.* **218**: 431–438.

528. **Crabbe, T., J. P. O'Connell, B. J. Smith, and A. J. P. Docherty.** 1994. Reciprocated matrix metalloproteinase activation: a process performed by interstitial collagenase and progelatinase A. *Biochemistry* **33**: 14419–14425.

529. **Crabbe, T., S. Zucker, M. I. Cockett, F. Willenbrock, S. Tickle, J. P. O'Connell, J. M. Scothern, G. Murphy, and A. J. P. Docherty.** 1994. Mutation of the active site glutamic acid of human gelatinase A: effects on

latency, catalysis, and the binding of tissue inhibitor of metalloproteinases-1. *Biochemistry* **33**: 6684–6690.

530. **de Souza, S. J. and R. Brentani.** 1993. Sequence homology between a bacterial metalloproteinase and eukaryotic matrix metalloproteinases. *J. Mol. Evol.* **36**: 596–598.

531. **de Souza, S. J., H. M. Pereira, S. Jacchieri, and R. R. Brentani.** 1996. Collagen/collagenase interaction: does the enzyme mimic the conformation of its own substrate? *FASEB J.* **10**: 927–930.

532. **Faber, H. R., C. R. Groom, H. M. Baker, W. T. Morgan, A. Smith, and E. N. Baker.** 1995. 1.8 Å crystal structure of the C-terminal domain of rabbit serum hemopexin. *Structure* **3**: 551–559.

533. **Fotouhi, N., A. Lugo, M. Visnick, L. Lusch, R. Walsky, J. W. Coffey, and A. C. Hanglow.** 1994. Potent peptide inhibitors of stromelysin based on the prodomain region of matrix metalloproteinases. *J. Biol. Chem.* **269**: 30227–30231.

534. **Fridman, R., T. R. Fuerst, R. E. Bird, M. Hoyhtya, M. Oelkuct, S. Kraus, D. Komarek, L. A. Liotta, M. L. Berman, and W. G. Stetler-Stevenson.** 1992. Domain structure of human 72-kDa gelatinase/type IV collagenase. Characterization of proteolytic activity and identification of the tissue inhibitor of metalloproteinase-2 (TIMP-2) binding regions. *J. Biol. Chem.* **267**: 15398–15405.

535. **Fujimoto, N., R. V. Ward, T. Shinya, K. Iwata, K. Yamashita, and T. Hayakawa.** 1996. Interaction between tissue inhibitor of metalloproteinases-2 and progelatinase A: immunoreactivity analyses. *Biochem. J.* **313**: 827–833.

536. **Gehring, M. R., B. Condon, S. A. Margosiak, and C. C. Kan.** 1995. Characterization of the Phe-81 and Val-82 human fibroblast collagenase catalytic domain purified from *Escherichia coli*. *J. Biol. Chem.* **270**: 22507–22513.

537. **Goldberg, G. I., S. M. Wilhelm, A. Kronberger, E. A. Bauer, G. A. Grant, and A. Z. Eisen.** 1986. Human fibroblast collagenase. Complete primary structure and homology to an oncogene transformation-induced rat protein. *J. Biol. Chem.* **261**: 6600–6605.

538. **Gomis-Rüth, F. X., U. Gohlke, M. Betz, V. Knäuper, G. Murphy, C. López-Otín, and W. Bode.** 1996. The helping hand of collagenase-3 (MMP-13): 2.7 Å crystal structure of its C-terminal haemopexin-like domain. *J. Mol. Biol.* **264**: 556–566.

539. **Gooley, P. R., B. A. Johnson, A. I. Marcy, G. C. Cuca, S. P. Salowe, W. K. Hagmann, C. K. Esser, and J. P. Springer.** 1993. Secondary structure and zinc ligation of human recombinant short-form stromelysin by multidimensional heteronuclear NMR. *Biochemistry* **32**: 13098–13108.

540. **Hanglow, A. C., A. Lugo, R. Walsky, M. Finch-Arietta, L. Lusch, M. Visnick, and N. Fotouhi.** 1993. Peptides based on the conserved predomain sequence of matrix metalloproteinases inhibit human stromelysin and collagenase. *Agents Actions* **39**: C148-C150

541. **Harrison, R. K., B. Chang, L. Niedzwiecki, and R. L. Stein.** 1992. Mechanistic studies on the human matrix metalloproteinase stromelysin. *Biochemistry* **31**: 10757–10762.

542. **Hirose, T., C. Patterson, T. Pourmotabbed, C. L. Mainardi, and K. A. Hasty.** 1993. Structure-function relationship of human neutrophil collagenase: identification of regions responsible for substrate specificity and general proteinase activity. *Proc. Natl. Acad. Sci. U.S.A* **90**: 2569–2573.

543. **Holz, R. C., S. P. Salowe, C. K. Smith, G. C. Cuca, and L. Que, Jr.** 1992. EXAFS evidence for a 'cysteine switch' in the activation of prostromelysin. *J. Am. Chem. Soc.* **114**: 9611–9614.

544. **Housley, T. J., A. P. Baumann, I. D. Braun, G. Davis, P. K. Seperack, and S. M. Wilhelm.** 1993. Recombinant Chinese hamster ovary cell matrix metalloprotease-3 (MMP-3, stromelysin-1). Role of calcium in promatrix metalloprotease-3 (pro-MMP-3, prostromelysin-1) activation and thermostability of the low mass catalytic domain of MMP-3. *J. Biol. Chem.* **268**: 4481–4487.

545. **Howard, E. W. and M. J. Banda.** 1991. Binding of tissue inhibitor of metalloproteinases 2 to two distinct sites on human 72-kDa gelatinase. Identification of a stabilization site. *J. Biol. Chem.* **266**: 17972–17977.

546. **Howard, E. W., E. C. Bullen, and M. J. Banda.** 1991. Regulation of the autoactivation of human 72-kDa progelatinase by tissue inhibitor of metalloproteinases-2. *J. Biol. Chem.* **266**: 13064–13069.

547. **Hu, P., Q. Z. Ye, and J. A. Loo.** 1994. Calcium stoichiometry determination for calcium binding proteins by electrospray ionization mass spectrometry. *Anal. Chem.* **66**: 4190–4194.

548. **Hunt, L. T., W. C. Barker, and H. R. Chen.** 1987. A domain structure common to hemopexin, vitronectin, interstitial collagenase, and a collagenase homolog. *Protein Seq. Data Anal.* **1**: 21–26.

549. **Imai, K., E. Ohuchi, T. Aoki, H. Nomura, Y. Fujii, H. Sato, M. Seiki, and Y. Okada.** 1996. Membrane-type matrix metalloproteinase 1 is a gelatinolytic enzyme and is secreted in a complex with tissue inhibitor of metalloproteinases 2. *Cancer Res.* **56**: 2707–2710.

550. **Jenne, D. and K. K. Stanley.** 1987. Nucleotide sequence and organization of the human S-protein gene: repeating peptide motifs in the 'pexin' family and a model for their evolution. *Biochemistry* **26**: 6735–6742.

551. **Knäuper, V., S. Cowell, B. Smith, C. López-Otín, M. O'Shea, H. Morris, L. Zardi, and G. Murphy.** 1997. The role of the C-terminal domain of human collagenase-3 (MMP-13) in the activation of procollagenase-3, substrate specificity, and tissue inhibitor of

metalloproteinase interaction. *J. Biol. Chem.* **272**: 7608–7616.

552. **Knäuper, V., A. J. P. Docherty, B. Smith, H. Tschesche, and G. Murphy.** 1997. Analysis of the contribution of the hinge region of human neutrophil collagenase (HNC, MMP-8) to stability and collagenolytic activity by alanine scanning mutagenesis. *FEBS Lett.* **405**: 60–64.

553. **Knäuper, V., A. Osthues, Y. A. DeClerck, K. E. Langley, J. Bläser, and H. Tschesche.** 1993. Fragmentation of human polymorphonuclear-leucocyte collagenase. *Biochem. J.* **291**: 847–854.

554. **Knäuper, V., S. M. Wilhelm, P. K. Seperack, Y. A. DeClerck, K. E. Langley, A. Osthues, and H. Tschesche.** 1993. Direct activation of human neutrophil procollagenase by recombinant stromelysin. *Biochem. J.* **295**: 581–586.

555. **Kröger, M. and H. Tschesche.** 1997. Cloning, expression and activation of a truncated 92-kDa gelatinase minienzyme. *Gene* **196**: 175–180.

556. **Lemaître, V., A. Jungbluth, and Y. Eeckhout.** 1997. The recombinant catalytic domain of mouse collagenase-3 depolymerizes type I collagen by cleaving its aminotelopeptides. *Biochem. Biophys. Res. Commun.* **230**: 202–205.

557. **Li, J., P. Brick, M. C. O'Hare, T. Skarzynski, L. F. Lloyd, V. A. Curry, I. M. Clark, H. F. Bigg, B. L. Hazleman, T. E. Cawston,** *et al.* 1995. Structure of full-length porcine synovial collagenase reveals a C-terminal domain containing a calcium-linked, four-bladed beta-propeller. *Structure* **3**: 541–549.

558. **Lichte, A., H. Kolkenbrock, and H. Tschesche.** 1996. The recombinant catalytic domain of membrane-type matrix metalloproteinase-1 (MT1-MMP) induces activation of progelatinase A and progelatinase A complexed with TIMP-2. *FEBS Lett.* **397**: 277–282.

559. **Lovejoy, B., A. Cleasby, A. M. Hassell, K. Longley, M. A. Luther, D. Weigl, G. McGeehan, A. B. McElroy, D. Drewry, M. H. Lambert,** *et al.* 1994. Structure of the catalytic domain of fibroblast collagenase complexed with an inhibitor. *Science* **263**: 375–377.

560. **Lovejoy, B., A. M. Hassell, M. A. Luther, D. Weigl, and S. R. Jordan.** 1994. Crystal structures of recombinant 19-kDa human fibroblast collagenase complexed to itself. *Biochemistry* **33**: 8207–8217.

561. **Lowry, C. L., G. McGeehan, and H. LeVine, 3rd.** 1992. Metal ion stabilization of the conformation of a recombinant 19-kDa catalytic fragment of human fibroblast collagenase. *Proteins* **12**: 42–48.

562. **Marcy, A. I., L. L. Eiberger, R. Harrison, H. K. Chan, N. I. Hutchinson, W. K. Hagmann, P. M. Cameron, D. A. Boulton, and J. D. Hermes.** 1991. Human fibroblast stromelysin catalytic domain: expression, purification, and characterization of a C-terminally truncated form. *Biochemistry* **30**: 6476–6483.

563. **Maruyama-Ohki, Y., T. Takazawa, T. Nakamura, and T. Kinoshita.** 1993. Assignment of disulfide bonds in type IV collagenase FN-II domain by mass spectrometry. *Nippon Iyo Masu Supekutoru Gakkai Koenshu* **18**: 189–192.

563a. **Murphy, G., J. A. Allan, F. Willenbrock, M. I. Cockett, J. P. O'Connell, and A. J. P. Docherty.** 1992. The role of the C-terminal domain in collagenase and stromelysin specificity. *J. Biol. Chem.* **267**: 9612–9618.

564. **Murphy, G., Q. Nguyen, M. I. Cockett, S. J. Atkinson, J. A. Allan, C. G. Knight, F. Willenbrock, and A. J. P. Docherty.** 1994. Assessment of the role of the fibronectin-like domain of gelatinase A by analysis of a deletion mutant. *J. Biol. Chem.* **269**: 6632–6636.

565. **Murphy, G., F. Willenbrock, R. V. Ward, M. I. Cockett, D. Eaton, and A. J. P. Docherty.** 1992. The C-terminal domain of 72 kDa gelatinase A is not required for catalysis, but is essential for membrane activation and modulates interactions with tissue inhibitors of metalloproteinases. *Biochem. J.* **283**: 637–641.

566. **Murphy, G. J., G. Murphy, and J. J. Reynolds.** 1991. The origin of matrix metalloproteinases and their familial relationships. *FEBS Lett.* **289**: 4–7.

567. **Nakahara, H., L. Howard, E. W. Thompson, H. Sato, M. Seiki, Y. Y. Yeh, and W. T. Chen.** 1997. Transmembrane/cytoplasmic domain-mediated membrane type 1-matrix metalloprotease docking to invadopodia is required for cell invasion. *Proc. Natl. Acad. Sci. USA* **94**: 7959–7964.

568. **Nguyen, Q., F. Willenbrock, M. I. Cockett, M. O'Shea, A. J. P. Docherty, and G. Murphy.** 1994. Different domain interactions are involved in the binding of tissue inhibitors of metalloproteinases to stromelysin-1 and gelatinase A. *Biochemistry* **33**: 2089–2095.

569. **Noël, A., M. Santavicca, I. Stoll, C. L'Hoir, A. Staub, G. Murphy, M.-C. Rio, and P. Basset.** 1995. Identification of structural determinants controlling human and mouse stromelysin-3 proteolytic activities. *J. Biol. Chem.* **270**: 22866–22872.

570. **Nomura, K., T. Shimizu, H. Kinoh, Y. Sendai, M. Inomata, and N. Suzuki.** 1997. Sea urchin hatching enzyme (envelysin)—cDNA cloning and deprivation of protein substrate specificity by autolytic degradation. *Biochemistry* **36**: 7225–7238.

571. **O'Connell, J. P., F. Willenbrock, A. J. P. Docherty, D. Eaton, and G. Murphy.** 1994. Analysis of the role of the COOH-terminal domain in the activation, proteolytic activity, and tissue inhibitor of metalloproteinase interactions of gelatinase B. *J. Biol. Chem.* **269**: 14967–14973.

572. **O'Hare, M. C., V. A. Curry, R. E. Mitchell, and T. E. Cawston.** 1995. Stabilisation of purified human

collagenase by site-directed mutagenesis. *Biochem. Biophys. Res. Commun.* **216**: 329–337.

573. **Ohuchi, E., K. Imai, Y. Fujii, H. Sato, M. Seiki, and Y. Okada.** 1997. Membrane type 1 matrix metalloproteinase digests interstitial collagens and other extracellular matrix macromolecules. *J. Biol. Chem.* **272**: 2446–2451.

574. **Okada, A., S. Saez, Y. Misumi, and P. Basset.** 1997. Rat stromelysin 3: cDNA cloning from healing skin wound, activation by furin and expression in rat tissues. *Gene* **185**: 187–193.

575. **Okumura, Y., H. Sato, M. Seiki, and H. Kido.** 1997. Proteolytic activation of the precursor of membrane type 1 matrix metalloproteinase by human plasmin. A possible cell surface activator. *FEBS Lett.* **402**: 181–184.

576. **Park, A. J., L. M. Matrisian, A. F. Kells, R. Pearson, Z. Y. Yuan, and M. Navre.** 1991. Mutational analysis of the transin (rat stromelysin) autoinhibitor region demonstrates a role for residues surrounding the 'cysteine switch'. *J. Biol. Chem.* **266**: 1584–1590.

577. **Pei, D. and S. J. Weiss.** 1995. Furin-dependent intracellular activation of the human stromelysin-3 zymogen. *Nature* **375**: 244–247.

578. **Pei, D. and S. J. Weiss.** 1996. Transmembrane-deletion mutants of the membrane-type matrix metalloproteinase-1 process progelatinase A and express intrinsic matrix-degrading activity. *J. Biol. Chem.* **271**: 9135–9140.

579. **Pendás, A. M., V. Knäuper, X. S. Puente, E. Llano, M. G. Mattei, S. Apte, G. Murphy, and C. López-Otín.** 1997. Identification and characterization of a novel human matrix metalloproteinase with unique structural characteristics, chromosomal location, and tissue distribution. *J. Biol. Chem.* **272**: 4281–4286.

580. **Pieper, M., M. Betz, N. Budisa, F. X. Gomis-Rüth, W. Bode, and H. Tschesche.** 1997. Expression, purification, characterization, and X-ray analysis of selenomethionine 215 variant of leukocyte collagenase. *J. Protein Chem.* **16**: 637–650.

581. **Pourmotabbed, T.** 1994. Relation between substrate specificity and domain structure of 92-kDa type IV collagenase. *Ann. N. Y. Acad. Sci.* **732**: 372–374.

582. **Pourmotabbed, T., J. A. Aelion, D. Tyrrell, K. A. Hasty, C. H. Bu, and C. L. Mainardi.** 1995. Role of the conserved histidine and aspartic acid residues in activity and stabilization of human gelatinase B: an example of matrix metalloproteinases. *J. Protein Chem.* **14**: 527–535.

583. **Puente, X. S., A. M. Pendás, E. Llano, G. Velasco, and C. López-Otín.** 1996. Molecular cloning of a novel membrane-type matrix metalloproteinase from a human breast carcinoma. *Cancer Res.* **56**: 944–949.

584. **Reinemer, P., F. Grams, R. Huber, T. Kleine, S. Schnierer, M. Piper, H. Tschesche, and W. Bode.** 1994. Structural implications for the role of the N terminus in the 'superactivation' of collagenases. A crystallographic study. *FEBS Lett.* **338**: 227–233.

585. **Rice, A. and M. J. Banda.** 1995. Neutrophil elastase processing of gelatinase A is mediated by extracellular matrix. *Biochemistry* **34**: 9249–9256.

586. **Ries, C., F. Lottspeich, K. H. Dittmann, and P. E. Petrides.** 1996. HL-60 leukemia cells produce an autocatalytically truncated form of matrix metalloproteinase-9 with impaired sensitivity to inhibition by tissue inhibitors of metalloproteinases. *Leukemia* **10**: 1520–1526.

587. **Salowe, S. P., A. I. Marcy, G. C. Cuca, C. K. Smith, I. E. Kopka, W. K. Hagmann, and J. D. Hermes.** 1992. Characterization of zinc-binding sites in human stromelysin-1: stoichiometry of the catalytic domain and identification of a cysteine ligand in the proenzyme. *Biochemistry* **31**: 4535–4540.

588. **Sanchez-Lopez, R., C. M. Alexander, O. Behrendtsen, R. Breathnach, and Z. Werb.** 1993. Role of zinc-binding- and hemopexin domain-encoded sequences in the substrate specificity of collagenase and stromelysin-2 as revealed by chimeric proteins. *J. Biol. Chem.* **268**: 7238–7247.

589. **Sanchez-Lopez, R., R. Nicholson, M. C. Gesnel, L. M. Matrisian, and R. Breathnach.** 1988. Structure-function relationships in the collagenase family member transin. *J. Biol. Chem.* **263**: 11892–11899.

590. **Santavicca, M., A. Noël, H. Angliker, I. Stoll, J. P. Segain, P. Anglard, M. Chretien, N. Seidah, and P. Basset.** 1996. Characterization of structural determinants and molecular mechanisms involved in pro-stromelysin-3 activation by 4-aminophenylmercuric acetate and furin-type convertases. *Biochem. J.* **315**: 953–958.

591. **Sato, H., T. Kinoshita, T. Takino, K. Nakayama, and M. Seiki.** 1996. Activation of a recombinant membrane type 1-matrix metalloproteinase (MT1-MMP) by furin and its interaction with tissue inhibitor of metalloproteinases (TIMP)-2. *FEBS Lett.* **393**: 101–104.

592. **Schnierer, S., T. Kleine, T. Gote, A. Hillemann, V. Knäuper, and H. Tschesche.** 1993. The recombinant catalytic domain of human neutrophil collagenase lacks type I collagen substrate specificity. *Biochem. Biophys. Res. Commun.* **191**: 319–326.

593. **Shipley, J. M., G. A. Doyle, C. J. Fliszar, Q. Z. Ye, L. L. Johnson, S. D. Shapiro, H. G. Welgus, and R. M. Senior.** 1996. The structural basis for the elastolytic activity of the 92-kDa and 72-kDa gelatinases. Role of the fibronectin type II-like repeats. *J. Biol. Chem.* **271**: 4335–4341.

594. **Shofuda, K., H. Yasumitsu, A. Nishihashi, K. Miki, and K. Miyazaki.** 1997. Expression of three membrane-type matrix metalloproteinases (MT-MMPs) in rat vascular smooth muscle cells and characterization of

MT3-MMPs with and without transmembrane domain. *J. Biol. Chem.* **272**: 9749–9754.

595. **Soler, D., T. Nomizu, W. E. Brown, M. Chen, Q. Z. Ye, H. E. Van Wart, and D. S. Auld.** 1994. Zinc content of promatrilysin, matrilysin and the stromelysin catalytic domain. *Biochem. Biophys. Res. Commun.* **201**: 917–923.

596. **Springman, E. B., E. L. Angleton, H. Birkedal-Hansen, and H. E. Van Wart.** 1990. Multiple modes of activation of latent human fibroblast collagenase: evidence for the role of a Cys^{73} active-site zinc complex in latency and a 'cysteine switch' mechanism for activation. *Proc. Natl. Acad. Sci. USA* **87**: 364–368.

597. **Spurlino, J. C., A. M. Smallwood, D. D. Carlton, T. M. Banks, K. J. Vavra, J. S. Johnson, E. R. Cook, J. Falvo, R. C. Wahl, T. A. Pulvino,** *et al.* 1994. 1.56 A structure of mature truncated human fibroblast collagenase. *Proteins* **19**: 98–109.

598. **Steffensen, B., U. M. Wallon, and C. M. Overall.** 1995. Extracellular matrix binding properties of recombinant fibronectin type II-like modules of human 72-kDa gelatinase/type IV collagenase. High affinity binding to native type I collagen but not native type IV collagen. *J. Biol. Chem.* **270**: 11555–11566.

599. **Stetler-Stevenson, W. G., J. A. Talano, M. E. Gallagher, H. C. Krutzsch, and L. A. Liotta.** 1991. Inhibition of human type IV collagenase by a highly conserved peptide sequence derived from its prosegment. *Am. J. Med. Sci.* **302**: 163–170.

600. **Stöcker, W., F. Grams, U. Baumann, P. Reinemer, F. X. Gomis-Rüth, D. B. McKay, and W. Bode.** 1995. The metzincins–topological and sequential relations between the astacins, adamalysins, serralysins, and matrixins (collagenases) define a superfamily of zinc-peptidases. *Protein Sci.* **4**: 823–840.

601. **Strongin, A. Y., I. E. Collier, P. A. Krasnov, L. T. Genrich, B. L. Marmer, and G. I. Goldberg.** 1993. Human 92 kDa type IV collagenase: functional analysis of fibronectin and carboxyl-end domains. *Kidney Int.* **43**: 158–162.

602. **Strongin, A. Y., B. L. Marmer, G. A. Grant, and G. I. Goldberg.** 1993. Plasma membrane-dependent activation of the 72-kDa type IV collagenase is prevented by complex formation with TIMP-2. *J. Biol. Chem.* **268**: 14033–14039.

603. **Suzuki, K., C. C. Kan, W. Hung, M. R. Gehring, K. Brew, and H. Nagase.** 1998. Expression of human pro-matrix metalloproteinase 3 that lacks the N-terminal 34 residues in *Escherichia coli*—autoactivation and interaction with tissue inhibitor of metalloproteinase 1 (TIMP-1). *Biol. Chem.* **379**: 185–191.

604. **Takino, T., H. Sato, E. Yamamoto, and M. Seiki.** 1995. Cloning of a human gene potentially encoding a novel matrix metalloproteinase having a C-terminal transmembrane domain. *Gene* **155**: 293–298.

605. **Taylor, K. B., L. J. Windsor, N. C. M. Caterina, M. K. Bodden, and J. A. Engler.** 1996. The mechanism of inhibition of collagenase by TIMP-1. *J. Biol. Chem.* **271**: 23938–23945.

606. **Vallee, B. L. and D. S. Auld.** 1990. Zinc coordination, function, and structure of zinc enzymes and other proteins. *Biochemistry* **29**: 5647–5659.

607. **Van Wart, H. E. and H. Birkedal-Hansen.** 1990. The cysteine switch: a principle of regulation of metalloproteinase activity with potential applicability to the entire matrix metalloproteinase gene family. *Proc. Natl. Acad. Sci. USA* **87**: 5578–5582.

608. **Wallon, U. M. and C. M. Overall.** 1997. The hemopexin-like domain (C domain) of human gelatinase A (matrix metalloproteinase-2) requires Ca^{2+} for fibronectin and heparin binding. Binding properties of recombinant gelatinase A C domain to extracellular matrix and basement membrane components. *J. Biol. Chem.* **272**: 7473–7481.

609. **Ward, R. V., S. J. Atkinson, J. J. Reynolds, and G. Murphy.** 1994. Cell surface-mediated activation of progelatinase A: demonstration of the involvement of the C-terminal domain of progelatinase A in cell surface binding and activation of progelatinase A by primary fibroblasts. *Biochem. J.* **304**: 263–269.

610. **Wetmore, D. R. and K. D. Hardman.** 1996. Roles of the propeptide and metal ions in the folding and stability of the catalytic domain of stromelysin (matrix metalloproteinase 3). *Biochemistry* **35**: 6549–6558.

611. **Wilhelm, S. M., I. E. Collier, B. L. Marmer, A. Z. Eisen, G. A. Grant, and G. I. Goldberg.** 1989. SV40-transformed human lung fibroblasts secrete a 92-kDa type IV collagenase which is identical to that secreted by normal human macrophages. *J. Biol. Chem.* **264**: 17213–17221.

612. **Will, H., S. J. Atkinson, G. S. Butler, B. Smith, and G. Murphy.** 1996. The soluble catalytic domain of membrane type 1 matrix metalloproteinase cleaves the propeptide of progelatinase A and initiates autoproteolytic activation. Regulation by TIMP-2 and TIMP-3. *J. Biol. Chem.* **271**: 17119–17123.

613. **Willenbrock, F., G. Murphy, I. R. Phillips, and K. Brocklehurst.** 1995. The second zinc atom in the matrix metalloproteinase catalytic domain is absent in the full-length enzymes: a possible role for the C-terminal domain. *FEBS Lett.* **358**: 189–192.

614. **Williams, D. H. and E. J. Murray.** 1994. Specific amino acid substitutions in human collagenase cause decreased autoproteolysis and reveal a requirement for a second zinc atom for catalytic activity. *FEBS Lett.* **354**: 267–270.

615. **Windsor, L. J., H. Birkedal-Hansen, B. Birkedal-Hansen, and J. A. Engler.** 1991. An internal cysteine plays a role in the maintenance of the latency of human fibroblast collagenase. *Biochemistry* **30**: 641–647.

616. **Windsor, L. J., M. K. Bodden, B. Birkedal-Hansen, J. A. Engler, and H. Birkedal-Hansen.** 1994. Mutational analysis of residues in and around the active site of human fibroblast-type collagenase. *J. Biol. Chem.* **269**: 26201–26207.

617. **Windsor, L. J., D. L. Steele, S. B. LeBlanc, and K. B. Taylor.** 1997. Catalytic domain comparisons of human fibroblast-type collagenase, stromelysin-1, and matrilysin. *Biochim. Biophys. Acta* **1334**: 261–272.

618. **Xia, Y., G. Garcia, S. Chen, C. B. Wilson, and L. Feng.** 1996. Cloning of rat 92-kDa type IV collagenase and expression of an active recombinant catalytic domain. *FEBS Lett.* **382**: 285–288.

619. **Yang, M. Z., M. T. Murray, and M. Kurkinen.** 1997. A novel matrix metalloproteinase gene (XMMP) encoding vitronectin-like motifs is transiently expressed in *Xenopus laevis* early embryo development. *J. Biol. Chem.* **272**: 13527–13533.

620. **Ye, Q. Z., L. L. Johnson, A. E. Yu, and D. Hupe.** 1995. Reconstructed 19 kDa catalytic domain of gelatinase A is an active proteinase. *Biochemistry* **34**: 4702–4708.

621. **Zhu, L., T. J. Hope, J. Hall, A. Davies, M. Stern, U. Muller-Eberhard, R. Stern, and T. G. Parslow.** 1994. Molecular cloning of a mammalian hyaluronidase reveals identity with hemopexin, a serum heme-binding protein. *J. Biol. Chem.* **269**: 32092–32097.

622. **Zucker, S., M. Drews, C. Conner, H. D. Foda, Y. A. De Clerck, K. E. Langley, W. F. Bahou, A. J. P. Docherty, and J. Cao.** 1998. Tissue inhibitor of metalloproteinase-2 (TIMP-2) binds to the catalytic domain of the cell surface receptor, membrane type 1 matrix metalloproteinase 1 (MT1-MMP). *J. Biol. Chem.* **273**: 1216–1222.

See also review articles: 18, 70, 82, 103, 104, 111.

Historical introduction and nomenclature

623. **Aer, J.** 1971. Purification of rat bone collagenolytic enzymes. *Ann. Med. Exp. Biol. Fenn.* **49**: 1–8.

624. **Apte, S. S., N. Fukai, D. R. Beier, and B. R. Olsen.** 1997. The matrix metalloproteinase-14 (MMP-14) gene is structurally distinct from other MMP genes and is co-expressed with the TIMP-2 gene during mouse embryogenesis. *J. Biol. Chem.* **272**: 25511–25517.

625. **Apte, S. S., K. Hayashi, M. F. Seldin, M. G. Mattei, M. Hayashi, and B. R. Olsen.** 1994. Gene encoding a novel murine tissue inhibitor of metalloproteinases (TIMP), TIMP-3, is expressed in developing mouse epithelia, cartilage, and muscle, and is located on mouse chromosome 10. *Dev. Dyn.* **200**: 177–197.

626. **Apte, S. S., M. G. Mattei, and B. R. Olsen.** 1994. Cloning of the cDNA encoding human tissue inhibitor of metalloproteinases-3 (TIMP-3) and mapping of the TIMP3 gene to chromosome 22. *Genomics* **19**: 86–90.

627. **Atkinson, S. J., T. Crabbe, S. Cowell, R. V. Ward, M. J. Butler, H. Sato, M. Seiki, J. J. Reynolds, and G. Murphy.** 1995. Intermolecular autolytic cleavage can contribute to the activation of progelatinase A by cell membranes. *J. Biol. Chem.* **270**: 30479–30485.

628. **Azzo, W. and J. F. Woessner, Jr.** 1986. Purification and characterization of an acid metalloproteinase from human articular cartilage. *J. Biol. Chem.* **261**: 5434–5441.

629. **Banda, M. J., E. J. Clark, and Z. Werb.** 1980. Limited proteolysis by macrophage elastase inactivates human α_1-proteinase inhibitor. *J. Exp. Med.* **152**: 1563–1570.

630. **Banda, M. J. and Z. Werb.** 1981. Mouse macrophage elastase. Purification and characterization as a metalloproteinase. *Biochem. J.* **193**: 589–605.

631. **Baramova, E. N., C. Munaut, A. Remacle, B. Pichot, and J. M. Foidart.** 1995. Correlation of tumor cells membrane activation of gelatinase A with the expression of MT-MMP mRNA. *Cell Biol. Int.* **19**: 248–248.

632. **Bartlett, J. D., J. P. Simmer, J. Xue, H. C. Margolis, and E. C. Moreno.** 1996. Molecular cloning and mRNA tissue distribution of a novel matrix metalloproteinase isolated from porcine enamel organ. *Gene* **183**: 123–128.

633. **Basset, P., J.P Bellocq, C. Wolf, I. Stoll, P. Hutin, J. M. Limacher, O. L. Podhajcer, M. P. Chenard, M. C. Rio, and P. Chambon.** 1990. A novel metalloproteinase gene specifically expressed in stromal cells of breast carcinomas. *Nature* **348**: 699–704.

634. **Bauer, E. A., A. Z. Eisen, and J. J. Jeffrey.** 1970. Immunologic relationship of a purified human skin collagenase to other human and animal collagenases. *Biochim. Biophys. Acta* **206**: 152–160.

635. **Bauer, E. A., G. P. Stricklin, J. J. Jeffrey, and A. Z. Eisen.** 1975. Collagenase production by human skin fibroblasts. *Biochem. Biophys. Res. Commun.* **64**: 232–240.

636. **Belaaouaj, A., J. M. Shipley, D. K. Kobayashi, D. B. Zimonjic, N. Popescu, G. A. Silverman, and S. D. Shapiro.** 1995. Human macrophage metalloelastase. Genomic organization, chromosomal location, gene linkage, and tissue-specific expression. *J. Biol. Chem.* **270**: 14568–14575.

637. **Beutner, E. H., C. Triftshauser, and S. P. Hazen.** 1966. Collagenase activity of gingival tissue from patients with periodontal disease. *Proc. Soc. Exp. Biol. Med.* **121**: 1082–1085.

638. **Bigg, H. F., Y. E. Shi, Y. L. E. Liu, B. Steffensen, and C. M. Overall.** 1997. Specific, high affinity binding of tissue inhibitor of metalloproteinases-4 (TIMP4) to the COOH-terminal hemopexin-like domain of human gelatinase A TIMP-4 binds progelatinase A and the COOH-terminal domain in a similar manner to TIMP-2. *J. Biol. Chem.* **272**: 15496–15500.

639. **Birkedal-Hansen, H., C. M. Cobb, R. E. Taylor, and H. M. Fullmer.** 1975. Activation of latent bovine gingival collagenase. *Arch. Oral Biol.* **20**: 681–685.

640. **Bjorn, S. F., N. Hastrup, L. R. Lund, K. Danø, J. F. Larsen, and C. Pyke.** 1997. Co-ordinated expression of MMP-2 and its putative activator, MT1-MMP, in human placentation. *Mol. Hum. Reprod.* **3**: 713–723.

641. **Black, R. A., C. T. Rauch, C. J. Kozlosky, J. J. Peschon, J. L. Slack, M. F. Wolfson, B. J. Castner, K. L. Stocking, P. Reddy, S. Srinivasan,** *et al.* 1997. A metalloproteinase disintegrin that releases tumour-necrosis factor-α from cells. *Nature* **385**: 729–733.

642. **Blenis, J. and S. P. Hawkes.** 1983. Transformation-sensitive protein associated with the cell substratum of chicken embryo fibroblasts. *Proc. Natl. Acad. Sci. USA* **80**: 770–774.

643. **Blenis, J. and S. P. Hawkes.** 1984. Characterization of a transformation-sensitive protein in the extracellular matrix of chicken embryo fibroblasts. *J. Biol. Chem.* **259**: 11563–11570.

644. **Boujrad, N., S. O. Ogwuegbu, M. Garnier, C. H. Lee, B. M. Martin, and V. Papadopoulos.** 1995. Identification of a stimulator of steroid hormone synthesis isolated from testis. *Science* **268**: 1609–1612.

645. **Breathnach, R., L. M. Matrisian, M.-C. Gesnel, A. Staub, and P. Leroy.** 1987. Sequences coding for part of oncogene-induced transin are highly conserved in a related rat gene. *Nucleic Acids Res.* **15**: 1139–1151.

646. **Buttner, F. H., S. Chubinskaya, D. Margerie, K. Huch, J. Flechtenmacher, A. A. Cole, K. E. Kuettner, and E. Bartnik.** 1997. Expression of membrane type 1 matrix metalloproteinase in human articular cartilage. *Arthritis Rheum.* **40**: 704–709.

647. **Cawston, T. E., W. A. Galloway, E. Mercer, G. Murphy, and J. J. Reynolds.** 1981. Purification of rabbit bone inhibitor of collagenase. *Biochem. J.* **195**: 159–165.

648. **Chin, J. R., G. Murphy, and Z. Werb.** 1985. Stromelysin, a connective tissue-degrading metalloendopeptidase secreted by stimulated rabbit synovial fibroblasts in parallel with collagenase. Biosynthesis, isolation, characterization, and substrates. *J. Biol. Chem.* **260**: 12367–12376.

649. **Collier, I. E., P. A. Krasnov, A. Y. Strongin, H. Birkedal-Hansen, and G. I. Goldberg.** 1992. Alanine scanning mutagenesis and functional analysis of the fibronectin-like collagen-binding domain from human 92-kDa type IV collagenase. *J. Biol. Chem.* **267**: 6776–6781.

650. **Collier, I. E., S. M. Wilhelm, A. Z. Eisen, B. L. Marmer, G. A. Grant, J. L. Seltzer, A. Kronberger, C. S. He, E. A. Bauer, and G. I. Goldberg.** 1988. H-ras oncogene-transformed human bronchial epithelial cells (TBE-1) secrete a single metalloprotease capable of degrading basement membrane collagen. *J. Biol. Chem.* **263**: 6579–6587.

651. **Cossins, J., T. J. Dudgeon, G. Catlin, A. J. Gearing, and J. M. Clements.** 1996. Identification of MMP-18, a putative novel human matrix metalloproteinase. *Biochem. Biophys. Res. Commun.* **228**: 494–498.

652. **d'Ortho, M. P., H. Will, S. Atkinson, G. Butler, A. Messent, J. Gavrilovic, B. Smith, R. Timpl, L. Zardi, and G. Murphy.** 1997. Membrane-type matrix metalloproteinases 1 and 2 exhibit broad-spectrum proteolytic capacities comparable to many matrix metalloproteinases. *Eur. J. Biochem.* **250**: 751–757.

653. **De Clerck, Y. A., T.-D. Yean, B. J. Ratzkin, H. S. Lu, and K. E. Langley.** 1989. Purification and characterization of two related but distinct metalloproteinase inhibitors secreted by bovine aortic endothelial cells. *J. Biol. Chem.* **264**: 17445–17453.

654. **DenBesten, P. K. and L. M. Heffernan.** 1989. Separation by polyacrylamide gel electrophoresis of multiple proteases in rat and bovine enamel. *Arch. Oral Biol.* **34**: 399–404.

655. **Deshmukh-Phadke, K., M. Lawrence, and S. Nanda.** 1978. Synthesis of collagenase and neutral proteases by articular chondrocytes: stimulation by a macrophage-derived factor. *Biochem. Biophys. Res. Commun.* **85**: 490–496.

656. **Docherty, A. J. P., A. Lyons, B. J. Smith, E. M. Wright, P. E. Stephens, T. J. R. Harris, G. Murphy, and J. J. Reynolds.** 1985. Sequence of human tissue inhibitor of metalloproteinases and its identity to erythroid-potentiating activity. *Nature* **318**: 66–69.

657. **Edwards, D. R., C. L. Parfett, and D. T. Denhardt.** 1985. Transcriptional regulation of two serum-induced RNAs in mouse fibroblasts: equivalence of one species to B2 repetitive elements. *Mol. Cell. Biol.* **5**: 3280–3288.

658. **Evanson, J. M., J. J. Jeffrey, and S. M. Krane.** 1967. Human collagenase: identification and characterization of an enzyme from rheumatoid synovium in culture. *Science* **158**: 499–502.

659. **Freije, J. M., I. Díez-Itza, M. Balbín, L. M. Sánchez, R. Blasco, J. Tolivia, and C. López-Otín.** 1994. Molecular cloning and expression of collagenase-3, a novel human matrix metalloproteinase produced by breast carcinomas. *J. Biol. Chem.* **269**: 16766–16773.

660. **Fullmer, H. M. and W. Gibson.** 1966. Collagenolytic activity in gingivae of man. *Nature* **209**: 728–729.

661. **Fullmer, H. M., W. A. Gibson, G. Lazarus, and A. C. Stam, Jr.** 1966. Collagenolytic activity of the skin associated with neuromuscular diseases including amyotrophic lateral sclerosis. *Lancet* **1**: 1007–1009.

662. **Fullmer, H. M. and G. Lazarus.** 1967. Collagenase in human goat and rat bone. *Israel J. Med. Sci.* **3**: 758–761.

663. **Galloway, W. A., G. Murphy, J. D. Sandy, J. Gavrilovic, T. E. Cawston, and J. J. Reynolds.** 1983.

Purification and characterization of a rabbit bone metalloproteinase that degrades proteoglycan and other connective-tissue components. *Biochem. J.* **209**: 741–752.

664. **Gasson, J. C., D. W. Golde, S. E. Kaufman, C. A. Westbrook, R. M. Hewick, R. J. Kaufman, G. G. Wong, P. A. Temple, A. C. Leary, E. L. Brown, *et al.*** 1985. Molecular characterization and expression of the gene encoding human erythroid-potentiating activity. *Nature* **315**: 768–771.

665. **Gauwerky, C. E., A. J. Lusis, and D. W. Golde.** 1980. Erythroid-potentiating activity: characterization and target cells. *Exp. Hematol.* **8**: 117–127.

666. **Gewert, D. R., B. Coulombe, M. Castelino, D. Skup, and B. R. G. Williams.** 1987. Characterization and expression of a murine gene homologous to human EPA/TIMP: a virus-induced gene in the mouse. *EMBO J.* **6**: 651–657.

667. **Goldberg, G. I., B. L. Marmer, G. A. Grant, A. Z. Eisen, S. Wilhelm, and C. S. He.** 1989. Human 72-kilodalton type IV collagenase forms a complex with a tissue inhibitor of metalloproteases designated TIMP-2. *Proc. Natl. Acad. Sci. USA* **86**: 8207–8211.

668. **Goldberg, G. I., A. Strongin, I. E. Collier, L. T. Genrich, and B. L. Marmer.** 1992. Interaction of 92-kDa type IV collagenase with the tissue inhibitor of metalloproteinases prevents dimerization, complex formation with interstitial collagenase, and activation of the proenzyme with stromelysin. *J. Biol. Chem.* **267**: 4583–4591.

669. **Goldberg, G. I., S. M. Wilhelm, A. Kronberger, E. A. Bauer, G. A. Grant, and A. Z. Eisen.** 1986. Human fibroblast collagenase. Complete primary structure and homology to an oncogene transformation-induced rat protein. *J. Biol. Chem.* **261**: 6600–6605.

670. **Graham, J. S., J. Xiong, and J. W. Gillikin.** 1991. Purification and developmental analysis of a metalloendoproteinase from the leaves of *Glycine max*. *Plant Physiol.* **97**: 786–792.

671. **Greene, J., M. Wang, Y. E. Liu, L. A. Raymond, C. Rosen, and Y. E. Shi.** 1996. Molecular cloning and characterization of human tissue inhibitor of metalloproteinase 4. *J. Biol. Chem.* **271**: 30375–30380.

672. **Gross, J. and C. M. Lapière.** 1962. Collagenolytic activity in amphibian tissues: a tissue culture assay. *Proc. Natl. Acad. Sci. USA* **48**: 1014–1022.

673. **Gross, J. and Y. Nagai.** 1965. Specific degradation of the collagen molecule by tadpole collagenolytic enzyme. *Proc. Natl. Acad. Sci. USA* **54**: 1197–1204.

674. **Guillem, J. G., M. F. Levy, L. L. Hsieh, M. D. Johnson, P. LoGerfo, K. A. Forde, and I. B. Weinstein.** 1990. Increased levels of phorbin, c-*myc*, and ornithine decarboxylase RNAs in human colon cancer. *Mol. Carcinog.* **3**: 68–74.

675. **Gunja-Smith, Z., H. Nagase, and J. F. Woessner, Jr.**

1989. Purification of the neutral proteoglycan-degrading metalloproteinase from human articular cartilage tissue and its identification as stromelysin matrix metalloproteinase-3. *Biochem. J.* **258**: 115–119.

676. **Halme, J., B. Tyree, and J. J. Jeffrey.** 1980. Collagenase production by primary cultures of rat uterine cells. Partial purification and characterization of the enzyme. *Arch. Biochem. Biophys.* **199**: 51–60.

677. **Hanemaaijer, R., T. Sorsa, Y. T. Konttinen, Y. L. Ding, M. Sutinen, H. Visser, V. W. M. Van Hinsbergh, T. Helaakoski, T. Kainulainen, H. Ronka, *et al.*** 1997. Matrix metalloproteinase-8 is expressed in rheumatoid synovial fibroblasts and endothelial cells—regulation by tumor necrosis factor-α and doxycycline. *J. Biol. Chem.* **272**: 31504–31509.

678. **Harper, E., K. J. Bloch, and J. Gross.** 1971. The zymogen of tadpole collagenase. *Biochemistry* **10**: 3035–3041.

679. **Harris, E. D., Jr. and S. M. Krane.** 1972. An endopeptidase from rheumatoid synovial tissue culture. *Biochim. Biophys. Acta* **258**: 566–576.

680. **Harrison, R. K., B. Chang, L. Niedzwiecki, and R. L. Stein.** 1992. Mechanistic studies on the human matrix metalloproteinase stromelysin. *Biochemistry* **31**: 10757–10762.

681. **Hasty, K. A., M. S. Hibbs, A. H. Kang, and C. L. Mainardi.** 1984. Heterogeneity among human collagenases demonstrated by monoclonal antibody that selectively recognizes and inhibits human neutrophil collagenase. *J. Exp. Med.* **159**: 1455–1463.

682. **Hasty, K. A., T. F. Pourmotabbed, G. I. Goldberg, J. P. Thompson, D. G. Spinella, R. M. Stevens, and C. L. Mainardi.** 1990. Human neutrophil collagenase. A distinct gene product with homology to other matrix metalloproteinases. *J. Biol. Chem.* **265**: 11421–11424.

683. **Herron, G. S., M. J. Banda, E. J. Clark, J. Gavrilovic, and Z. Werb.** 1986. Secretion of metalloproteinases by stimulated capillary endothelial cells. II. Expression of collagenase and stromelysin activities is regulated by endogenous inhibitors. *J. Biol. Chem.* **261**: 2814–2818.

684. **Jeffrey, J. J. and J. Gross.** 1970. Collagenase from rat uterus. Isolation and partial characterization. *Biochemistry* **9**: 268–273.

685. **Johnson, M. D., G. M. Housey, P. T. Kirschmeier, and I. B. Weinstein.** 1987. Molecular cloning of gene sequences regulated by tumor promoters and mitogens through protein kinase C. *Mol. Cell. Biol.* **7**: 2821–2829.

686. **Kettner, C., E. Shaw, R. White, and A. Janoff.** 1981. The specificity of macrophage elastase on the insulin B-chain. *Biochem. J.* **195**: 369–372.

687. **Kinoh, H., H. Sato, Y. Tsunezuka, T. Takino, A. Kawashima, Y. Okada, and M. Seiki.** 1996. MT-MMP, the cell surface activator of proMMP-2 (pro-

gelatinase A), is expressed with its substrate in mouse tissue during embryogenesis. *J. Cell Sci.* **109**: 953–959.

688. **Kjeldsen, L., A. H. Johnsen, H. Sengelov, and N. Borregaard.** 1993. Isolation and primary structure of NGAL, a novel protein associated with human neutrophil gelatinase. *J. Biol. Chem.* **268**: 10425–10432.

689. **Kobayashi, S. and Y. Nagai.** 1978. Human leucocyte neutral proteases, with special reference to collagen metabolism. *J. Biochem. (Tokyo)* **84**: 559–567.

690. **Kolb, C., S. Mauch, H. H. Peter, U. Krawinkel, and R. Sedlacek.** 1997. The matrix metalloproteinase RASI-1 is expressed in synovial blood vessels of a rheumatoid arthritis patient. *Immunol. Lett.* **57**: 83–88.

691. **Krane, S. M., M. H. Byrne, V. Lemaître, P. Henriet, J. J. Jeffrey, J. P. Witter, X. Liu, H. Wu, R. Jaenisch, and Y. Eeckhout.** 1996. Different collagenase gene products have different roles in degradation of type I collagen. *J. Biol. Chem.* **271**: 28509–28515.

692. **Kruze, D. and E. Wojtecka.** 1972. Activation of leucocyte collagenase proenzyme by rheumatoid synovial fluid. *Biochim. Biophys. Acta* **285**: 436–446.

693. **Kuettner, K. E., J. Hiti, R. Eisenstein, and E. Harper.** 1976. Collagenase inhibition by cationic proteins derived from cartilage and aorta. *Biochem. Biophys. Res. Commun.* **72**: 40–46.

694. **Lawson, N. D., A. Khannagupta, and N. Berliner.** 1998. Isolation and characterization of the cDNA for mouse neutrophil collagenase—demonstration of shared negative regulatory pathways for neutrophil secondary granule protein gene expression. *Blood* **91**: 2517–2524.

695. **Lazarus, G. S., R. S. Brown, J. R. Daniels, and H. M. Fullmer.** 1968. Human granulocyte collagenase. *Science* **159**: 1483–1485.

696. **Lazarus, G. S., J. R. Daniels, R. S. Brown, H. A. Bladen, and H. M. Fullmer.** 1968. Degradation of collagen by a human granulocyte collagenolytic system. *J. Clin. Invest.* **47**: 2622–2629.

697. **Leco, K. J., S. S. Apte, G. T. Taniguchi, S. P. Hawkes, R. Khokha, G. A. Schultz, and D. R. Edwards.** 1997. Murine tissue inhibitor of metalloproteinases-4 (TIMP-4): cDNA isolation and expression in adult mouse tissues. *FEBS Lett.* **401**: 213–217.

698. **Lefebvre, O., C. Regnier, M. P. Chenard, C. Wendling, P. Chambon, P. Basset, and M. C. Rio.** 1995. Developmental expression of mouse stromelysin-3 mRNA. *Development* **121**: 947–955.

699. **Lepage, T. and C. Gache.** 1989. Purification and characterization of the sea urchin embryo hatching enzyme. *J. Biol. Chem.* **264**: 4787–4793.

700. **Lepage, T. and C. Gache.** 1990. Early expression of a collagenase-like hatching enzyme gene in the sea urchin embryo. *EMBO J.* **9**: 3003–3012.

701. **Liotta, L. A., S. Abe, P. G. Robey, and G. R. Martin.** 1979. Preferential digestion of basement membrane collagen by an enzyme derived from a metastatic murine tumor. *Proc. Natl. Acad. Sci. USA* **76**: 2268–2272.

702. **Liu, Y. L. E., M. S. Wang, J. Greene, J. Su, S. Ullrich, H. Li, S. J. Sheng, P. Alexander, Q. X. A. Sang, and Y. E. Shi.** 1997. Preparation and characterization of recombinant tissue inhibitor of metalloproteinase 4 (TIMP-4). *J. Biol. Chem.* **272**: 20479–20483.

703. **Llano, E., A. M. Pendás, V. Knäuper, T. Sorsa, T. Salo, E. Salido, G. Murphy, J. P. Simmer, J. D. Bartlett, and C. López-Otín.** 1997. Identification and structural and functional characterization of human enamelysin (MMP-20). *Biochemistry* **36**: 15101–15108.

704. **Macartney, H. W. and H. Tschesche.** 1983. The collagenase inhibitor from human polymorphonuclear leukocytes. Isolation, purification and characterisation. *Eur. J. Biochem.* **130**: 79–83.

705. **Mainardi, C. L., J. M. Seyer, and A. H. Kang.** 1980. Type-specific collagenolysis: a type V collagen-degrading enzyme from macrophages. *Biochem. Biophys. Res. Commun.* **97**: 1108–1115.

706. **Marcotte, P. A., I. M. Kozan, S. A. Dorwin, and J. M. Ryan.** 1992. The matrix metalloproteinase pump-1 catalyzes formation of low molecular weight (pro)urokinase in cultures of normal human kidney cells. *J. Biol. Chem.* **267**: 13803–13806.

707. **Matrisian, L. M., N. Glaichenhaus, M.-C. Gesnel, and R. Breathnach.** 1985. Epidermal growth factor and oncogenes induce transcription of the same cellular mRNA in rat fibroblasts. *EMBO J.* **4**: 1435–1440.

708. **Matsumoto, H., Y. Ishibashi, T. Ohtaki, Y. Hasegawa, C. Koyama, and K. Inoue.** 1993. Newly established murine pituitary folliculo-stellate-like cell line (TtT/GF) secretes potent pituitary glandular cell survival factors, one of which corresponds to metalloproteinase inhibitor. *Biochem. Biophys. Res. Commun.* **194**: 909–915.

709. **Matsumoto, S. I., M. Katoh, S. Saito, T. Watanabe, and Y. Masuho.** 1997. Identification of soluble type of membrane-type matrix metalloproteinase-3 formed by alternatively spliced mRNA. *Biochim. Biophys. Acta—Gene Struct. Express.* **1354**: 159–170.

710. **McGeehan, G., W. Burkhart, R. Anderegg, J. D. Becherer, J. W. Gillikin, and J. S. Graham.** 1992. Sequencing and characterization of the soybean leaf metalloproteinase. Structural and functional similarity to the matrix metalloproteinase family. *Plant Physiol.* **99**: 1179–1183.

711. **Miyazaki, K., Y. Hattori, F. Umenishi, H. Yasumitsu, and M. Umeda.** 1990. Purification and characterization of extracellular matrix-degrading metalloproteinase, matrin (pump-1), secreted from human rectal carcinoma cell line. *Cancer Res.* **50**: 7758–7764.

712. **Moncrief, J. S., R. Obiso, Jr., L. A. Barroso, J. J. Kling, R. L. Wright, R. L. van Tassell, D. M. Lyerly, and T. D. Wilkins.** 1995. The enterotoxin of *Bacteroides fragilis* is a metalloprotease. *Infect. Immun.* **63**: 175–181.

713. **Moradian-Oldak, J., J. P. Simmer, P. E. Sarte, M. Zeichner-David, and A. G. Fincham.** 1994. Specific cleavage of a recombinant murine amelogenin at the carboxy-terminal region by a proteinase fraction isolated from developing bovine tooth enamel. *Arch. Oral Biol.* **39**: 647–656.

714. **Moss, M. L., S. L. C. Jin, M. E. Milla, W. Burkhart, H. L. Carter, W. J. Chen, W. C. Clay, J. R. Didsbury, D. Hassler, C. R. Hoffman,** *et al.* 1997. Cloning of a disintegrin metalloproteinase that processes precursor tumour-necrosis factor-α. *Nature* **385**: 733–736.

715. **Muller, D., R. Breathnach, A. Engelmann, R. Millon, G. Bronner, H. Flesch, P. Dumont, M. Eber, and J. Abecassis.** 1991. Expression of collagenase-related metalloproteinase genes in human lung or head and neck tumours. *Int. J. Cancer* **48**: 550–556.

716. **Muller, D., B. Quantin, M. C. Gesnel, R. Millon-Collard, J. Abecassis, and R. Breathnach.** 1988. The collagenase gene family in humans consists of at least four members. *Biochem. J.* **253**: 187–192.

717. **Murphy, G., U. Bretz, M. Baggiolini, and J. J. Reynolds.** 1980. The latent collagenase and gelatinase of human polymorphonuclear neutrophil leucocytes. *Biochem. J.* **192**: 517–525.

718. **Murphy, G., E. C. Cartwright, A. Sellers, and J. J. Reynolds.** 1977. The detection and characterisation of collagenase inhibitors from rabbit tissues in culture. *Biochim. Biophys. Acta* **483**: 493–498.

719. **Murphy, G., C. G. McAlpine, C. T. Poll, and J. J. Reynolds.** 1985. Purification and characterization of a bone metalloproteinase that degrades gelatin and types IV and V collagen. *Biochim. Biophys. Acta* **831**: 49–58.

720. **Murphy, G., J. J. Reynolds, U. Bretz, and M. Baggiolini.** 1977. Collagenase is a component of the specific granules of human neutrophil leucocytes. *Biochem. J.* **162**: 195–197.

721. **Murphy, G., J. P. Segain, M. O'Shea, M. Cockett, C. Ioannou, O. Lefebvre, P. Chambon, and P. Basset.** 1993. The 28-kDa N-terminal domain of mouse stromelysin-3 has the general properties of a weak metalloproteinase. *J. Biol. Chem.* **268**: 15435–15441.

722. **Nagase, H., A. J. Barrett, and J. F. Woessner, Jr.** 1992. Nomenclature and glossary of the matrix metalloproteinases. *Matrix Suppl.* **1**: 421–424.

723. **Nakano, T. and P. G. Scott.** 1986. Purification and characterization of a gelatinase produced by fibroblasts from human gingiva. *Biochem. Cell. Biol.* **64**: 387–393.

724. **Nakano, T. and P. G. Scott.** 1987. Partial purification and characterization of a neutral proteinase with collagen telopeptidase activity produced by human gingival fibroblasts. *Biochem. Cell. Biol.* **65**: 286–292.

725. **Nawrocki, B., M. Polette, C. Clavel, A. Morrone, J. P. Eschard, J. C. Etienne, and P. Birembaut.** 1994. Expression of stromelysin 3 and tissue inhibitors of matrix metallo-proteinases, TIMP-1 and TIMP-2, in rheumatoid arthritis. *Pathol. Res. Pract.* **190**: 690–696.

726. **Nemeth, J. A. and C. L. Goolsby.** 1993. TIMP-2, a growth-stimulatory protein from SV40-transformed human fibroblasts. *Exp. Cell Res.* **207**: 376–382.

727. **Nicholson, R., G. Murphy, and R. Breathnach.** 1989. Human and rat malignant-tumor-associated mRNAs encode stromelysin-like metalloproteinases. *Biochemistry* **28**: 5195–5203.

728. **Nolan, J. C., S. Ridge, A. L. Oronsky, L. L. Slakey, and S. S. Kerwar.** 1978. Synthesis of a collagenase inhibitor by smooth muscle cells in culture. *Biochem. Biophys. Res. Commun.* **83**: 1183–1190.

729. **Nolan, J. C., S. C. Ridge, A. L. Oronsky, and S. S. Kerwar.** 1980. Purification and properties of a collagenase inhibitor from cultures of bovine aorta. *Atherosclerosis* **35**: 93–102.

730. **Nomura, H., H. Sato, M. Seiki, M. Mai, and Y. Okada.** 1995. Expression of membrane-type matrix metalloproteinase in human gastric carcinomas. *Cancer Res.* **55**: 3263–3266.

731. **Nomura, K., T. Shimizu, H. Kinoh, Y. Sendai, M. Inomata, and N. Suzuki.** 1997. Sea urchin hatching enzyme (envelysin)—cDNA cloning and deprivation of protein substrate specificity by autolytic degradation. *Biochemistry* **36**: 7225–7238.

732. **Obiso, R. J., D. M. Lyerly, R. L. van Tassell, and T. D. Wilkins.** 1995. Proteolytic activity of the *Bacteroides fragilis* enterotoxin causes fluid secretion and intestinal damage in vivo. *Infect. Immun.* **63**: 3820–3826.

733. **Ohlsson, K.** 1980. Polymorphonuclear leukocyte collagenase. In *Collagenase in Normal and Pathological Connective Tissues.* D. E. Woolley and J. M. Evanson (eds), pp. 209–222. Wiley, Chichester, England.

734. **Okada, A., J. P. Bellocq, N. Rouyer, M. P. Chenard, M. C. Rio, P. Chambon, and P. Basset.** 1995. Membrane-type matrix metalloproteinase (MT-MMP) gene is expressed in stromal cells of human colon, breast, and head and neck carcinomas. *Proc. Natl. Acad. Sci. U.S.A* **92**: 2730–2734.

735. **Oronsky, A. L., R. J. Perper, and H. C. Schroder.** 1973. Phagocytic release and activation of human leukocyte procollagenase. *Nature* **246**: 417–419.

736. **Osawa, K., Shirai, T., Yanagawa, M., Yamada, K., Nishikawa, K., and Tanaka, H.** 1994. Purification and characterization of growth and differentiation factors from human osteosarcoma cell line, OST-1-PF. In *Anim. Cell Technol.: Basic Appl. Aspects, Proc. Int. Meet. Jpn. Assoc. Anim. Cell Technol. ,6th.* T. Kobayashi, Y.

Kitagawa, and K. Okumura (eds), pp. 589–593. Dordrecht, Kluwer.

737. **Overall, C. M. and J. Sodek.** 1987. Initial characterization of a neutral metalloproteinase, active on native 3/4-collagen fragments, synthesized by ROS 17/2.8 osteoblastic cells, periodontal fibroblasts, and identified in gingival crevicular fluid. *J. Dent. Res.* **66**: 1271–1282.

738. **Pak, J. H., C. Y. Liu, J. Huangpu, and J. S. Graham.** 1997. Construction and characterization of the soybean leaf metalloproteinase cDNA. *FEBS Lett.* **404**: 283–288.

739. **Pavloff, N., P. W. Staskus, N. S. Kishnani, and S. P. Hawkes.** 1992. A new inhibitor of metalloproteinases from chicken: ChIMP-3. A third member of the TIMP family. *J. Biol. Chem.* **267**: 17321–17326.

740. **Pei, D., G. Majmudar, and S. J. Weiss.** 1994. Hydrolytic inactivation of a breast carcinoma cell-derived serpin by human stromelysin-3. *J. Biol. Chem.* **269**: 25849–25855.

741. **Pei, D. and S. J. Weiss.** 1995. Furin-dependent intracellular activation of the human stromelysin-3 zymogen. *Nature* **375**: 244–247.

742. **Pelletier, J. P., J. Martel-Pelletier, J. M. Cloutier, and J. F. Woessner, Jr.** 1987. Proteoglycan-degrading acid metalloprotease activity in human osteoarthritic cartilage, and the effect of intraarticular steroid injections. *Arthritis Rheum.* **30**: 541–548.

743. **Pendás, A. M., V. Knäuper, X. S. Puente, E. Llano, M. G. Mattei, S. Apte, G. Murphy, and C. López-Otín.** 1997. Identification and characterization of a novel human matrix metalloproteinase with unique structural characteristics, chromosomal location, and tissue distribution. *J. Biol. Chem.* **272**: 4281–4286.

744. **Puente, X. S., A. M. Pendás, E. Llano, G. Velasco, and C. López-Otín.** 1996. Molecular cloning of a novel membrane-type matrix metalloproteinase from a human breast carcinoma. *Cancer Res.* **56**: 944–949.

745. **Quantin, B., G. Murphy, and R. Breathnach.** 1989. Pump-1 cDNA codes for a protein with characteristics similar to those of classical collagenase family members. *Biochemistry* **28**: 5327–5334.

746. **Quinn, C. O., D. K. Scott, C. E. Brinckerhoff, L. M. Matrisian, J. J. Jeffrey, and N. C. Partridge.** 1990. Rat collagenase. Cloning, amino acid sequence comparison, and parathyroid hormone regulation in osteoblastic cells. *J. Biol. Chem.* **265**: 22342–22347.

747. **Ragster, L. and M. J. Chrispeels.** 1979. Azocoll-digesting proteinases in soybean leaves: characteristics and changes during leaf maturation. *Plant Physiol.* **64**: 857–862.

748. **Rantala-Ryhänen, S., L. Ryhänen, F. V. Nowak, and J. Uitto.** 1983. Proteinases in human polymorphonuclear leukocytes. Purification and characterization of an enzyme which cleaves denatured collagen and a synthetic peptide with a Gly-Ile sequence. *Eur. J. Biochem.* **134**: 129–137.

749. **Raynes, J. G., F. A. Clarke, J. C. Anderson, R. J. Fitzpatrick, and H. Dobson.** 1988. Collagenase inhibitor concentration in cultured cervical tissue of sheep is increased in late pregnancy. *J. Reprod. Fertil.* **83**: 893–900.

750. **Rodgers, W. H., L. M. Matrisian, L. C. Giudice, B. Dsupin, P. Cannon, C. Svitek, F. Gorstein, and K. G. Osteen.** 1994. Patterns of matrix metalloproteinase expression in cycling endometrium imply differential functions and regulation by steroid hormones. *J. Clin. Invest.* **94**: 946–953.

751. **Roswit, W. T., J. Halme, and J. J. Jeffrey.** 1983. Purification and properties of rat uterine procollagenase. *Arch. Biochem. Biophys.* **225**: 285–295.

752. **Roughley, P. J., G. Murphy, and A. J. Barrett.** 1978. Proteinase inhibitors of bovine nasal cartilage. *Biochem. J.* **169**: 721–724.

753. **Sakamoto, S., P. Goldhaber, and M. J. Glimcher.** 1972. The further purification and characterization of mouse bone collagenase. *Calcif. Tissue Res.* **10**: 142–151.

754. **Salamonsen, L. A. and D. E. Woolley.** 1996. Matrix metalloproteinases and their tissue inhibitors in endometrial remodelling and menstruation. *Reprod. Med. Rev.* **5**: 185–203.

755. **Salo, T., L. A. Liotta, and K. Tryggvason.** 1983. Purification and characterization of a murine basement membrane collagen-degrading enzyme secreted by metastatic tumor cells. *J. Biol. Chem.* **258**: 3058–3063.

756. **Sapolsky, A. I., D. S. Howell, and J. F. Woessner, Jr.** 1974. Neutral proteases and cathepsin D in human articular cartilage. *J. Clin. Invest.* **53**: 1044–1053.

757. **Sapolsky, A. I., H. Keiser, D. S. Howell, and J. F. Woessner, Jr.** 1976. Metalloproteases of human articular cartilage that digest cartilage proteoglycan at neutral and acid pH. *J. Clin. Invest.* **58**: 1030–1041.

758. **Sato, H., T. Takino, Y. Okada, J. Cao, A. Shinagawa, E. Yamamoto, and M. Seiki.** 1994. A matrix metalloproteinase expressed on the surface of invasive tumour cells. *Nature* **370**: 61–65.

759. **Satoh, T., K. Kobayashi, S. Yamashita, M. Kikuchi, Y. Sendai, and H. Hoshi.** 1994. Tissue inhibitor of metalloproteinases (TIMP-1) produced by granulosa and oviduct cells enhances in vitro development of bovine embryo. *Biol. Reprod.* **50**: 835–844.

760. **Scott, P. G. and H. A. Goldberg.** 1983. Cleavage of the carboxy-terminal cross-linking region of type I collagen by proteolytic activity from cultured porcine gingival explants. *Coll. Relat. Res.* **3**: 295–304.

761. **Scott, P. G., H. A. Goldberg, and C. M. Dodd.** 1983. A neutral proteinase from human gingival fibroblasts active against the C-terminal cross-linking region of type I collagen. *Biochem. Biophys. Res. Commun.* **114**: 1064–1070.

762. **Sedlacek, R., S. Mauch, B. Kolb, C. Schatzlein, H. Eibel, H. H. Peter, J. Schmitt, and U. Krawinkel.** 1998. Matrix metalloproteinase MMP-19 (RASI 1) is expressed on the surface of activated peripheral blood mononuclear cells and is detected as an autoantigen in rheumatoid arthritis. *Immunobiology* **198**: 408–423.

763. **Sellers, A., J. J. Reynolds, and M. C. Meikle.** 1978. Neutral metallo-proteinases of rabbit bone. Separation in latent forms of distinct enzymes that when activated degrade collagen, gelatin and proteoglycans. *Biochem. J.* **171**: 493–496.

764. **Sellers, A. and J. F. Woessner, Jr.** 1980. The extraction of a neutral metalloproteinase from the involuting rat uterus, and its action on cartilage proteoglycan. *Biochem. J.* **189**: 521–531.

765. **Seltzer, J. L., M. L. Eschbach, and A. Z. Eisen.** 1985. Purification of gelatin-specific neutral protease from human skin by conventional and high-performance liquid chromatography. *J. Chromatogr.* **326**: 147–155.

766. **Shapiro, S. D., G. L. Griffin, D. J. Gilbert, N. A. Jenkins, N. G. Copeland, H. G. Welgus, R. M. Senior, and T. J. Ley.** 1992. Molecular cloning, chromosomal localization, and bacterial expression of a murine macrophage metalloelastase. *J. Biol. Chem.* **267**: 4664–4671.

767. **Sharpe-Timms, K. L., L. L. Penney, R. L. Zimmer, J. A. Wright, Y. Zhang, and K. Surewicz.** 1995. Partial purification and amino acid sequence analysis of endometriosis protein-II (ENDO-II) reveals homology with tissue inhibitor of metalloproteinases-1 (TIMP-1). *J. Clin. Endocrinol. Metab.* **80**: 3784–3787.

768. **Shimizu, M., M. J. Glimcher, D. Travis, and P. Goldhaber.** 1969. Mouse bone collagenase: isolation, partial purification, and mechanism of action. *Proc. Soc. Exp. Biol. Med.* **130**: 1175–1180.

769. **Shlopov, B. V., W. R. Lie, C. L. Mainardi, A. A. Cole, S. Chubinskaya, and K. A. Hasty.** 1997. Osteoarthritic lesions: involvement of three different collagenases. *Arthritis Rheum.* **40**: 2065–2074.

770. **Shofuda, K., H. Yasumitsu, A. Nishihashi, K. Miki, and K. Miyazaki.** 1997. Expression of three membrane-type matrix metalloproteinases (MT-MMPs) in rat vascular smooth muscle cells and characterization of MT3-MMPs with and without transmembrane domain. *J. Biol. Chem.* **272**: 9749–9754.

771. **Sopata, I. and A. M. Dancewicz.** 1974. Presence of a gelatin-specific proteinase and its latent form in human leucocytes. *Biochim. Biophys. Acta* **370**: 510–523.

772. **Sopata, I. and J. Wize.** 1979. A latent gelatin specific proteinase of human leucocytes and its activation. *Biochim. Biophys. Acta* **571**: 305–312.

773. **Staskus, P. W., F. R. Masiarz, L. J. Pallanck, and S. P. Hawkes.** 1991. The 21-kDa protein is a transformation-sensitive metalloproteinase inhibitor of chicken fibroblasts. *J. Biol. Chem.* **266**: 449–454.

774. **Stetler-Stevenson, W. G., P. D. Brown, M. Onisto, A. T. Levy, and L. A. Liotta.** 1990. Tissue inhibitor of metalloproteinases-2 (TIMP-2) mRNA expression in tumor cell lines and human tumor tissues. *J. Biol. Chem.* **265**: 13933–13938.

775. **Stetler-Stevenson, W. G., H. C. Krutzsch, and L. A. Liotta.** 1989. Tissue inhibitor of metalloproteinase (TIMP-2). A new member of the metalloproteinase inhibitor family. *J. Biol. Chem.* **264**: 17374–17378.

776. **Stolow, M. A., D. D. Bauzon, J. Li, T. Sedgwick, V. C. Liang, Q. A. Sang, and Y. B. Shi.** 1996. Identification and characterization of a novel collagenase in *Xenopus laevis*: possible roles during frog development. *Mol. Biol. Cell.* **7**: 1471–1483.

777. **Stricklin, G. P. and H. G. Welgus.** 1986. Physiological relevance of erythroid-potentiating activity of TIMP. *Nature* **321**: 628–628.

778. **Strongin, A. Y., I. Collier, G. Bannikov, B. L. Marmer, G. A. Grant, and G. I. Goldberg.** 1995. Mechanism of cell surface activation of 72-kDa type IV collagenase. Isolation of the activated form of the membrane metalloprotease. *J. Biol. Chem.* **270**: 5331–5338.

779. **Takino, T., H. Sato, A. Shinagawa, and M. Seiki.** 1995. Identification of the second membrane-type matrix metalloproteinase (MT-MMP-2) gene from a human placenta cDNA library. MT-MMPs form a unique membrane-type subclass in the MMP family. *J. Biol. Chem.* **270**: 23013–23020.

780. **Takino, T., H. Sato, E. Yamamoto, and M. Seiki.** 1995. Cloning of a human gene potentially encoding a novel matrix metalloproteinase having a C-terminal transmembrane domain. *Gene* **155**: 293–298.

781. **Tanabe, T., M. Fukae, T. Uchida, and M. Shimizu.** 1992. The localization and characterization of proteinases for the initial cleavage of porcine amelogenin. *Calcif. Tissue Int.* **51**: 213–217.

782. **Treadwell, B. V., J. Neidel, M. Pavia, C. A. Towle, M. E. Trice, and H. J. Mankin.** 1986. Purification and characterization of collagenase activator protein synthesized by articular cartilage. *Arch. Biochem. Biophys.* **251**: 715–723.

783. **Ueno, H., H. Nakamura, M. Inoue, K. Imai, M. Noguchi, H. Sato, M. Seiki, and Y. Okada.** 1997. Expression and tissue localization of membrane-types 1, 2, and 3 matrix metalloproteinases in human invasive breast carcinomas. *Cancer Res.* **57**: 2055–2060.

784. **Vaes, G., Y. Eeckhout, G. Lenaers-Claeys, C. François-Gillet, and J. E. Druetz.** 1978. The simultaneous release by bone explants in culture and the parallel activation of procollagenase and of a latent

neutral proteinase that degrades cartilage proteoglycans and denatured collagen. *Biochem. J.* **172**: 261–274.

785. **van Tassell, R. L., D. M. Lyerly, and T. D. Wilkins.** 1992. Purification and characterization of an enterotoxin from *Bacteroides fragilis. Infect. Immun.* **60**: 1343–1350.

786. **Vartio, T.** 1985. The 95000-Mr gelatin-binding protein in human serum and plasma. *Biochem. J.* **228**: 605–608.

787. **Vartio, T. and M. Baumann.** 1989. Human gelatinase/type IV procollagenase is a regular plasma component. *FEBS Lett.* **255**: 285–289.

788. **Vartio, T., K. Hedman, S. E. Jansson, and T. Hovi.** 1985. The Mr 95,000 gelatin-binding protein in differentiated human macrophages and granulocytes. *Blood* **65**: 1175–1180.

789. **Vartio, T. and A. Vaheri.** 1981. A gelatin-binding 70,000-dalton glycoprotein synthesized distinctly from fibronectin by normal and malignant adherent cells. *J. Biol. Chem.* **256**: 13085–13090.

790. **Vater, C. A., C. L. Mainardi, and E. D. Harris, Jr.** 1978. Activation in vitro of rheumatoid synovial collagenase from cell cultures. *J. Clin. Invest.* **62**: 987–992.

791. **Vater, C. A., C. L. Mainardi, and E. D. Harris, Jr.** 1979. Inhibitor of human collagenase from cultures of human tendon. *J. Biol. Chem.* **254**: 3045–3053.

792. **Wada, K., H. Sato, H. Kinoh, M. Kajita, H. Yamamoto, and M. Seiki.** 1998. Cloning of three *Caenorhabditis elegans* genes potentially encoding novel matrix metalloproteinases. *Gene* **211**: 57–62.

793. **Weber, B. H. F., G. Vogt, R. C. Pruett, H. Stöhr, and U. Felbor.** 1994. Mutations in the tissue inhibitor of metalloproteinases-3 (TIMP3) in patients with Sorsby's fundus dystrophy. *Nature Genet.* **8**: 352–356.

794. **Weeks, J. G., J. Halme, and J. F. Woessner, Jr.** 1976. Extraction of collagenase from the involuting rat uterus. *Biochim. Biophys. Acta* **445**: 205–214.

795. **Welgus, H. G., G. P. Stricklin, A. Z. Eisen, E. A. Bauer, R. V. Cooney, and J. J. Jeffrey.** 1979. A specific inhibitor of vertebrate collagenase produced by human skin fibroblasts. *J. Biol. Chem.* **254**: 1938–1943.

796. **Werb, Z. and M. C. Burleigh.** 1974. A specific collagenase from rabbit fibroblasts in monolayer culture. *Biochem. J.* **137**: 373–385.

797. **Werb, Z., J. T. Dingle, J. J. Reynolds, and A. J. Barrett.** 1978. Proteoglycan-degrading enzymes of rabbit fibroblasts and granulocytes. *Biochem. J.* **173**: 949–958.

798. **Werb, Z. and S. Gordon.** 1975. Elastase secretion by stimulated macrophages. Characterization and regulation. *J. Exp. Med.* **142**: 361–377.

799. **Werb, Z. and J. J. Reynolds.** 1974. Stimulation by endocytosis of the secretion of collagenase and neutral proteinase from rabbit synovial fibroblasts. *J. Exp. Med.* **140**: 1482–1497.

800. **Wick, M., C. Bürger, S. Brüsselbach, F. C. Lucibello, and R. Müller.** 1994. A novel member of

human tissue inhibitor of metalloproteinases (TIMP) gene family is regulated during G1 progression, mitogenic stimulation, differentiation, and senescence. *J. Biol. Chem.* **269**: 18953–18960.

801. **Wilhelm, S. M., I. E. Collier, B. L. Marmer, A. Z. Eisen, G. A. Grant, and G. I. Goldberg.** 1989. SV40-transformed human lung fibroblasts secrete a 92-kDa type IV collagenase which is identical to that secreted by normal human macrophages. *J. Biol. Chem.* **264**: 17213–17221.

802. **Wilhelm, S. M., Z. H. Shao, T. J. Housley, P. K. Seperack, A. P. Baumann, Z. Gunja-Smith, and J. F. Woessner, Jr.** 1993. Matrix metalloproteinase-3 (stromelysin-1). Identification as the cartilage acid metalloprotease and effect of pH on catalytic properties and calcium affinity. *J. Biol. Chem.* **268**: 21906–21913.

803. **Will, H. and B. Hinzmann.** 1995. cDNA sequence and mRNA tissue distribution of a novel human matrix metalloproteinase with a potential transmembrane segment. *Eur. J. Biochem.* **231**: 602–608.

804. **Woessner, J. F., Jr. and M. G. Selzer.** 1984. Two latent metalloproteases of human articular cartilage that digest proteoglycan. *J. Biol. Chem.* **259**: 3633–3638.

805. **Woessner, J. F., Jr. and C. J. Taplin.** 1988. Purification and properties of a small latent matrix metalloproteinase of the rat uterus. *J. Biol. Chem.* **263**: 16918–16925.

806. **Woolley, D. E., D. R. Roberts, and J. M. Evanson.** 1975. Inhibition of human collagenase activity by a small molecular weight serum protein. *Biochem. Biophys. Res. Commun.* **66**: 747–754.

807. **Woolley, D. E., D. R. Roberts, and J. M. Evanson.** 1976. Small molecular weight β1 serum protein which specifically inhibits human collagenases. *Nature* **261**: 325–327.

808. **Wu, I. M. and M. A. Moses.** 1998. Molecular cloning and expression analysis of the cDNA encoding rat tissue inhibitor of metalloproteinase-4. *Matrix Biol.* **16**: 339–342.

809. **Wu, J. J., M. W. Lark, L. E. Chun, and D. R. Eyre.** 1991. Sites of stromelysin cleavage in collagen types II, IX, X, and XI of cartilage. *J. Biol. Chem.* **266**: 5625–5628.

810. **Yang, M. Z. and M. Kurkinen.** 1998. Cloning and characterization of a novel matrix metalloproteinase (MMP), CMMP, from chicken embryo fibroblasts—CMMP, *Xenopus* XMMP, and human MMP-19 have a conserved unique cysteine in the catalytic domain. *J. Biol. Chem.* **273**: 17893–17900.

811. **Yang, M. Z., M. T. Murray, and M. Kurkinen.** 1997. A novel matrix metalloproteinase gene (XMMP) encoding vitronectin-like motifs is transiently expressed in *Xenopus laevis* early embryo development. *J. Biol. Chem.* **272**: 13527–13533.

See also review articles: 10, 16, 28, 34, 36, 44, 49, 51, 55, 59, 60, 63, 73, 74, 80, 81, 84, 86, 89, 90, 91, 95, 100, 114, 115, 117, 118, 129, 130, 135, 140, 141, 142, 145, 147, 150.

Inhibitors of MMPs

812. **Abe, S. and Y. Nagai.** 1972. Interaction between tadpole collagenase and human α_2-macroglobulin. *Biochim. Biophys. Acta* **278**: 125–132.

813. **Abe, S. and Y. Nagai.** 1973. Evidence for the presence of a complex of collagenase with α_2-macroglobulin in human rheumatoid synovial fluid: a possible regulatory mechanism of collagenase activity in vivo. *J. Biochem. (Tokyo)* **73**: 897–900.

814. **Abe, S., M. Shinmei, and Y. Nagai.** 1973. Synovial collagenase and joint diseases: the significancy of latent collagenase with special reference to rheumatoid arthritis. *J. Biochem. (Tokyo)* **73**: 1007–1011.

815. **Ackerman, N. R., S. Jubb, B. Trimble, S. Marlowe, L. Miram, and P. Maloney.** 1983. Divergent effects of suramin on in vitro and in vivo assays of cartilage degradation. *J. Pharmacol. Exp. Ther.* **225**: 243–250.

816. **Albin, R. J., R. M. Senior, H. G. Welgus, N. L. Connolly, and E. J. Campbell.** 1987. Human alveolar macrophages secrete an inhibitor of metalloproteinase elastase. *Am. Rev. Respir. Dis.* **135**: 1281–1285.

817. **Altman, R. D., D. D. Dean, O. E. Muniz, and D. S. Howell.** 1987. Therapeutic treatment of canine osteoarthritis with glycosaminoglycan polysulfuric acid ester. *Arthritis Rheum.* **32**: 1300–1307.

818. **Andersen, G. R., T. J. Koch, K. Dolmer, L. Sottrup-Jensen, and J. Nyborg.** 1995. Low resolution X-ray structure of human methylamine-treated α_2-macroglobulin. *J. Biol. Chem.* **270**: 25133–25141.

819. **Arbelaez, L. F., U. Bergmann, A. Tuuttila, V. P. Shanbhag, and T. Stigbrand.** 1997. Interaction of matrix metalloproteinases-2 and -9 with pregnancy zone protein and α_2-macroglobulin. *Arch. Biochem. Biophys.* **347**: 62–68.

820. **Ata, N., T. Oku, M. Hattori, H. Fujii, M. Nakajima, and I. Saiki.** 1996. Inhibition by galloylglucose (GG6-10) of tumor invasion through extracellular matrix and gelatinase-mediated degradation of type IV collagens by metastatic tumor cells. *Oncol. Res.* **8**: 503–511.

821. **Azzo, W. and J. F. Woessner, Jr.** 1986. Purification and characterization of an acid metalloproteinase from human articular cartilage. *J. Biol. Chem.* **261**: 5434–5441.

822. **Bailey, S., B. Bolognese, D. R. Buckle, A. Faller, S. Jackson, P. Louis-Flamberg, M. McCord, R. J. Mayer, L. A. Marshall, and D. G. Smith.** 1998. Hydroxamate-based inhibitors of low affinity IgE receptor (CD23) processing. *Bioorg. Med. Chem. Lett.* **8**: 23–28.

823. **Bailey, S., B. Bolognese, D. R. Buckle, A. Faller, S. Jackson, P. Louis-Flamberg, M. McCord, R. J. Mayer, L. A. Marshall, and D. G. Smith.** 1998. Selective inhibition of low affinity IgE receptor (CD23) processing. *Bioorg. Med. Chem. Lett.* **8**: 29–34.

824. **Banda, M. J., E. J. Clark, and Z. Werb.** 1980. Limited proteolysis by macrophage elastase inactivates human α_1-proteinase inhibitor. *J. Exp. Med.* **152**: 1563–1570.

825. **Banda, M. J. and Z. Werb.** 1981. Mouse macrophage elastase. Purification and characterization as a metalloproteinase. *Biochem. J.* **193**: 589–605.

826. **Barrett, A. J. and P. M. Starkey.** 1973. The interaction of α_2-macroglobulin with proteinases. Characteristics and specificity of the reaction, and a hypothesis concerning its molecular mechanism. *Biochem. J.* **133**: 709–724.

827. **Bauer, E. A., G. P. Stricklin, J. J. Jeffrey, and A. Z. Eisen.** 1975. Collagenase production by human skin fibroblasts. *Biochem. Biophys. Res. Commun.* **64**: 232–240.

828. **Baxter, A. D., R. Bhogal, J. B. Bird, G. M. Buckley, D. S. Gregory, P. C. Hedger, D. T. Manallack, T. Massil, K. J. Minton, J. G. Montana,** *et al.* 1997. Mercaptoacyl matrix metalloproteinase inhibitors—the effect of substitution at the mercaptoacyl moiety. *Bioorg. Med. Chem. Lett.* **7**: 2765–2770.

829. **Berman, M., J. Gordon, L. A. Garcia, and L. Gage.** 1975. Corneal ulceration and the serum antiproteases. II. Complexes of corneal collagenases and alpha-macroglobulins. *Exp. Eye Res.* **20**: 231–244.

830. **Berman, M. B., J. C. Barber, R. C. Talamo, and C. E. Langley.** 1973. Corneal ulceration and the serum antiproteases. I. α_1-Antitrypsin. *Invest. Ophthalmol.* **12**: 759–770.

831. **Berman, M. B., A. P. Kerza-Kwiatecki, and P. F. Davison.** 1973. Characterization of human corneal collagenase. *Exp. Eye Res.* **15**: 367–373.

832. **Berman, M. B. and R. Manabe.** 1973. Corneal collagenases: evidence for zinc metalloenzymes. *Ann. Ophthalmol.* **5**: 1193–1209.

833. **Beszant, B., J. Bird, L. M. Gaster, G. P. Harper, I. Hughes, E. H. Karran, R. E. Markwell, A. J. Miles-Williams, and S. A. Smith.** 1993. Synthesis of novel modified dipeptide inhibitors of human collagenase: β-mercapto carboxylic acid derivatives. *J. Med. Chem.* **36**: 4030–4039.

834. **Betz, M., P. Huxley, S. J. Davies, Y. Mushtaq, M. Pieper, H. Tschesche, W. Bode, and F. X. Gomis-Rüth.** 1997. 1.8-Å crystal structure of the catalytic domain of human neutrophil collagenase (matrix metalloproteinase-8) complexed with a peptidomimetic hydroxamate prime-side inhibitor with a distinct selectivity profile. *Eur. J. Biochem.* **247**: 356–363.

835. **Bird, J., R. C. De Mello, G. P. Harper, D. J. Hunter, E. H. Karran, R. E. Markwell, A. J. Miles-Williams, S. S. Rahman, and R. W. Ward.** 1994. Synthesis of novel N-phosphonoalkyl dipeptide inhibitors of human collagenase. *J. Med. Chem.* **37**: 158–169.

836. **Bird, J., G. P. Harper, I. Hughes, D. J. Hunter, E. H. Karran, R. E. Markwell, A. J. Miles-Williams, S. S. Rahman, and R. W. Ward.** 1995. Inhibitors of human

collagenase: dipeptide mimetics with lactam and azalactam moieties at the P-2′ P-3′ position. *Bioorg. Med. Chem. Lett.* **5**: 2593–2598.

837. **Birkedal-Hansen, H., C. M. Cobb, R. E. Taylor, and H. M. Fullmer.** 1974. Serum inhibition of gingival collagenase. *J. Oral Pathol.* **3**: 284–290.

838. **Bode, W., P. Reinemer, R. Huber, T. Kleine, S. Schnierer, and H. Tschesche.** 1994. The X-ray crystal structure of the catalytic domain of human neutrophil collagenase inhibited by a substrate analogue reveals the essentials for catalysis and specificity. *EMBO J.* **13**: 1263–1269.

839. **Boisset, N., J.-C. Taveau, F. Pochon, A. Tardieu, M. Barray, J. N. Lamy, and E. Delain.** 1989. Image processing of proteinase- and methylamine-transformed human α2-macroglobulin. Localization of the proteinases. *J. Biol. Chem.* **264**: 12046–12052.

840. **Bramhall, S. R.** 1997. The matrix metalloproteinases and their inhibitors in pancreatic cancer. From molecular science to a clinical application. *Int. J. Pancreatol.* **21**: 1–12

841. **Brandstetter, H., R. A. Engh, E. Graf von Roedern, L. Moroder, R. Huber, W. Bode, and F. Grams.** 1998. Structure of malonic acid-based inhibitors bound to human neutrophil collagenase—a new binding mode explains apparently anomalous data. *Prot. Sci.* **7**: 1303–1309.

842. **Brännström, M., J. F. Woessner, Jr., R. D. Koos, C. H. Sear, and W. J. LeMaire.** 1988. Inhibitors of mammalian tissue collagenase and metalloproteinases suppress ovulation in the perfused rat ovary. *Endocrinology* 122: 1715–1721.

843. **Brown, F. K., P. J. Brown, D. M. Bickett, C. L. Chambers, H. G. Davies, D. N. Deaton, D. Drewry, M. Foley, A. B. McElroy, M. Gregson,** *et al.* 1994. Matrix metalloproteinase inhibitors containing a (carboxyalkyl)amino zinc ligand: modification of the P1 and P2′ residues. *J. Med. Chem.* **37**: 674–688.

844. **Brown, S. I., S. Akiya, and C. A. Weller.** 1969. Prevention of the ulcers of the alkali-burned cornea. Preliminary studies with collagenase inhibitors. *Arch. Ophthalmol.* **82**: 95–97.

845. **Browner, M. F., W. W. Smith, and A. L. Castelhano.** 1995. Matrilysin-inhibitor complexes: common themes among metalloproteases. *Biochemistry* **34**: 6602–6610.

846. **Burns, F. R., M. S. Stack, R. D. Gray, and C. A. Paterson.** 1989. Inhibition of purified collagenase from alkali-burned rabbit corneas. *Invest. Ophthalmol. Vis. Sci.* **30**: 1569–1575.

847. **Butler, T. A., C. Zhu, R. A. Mueller, G. C. Fuller, W. J. LeMaire, and J. F. Woessner, Jr.** 1991. Inhibition of ovulation in the perfused rat ovary by the synthetic collagenase inhibitor SC 44463. *Biol. Reprod.* **44**: 1183–1188.

848. **Caccese, R. G., J. F. DiJoseph, J. S. Skotnicki, L. E. Borella, and L. M. Adams.** 1991. Inhibition of interleukin-1 (IL-1) induced neutral proteases from rabbit articular chondrocytes by WY-46,135 and WY-48,989. *Agents Actions* **34**: 223–225.

849. **Caldwell, C. G., S. P. Sahoo, S. A. Polo, R. R. Eversole, T. J. Lanza, S. G. Mills, L. M. Niedzwiecki, M. Izquierdo-Martin, B. C. Chang, R. K. Harrison,** *et al.* 1996. Phosphinic acid inhibitors of matrix metalloproteinases. *Bioorg. Med. Chem. Lett.* **6**: 323–328.

850. **Caputo, C. B., L. A. Sygowski, S. P. Patton, R. F. Piehl, R. G. Caccese, and G. DiPasquale.** 1988. Degradation of rat chondrosarcoma proteoglycans by a neutral metalloprotease from rabbit chondrocytes. *Connect. Tissue Res.* **18**: 191–203.

851. **Caputo, C. B., L. A. Sygowski, D. J. Wolanin, S. P. Patton, R. G. Caccese, A. Shaw, R. A. Roberts, and G. DiPasquale.** 1987. Effect of synthetic metalloprotease inhibitors on cartilage autolysis in vitro. *J. Pharmacol. Exp. Ther.* **240**: 460–465.

852. **Caputo, C. B., D. J. Wolanin, R. A. Roberts, L. A. Sygowski, S. P. Patton, R. G. Caccese, A. Shaw, and G. DiPasquale.** 1987. Proteoglycan degradation by a chondrocyte metalloprotease. Effects of synthetic protease inhibitors. *Biochem. Pharmacol.* **36**: 995–1002.

853. **Castelhano, A. L., R. Billedeau, N. Dewdney, S. Donnelly, S. Horne, L. J. Kurz, T. J. Liak, R. Martin, R. Uppington, Z. Y. Yuan, and A. Krantz.** 1995. Novel indolactam-based inhibitors of matrix metalloproteinases. *Bioorg. Med. Chem. Lett.* **5**: 1415–1420.

854. **Cawston, T. E. and E. Mercer.** 1986. Preferential binding of collagenase to α₂-macroglobulin in the presence of the tissue inhibitor of metalloproteinases. *FEBS Lett.* **209**: 9–12.

855. **Chapman, K. T., P. L. Durette, C. G. Caldwell, K. M. Sperow, L. M. Niedzwiecki, R. K. Harrison, C. Saphos, A. J. Christen, J. M. Olszewski, V. L. Moore,** *et al.* 1996. Orally active inhibitors of stromelysin-1 (MMP-3). *Bioorg. Med. Chem. Lett.* **6**: 803–806.

856. **Chapman, K. T., I. E. Kopka, P. L. Durette, C. K. Esser, T. J. Lanza, M. Izquierdo-Martin, L. Niedzwiecki, B. Chang, R. K. Harrison, D. W. Kuo,** *et al.* 1993. Inhibition of matrix metalloproteinases by N-carboxyalkyl peptides. *J. Med. Chem.* **36**: 4293–4301.

857. **Chapman, K. T., J. Wales, S. P. Sahoo, L. M. Niedzwiecki, M. Izquierdo-Martin, B. C. Chang, R. K. Harrison, R. L. Stein, and W. K. Hagmann.** 1996. Inhibition of matrix metalloproteinases by P-1 substituted N-carboxyalkyl dipeptides. *Bioorg. Med. Chem. Lett.* **6**: 329–332.

858. **Chen, J. J., Y. P. Zhang, S. Hammond, N. Dewdney, T. Ho, X. H. Lin, M. F. Browner, and A. L. Castelhano.** 1996. Design, synthesis, activity, and structure of a novel class of matrix metalloproteinase

inhibitors containing a heterocyclic P_2'-P_3' amide bond isostere. *Bioorg. Med. Chem. Lett.* **6**: 1601–1606.

859. **Cherney, R. J., C. P. Decicco, D. J. Nelson, L. Wang, D. T. Meyer, K. D. Hardman, R. A. Copeland, and E. C. Arner.** 1997. Potent carboxylate inhibitors of stromelysin containing P2' piperazic acids and P1' biaryl moeities. *Bioorg. Med. Chem. Lett.* **7**: 1757–1762.

860. **Cherney, R. J., L. Wang, D. T. Meyer, C. B. Xue, Z. R. Wasserman, K. D. Hardman, P. K. Welch, M. B. Covington, R. A. Copeland, E. C. Arner, *et al.* 1998. Macrocyclic amino carboxylates as selective MMP-8 inhibitors. *J. Med. Chem.* **41**: 1749–1751.

861. **Clark, D. E., P. Wei, and N. H. Grant.** 1985. A novel inhibitor of mammalian collagenase. *Life Sci.* **37**: 575–578.

862. **Conway, J. G., S. J. Trexler, J. A. Wakefield, B. E. Marron, D. L. Emerson, D. M. Bickett, D. N. Deaton, D. Garrison, M. Elder, A. McElroy, *et al.* 1996. Effect of matrix metalloproteinase inhibitors on tumor growth and spontaneous metastasis. *Clin. Exp. Metastasis* **14**: 115–124.

863. **Crossley, M. J. and I. M. Hunneyball.** 1982. Biochemical and pharmacological studies on synovium-cartilage interactions in organ culture. *Eur. J. Rheumatol. Inflamm.* **5**: 15–29.

864. **Darlak, K., R. B. Miller, M. S. Stack, A. F. Spatola, and R. D. Gray.** 1990. Thiol-based inhibitors of mammalian collagenase. Substituted amide and peptide derivatives of the leucine analogue, 2-[(R,S)-mercaptomethyl]-4-methylpentanoic acid. *J. Biol. Chem.* **265**: 5199–5205.

864a. **Davies, B., P. D. Brown, N. East, M. J. Crimmin, and F. R. Balkwill.** 1993. A synthetic matrix metalloproteinase inhibitor decreases tumor burden and prolongs survival of mice bearing human ovarian carcinoma xenografts. *Cancer Res.* **53**: 2087–2091.

865. **Decicco, C. P., J. L. Seng, K. E. Kennedy, M. B. Covington, P. K. Welch, E. C. Arner, R. L. Magolda, and D. J. Nelson.** 1997. Amide surrogates of matrix metalloproteinase inhibitors – urea and sulfonamide mimics. *Bioorg. Med. Chem. Lett.* **7**: 2331–2336.

866. **Delaissé, J. M., Y. Eeckhout, C. Sear, A. Galloway, K. McCullagh, and G. Vaes.** 1985. A new synthetic inhibitor of mammalian tissue collagenase inhibits bone resorption in culture. *Biochem. Biophys. Res. Commun.* **133**: 483–490.

867. **Dhanaraj, V., Q. Z. Ye, L. L. Johnson, D. J. Hupe, D. F. Ortwine, J. B. Dunbar, Jr., J. R. Rubin, A. Pavlovsky, C. Humblet, and T. L. Blundell.** 1996. Designing inhibitors of the metalloproteinase superfamily: comparative analysis of representative structures. *Drug Des. Discov.* **13**: 3–14.

868. **Di Giulio, A., A. Barracchini, L. Cellai, C. Martuccio, G. Amicosante, A. Oratore, and G. Pantaleoni.** 1996. Rifamycins as inhibitors of collagenase activity: their possible pharmacological role in collagen degradative diseases. *Int. J. Pharm.* **144**: 27–35.

869. **DiPasquale, G., R. Caccese, R. Pasternak, J. Conaty, S. Hubbs, and K. Perry.** 1986. Proteoglycan-and collagen-degrading enzymes from human interleukin 1-stimulated chondrocytes from several species: proteoglycanase and collagenase inhibitors as potentially new disease-modifying antiarthritic agents. *Proc. Soc. Exp. Biol. Med.* **183**: 262–267.

870. **Eisen, A. Z., E. A. Bauer, and J. J. Jeffrey.** 1971. Human skin collagenase. The role of serum alpha-globulins in the control of activity in vivo and in vitro. *Proc. Natl. Acad. Sci. USA* **68**: 248–251.

871. **Eisen, A. Z., K. J. Bloch, and T. Sakai.** 1970. Inhibition of human skin collagenase by human serum. *J. Lab. Clin. Med.* **75**: 258–263.

872. **Eisen, A. Z., J. J. Jeffrey, and J. Gross.** 1968. Human skin collagenase. Isolation and mechanism of attack on the collagen molecule. *Biochim. Biophys. Acta* **151**: 637–645.

873. **Enghild, J. J., G. Salvesen, K. Brew, and H. Nagase.** 1989. Interaction of human rheumatoid synovial collagenase (matrix metalloproteinase 1) and stromelysin (matrix metalloproteinase 3) with human α_2-macroglobulin and chicken ovostatin. Binding kinetics and identification of matrix metalloproteinase cleavage sites. *J. Biol. Chem.* **264**: 8779–8785.

874. **Esnard, F., N. Gutman, A. el Moujahed, and F. Gauthier.** 1985. Rat plasma α_1-inhibitor$_3$: a member of the α-macroglobulin family. *FEBS Lett.* **182**: 125–129.

875. **Esser, C. K., R. L. Bugianesi, C. G. Caldwell, K. T. Chapman, P. L. Durette, N. N. Girotra, I. E. Kopka, T. J. Lanza, D. A. Levorse, M. MacCoss, *et al.* 1997. Inhibition of stromelysin-1 (MMP-3) by P1'-biphenylylethyl carboxyalkyl dipeptides. *J. Med. Chem.* **40**: 1026–1040.

876. **Esser, C. K., N. J. Kevin, N. A. Yates, and K. T. Chapman.** 1997. Solid-phase synthesis of a N-carboxyalkyl tripeptide combinatorial library. *Bioorg. Med. Chem. Lett.* **7**: 2639–2644.

877. **Esser, C. K., I. E. Kopka, P. L. Durette, R. K. Harrison, L. M. Niedzwiecki, M. Izquierdo-Martin, R. L. Stein, and W. K. Hagmann.** 1995. Inhibition of matrix metalloproteinases by N-carboxyalkyl peptides containing extended alkyl residues at P1'. *Bioorg. Med. Chem. Lett.* **5**: 539–542.

878. **Evans, C. H. and J. D. Ridella.** 1985. Inhibition, by lanthanides, of neutral proteinases secreted by human, rheumatoid synovium. *Eur. J. Biochem.* **151**: 29–32.

879. **Evanson, J. M., J. J. Jeffrey, and S. M. Krane.** 1967. Human collagenase: identification and characterization of an enzyme from rheumatoid synovium in culture. *Science* **158**: 499–502.

880. **Faucher, D. C., Y. Lelièvre, and T. Cartwright.** 1987. An inhibitor of mammalian collagenase active at

micromolar concentrations from an actinomycete culture broth. *J. Antibiot. (Tokyo)* **40**: 1757–1761.

881. Ferry, G., J. A. Boutin, G. Atassi, J. L. Fauchère, and G. C. Tucker. 1996. Selection of a histidine-containing inhibitor of gelatinases through deconvolution of combinatorial tetrapeptide libraries. *Mol. Diversity* **2**: 135–146.

882. Foley, M. A., A. S. Hassman, D. H. Drewry, D. G. Greer, C. D. Wagner, P. L. Feldman, J. Berman, D. M. Bickett, G. M. McGeehan, M. H. Lambert, and M. Green. 1996. Rapid synthesis of novel dipeptide inhibitors of human collagenase and gelatinase using solid phase chemistry. *Bioorg. Med. Chem. Lett.* **6**: 1905–1910.

883. Folkersen, J., B. Teisner, N. Grunnet, J. G. Grudzinskas, J. G. Westergaard, and P. Hindersson. 1981. Circulating levels of pregnancy zone protein: normal range and the influence of age and gender. *Clin. Chim. Acta* **110**: 139–145.

884. Fotouhi, N., A. Lugo, M. Visnick, L. Lusch, R. Walsky, J. W. Coffey, and A. C. Hanglow. 1994. Potent peptide inhibitors of stromelysin based on the prodomain region of matrix metalloproteinases. *J. Biol. Chem.* **269**: 30227–30231.

885. Fullmer, H. M. and G. Lazarus. 1967. Collagenase in human goat and rat bone. *Israel J. Med. Sci.* **3**: 758–761.

886. Gearing, A. J., P. Beckett, M. Christodoulou, M. Churchill, J. Clements, A. H. Davidson, A. H. Drummond, W. A. Galloway, R. Gilbert, J. L. Gordon, *et al.* 1994. Processing of tumour necrosis factor-α precursor by metalloproteinases. *Nature* **370**: 555–557.

887. Geiger, T., Y. Lamri, T. A. Tran-Thi, F. Gauthier, G. Feldmann, K. Decker, and P. C. Heinrich. 1987. Biosynthesis and regulation of rat α_1-inhibitor3, a negative acute-phase reactant of the macroglobulin family. *Biochem. J.* **245**: 493–500.

888. Georgopapadakou, N. H. 1996. Does cephalothin inhibit matrix metalloproteinases? *Antimicrob. Agents Chemother.* **40**: 1969–1970.

889. Ghose, A. K., M. E. Logan, A. M. Treasurywala, H. Wang, R. C. Wahl, B. E. Tomczuk, M. R. Gowravaram, E. P. Jaeger, and J. J. Wendoloski. 1995. Determination of pharmacophoric geometry for collagenase inhibitors using a novel computational method and its verification using molecular dynamics, NMR, and x-ray crystallography. *J. Am. Chem. Soc.* **117**: 4671–4682.

890. Golub, L. M., H. M. Lee, G. Lehrer, A. Nemiroff, T. F. McNamara, R. Kaplan, and N. S. Ramamurthy. 1983. Minocycline reduces gingival collagenolytic activity during diabetes. Preliminary observations and a proposed new mechanism of action. *J. Periodontal Res.* **18**: 516–526.

891. Golub, L. M., T. F. McNamara, G. D'Angelo, R. A. Greenwald, and N. S. Ramamurthy. 1987. A non-antibacterial chemically-modified tetracycline inhibits mammalian collagenase activity. *J. Dent. Res.* **66**: 1310–1314.

892. Golub, L. M., M. Wolff, H. M. Lee, T. F. McNamara, N. S. Ramamurthy, J. Zambon, and S. Ciancio. 1985. Further evidence that tetracyclines inhibit collagenase activity in human crevicular fluid and from other mammalian sources. *J. Periodontal Res.* **20**: 12–23.

893. Gonnella, N. C., R. Bohacek, X. Zhang, I. Kolossvary, C. G. Paris, R. Melton, C. Winter, S. I. Hu, and V. Ganu. 1995. Bioactive conformation of stromelysin inhibitors determined by transferred nuclear Overhauser effects. *Proc. Natl. Acad. Sci. USA* **92**: 462–466.

894. Gonnella, N. C., Y. C. Li, X. L. Zhang, and C. G. Paris. 1997. Bioactive conformation of a potent stromelysin inhibitor determined by X-nucleus filtered and multidimensional NMR spectroscopy. *Bioorg. Med. Chem.* **5**: 2193–2201.

895. Gooley, P. R., B. A. Johnson, A. I. Marcy, G. C. Cuca, S. P. Salowe, W. K. Hagmann, C. K. Esser, and J. P. Springer. 1993. Secondary structure and zinc ligation of human recombinant short-form stromelysin by multidimensional heteronuclear NMR. *Biochemistry* **32**: 13098–13108.

896. Gooley, P. R., J. F. O'Connell, A. I. Marcy, G. C. Cuca, S. P. Salowe, B. L. Bush, J. D. Hermes, C. K. Esser, W. K. Hagmann, J. P. Springer, *et al.* 1994. The NMR structure of the inhibited catalytic domain of human stromelysin-1. *Nature Struct. Biol.* **1**: 111–118.

897. Gordon, A. H. 1976. The α macroglobulins of rat serum. *Biochem. J.* **159**: 643–650.

898. Goulet, J. L., J. F. Kinneary, P. L. Durette, R. L. Stein, R. K. Harrison, M. Izquierdo-Martin, D. W. Kuo, T. Y. Lin, and W. K. Hagmann. 1994. Inhibition of stromelysin-1 (MMP-3) by peptidyl phosphinic acids. *Bioorg. Med. Chem. Lett.* **4**: 1221–1224.

899. Gowravaram, M. R., B. E. Tomczuk, J. S. Johnson, D. Delecki, E. R. Cook, A. K. Ghose, A. M. Mathiowetz, J. C. Spurlino, B. Rubin, D. L. Smith, *et al.* 1995. Inhibition of matrix metalloproteinases by hydroxamates containing heteroatom-based modifications of the P1′ group. *J. Med. Chem.* **38**: 2570–2581.

900. Graf von Roedern, E., F. Grams, H. Brandstetter, and L. Moroder. 1998. Design and synthesis of malonic acid-based inhibitors of human neutrophil collagenase (MMP8). *J. Med. Chem.* **41**: 339–345.

901. Grams, F., M. Crimmin, L. Hinnes, P. Huxley, M. Pieper, H. Tschesche, and W. Bode. 1995. Structure determination and analysis of human neutrophil collagenase complexed with a hydroxamate inhibitor. *Biochemistry* **34**: 14012–14020.

902. **Grams, F., P. Reinemer, J. C. Powers, T. Kleine, M. Pieper, H. Tschesche, R. Huber, and W. Bode.** 1995. X-ray structures of human neutrophil collagenase complexed with peptide hydroxamate and peptide thiol inhibitors. Implications for substrate binding and rational drug design. *Eur. J. Biochem.* **228**: 830–841.

903. **Gray, R. D., R. B. Miller, and A. F. Spatola.** 1986. Inhibition of mammalian collagenases by thiol-containing peptides. *J. Cell. Biochem.* **32**: 71–77.

904. **Gray, R. D., H. H. Saneii, and A. F. Spatola.** 1981. Metal binding peptide inhibitors of vertebrate collagenase. *Biochem. Biophys. Res. Commun.* **101**: 1251–1258.

905. **Greenwald, R. A., L. M. Golub, B. Lavietes, N. S. Ramamurthy, B. Gruber, R. S. Laskin, and T. F. McNamara.** 1987. Tetracyclines inhibit human synovial collagenase *in vivo* and *in vitro*. *J. Rheumatol.* **14**: 28–32.

906. **Greenwald, R. A., L. M. Golub, N. S. Ramamurthy, M. Chowdhury, S. A. Moak, and T. Sorsa.** 1998. In vitro sensitivity of the three mammalian collagenases to tetracycline inhibition—relationship to bone and cartilage degradation. *Bone* **22**: 33–38.

907. **Grobelny, D., L. Poncz, and R. E. Galardy.** 1992. Inhibition of human skin fibroblast collagenase, thermolysin, and *Pseudomonas aeruginosa* elastase by peptide hydroxamic acids. *Biochemistry* **31**: 7152–7154.

908. **Hajduk, P. J., E. T. Olejniczak, and S. W. Fesik.** 1997. One-dimensional relaxation- and diffusion-edited NMR methods for screening compounds that bind to macromolecules. *J. Am. Chem. Soc.* **119**: 12257–12261.

909. **Hajduk, P. J., G. Sheppard, D. G. Nettesheim, E. T. Olejniczak, S. B. Shuker, R. P. Meadows, D. H. Steinman, G. M. Carrera, P. A. Marcotte, J. Severin,** *et al.* 1997. Discovery of potent nonpeptide inhibitors of stromelysin using SAR by NMR. *J. Am. Chem. Soc.* **119**: 5818–5827.

910. **Hanemaaijer, R., T. Sorsa, Y. T. Konttinen, Y. L. Ding, M. Sutinen, H. Visser, V. W. M. Van Hinsbergh, T. Helaakoski, T. Kainulainen, H. Ronka,** *et al.* 1997. Matrix metalloproteinase-8 is expressed in rheumatoid synovial fibroblasts and endothelial cells—regulation by tumor necrosis factor-α and doxycycline. *J. Biol. Chem.* **272**: 31504–31509.

911. **Hanglow, A. C., A. Lugo, R. Walsky, M. Finch-Arietta, L. Lusch, M. Visnick, and N. Fotouhi.** 1993. Peptides based on the conserved predomain sequence of matrix metalloproteinases inhibit human stromelysin and collagenase. *Agents Actions* **39**: C148–C150

912. **Harris, G. H., K. Hoogsteen, K. C. Silverman, S. L. Raghoobar, G. F. Bills, R. B. Lingham, J. L. Smith, H. W. Dougherty, C. Cascales, and F. Pelaez.** 1993. Isolation and structure determination of pycnidione, a novel bistropolone stromelysin inhibitor from a *Phoma* sp. *Tetrahedron* **49**: 2139–2144.

913. **Haruyama, H., Y. Ohkuma, H. Nagaki, T. Ogita, K. Tamaki, and T. Kinoshita.** 1994. Matlystatins, new inhibitors of type IV collagenases from *Actinomadura atramentaria*. III. Structure elucidation of matlystatins A to F. *J. Antibiot. (Tokyo)* **47**: 1473–1480.

914. **Hawley, P. R. and W. P. Faulk.** 1970. A circulatory collagenase inhibitor. *Br. J. Surg.* **57**: 900–904.

915. **Heard, S. B.** 1976. Collagenase studies in bones of guinea pigs. *J. Oral Pathol.* **5**: 17–32.

916. **Henderson, B., A. J. P. Docherty, and N. R. A. Beeley.** 1990. Design of inhibitors of articular cartilage destruction. *Drugs of Future* **15**: 495–508.

917. **Hill, P. A., A. J. P. Docherty, K. M. Bottomley, J. P. O'Connell, J. R. Morphy, J. J. Reynolds, and M. C. Meikle.** 1995. Inhibition of bone resorption in vitro by selective inhibitors of gelatinase and collagenase. *Biochem. J.* **308**: 167–175.

918. **Hirayama, R., M. Yamamoto, T. Tsukida, K. Matsuo, Y. Obata, F. Sakamoto, and S. Ikeda.** 1997. Synthesis and biological evaluation of orally active matrix metalloproteinase inhibitors. *Bioorg. Med. Chem.* **5**: 765–778.

919. **Hughes, I., G. P. Harper, E. H. Karran, R. E. Markwell, and A. J. Miles-Williams.** 1995. Synthesis of thiophenol derivatives as inhibitors of human collagenase. *Bioorg. Med. Chem. Lett.* **5**: 3039–3042.

920. **Hunter, D. J., J. Bird, F. Cassidy, R. C. De Mello, G. P. Harper, E. H. Karran, R. E. Markwell, A. J. Miles-Williams, and R. W. Ward.** 1994. Aminophosphonic acid containing inhibitors of human collagenase: modification of the P_1 residue. *Bioorg. Med. Chem. Lett.* **4**: 1833–1836.

921. **Ikai, A., T. Kitamoto, and M. Nishigai.** 1982. α_2-Macroglobulin-like protease inhibitor from the egg white of Cuban crocodile (*Crocodylus rhombifer*). *J. Biochem. (Tokyo)* **93**: 121–127.

922. **Izquierdo-Martin, M., K. T. Chapman, W. K. Hagmann, and R. L. Stein.** 1994. Studies on the kinetic and chemical mechanism of inhibition of stromelysin by an N-(carboxyalkyl)dipeptide. *Biochemistry* **33**: 1356–1365.

923. **Jacobson, I. C., P. G. Reddy, Z. R. Wasserman, K. D. Hardman, M. B. Covington, E. C. Arner, R. A. Copeland, C. P. Decicco, and R. L. Magolda.** 1998. Structure-based design and synthesis of a series of hydroxamic acids with a quaternary-hydroxy group in P1 as inhibitors of matrix metalloproteinases. *Bioorg. Med. Chem. Lett.* **8**: 837–842.

924. **Jeng, A. Y., M. Chou, and D. T. Parker.** 1998. Sulfonamide-based hydroxamic acids as potent inhibitors of mouse macrophage metalloelastase. *Bioorg. Med. Chem. Lett.* **8**: 897–902.

925. **Jonat, C., F. Z. Chung, and V. M. Baragi.** 1996. Transcriptional downregulation of stromelysin by tetracycline. *J. Cell. Biochem.* **60**: 341–347.

926. **Karakiulakis, G., E. Missirlis, and M. E. Maragoudakis.** 1989. Mode of action of razoxane: inhibition of basement membrane collagen-degradation by a malignant tumor enzyme. *Meth. Find. Exp. Clin. Pharmacol.* **11**: 255–261.

927. **Karakiulakis, G., E. Missirlis, and M. E. Maragoudakis.** 1990. Basement membrane collagen-degrading activity from a malignant tumor is inhibited by anthracycline antibiotics. *Biochim. Biophys. Acta* **1035**: 218–222.

928. **Kobayashi, S. and Y. Nagai.** 1978. Human leucocyte neutral proteases, with special reference to collagen metabolism. *J. Biochem. (Tokyo)* **84**: 559–567.

929. **Kortylewicz, Z. P. and R. E. Galardy.** 1989. Phthaloyl-glycyl^P-isoleucyl-tryptophan benzylamide is a potent inhibitor of human skin fibroblast collagenase with a Ki of 25 nM. *J. Enz. Inhib.* **3**: 159–162.

930. **Kortylewicz, Z. P. and R. E. Galardy.** 1990. Phosphoramidate peptide inhibitors of human skin fibroblast collagenase. *J. Med. Chem.* **33**: 263–273.

931. **Kuttan, R., P. V. Donnelly, and N. Di Ferrante.** 1981. Collagen treated with (+)-catechin becomes resistant to the action of mammalian collagenase. *Experientia* **37**: 221–223.

932. **Lee, H.-J., M.-C. Chung, C.-H. Lee, H.-K. Chun, H.-M. Kim, and Y.-H. Kho.** 1996. Pyridoxatin, an inhibitor of gelatinase A with cytotoxic activity. *J. Microbiol. Biotechnol.* **6**: 445–450.

933. **Lee, H. J., M. C. Chung, C. H. Lee, B. S. Yun, H. K. Chun, and Y. H. Kho.** 1997. Gelastatins A and B, new inhibitors of gelatinase A from *Westerdykella multispora* F50733. *J. Antibiot.* **50**: 357–359.

934. **Lelièvre, Y., R. Bouboutou, J. Boiziau, and T. Cartwright.** 1989. [Inhibition of synovial collagenase by actinonin. Study of structure/activity relationship]. [French]. *Pathol. Biol. (Paris)* **37**: 43–46.

935. **Lelièvre, Y., R. Bouboutou, J. Boiziau, D. Faucher, D. Achard, and T. Cartwright.** 1990. Low molecular weight, sequence based, collagenase inhibitors selectively block the interaction between collagenase and TIMP (tissue inhibitor of metalloproteinases). *Matrix* **10**: 292–299.

936. **Levin, J. I., J. F. DiJoseph, L. M. Killar, M. A. Sharr, J. S. Skotnicki, D. V. Patel, X. Y. Xiao, L. H. Shi, M. Navre, and D. A. Campbell.** 1998. The asymmetric synthesis and in vitro characterization of succinyl mercaptoalcohol and mercaptoketone inhibitors of matrix metalloproteinases. *Bioorg. Med. Chem. Lett.* **8**: 1163–1168.

937. **Levy, D. E., F. Lapierre, W. S. Liang, W. Q. Ye, C. W. Lange, X. Y. Li, D. Grobelny, M. Casabonne, D. Tyrrell, K. Holme, A. Nadzan, and R. E. Galardy.** 1998. Matrix metalloproteinase inhibitors—a structure-activity study. *J. Med. Chem.* **41**: 199–223.

938. **Lewis, E. J., J. Bishop, K. M. K. Bottomley, D. Bradshaw, M. Brewster, M. J. Broadhurst, P. A. Brown, J. M. Budd, L. Elliott, A. K. Greenham,** *et al.* 1997. Ro 32-3555, an orally active collagenase inhibitor, prevents cartilage breakdown *in vitro* and *in vivo*. *Br. J. Pharmacol.* **121**: 540–546.

939. **Lindy, O., Y. T. Konttinen, T. Sorsa, Y. L. Ding, S. Santavirta, A. Ceponis, and C. López-Otín.** 1997. Matrix metalloproteinase 13 (collagenase 3) in human rheumatoid synovium. *Arthritis Rheum.* **40**: 1391–1399.

940. **Lonberg-Holm, K., D. L. Reed, R. C. Roberts, R. R. Hebert, M. C. Hillman, and R. M. Kutney.** 1987. Three high molecular weight protease inhibitors of rat plasma. Isolation, characterization, and acute phase changes. *J. Biol. Chem.* **262**: 438–445.

941. **Lubec, G. and E. Ratzenhofer.** 1979. Collagenase activity of rat kidney with glomerulonephritis is inhibited by erythromycin. *Nephron* **24**: 93–95.

942. **MacPherson, L. J., E. K. Bayburt, M. P. Capparelli, B. J. Carroll, R. Goldstein, M. R. Justice, L. J. Zhu, S. I. Hu, R. A. Melton, L. Fryer,** *et al.* 1997. Discovery of CGS 27023A, a non-peptidic, potent, and orally active stromelysin inhibitor that blocks cartilage degradation in rabbits. *J. Med. Chem.* **40**: 2525–2532.

943. **Makimura, M., M. Hirasawa, K. Kobayashi, J. Indo, S. Sakanaka, T. Taguchi, and S. Otake.** 1993. Inhibitory effect of tea catechins on collagenase activity. *J. Periodontol.* **64**: 630–636.

944. **Mallya, S. K. and H. E. Van Wart.** 1987. Inhibition of human neutrophil collagenase by gold(I) salts used in chrysotherapy. *Biochem. Biophys. Res. Commun.* **144**: 101–108.

945. **Mallya, S. K. and H. E. Van Wart.** 1989. Mechanism of inhibition of human neutrophil collagenase by gold(I) chrysotherapeutic compounds. Interaction at a heavy metal binding site. *J. Biol. Chem.* **264**: 1594–1601.

946. **Martin, S. F., C. J. Oalmann, and S. Liras.** 1993. Cyclopropanes as conformationally restricted peptide isosteres—design and synthesis of novel collagenase inhibitors. *Tetrahedron* **49**: 3521–3532.

947. **Masui, Y., T. Takemoto, S. Sakakibara, H. Hori, and Y. Nagai.** 1977. Synthetic substrates for vertebrate collagenase. *Biochem. Med.* **17**: 215–221.

948. **McCroskery, P. A., J. F. Richards, and E. D. Harris, Jr.** 1975. Purification and characterization of a collagenase extracted from rabbit tumours. *Biochem. J.* **152**: 131–142.

949. **McCullagh, K., Wadsworth, H., and Hann, M.** 1984. Carboxyalkyl peptide derivatives. Eur.Pat.Appl. EP 126,974, pp. 1–111.

950. **McCulloch, C. A., P. Birek, C. Overall, S. Aitken, W. Lee, and G. Kulkarni.** 1990. Randomized controlled trial of doxycycline in prevention of recurrent periodontitis in high-risk patients: antimicrobial activity

and collagenase inhibition. *J. Clin. Periodontol.* **17**: 616–622.

951. **McGeehan, G. M., J. D. Becherer, R. C. Bast, C. M. Boyer, B. Champion, K. M. Connolly, J. G. Conway, P. Furdon, S. Karp, S. Kidao,** *et al.* 1994. Regulation of tumour necrosis factor-α processing by a metalloproteinase inhibitor. *Nature* **370**: 558–561.

952. **Melchiori, A., A. Albini, J. M. Ray, and W. G. Stetler-Stevenson.** 1992. Inhibition of tumor cell invasion by a highly conserved peptide sequence from the matrix metalloproteinase enzyme prosegment. *Cancer Res.* **52**: 2353–2356.

953. **Menzel, J. and W. Borth.** 1983. Influence of plasma fibronectin on collagen cleavage by collagenase. *Coll. Relat. Res.* **3**: 217–230.

954. **Miller, A., M. Askew, R. P. Beckett, C. L. Bellamy, E. A. Bone, R. E. Coates, A. H. Davidson, A. H. Drummond, P. Huxley, F. M. Martin,** *et al.* 1997. Inhibition of matrix metalloproteinases: an examination of the S1' pocket. *Bioorg. Med. Chem. Lett.* **7**: 193–198.

955. **Mookhtiar, K. A., C. K. Marlowe, P. A. Bartlett, and H. E. Van Wart.** 1987. Phosphonamidate inhibitors of human neutrophil collagenase. *Biochemistry* **26**: 1962–1965.

956. **Moore, W. M. and C. A. Spilburg.** 1986. Peptide hydroxamic acids inhibit skin collagenase. *Biochem. Biophys. Res. Commun.* **136**: 390–395.

957. **Morodomi, T., Y. Ogata, Y. Sasaguri, M. Morimatsu, and H. Nagase.** 1992. Purification and characterization of matrix metalloproteinase 9 from U937 monocytic leukaemia and HT1080 fibrosarcoma cells. *Biochem. J.* **285**: 603–611.

958. **Morphy, J. R., N. R. A. Beeley, B. A. Boyce, J. Leonard, B. Mason, A. Millican, K. Millar, J. P. Oconnell, and J. Porter.** 1994. Potent and selective inhibitors of gelatinase A. 2. Carboxylic and phosphonic acid derivatives. *Bioorg. Med. Chem. Lett.* **4**: 2747–2752.

959. **Morris, A. D., S. Leonce, N. Guilbaud, G. C. Tucker, V. Perez, M. Jan, A. A. Cordi, A. Pierre, and G. Atassi.** 1997. Eriochrome black T structurally related to suramin, inhibits angiogenesis and tumor growth in vivo. *Anticancer Drugs* **8**: 746–755.

960. **Müller, J. C. D., E. Graf von Roedern, F. Grams, H. Nagase, and L. Moroder.** 1997. Non-peptidic cysteine derivatives as inhibitors of matrix metalloproteinases. *Biol. Chem.* **378**: 1475–1480.

961. **Nagai, Y., C. M. Lapiere, and J. Gross.** 1966. Tadpole collagenase. Preparation and purification. *Biochemistry* **5**: 3123–3130.

962. **Nagai, Y., H. Shinkai, and Y. Ninomiya.** 1978. The release of collagenase inhibitors from procollagen additional peptides by pepsin treatment. *Proc. Jap. Acad. Ser. B* **54**: 140–144.

963. **Nagase, H. and E. D. Harris, Jr.** 1983. Ovostatin: a novel proteinase inhibitor from chicken egg white. II. Mechanism of inhibition studied with collagenase and thermolysin. *J. Biol. Chem.* **258**: 7490–7498.

964. **Nagase, H., E. D. Harris, Jr., J. F. Woessner, Jr., and K. Brew.** 1983. Ovostatin: a novel proteinase inhibitor from chicken egg white. I. Purification, physicochemical properties, and tissue distribution of ovostatin. *J. Biol. Chem.* **258**: 7481–7489.

965. **Nagase, H., Y. Itoh, and S. Binner.** 1994. Interaction of α₂-macroglobulin with matrix metalloproteinases and its use for identification of their active forms. *Ann. N. Y. Acad. Sci.* **732**: 294–302.

966. **Naito, K., S. Nakajima, N. Kanbayashi, A. Okuyama, and M. Goto.** 1993. Inhibition of metalloproteinase activity of rheumatoid arthritis synovial cells by a new inhibitor [BE16627B; L-N-(N-hydroxy-2-isobutylsuccinamoyl)-seryl-L-valine]. *Agents Actions* **39**: 182–186.

967. **Nakagawa, T., T. Kubota, M. Kabuto, and T. Kodera.** 1995. Captopril inhibits glioma cell invasion in vitro: involvement of matrix metalloproteinases. *Anticancer Res.* **15**: 1985–1989.

968. **Nielsen, K. L., L. Sottrup-Jensen, H. Nagase, and M. Etzerodt.** 1994. The primary structure of ovomacroglobulin. *Ann. NY Acad. Sci.* **737**: 476–479.

969. **Nip, L. H., V.-J. Uitto, and L. M. Golub.** 1993. Inhibition of epithelial cell matrix metalloproteinases by tetracyclines. *J. Periodontal Res.* **28**: 379–385.

970. **Nixon, J. S., K. M. Bottomley, M. J. Broadhurst, P. A. Brown, W. H. Johnson, G. Lawton, J. Marley, A. D. Sedgwick, and S. E. Wilkinson.** 1991. Potent collagenase inhibitors prevent interleukin-1-induced cartilage degradation in vitro. *Int. J. Tissue React.* **13**: 237–241.

971. **Norga, K., B. Grillet, S. Masure, L. Paemen, and G. Opdenakker.** 1996. Human gelatinase B, a marker enzyme in rheumatoid arthritis, is inhibited by D-penicillamine: anti-rheumatic activity by protease inhibition. *Clin. Rheumatol.* **15**: 31–34.

972. **Norga, K., L. Paemen, S. Masure, C. Dillen, H. Heremans, A. Billiau, H. Carton, L. Cuzner, T. Olsson, J. Van Damme,** *et al.* 1995. Prevention of acute autoimmune encephalomyelitis and abrogation of relapses in murine models of multiple sclerosis by the protease inhibitor D-penicillamine. *Inflamm. Res.* **44**: 529–534.

973. **Nugiel, D. A., K. Jacobs, C. P. Decicco, D. J. Nelson, R. A. Copeland, and K. D. Hardman.** 1995. Probing the P3' pocket of stromelysin with piperazic acid analogs. *Bioorg. Med. Chem. Lett.* **5**: 3053–3056.

974. **Odake, S., Y. Morita, T. Morikawa, N. Yoshida, H. Hori, and Y. Nagai.** 1994. Inhibition of matrix metalloproteinases by peptidyl hydroxamic acids. *Biochem. Biophys. Res. Commun.* **199**: 1442–1446.

975. **Odake, S., T. Okayama, M. Obata, T. Morikawa, S. Hattori, H. Hori, and Y. Nagai.** 1990. Vertebrate collagenase inhibitor. I. Tripeptidyl hydroxamic acids. *Chem. Pharm. Bull. (Tokyo)* **38**: 1007–1011.

976. **Odake, S., T. Okayama, M. Obata, T. Morikawa, S. Hattori, H. Hori, and Y. Nagai.** 1991. Vertebrate collagenase inhibitor. II. Tetrapeptidyl hydroxamic acids. *Chem. Pharm. Bull. (Tokyo)* **39**: 1489–1494.

977. **Ogita, T., A. Sato, R. Enokita, K. Suzuki, M. Ishii, T. Negishi, T. Okazaki, K. Tamaki, and K. Tanzawa.** 1992. Matlystatins, new inhibitors of typeIV collagenases from *Actinomadura atramentaria*. I. Taxonomy, fermentation, isolation, and physico-chemical properties of matlystatin-group compounds. *J. Antibiot. (Tokyo)* **45**: 1723–1732.

978. **Ohta, A., J. S. Louie, and J. Uitto.** 1986. Collagenase production by human mononuclear cells in culture: inhibition by gold containing compounds and other antirheumatic agents. *Ann. Rheum. Dis.* **45**: 996–1003.

979. **Ohuchi, E., K. Imai, Y. Fujii, H. Sato, M. Seiki, and Y. Okada.** 1997. Membrane type 1 matrix metalloproteinase digests interstitial collagens and other extracellular matrix macromolecules. *J. Biol. Chem.* **272**: 2446–2451.

980. **Olejniczak, E. T., P. J. Hajduk, P. A. Marcotte, D. G. Nettesheim, R. P. Meadows, R. Edalji, T. F. Holzman, and S. W. Fesik.** 1997. Stromelysin inhibitors designed from weakly bound fragments—effects of linking and cooperativity. *J. Am. Chem. Soc.* **119**: 5828–5832.

981. **Paemen, L., E. Martens, K. Norga, S. Masure, E. Roets, J. Hoogmartens, and G. Opdenakker.** 1996. The gelatinase inhibitory activity of tetracyclines and chemically modified tetracycline analogues as measured by a novel microtiter assay for inhibitors. *Biochem. Pharmacol.* **52**: 105–111.

982. **Park, A. J., L. M. Matrisian, A. F. Kells, R. Pearson, Z. Y. Yuan, and M. Navre.** 1991. Mutational analysis of the transin (rat stromelysin) autoinhibitor region demonstrates a role for residues surrounding the 'cysteine switch'. *J. Biol. Chem.* **266**: 1584–1590.

983. **Parsons, S. L., S. A. Watson, and R. J. Steele.** 1997. Phase I/II trial of batimastat, a matrix metalloproteinase inhibitor, in patients with malignant ascites. *Eur. J. Surg. Oncol.* **23**: 526–531.

984. **Pei, D., G. Majmudar, and S. J. Weiss.** 1994. Hydrolytic inactivation of a breast carcinoma cell-derived serpin by human stromelysin-3. *J. Biol. Chem.* **269**: 25849–25855.

985. **Porter, J. R., N. R. A. Beeley, B. A. Boyce, B. Mason, A. Millican, K. Millar, J. Leonard, J. R. Morphy, and J. P. O'Connell.** 1994. Potent and selective inhibitors of gelatinase A .1. Hydroxamic acid derivatives. *Bioorg. Med. Chem. Lett.* **4**: 2741–2746.

986. **Ratzenhofer, E., G. Lubec, and E. M. Kokoschka.** 1978. [In-vitro inhibition of collagenase activity in epidermolysis bullosa hereditaria dystrophica]. [German]. *Z. Hautkr.* **53**: 929–934.

987. **Reich, R., E. W. Thompson, Y. Iwamoto, G. R. Martin, J. R. Deason, G. C. Fuller, and R. Miskin.** 1988. Effects of inhibitors of plasminogen activator, serine proteinases, and collagenase IV on the invasion of basement membranes by metastatic cells. *Cancer Res.* **48**: 3307–3312.

988. **Rockwell, A., M. Melden, R. A. Copeland, K. Hardman, C. P. Decicco, and W. F. Degrado.** 1996. Complementarity of combinatorial chemistry and structure-based ligand design: application to the discovery of novel inhibitors of matrix metalloproteinases. *J. Am. Chem. Soc.* **118**: 10337–10338.

989. **Sahoo, S. P., C. G. Caldwell, K. T. Chapman, P. L. Durette, C. K. Esser, I. E. Kopka, S. A. Polo, K. M. Sperow, L. M. Niedzwiecki, M. Izquierdo-Martin, *et al.** 1995. Inhibition of matrix metalloproteinases by N-carboxyalkyl dipeptides: enhanced potency and selectivity with substituted P1′ homophenylalanines. *Bioorg. Med. Chem. Lett.* **5**: 2441–2446.

990. **Saito, A. and H. Sinohara.** 1985. Murinoglobulin, a novel protease inhibitor from murine plasma. Isolation, characterization, and comparison with murine α-macroglobulin and human α_2-macroglobulin. *J. Biol. Chem.* **260**: 775–781.

991. **Sakamoto, S., P. Goldhaber, and M. J. Glimcher.** 1972. Further studies on the nature of the components in serum which inhibit mouse bone collagenase. *Calcif. Tissue Res.* **10**: 280–288.

992. **Sakamoto, S., P. Goldhaber, and M. J. Glimcher.** 1972. Maintenance of mouse bone collagenase activity in the presence of serum protein by addition of trypsin. *Proc. Soc. Exp. Biol. Med.* **139**: 1038–1041.

993. **Salvesen, G. S., C. A. Sayers, and A. J. Barrett.** 1981. Further characterization of the covalent linking reaction of α_2-macroglobulin. *Biochem. J.* **195**: 453–461.

994. **Sand, O., J. Folkersen, J. G. Westergaard, and L. Sottrup-Jensen.** 1985. Characterization of human pregnancy zone protein. Comparison with human α_2-macroglobulin. *J. Biol. Chem.* **260**: 15723–15735.

995. **Santavirta, S., M. Takagi, Y. T. Konttinen, T. Sorsa, and A. Suda.** 1996. Inhibitory effect of cephalothin on matrix metalloproteinase activity around loose hip prostheses. *Antimicrob. Agents Chemother.* **40**: 244–246.

996. **Sato, T., Y. Takebayashi, T. Tokunaga, and T. Ozasa.** 1996. YM-24074, a new peptide antibiotic. II. Structural elucidation. *J. Antibiot. (Tokyo)* **49**: 811–814.

997. **Sazuka, M., H. Imazawa, Y. Shoji, T. Mita, Y. Hara, and M. Isemura.** 1997. Inhibition of collagenases from mouse lung carcinoma cells by green tea catechins and

black tea theaflavins. *Biosci. Biotechnol. Biochem.* **61**: 1504–1506.

998. **Schaefer, R. M., L. Paczek, S. Huang, M. Teschner, L. Schaefer, and A. Heidland.** 1992. Role of glomerular proteinases in the evolution of glomerulosclerosis. *Eur. J. Clin. Chem. Clin. Biochem.* **30**: 641–646.

999. **Schwartz, M. A., S. Venkataraman, M. A. Ghaffari, A. Libby, K. A. Mookhtiar, S. K. Mallya, H. Birkedal-Hansen, and H. E. Van Wart.** 1991. Inhibition of human collagenases by sulfur-based substrate analogs. *Biochem. Biophys. Res. Commun.* **176**: 173–179.

1000. **Seltzer, J. L., M. L. Eschbach, J. O. Winberg, E. A. Bauer, A. Z. Eisen, and H. Weingarten.** 1987. Eriochrome black T inhibition of human skin collagenase, but not gelatinase, using both protein and synthetic substrates. *Coll. Relat. Res.* **7**: 399–407.

1001. **Smith, G. N., Jr., K. D. Brandt, and K. A. Hasty.** 1994. Procollagenase is reduced to inactive fragments upon activation in the presence of doxycycline. *Ann. N. Y. Acad. Sci.* **732**: 436–438.

1002. **Smith, G. N., Jr., K. D. Brandt, and K. A. Hasty.** 1996. Activation of recombinant human neutrophil procollagenase in the presence of doxycycline results in fragmentation of the enzyme and loss of enzyme activity. *Arthritis Rheum.* **39**: 235–244.

1003. **Sorbi, D., M. Fadly, R. Hicks, S. Alexander, and L. Arbeit.** 1993. Captopril inhibits the 72 kDa and 92 kDa matrix metalloproteinases. *Kidney Int.* **44**: 1266–1272.

1004. **Sorsa, T., T. Ingman, K. Suomalainen, S. Halinen, H. Saari, Y. T. Konttinen, V.-J. Uitto, and L. M. Golub.** 1992. Cellular source and tetracycline-inhibition of gingival crevicular fluid collagenase of patients with labile diabetes mellitus. *J. Clin. Periodontol.* **19**: 146–149.

1005. **Sorsa, T., N. S. Ramamurthy, A. T. Vernillo, X. Zhang, Y. T. Konttinen, B. R. Rifkin, and L. M. Golub.** 1998. Functional sites of chemically modified tetracyclines—inhibition of the oxidative activation of human neutrophil and chicken osteoclast pro-matrix metalloproteinases. *J. Rheumatol.* **25**: 975–982.

1006. **Sottrup-Jensen, L. and H. Birkedal-Hansen.** 1989. Human fibroblast collagenase-α-macroglobulin interactions. Localization of cleavage sites in the bait regions of five mammalian α-macroglobulins. *J. Biol. Chem.* **264**: 393–401.

1007. **Sottrup-Jensen, L., J. Gliemann, and F. Van Leuven.** 1986. Domain structure of human α_2-macroglobulin. Characterization of a receptor-binding domain obtained by digestion with papain. *FEBS Lett.* **205**: 20–24.

1008. **Springman, E. B., H. Nagase, H. Birkedal-Hansen, and H. E. Van Wart.** 1995. Zinc content and function in human fibroblast collagenase. *Biochemistry* **34**: 15713–15720.

1009. **Stack, M. S., C. G. Emberts, and R. D. Gray.** 1991. Application of N-carboxyalkyl peptides to the inhibition and affinity purification of the porcine matrix metalloproteinases collagenase, gelatinase, and stromelysin. *Arch. Biochem. Biophys.* **287**: 240–249.

1010. **Stetler-Stevenson, W. G., J. A. Talano, M. E. Gallagher, H. C. Krutzsch, and L. A. Liotta.** 1991. Inhibition of human type IV collagenase by a highly conserved peptide sequence derived from its prosegment. *Am. J. Med. Sci.* **302**: 163–170.

1011. **Sun, H. H., P. V. Kaplita, D. R. Houck, M. B. Stawicki, R. McGarry, R. C. Wahl, A. M. Gillum, and R. Cooper.** 1996. A metalloproteinase inhibitor from *Doliocarpus verruculosus. Phytother. Res.* **10**: 194–197.

1012. **Suomalainen, K., T. Sorsa, L. M. Golub, N. Ramamurthy, H. M. Lee, V.-J. Uitto, H. Saari, and Y. T. Konttinen.** 1992. Specificity of the anticollagenase action of tetracyclines: relevance to their anti-inflammatory potential. *Antimicrob. Agents Chemother.* **36**: 227–229.

1013. **Suzuki, K., K. Shimada, S. Nozoe, K. Tanzawa, and T. Ogita.** 1996. Isolation of nicotianamine as a gelatinase inhibitor. *J. Antibiot. (Tokyo)* **49**: 1284–1285.

1014. **Tamaki, K., S. Kurihara, T. Oikawa, K. Tanzawa, and Y. Sugimura.** 1994. Matlystatins, new inhibitors of type IV collagenases from *Actinomadura atramentaria.* IV. Synthesis and structure-activity relationships of matlystatin B and its stereoisomers. *J. Antibiot. (Tokyo)* **47**: 1481–1492.

1015. **Tamaki, K., A. Matsubara, K. Tanzawa, and Y. Sugimura.** 1995. Total synthesis and inhibitory activity against gelatinase B of YL-01869P. *J. Antibiot. (Tokyo)* **48**: 87–88.

1016. **Tamaki, K., K. Tanzawa, S. Kurihara, T. Oikawa, S. Monma, K. Shimada, and Y. Sugimura.** 1995. Synthesis and structure-activity relationships of gelatinase inhibitors derived from matlystatins. *Chem. Pharm. Bull. (Tokyo)* **43**: 1883–1893.

1017. **Tamura, Y., F. Watanabe, T. Nakatani, K. Yasui, M. Fuji, T. Komurasaki, H. Tsuzuki, R. Maekawa, T. Yoshioka, K. Kawada, K. Sugita, and M. Ohtani.** 1998. Highly selective and orally active inhibitors of type IV collagenase (MMP-9 and MMP-2)—N-sulfonylamino acid derivatives. *J. Med. Chem.* **41**: 640–649.

1018. **Tanzawa, K., M. Ishii, T. Ogita, and K. Shimada.** 1992. Matlystatins, new inhibitors of type IV collagenases from *Actinomadura atramentaria.* II. Biological activities. *J. Antibiot. (Tokyo)* **45**: 1733–1737.

1019. **Tapon-Bretaudière, J., A. Bros, E. Couture-Tosi, and E. Delaine.** 1985. Electron microscopy of the conformational changes of α_2-macroglobulin from human placents. *EMBO J.* **4**: 85–89.

1020. **Teronen, O., Y. T. Konttinen, C. Lindqvist, T. Salo, T. Ingman, A. Lauhio, Y. Ding, S. Santavirta, and T.**

Sorsa. 1997. Human neutrophil collagenase MMP-8 in peri-implant sulcus fluid and its inhibition by clodronate. *J. Dent. Res.* **76**: 1529–1537.

1021. **Teronen, O., Y. T. Konttinen, C. Lindqvist, T. Salo, T. Ingman, A. Lauhio, Y. Ding, S. Santavirta, H. Valleala, and T. Sorsa.** 1997. Inhibition of matrix metalloproteinase-1 by dichloromethylene bisphosphonate (clodronate). *Calcif. Tissue Int.* **61**: 59–61.

1022. **Uitto, V.-J., J. D. Firth, L. Nip, and L. M. Golub.** 1994. Doxycycline and chemically modified tetracyclines inhibit gelatinase A (MMP-2) gene expression in human skin keratinocytes. *Ann. N. Y. Acad. Sci.* **732**: 140–151.

1023. **Van Leuven, F., J. J. Cassiman, and H. Van Den Berghe.** 1986. Human pregnancy zone protein and $\alpha2$-macroglobulin. High affinity binding of complexes to the same receptor on fibroblasts and characterization by monoclonal antibodies. *J. Biol. Chem.* **261**: 16622–16625.

1024. **Wallace, D. A., S. R. Bates, B. Walker, G. Kay, J. White, D. J. Guthrie, N. L. Blumson, and D. T. Elmore.** 1986. Competitive inhibition of human skin collagenase by N-benzyloxycarbonyl-L-prolyl-L-alanyl-3-amino-2-oxopropyl-L-leucyl-L-alanylglycine ethyl ester. *Biochem. J.* **239**: 797–799.

1025. **Wang, X., X. Fu, P. D. Brown, M. J. Crimmin, and R. M. Hoffman.** 1994. Matrix metalloproteinase inhibitor BB-94 (batimastat) inhibits human colon tumor growth and spread in a patient-like orthotopic model in nude mice. *Cancer Res.* **54**: 4726–4728.

1026. **Wauters, P., Y. Eeckhout, and G. Vaes.** 1986. Oxidation products are responsible for the resistance to the action of collagenase conferred on collagen by (+)-catechin. *Biochem. Pharmacol.* **35**: 2971–2973.

1027. **Welgus, H. G., J. J. Jeffrey, and A. Z. Eisen.** 1981. The collagen substrate specificity of human skin fibroblast collagenase. *J. Biol. Chem.* **256**: 9511–9515.

1028. **Werb, Z., M. C. Burleigh, A. J. Barrett, and P. M. Starkey.** 1974. The interaction of α_2-macroglobulin with proteinases. Binding and inhibition of mammalian collagenases and other metal proteinases. *Biochem. J.* **139**: 359–368.

1029. **Winwood, P. J., D. Schuppan, J. P. Iredale, C. A. Kawser, A. J. P. Docherty, and M. J. Arthur.** 1995. Kupffer cell-derived 95-kD type IV collagenase/gelatinase B: characterization and expression in cultured cells. *Hepatology* **22**: 304–315.

1030. **Woessner, J. F.** 1996. Regulation of matrilysin in the rat uterus. *Biochem. Cell. Biol.* **74**: 777–784.

1031. **Woessner, J. F., Jr. and C. J. Taplin.** 1988. Purification and properties of a small latent matrix metalloproteinase of the rat uterus. *J. Biol. Chem.* **263**: 16918–16925.

1032. **Wojtowicz-Praga, S., J. Low, J. Marshall, E. Ness, R. Dickson, J. Barter, M. Sale, P. McCann, J. Moore, A. Cole, and M. J. Hawkins.** 1996. Phase I trial of a novel matrix metalloproteinase inhibitor batimastat (BB-94) in patients with advanced cancer. *Invest. New Drugs* **14**: 193–202.

1033. **Wojtowicz-Praga, S., J. Torri, M. Johnson, V. Steen, J. Marshall, E. Ness, R. Dickson, M. Sale, H. S. Rasmussen, T. A. Chiodo, and M. J. Hawkins.** 1998. Phase I trial of marimastat, a novel matrix metalloproteinase inhibitor, administered orally to patients with advanced lung cancer. *J. Clin. Oncol.* **16**: 2150–2156.

1034. **Woolley, D. E., R. W. Glanville, D. R. Roberts, and J. M. Evanson.** 1978. Purification, characterization and inhibition of human skin collagenase. *Biochem. J.* **169**: 265–276.

1035. **Woolley, D. E., D. R. Roberts, and J. M. Evanson.** 1976. Small molecular weight β_1 serum protein which specifically inhibits human collagenases. *Nature* **261**: 325–327.

1036. **Xue, C. B., X. H. He, J. Roderick, W. F. DeGrado, C. Decicco, and R. A. Copeland.** 1996. Potent matrix metalloproteinase inhibitors: amino-carboxylate compounds containing modifications of the P1 residue. *Bioorg. Med. Chem. Lett.* **6**: 379–384.

1037. **Yamamoto, M., H. Tsujishita, N. Hori, Y. Ohishi, S. Inoue, S. Ikeda, and Y. Okada.** 1998. Inhibition of membrane-type 1 matrix metalloproteinase by hydroxamate inhibitors—an examination of the subsite pocket. *J. Med. Chem.* **41**: 1209–1217.

1038. **Yankeelov, J. A., Jr., H. A. Parish, Jr., and A. F. Spatola.** 1978. Mercaptoimidazolylpropionic acid hydrobromide. Inhibition of tadpole collagenase and related properties. *J. Med. Chem.* **21**: 701–704.

1039. **Zhou, N. M., L. Paemen, G. Opdenakker, and G. Froyen.** 1997. Cloning and expression in *Escherichia coli* of a human gelatinase B-inhibitory single-chain immunoglobulin variable fragment (scFv). *FEBS Lett.* **414**: 562–566.

1040. **Zhu, C. and J. F. Woessner, Jr.** 1991. A tissue inhibitor of metalloproteinases and α-macroglobulins in the ovulating rat ovary: possible regulators of collagen matrix breakdown. *Biol. Reprod.* **45**: 334–342.

See also review articles: 7, 12, 13, 14, 15, 20, 24, 53, 61, 77, 101, 121, 151.

Sequences of MMPs

1041. **Abramson, S. R., G. E. Conner, H. Nagase, I. Neuhaus, and J. F. Woessner, Jr.** 1995. Characterization of rat uterine matrilysin and its cDNA. Relationship to human pump-1 and activation of procollagenases. *J. Biol. Chem.* **270**: 16016–16022.

1042. **Aimes, R. T., D. L. French, and J. P. Quigley.** 1994. Cloning of a 72 kDa matrix metalloproteinase (gelatinase) from chicken embryo fibroblasts using gene family PCR: expression of the gelatinase increases upon malignant transformation. *Biochem. J.* **300**: 729–736.

1043. **Angel, P., I. Baumann, B. Stein, H. Delius, H. J. Rahmsdorf, and P. Herrlich.** 1987. 12–O-Tetradecanoyl-phorbol-13-acetate induction of the human collagenase gene is mediated by an inducible enhancer element located in the 5′-flanking region. *Mol. Cell. Biol.* **7**: 2256–2266.

1044. **Apte, S. S., N. Fukai, D. R. Beier, and B. R. Olsen.** 1997. The matrix metalloproteinase-14 (MMP-14) gene is structurally distinct from other MMP genes and is co-expressed with the TIMP-2 gene during mouse embryogenesis. *J. Biol. Chem.* **272**: 25511–25517.

1045. **Balbín, M., A. Fueyo, V. Knäuper, A. M. Pendás, J. M. Lopez, M. G. Jiménez, G. Murphy, and C. López-Otín.** 1998. Collagenase 2 (MMP-8) expression in murine tissue-remodeling processes—analysis of its potential role in postpartum involution of the uterus. *J. Biol. Chem.* **273**: 23959–23968.

1046. **Balkman, C. E. and A. J. Nixon.** 1998. Molecular cloning and cartilage gene expression of equine stromelysin 1 (matrix metalloproteinase 3). *Am. J. Vet. Res.* **59**: 30–36.

1047. **Bartlett, J. D., J. P. Simmer, J. Xue, H. C. Margolis, and E. C. Moreno.** 1996. Molecular cloning and mRNA tissue distribution of a novel matrix metalloproteinase isolated from porcine enamel organ. *Gene* **183**: 123–128.

1048. **Basset, P., J. P. Bellocq, C. Wolf, I. Stoll, P. Hutin, J. M. Limacher, O. L. Podhajcer, M. P. Chenard, M. C. Rio, and P. Chambon.** 1990. A novel metalloproteinase gene specifically expressed in stromal cells of breast carcinomas. *Nature* **348**: 699–704.

1049. **Baylis, H. A., A. Megson, and R. Hall.** 1995. Infection with *Theileria annulata* induces expression of matrix metalloproteinase 9 and transcription factor AP-1 in bovine leucocytes. *Mol. Biochem. Parasitol.* **69**: 211–222.

1050. **Bayne, E. K., N. I. Hutchinson, L. A. Walakovits, S. Donatelli, K. L. MacNaul, C. F. Harper, P. Cameron, V. L. Moore, and M. W. Lark.** 1992. Production, purification and characterization of canine prostromelysin. *Matrix* **12**: 173–184.

1051. **Breathnach, R., L. M. Matrisian, M.-C. Gesnel, A. Staub, and P. Leroy.** 1987. Sequences coding for part of oncogene-induced transin are highly conserved in a related rat gene. *Nucleic Acids Res.* **15**: 1139–1151.

1052. **Brinckerhoff, C. E., P. L. Ruby, S. D. Austin, M. E. Fini, and H. D. White.** 1987. Molecular cloning of human synovial cell collagenase and selection of a single gene from genomic DNA. *J. Clin. Invest.* **79**: 542–546.

1053. **Brown, D. D., Z. Wang, J. D. Furlow, A. Kanamori, R. A. Schwartzman, B. F. Remo, and A. Pinder.** 1996. The thyroid hormone-induced tail resorption program during *Xenopus laevis* metamorphosis. *Proc. Natl. Acad. Sci. USA* **93**: 1924–1929.

1054. **Chan, J. C., M. Scanlon, H. Z. Zhang, L. B. Jia, D. H. Yu, M. C. Hung, M. French, and E. M. Eastman.** 1992. Molecular cloning and characterization of v-mos-activated transformation-associated proteins. *J. Biol. Chem.* **267**: 1099–1103.

1055. **Chen, J. M., R. T. Aimes, G. R. Ward, G. L. Youngleib, and J. P. Quigley.** 1991. Isolation and characterization of a 70-kDa metalloprotease (gelatinase) that is elevated in Rous sarcoma virus-transformed chicken embryo fibroblasts. *J. Biol. Chem.* **266**: 5113–5121.

1056. **Chowdhury, S. K., K. J. Vavra, P. G. Brake, T. Banks, J. Falvo, R. Wahl, J. Eshraghi, G. Gonyea, B. T. Chait, and C. H. Vestal.** 1995. Examination of recombinant truncated mature human fibroblast collagenase by mass spectrometry: identification of differences with the published sequence and determination of stable isotope incorporation. *Rapid. Commun. Mass. Spectrom.* **9**: 563–569.

1057. **Clarke, N. J., M. C. O'Hare, T. E. Cawston, and G. P. Harper.** 1990. Nucleotide sequence of a cDNA for porcine type I collagenase, obtained by PCR. *Nucleic Acids Res.* **18**: 6703–6703.

1058. **Collier, I. E., S. M. Wilhelm, A. Z. Eisen, B. L. Marmer, G. A. Grant, J. L. Seltzer, A. Kronberger, C. He, E. A. Bauer, and G. I. Goldberg.** 1988. H-ras oncogene-transformed human bronchial epithelial cells (TBE-1) secrete a single metalloprotease capable of degrading basement membrane collagen. *J. Biol. Chem.* **263**: 6579–6587.

1059. **Cossins, J., T. J. Dudgeon, G. Catlin, A. J. Gearing, and J. M. Clements.** 1996. Identification of MMP-18, a putative novel human matrix metalloproteinase. *Biochem. Biophys. Res. Commun.* **228**: 494–498.

1060. **DenBesten, P. K., J. S. Punzi, and W. Li.** 1998. Purification and sequencing of a 21 kDa and 25 kDa bovine enamel metalloproteinase. *Eur. J. Oral Sci.* **106** (Suppl. 1): 345–349.

1061. **Devarajan, P., J. J. Johnston, S. S. Ginsberg, H. E. Van Wart, and N. Berliner.** 1992. Structure and expression of neutrophil gelatinase cDNA. Identity with type IV collagenase from HT1080 cells. *J. Biol. Chem.* **267**: 25228–25232.

1062. **Devarajan, P., K. Mookhtiar, H. Van Wart, and N. Berliner.** 1991. Structure and expression of the cDNA encoding human neutrophil collagenase. *Blood* **77**: 2731–2738.

1063. **Fang, K. C., W. W. Raymond, J. L. Blount, and G. H. Caughey.** 1997. Dog mast cell α-chymase activates progelatinase B by cleaving the Phe[88]-Gln[89] and Phe[91]-Glu[92] bonds of the catalytic domain. *J. Biol. Chem.* **272**: 25628–25635.

1064. **Fini, M. E., J. D. Bartlett, M. Matsubara, W. B. Rinehart, M. K. Mody, M. T. Girard, and M. Rainville.** 1994. The rabbit gene for 92-kDa matrix metalloproteinase. Role of AP1 and AP2 in cell type-specific transcription. *J. Biol. Chem.* **269**: 28620–28628.

1065. **Fini, M. E., R. H. Gross, and C. E. Brinckerhoff.** 1986. Characterization of rabbit genes for synovial cell collagenase. *Arthritis Rheum.* **29**: 1301–1315.

1066. **Fini, M. E., M. J. Karmilowicz, P. L. Ruby, A. M. Beeman, K. A. Borges, and C. E. Brinckerhoff.** 1987. Cloning of a complementary DNA for rabbit proactivator. A metalloproteinase that activates synovial cell collagenase, shares homology with stromelysin and transin, and is coordinately regulated with collagenase. *Arthritis Rheum.* **30**: 1254–1264.

1067. **Fini, M. E., I. M. Plucinska, A. S. Mayer, R. H. Gross, and C. E. Brinckerhoff.** 1987. A gene for rabbit synovial cell collagenase: member of a family of metalloproteinases that degrade the connective tissue matrix. *Biochemistry* **26**: 6156–6165.

1068. **Franco, A. A., L. M. Mundy, M. Trucksis, S. Wu, J. B. Kaper, and C. L. Sears.** 1997. Cloning and characterization of the *Bacteroides fragilis* metalloprotease toxin gene. *Infect. Immun.* **65**: 1007–1013.

1069. **Freije, J. M., I. Díez-Itza, M. Balbin, L. M. Sanchez, R. Blasco, J. Tolivia, and C. López-Otín.** 1994. Molecular cloning and expression of collagenase-3, a novel human matrix metalloproteinase produced by breast carcinomas. *J. Biol. Chem.* **269**: 16766–16773.

1070. **Gack, S., R. Vallon, J. Schaper, U. Rüther, and P. Angel.** 1994. Phenotypic alterations in fos-transgenic mice correlate with changes in Fos/Jun-dependent collagenase type I expression. Regulation of mouse metalloproteinases by carcinogens, tumor promoters, cAMP, and Fos oncoprotein. *J. Biol. Chem.* **269**: 10363–10369.

1071. **Gaire, M., Z. Magbanua, S. McDonnell, L. McNeil, D. H. Lovett, and L. M. Matrisian.** 1994. Structure and expression of the human gene for the matrix metalloproteinase matrilysin. *J. Biol. Chem.* **269**: 2032–2040.

1072. **Ghiglione, C., G. Lhomond, T. Lepage, and C. Gache.** 1994. Structure of the sea urchin hatching enzyme gene. *Eur. J. Biochem.* **219**: 845–854.

1073. **Goldberg, G. I., S. M. Wilhelm, A. Kronberger, E. A. Bauer, G. A. Grant, and A. Z. Eisen.** 1986. Human fibroblast collagenase. Complete primary structure and homology to an oncogene transformation-induced rat protein. *J. Biol. Chem.* **261**: 6600–6605.

1074. **Graubert, T., J. Johnston, and N. Berliner.** 1993. Cloning and expression of the cDNA encoding mouse neutrophil gelatinase: demonstration of coordinate secondary granule protein gene expression during terminal neutrophil maturation. *Blood* **82**: 3192–3197.

1075. **Hammani, K., P. Henriet, and Y. Eeckhout.** 1992. Cloning and sequencing of a cDNA encoding mouse stromelysin 1. *Gene* **120**: 321–322.

1076. **Hasty, K. A., T. F. Pourmotabbed, G. I. Goldberg, J. P. Thompson, D. G. Spinella, R. M. Stevens, and C. L. Mainardi.** 1990. Human neutrophil collagenase. A distinct gene product with homology to other matrix metalloproteinases. *J. Biol. Chem.* **265**: 11421–11424.

1077. **Henriet, P., G. G. Rousseau, and Y. Eeckhout.** 1992. Cloning and sequencing of mouse collagenase cDNA. Divergence of mouse and rat collagenases from the other mammalian collagenases. *FEBS Lett.* **310**: 175–178.

1078. **Huhtala, P., L. T. Chow, and K. Tryggvason.** 1990. Structure of the human type IV collagenase gene. *J. Biol. Chem.* **265**: 11077–11082.

1079. **Huhtala, P., R. L. Eddy, Y. S. Fan, M. G. Byers, T. B. Shows, and K. Tryggvason.** 1990. Completion of the primary structure of the human type IV collagenase preproenzyme and assignment of the gene (*CLG4*) to the q21 region of chromosome 16. *Genomics* **6**: 554–559.

1080. **Huhtala, P., A. Tuuttila, L. T. Chow, J. Lohi, J. Keski-Oja, and K. Tryggvason.** 1991. Complete structure of the human gene for 92-kDa type IV collagenase. Divergent regulation of expression for the 92- and 72-kilodalton enzyme genes in HT-1080 cells. *J. Biol. Chem.* **266**: 16485–16490.

1081. **Knäuper, V., S. Kramer, H. Reinke, and H. Tschesche.** 1990. Partial amino acid sequence of human PMN leukocyte procollagenase. *Biol. Chem. Hoppe Seyler* **371**: 295–304.

1082. **Lawson, N. D., A. Khannagupta, and N. Berliner.** 1998. Isolation and characterization of the cDNA for mouse neutrophil collagenase—demonstration of shared negative regulatory pathways for neutrophil secondary granule protein gene expression. *Blood* **91**: 2517–2524.

1083. **Lefebvre, O., C. Wolf, J. M. Limacher, P. Hutin, C. Wendling, M. LeMeur, P. Basset, and M. C. Rio.** 1992. The breast cancer-associated stromelysin-3 gene is expressed during mouse mammary gland apoptosis. *J. Cell Biol.* **119**: 997–1002.

1084. **Lepage, T. and C. Gache.** 1990. Early expression of a collagenase-like hatching enzyme gene in the sea urchin embryo. *EMBO J.* **9**: 3003–3012.

1085. **Li, F., R. Strange, R. R. Friis, V. Djonov, H. J. Altermatt, S. Saurer, H. Niemann, and A. C. Andres.** 1994. Expression of stromelysin-1 and TIMP-1 in the involuting mammary gland and in early invasive tumors of the mouse. *Int. J. Cancer* **59**: 560–568.

1086. **Llano, E., A. M. Pendás, V. Knäuper, T. Sorsa, T. Salo, E. Salido, G. Murphy, J. P. Simmer, J. D. Bartlett, and C. López-Otín.** 1997. Identification and structural and functional characterization of human enamelysin (MMP-20). *Biochemistry* **36**: 15101–15108.

1087. **Lohi, J., K. Lehti, J. Westermarck, V. M. Kahari, and J. Keski-Oja.** 1996. Regulation of membrane-type matrix metalloproteinase-1 expression by growth factors and phorbol 12-myristate 13-acetate. *Eur. J. Biochem.* **239**: 239–247.

1088. **Madlener, H. and S. Werner.** 1997. cDNA cloning and expression of the gene encoding murine stromelysin-2 (MMP-10). *Gene* **202**: 75–81.

1089. **Marti, H. P., L. McNeil, M. Davies, J. Martin, and D. H. Lovett.** 1993. Homology cloning of rat 72 kDa type IV collagenase: cytokine and second-messenger inducibility in glomerular mesangial cells. *Biochem. J.* **291**: 441–446.

1090. **Marti, H. P., L. McNeil, G. Thomas, M. Davies, and D. H. Lovett.** 1992. Molecular characterization of a low-molecular-mass matrix metalloproteinase secreted by glomerular mesangial cells as PUMP-1. *Biochem. J.* **285**: 899–905.

1091. **Masure, S., G. Nys, P. Fiten, J. Van Damme, and G. Opdenakker.** 1993. Mouse gelatinase B. cDNA cloning, regulation of expression and glycosylation in WEHI-3 macrophages and gene organisation. *Eur. J. Biochem.* **218**: 129–141.

1092. **Matrisian, L. M., N. Glaichenhaus, M.-C. Gesnel, and R. Breathnach.** 1985. Epidermal growth factor and oncogenes induce transcription of the same cellular mRNA in rat fibroblasts. *EMBO J.* **4**: 1435–1440.

1093. **Matsumoto, S., M. Katoh, T. Watanabe, and Y. Masuho.** 1996. Molecular cloning of rabbit matrix metalloproteinase-2 and its broad expression at several tissues. *Biochim. Biophys. Acta* **1307**: 137–139.

1094. **Matsumoto, S. I., M. Katoh, S. Saito, T. Watanabe, and Y. Masuho.** 1997. Identification of soluble type of membrane-type matrix metalloproteinase-3 formed by alternatively spliced mRNA. *Biochim. Biophys. Acta - Gene Struct. Express.* **1354**: 159–170.

1095. **McGeehan, G., W. Burkhart, R. Anderegg, J. D. Becherer, J. W. Gillikin, and J. S. Graham.** 1992. Sequencing and characterization of the soybean leaf metalloproteinase. Structural and functional similarity to the matrix metalloproteinase family. *Plant Physiol.* **99**: 1179–1183.

1096. **Miyazaki, K., K. Uchiyama, Y. Imokawa, and K. Yoshizato.** 1996. Cloning and characterization of cDNAs for matrix metalloproteinases of regenerating newt limbs. *Proc. Natl. Acad. Sci. U.S.A* **93**: 6819–6824.

1097. **Moncrief, J. S., R. Obiso, Jr., L. A. Barroso, J. J. Kling, R. L. Wright, R. L. van Tassell, D. M. Lyerly, and T. D. Wilkins.** 1995. The enterotoxin of *Bacteroides fragilis* is a metalloprotease. *Infect. Immun.* **63**: 175–181.

1098. **Muller, D., B. Quantin, M. C. Gesnel, R. Millon-Collard, J. Abecassis, and R. Breathnach.** 1988. The collagenase gene family in humans consists of at least four members. *Biochem. J.* **253**: 187–192.

1099. **Nomura, K., T. Shimizu, H. Kinoh, Y. Sendai, M. Inomata, and N. Suzuki.** 1997. Sea urchin hatching enzyme (envelysin) cDNA cloning and deprivation of protein substrate specificity by autolytic degradation. *Biochemistry* **36**: 7225–7238.

1100. **Okada, A., J. P. Bellocq, N. Rouyer, M. P. Chenard, M. C. Rio, P. Chambon, and P. Basset.** 1995. Membrane-type matrix metalloproteinase (MT-MMP) gene is expressed in stromal cells of human colon, breast, and head and neck carcinomas. *Proc. Natl. Acad. Sci. U.S.A* **92**: 2730–2734.

1101. **Okada, A., S. Saez, Y. Misumi, and P. Basset.** 1997. Rat stromelysin 3: cDNA cloning from healing skin wound, activation by furin and expression in rat tissues. *Gene* **185**: 187–193.

1102. **Okada, A., M. Santavicca, and P. Basset.** 1995. The cDNA cloning and expression of the gene encoding rat gelatinase B. *Gene* **164**: 317–321.

1103. **Okada, A., C. Tomasetto, Y. Lutz, J. P. Bellocq, M. C. Rio, and P. Basset.** 1997. Expression of matrix metalloproteinases during rat skin wound healing: evidence that membrane type-1 matrix metalloproteinase is a stromal activator of pro-gelatinase A. *J. Cell Biol.* **137**: 67–77.

1104. **Oofusa, K. and K. Yoshizato.** 1996. Thyroid hormone-dependent expression of bullfrog tadpole collagenase gene. *Roux's Arch. Dev. Biol.* **205**: 243–251.

1105. **Ota, K., W. G. Stetler-Stevenson, Q. W. Yang, A. Kumar, J. Wada, N. Kashihara, E. I. Wallner, and Y. S. Kanwar.** 1998. Cloning of murine membrane-type-1-matrix metalloproteinase (MT-1-MMP) and its metanephric developmental regulation with respect to MMP-2 and its inhibitor. *Kidney Int.* **54**: 131–142.

1106. **Pak, J. H., C. Y. Liu, J. Huangpu, and J. S. Graham.** 1997. Construction and characterization of the soybean leaf metalloproteinase cDNA. *FEBS Lett.* **404**: 283–288.

1107. **Patterton, D., W. P. Hayes, and Y. B. Shi.** 1995. Transcriptional activation of the matrix metalloproteinase gene stromelysin-3 coincides with thyroid hormone-induced cell death during frog metamorphosis. *Dev. Biol.* **167**: 252–262.

1108. **Pendás, A. M., V. Knäuper, X. S. Puente, E. Llano, M. G. Mattei, S. Apte, G. Murphy, and C. López-Otín.** 1997. Identification and characterization of a novel human matrix metalloproteinase with unique structural characteristics, chromosomal location, and tissue distribution. *J. Biol. Chem.* **272**: 4281–4286.

1109. **Pourmotabbed, T., T. L. Solomon, K. A. Hasty, and C. L. Mainardi.** 1994. Characteristics of 92 kDa type IV collagenase/gelatinase produced by granulocytic leukemia cells: structure, expression of cDNA in *E. coli* and enzymic properties. *Biochim. Biophys. Acta* **1204**: 97–107.

1110. **Puente, X. S., A. M. Pendás, E. Llano, G. Velasco, and C. López-Otín.** 1996. Molecular cloning of a novel membrane-type matrix metalloproteinase from a human breast carcinoma. *Cancer Res.* **56**: 944–949.

1111. **Quinn, C. O., D. K. Scott, C. E. Brinckerhoff, L. M.**

Matrisian, J. J. Jeffrey, and N. C. Partridge. 1990. Rat collagenase. Cloning, amino acid sequence comparison, and parathyroid hormone regulation in osteoblastic cells. *J. Biol. Chem.* **265**: 22342–22347.

1112. **Rawlings, N. D. and A. J. Barrett.** 1998. Introduction: family M10 of interstitial collagenase (clan MB). In *Handbook of Proteolytic Enzymes.* A. J. Barrett, N. D. Rawlings, and J. F. Woessner (eds),pp. 1144–1147. Academic Press, London.

1113. **Reponen, P., C. Sahlberg, P. Huhtala, T. Hurskainen, I. Thesleff, and K. Tryggvason.** 1992. Molecular cloning of murine 72-kDa type IV collagenase and its expression during mouse development. *J. Biol. Chem.* **267**: 7856–7862.

1114. **Reponen, P., C. Sahlberg, C. Munaut, I. Thesleff, and K. Tryggvason.** 1994. High expression of 92-kD type IV collagenase (gelatinase B) in the osteoclast lineage during mouse development. *J. Cell Biol.* **124**: 1091–1102.

1115. **Richards, C. D., J. A. Rafferty, J. J. Reynolds, and J. Saklatvala.** 1991. Porcine collagenase from synovial fibroblasts: cDNA sequence and modulation of expression of RNA in vitro by various cytokines. *Matrix* **11**: 161–167.

1116. **Sato, H., T. Takino, Y. Okada, J. Cao, A. Shinagawa, E. Yamamoto, and M. Seiki.** 1994. A matrix metalloproteinase expressed on the surface of invasive tumour cells. *Nature* **370**: 61–65.

1117. **Sato, T., M. del Carmen Ovejero, P. Hou, A. M. Heegaard, M. Kumegawa, N. T. Foged, and J. M. Delaissé.** 1997. Identification of the membrane-type matrix metalloproteinase MT1-MMP in osteoclasts. *J. Cell Sci.* **110**: 589–596.

1118. **Saus, J., S. Quinones, Y. Otani, H. Nagase, E. D. Harris, Jr., and M. Kurkinen.** 1988. The complete primary structure of human matrix metalloproteinase-3. Identity with stromelysin. *J. Biol. Chem.* **263**: 6742–6745.

1119. **Scalzo, C. M., H. G. Verhage, and R. C. Jaffe.** 1994. Expression and estrogen control of PUMP-1 mRNA in the cat uterus. *Endocr. J. -U. K.* **2**: 229–235.

1120. **Sedlacek, R., S. Mauch, B. Kolb, C. Schätzlein, H. Eibel, H. H. Peter, J. Schmitt, and U. Krawinkel.** 1998. Matrix metalloproteinase MMP-19 (RASI 1) is expressed on the surface of activated peripheral blood mononuclear cells and is detected as an autoantigen in rheumatoid arthritis. *Immunobiology* **198**: 408–423.

1121. **Shapiro, S. D., G. L. Griffin, D. J. Gilbert, N. A. Jenkins, N. G. Copeland, H. G. Welgus, R. M. Senior, and T. J. Ley.** 1992. Molecular cloning, chromosomal localization, and bacterial expression of a murine macrophage metalloelastase. *J. Biol. Chem.* **267**: 4664–4671.

1122. **Shapiro, S. D., D. K. Kobayashi, and T. J. Ley.** 1993. Cloning and characterization of a unique elastolytic metalloproteinase produced by human alveolar macrophages. *J. Biol. Chem.* **268**: 23824–23829.

1123. **Shofuda, K., H. Yasumitsu, A. Nishihashi, K. Miki, and K. Miyazaki.** 1997. Expression of three membrane-type matrix metalloproteinases (MT-MMPs) in rat vascular smooth muscle cells and characterization of MT3-MMPs with and without transmembrane domain. *J. Biol. Chem.* **272**: 9749–9754.

1124. **Stolow, M. A., D. D. Bauzon, J. Li, T. Sedgwick, V. C. Liang, Q. A. Sang, and Y. B. Shi.** 1996. Identification and characterization of a novel collagenase in *Xenopus laevis*: possible roles during frog development. *Mol. Biol. Cell.* **7**: 1471–1483.

1125. **Takino, T., H. Sato, A. Shinagawa, and M. Seiki.** 1995. Identification of the second membrane-type matrix metalloproteinase (MT-MMP-2) gene from a human placenta cDNA library. MT-MMPs form a unique membrane-type subclass in the MMP family. *J. Biol. Chem.* **270**: 23013–23020.

1126. **Takino, T., H. Sato, E. Yamamoto, and M. Seiki.** 1995. Cloning of a human gene potentially encoding a novel matrix metalloproteinase having a C-terminal transmembrane domain. *Gene* **155**: 293–298.

1127. **Tamura, M., H. Shimokawa, and S. Sasaki.** 1994. Primary structure of bovine interstitial collagenase deduced from cDNA sequence. *DNA Seq.* **5**: 63–66.

1128. **Tanaka, H., K. Hojo, H. Yoshida, T. Yoshioka, and K. Sugita.** 1993. Molecular cloning and expression of the mouse 105-kDa gelatinase cDNA. *Biochem. Biophys. Res. Commun.* **190**: 732–740.

1129. **Tanaka, M., H. Sato, T. Takino, K. Iwata, M. Inoue, and M. Seiki.** 1997. Isolation of a mouse MT2-MMP gene from a lung cDNA library and identification of its product. *FEBS Lett.* **402**: 219–222.

1130. **Templeton, N. S., P. D. Brown, A. T. Levy, I. M. Margulies, L. A. Liotta, and W. G. Stetler-Stevenson.** 1990. Cloning and characterization of human tumor cell interstitial collagenase. *Cancer Res.* **50**: 5431–5437.

1131. **Tezuka, K., K. Nemoto, Y. Tezuka, T. Sato, Y. Ikeda, M. Kobori, H. Kawashima, H. Eguchi, Y. Hakeda, and M. Kumegawa.** 1994. Identification of matrix metalloproteinase 9 in rabbit osteoclasts. *J. Biol. Chem.* **269**: 15006–15009.

1132. **Tyagi, S. C. and J. P. Cleutjens.** 1996. Myocardial collagenase: purification and structural characterization. *Can. J. Cardiol.* **12**: 165–171.

1133. **Umenishi, F., H. Yasumitsu, Y. Ashida, J. Yamauti, M. Umeda, and K. Miyazaki.** 1990. Purification and properties of extracellular matrix-degrading metalloproteinase overproduced by Rous sarcoma virus-transformed rat liver cell line, and its identification as transin. *J. Biochem. (Tokyo)* **108**: 537–543.

1134. **Vincenti, M. P., C. I. Coon, J. A. Mengshol, S. Yocum, P. Mitchell, and C. E. Brinckerhoff.** 1998.

Cloning of the gene for interstitial collagenase-3 (matrix metalloproteinase-13) from rabbit synovial fibroblasts — differential expression with collagenase-1 (matrix metalloproteinase-1). *Biochem. J.* **331**: 341–346.

1135. **Vincenti, M. P., C. I. Coon, L. A. White, A. Barchowsky, and C. E. Brinckerhoff.** 1996. *src*-Related tyrosine kinases regulate transcriptional activation of the interstitial collagenase gene, MMP-1, in interleukin-1-stimulated synovial fibroblasts. *Arthritis Rheum.* **39**: 574–582.

1136. **Wada, K., H. Sato, H. Kinoh, M. Kajita, H. Yamamoto, and M. Seiki.** 1998. Cloning of three *Caenorhabditis elegans* genes potentially encoding novel matrix metalloproteinases. *Gene* **211**: 57–62.

1137. **Wernicke, D., C. Seyfert, B. Hinzmann, and E. Gromnica-Ihle.** 1996. Cloning of collagenase 3 from the synovial membrane and its expression in rheumatoid arthritis and osteoarthritis. *J. Rheumatol.* **23**: 590–595.

1138. **Whitham, S. E., G. Murphy, P. Angel, H. J. Rahmsdorf, B. J. Smith, A. Lyons, T. J. Harris, J. J. Reynolds, P. Herrlich, and A. J. P. Docherty.** 1986. Comparison of human stromelysin and collagenase by cloning and sequence analysis. *Biochem. J.* **240**: 913–916.

1139. **Wilhelm, S. M., I. E. Collier, A. Kronberger, A. Z. Eisen, B. L. Marmer, G. A. Grant, E. A. Bauer, and G. I. Goldberg.** 1987. Human skin fibroblast stromelysin: structure, glycosylation, substrate specificity, and differential expression in normal and tumorigenic cells. *Proc. Natl. Acad. Sci. USA* **84**: 6725–6729.

1140. **Wilhelm, S. M., I. E. Collier, B. L. Marmer, A. Z. Eisen, G. A. Grant, and G. I. Goldberg.** 1989. SV40-transformed human lung fibroblasts secrete a 92-kDa type IV collagenase which is identical to that secreted by normal human macrophages. *J. Biol. Chem.* **264**: 17213–17221.

1141. **Will, H. and B. Hinzmann.** 1995. cDNA sequence and mRNA tissue distribution of a novel human matrix metalloproteinase with a potential transmembrane segment. *Eur. J. Biochem.* **231**: 602–608.

1142. **Willmroth, F., H. H. Peter, and W. Conca.** 1998. A matrix metalloproteinase gene expressed in human T lymphocytes is identical with collagenase 3 from breast carcinomas. *Immunobiology* **198**: 375–384.

1143. **Wilson, C. L., K. J. Heppner, L. A. Rudolph, and L. M. Matrisian.** 1995. The metalloproteinase matrilysin is preferentially expressed by epithelial cells in a tissue-restricted pattern in the mouse. *Mol. Biol. Cell.* **6**: 851–869.

1144. **Xia, Y., G. Garcia, S. Chen, C. B. Wilson, and L. Feng.** 1996. Cloning of rat 92-kDa type IV collagenase and expression of an active recombinant catalytic domain. *FEBS Lett.* **382**: 285–288.

1145. **Yang, M., K. Hayashi, M. Hayashi, J. T. Fujii, and M. Kurkinen.** 1996. Cloning and developmental expression of a membrane-type matrix metalloproteinase from chicken. *J. Biol. Chem.* **271**: 25548–25554.

1146. **Yang, M. Z. and M. Kurkinen.** 1998. Cloning and characterization of a novel matrix metalloproteinase (MMP), CMMP, from chicken embryo fibroblasts—CMMP, *Xenopus* XMMP, and human MMP-19 have a conserved unique cysteine in the catalytic domain. *J. Biol. Chem.* **273**: 17893–17900.

1147. **Yang, M. Z., M. T. Murray, and M. Kurkinen.** 1997. A novel matrix metalloproteinase gene (XMMP) encoding vitronectin-like motifs is transiently expressed in *Xenopus laevis* early embryo development. *J. Biol. Chem.* **272**: 13527–13533.

1148. **Ye, Q. Z., L. L. Johnson, D. J. Hupe, and V. Baragi.** 1992. Purification and characterization of the human stromelysin catalytic domain expressed in Escherichia coli. *Biochemistry* **31**: 11231–1123.

See also review articles: 18, 70, 82, 103, 104, 111.

Sequences of TIMPs

1149. **Aimes, R. T., L. H. Li, B. Weaver, S. Hawkes, E. A. Hahn-Dantona, and J. P. Quigley.** 1998. Cloning, expression, and characterization of chicken tissue inhibitor of metalloproteinase-2 (TIMP-2) in normal and transformed chicken embryo fibroblasts. *J. Cell. Physiol.* **174**: 342–352.

1150. **Apte, S. S., K. Hayashi, M. F. Seldin, M. G. Mattei, M. Hayashi, and B. R. Olsen.** 1994. Gene encoding a novel murine tissue inhibitor of metalloproteinases (TIMP), TIMP-3, is expressed in developing mouse epithelia, cartilage, and muscle, and is located on mouse chromosome 10. *Dev. Dyn.* **200**: 177–197.

1151. **Apte, S. S., M. G. Mattei, and B. R. Olsen.** 1994. Cloning of the cDNA encoding human tissue inhibitor of metalloproteinases-3 (TIMP-3) and mapping of the TIMP3 gene to chromosome 22. *Genomics* **19**: 86–90.

1152. **Apte, S. S., B. R. Olsen, and G. Murphy.** 1995. The gene structure of tissue inhibitor of metalloproteinases (TIMP)-3 and its inhibitory activities define the distinct TIMP gene family. *J. Biol. Chem.* **270**: 14313–14318.

1153. **Banda, M. J., E. W. Howard, G. S. Herron, and G. Apodaca.** 1992. Secreted inhibitors of metalloproteinases (IMPs) that are distinct from TIMP. *Matrix Suppl.* **1**: 294–298.

1154. **Boone, T. C., M. J. Johnson, Y. A. De Clerck, and K. E. Langley.** 1990. cDNA cloning and expression of a metalloproteinase inhibitor related to tissue inhibitor of metalloproteinases. *Proc. Natl. Acad. Sci. USA* **87**: 2800–2804.

1155. **Boujrad, N., S. O. Ogwuegbu, M. Garnier, C. H. Lee, B. M. Martin, and V. Papadopoulos.** 1995.

Identification of a stimulator of steroid hormone synthesis isolated from testis. *Science* **268**: 1609–1612.

1156. **Byrne, J. A., C. Tomasetto, N. Rouyer, J. P. Bellocq, M. C. Rio, and P. Basset.** 1995. The tissue inhibitor of metalloproteinases-3 gene in breast carcinoma: identification of multiple polyadenylation sites and a stromal pattern of expression. *Mol. Med.* **1**: 418–427.

1157. **Carmichael, D. F., A. Sommer, R. C. Thompson, D. C. Anderson, C. G. Smith, H. G. Welgus, and G. P. Stricklin.** 1986. Primary structure and cDNA cloning of human fibroblast collagenase inhibitor. *Proc. Natl. Acad. Sci. USA* **83**: 2407–2411.

1158. **Chopra, R., P. A. Koklitis, S. Bergin, J. Rowe, and S. Angal.** 1991. Purification of recombinant dog tissue inhibitor of metalloproteinases. *Biochem. Soc. Trans.* **19**: 372S

1159. **Cook, T. F., J. S. Burke, K. D. Bergman, C. O. Quinn, J. J. Jeffrey, and N. C. Partridge.** 1994. Cloning and regulation of rat tissue inhibitor of metalloproteinases-2 in osteoblastic cells. *Arch. Biochem. Biophys.* **311**: 313–320.

1160. **DeClerck, Y. A., T. D. Yean, Y. Lee, J. M. Tomich, and K. E. Langley.** 1993. Characterization of the functional domain of tissue inhibitor of metalloproteinases-2 (TIMP-2). *Biochem. J.* **289**: 65–69.

1161. **Docherty, A. J. P., A. Lyons, B. J. Smith, E. M. Wright, P. E. Stephens, T. J. R. Harris, G. Murphy, and J. J. Reynolds.** 1985. Sequence of human tissue inhibitor of metalloproteinases and its identity to erythroid-potentiating activity. *Nature* **318**: 66–69.

1162. **Edwards, D. R., P. Waterhouse, M. I. Holman, and D. T. Denhardt.** 1986. A growth-responsive gene (16C8) in normal mouse fibroblasts homologous to a human collagenase inhibitor with erythroid-potentiating activity: evidence for inducible and constitutive transcripts. *Nucleic Acids Res.* **14**: 8863–8878.

1163. **Faucher, D., Y. Lelièvre, J. Boiziau, P. Cornet, and T. Cartwright.** 1989. [Natural inhibitor of metalloproteinases: structural and functional study]. [French]. *Pathol. Biol. (Paris)* **37**: 199–205.

1164. **Forough, R., S. T. Nikkari, D. Hasenstab, H. Lea, and A. W. Clowes.** 1995. Cloning and characterization of a cDNA encoding the baboon tissue inhibitor of matrix metalloproteinase-1 (TIMP-1). *Gene* **163**: 267–271.

1165. **Freudenstein, J., S. Wagner, R. M. Luck, R. Einspanier, and K. H. Scheit.** 1990. mRNA of bovine tissue inhibitor of metalloproteinase: sequence and expression in bovine ovarian tissue. *Biochem. Biophys. Res. Commun.* **171**: 250–256.

1166. **Gasson, J. C., D. W. Golde, S. E. Kaufman, C. A. Westbrook, R. M. Hewick, R. J. Kaufman, G. G. Wong, P. A. Temple, A. C. Leary, E. L. Brown,** *et al.* 1985. Molecular characterization and expression of the gene encoding human erythroid-potentiating activity. *Nature* **315**: 768–771.

1167. **Gewert, D. R., B. Coulombe, M. Castelino, D. Skup, and B. R. G. Williams.** 1987. Characterization and expression of a murine gene homologous to human EPA/TIMP: a virus-induced gene in the mouse. *EMBO J.* **6**: 651–657.

1168. **Goldberg, G. I., B. L. Marmer, G. A. Grant, A. Z. Eisen, S. Wilhelm, and C. S. He.** 1989. Human 72-kilodalton type IV collagenase forms a complex with a tissue inhibitor of metalloproteases designated TIMP-2. *Proc. Natl. Acad. Sci. USA* **86**: 8207–8211.

1169. **Greene, J. M. and Rosen, C. A.** 1996. Cloning and expression of cDNA for human tissue inhibitor of metalloproteinase-4 and clinical applications. Patent WO 96 18,725), 50 pp.

1170. **Grima, J., K. Calcagno, and C. Y. Cheng.** 1996. Purification, cDNA cloning, and developmental changes in the steady-state mRNA level of rat testicular tissue inhibitor of metalloproteases-2 (TIMP-2). *J. Androl.* **17**: 263–275.

1171. **Hammani, K., A. Blakis, D. Morsette, A. M. Bowcock, C. Schmutte, P. Henriet, and Y. A. DeClerck.** 1996. Structure and characterization of the human tissue inhibitor of metalloproteinases-2 gene. *J. Biol. Chem.* **271**: 25498–25505.

1172. **Hammani, K., P. Henriet, S. M. Silbiger, and Y. A. DeClerck.** 1996. Cloning and partial structure of the gene encoding human tissue inhibitor of metalloproteinases-3. *Gene* **170**: 287–288.

1173. **Hardcastle, A. J., D. L. Thiselton, M. Nayudu, R. M. Hampson, and S. S. Bhattacharya.** 1997. Genomic organization of the human TIMP-1 gene—investigation of a causative role in the pathogenesis of X-linked retinitis pigmentosa 2. *Invest. Ophthalmol. Vis. Sci.* **38**: 1893–1896.

1174. **Higuchi, T., H. Kanzaki, H. Nakayama, M. Fujimoto, H. Hatayama, K. Kojima, M. Iwai, T. Mori, and J. Fujita.** 1995. Induction of tissue inhibitor of metalloproteinase 3 gene expression during *in vitro* decidualization of human endometrial stromal cells. *Endocrinology* **136**: 4973–4981.

1175. **Horowitz, S., N. Dafni, D. L. Shapiro, B. A. Holm, R. H. Notter, and D. J. Quible.** 1989. Hyperoxic exposure alters gene expression in the lung. Induction of the tissue inhibitor of metalloproteinases mRNA and other mRNAs. *J. Biol. Chem.* **264**: 7092–7095.

1176. **Iredale, J. P., R. C. Benyon, M. J. Arthur, W. F. Ferris, R. Alcolado, P. J. Winwood, N. Clark, and G. Murphy.** 1996. Tissue inhibitor of metalloproteinase-1 messenger RNA expression is enhanced relative to interstitial collagenase messenger RNA in experimental liver injury and fibrosis. *Hepatology* **24**: 176–184.

1177. **Johnson, M. D., G. M. Housey, P. T. Kirschmeier,**

and I. B. Weinstein. 1987. Molecular cloning of gene sequences regulated by tumor promoters and mitogens through protein kinase C. *Mol. Cell. Biol.* **7**: 2821–2829.

1178. **Kaczorek, M., N. Honore, V. Ribes, P. Dehoux, P. Cornet, T. Cartwright, and R. E. Streeck.** 1987. Molecular cloning and synthesis of biological active human tissue inhibitor of metalloproteinases in yeast. *Bio/Technology* **5**: 595–598.

1179. **Kishi, J., K. Ogawa, S. Yamamoto, and T. Hayakawa.** 1991. Purification and characterization of a new tissue inhibitor of metalloproteinases (TIMP-2) from mouse colon 26 tumor cells. *Matrix* **11**: 10–16.

1180. **Leco, K. J., S. S. Apte, G. T. Taniguchi, S. P. Hawkes, R. Khokha, G. A. Schultz, and D. R. Edwards.** 1997. Murine tissue inhibitor of metalloproteinases-4 (TIMP-4): cDNA isolation and expression in adult mouse tissues. *FEBS Lett.* **401**: 213–217.

1181. **Leco, K. J., L. J. Hayden, R. R. Sharma, H. Rocheleau, A. H. Greenberg, and D. R. Edwards.** 1992. Differential regulation of TIMP-1 and TIMP-2 mRNA expression in normal and Ha-ras-transformed murine fibroblasts. *Gene* **117**: 209–217.

1182. **Leco, K. J., R. Khokha, N. Pavloff, S. P. Hawkes, and D. R. Edwards.** 1994. Tissue inhibitor of metalloproteinases-3 (TIMP-3) is an extracellular matrix-associated protein with a distinctive pattern of expression in mouse cells and tissues. *J. Biol. Chem.* **269**: 9352–9360.

1183. **Matsumoto, H., Y. Ishibashi, T. Ohtaki, Y. Hasegawa, C. Koyama, and K. Inoue.** 1993. Newly established murine pituitary folliculo-stellate-like cell line (TtT/GF) secretes potent pituitary glandular cell survival factors, one of which corresponds to metalloproteinase inhibitor. *Biochem. Biophys. Res. Commun.* **194**: 909–915.

1184. **Murphy, G., A. Houbrechts, M. I. Cockett, R. A. Williamson, M. O'Shea, and A. J. P. Docherty.** 1991. The N-terminal domain of tissue inhibitor of metalloproteinases retains metalloproteinase inhibitory activity. *Biochemistry* **30**: 8097–8102.

1185. **Okada, A., J. M. Garnier, S. Vicaire, and P. Basset.** 1994. Cloning of the cDNA encoding rat tissue inhibitor of metalloproteinase 1 (TIMP-1), amino acid comparison with other TIMPs, and gene expression in rat tissues. *Gene* **147**: 301–302.

1186. **Opbroek, A., M. C. Kenney, and D. Brown.** 1993. Characterization of a human corneal metalloproteinase inhibitor (TIMP-1). *Curr. Eye Res.* **12**: 877–883.

1187. **Pavloff, N., P. W. Staskus, N. S. Kishnani, and S. P. Hawkes.** 1992. A new inhibitor of metalloproteinases from chicken: ChIMP-3. A third member of the TIMP family. *J. Biol. Chem.* **267**: 17321–17326.

1188. **Rapp, G., J. Freudenstein, J. Klaudiny, J. Mucha, F. Wempe, M. Zimmer, and K. H. Scheit.** 1990. Characterization of three abundant mRNAs from human ovarian granulosa cells. *DNA Cell Biol.* **9**: 479–485.

1189. **Roswit, W. T., D. W. McCourt, N. C. Partridge, and J. J. Jeffrey.** 1992. Purification and sequence analysis of two rat tissue inhibitors of metalloproteinases. *Arch. Biochem. Biophys.* **292**: 402–410.

1190. **Santoro, M., C. Battaglia, L. Zhang, F. Carlomagno, M. L. Martelli, D. Salvatore, and A. Fusco.** 1994. Cloning of the rat tissue inhibitor of metalloproteinases type 2 (TIMP-2) gene: analysis of its expression in normal and transformed thyroid cells. *Exp. Cell Res.* **213**: 398–403.

1191. **Satoh, T., K. Kobayashi, S. Yamashita, M. Kikuchi, Y. Sendai, and H. Hoshi.** 1994. Tissue inhibitor of metalloproteinases (TIMP-1) produced by granulosa and oviduct cells enhances in vitro development of bovine embryo. *Biol. Reprod.* **50**: 835–844.

1192. **Shimizu, S., K. Malik, H. Sejima, J. Kishi, T. Hayakawa, and O. Koiwai.** 1992. Cloning and sequencing of the cDNA encoding a mouse tissue inhibitor of metalloproteinase-2. *Gene* **114**: 291–292.

1193. **Silbiger, S. M., V. L. Jacobsen, R. L. Cupples, and R. A. Koski.** 1994. Cloning of cDNAs encoding human TIMP-3, a novel member of the tissue inhibitor of metalloproteinase family. *Gene* **141**: 293–297.

1194. **Skup, D., J. D. Windass, F. Sor, H. George, B. R. Williams, H. Fukuhara, J. De Maeyer-Guignard, and E. De Maeyer.** 1982. Molecular cloning of partial cDNA copies of two distinct mouse IFN-β mRNAs. *Nucleic Acids Res.* **10**: 3069–3084.

1195. **Smith, G. W., T. L. Goetz, R. V. Anthony, and M. F. Smith.** 1994. Molecular cloning of an ovine ovarian tissue inhibitor of metalloproteinases: ontogeny of messenger ribonucleic acid expression and *in situ* localization within preovulatory follicles and luteal tissue. *Endocrinology* **134**: 344–352.

1196. **Staskus, P. W., F. R. Masiarz, L. J. Pallanck, and S. P. Hawkes.** 1991. The 21-kDa protein is a transformation-sensitive metalloproteinase inhibitor of chicken fibroblasts. *J. Biol. Chem.* **266**: 449–454.

1197. **Stetler-Stevenson, W. G., P. D. Brown, M. Onisto, A. T. Levy, and L. A. Liotta.** 1990. Tissue inhibitor of metalloproteinases-2 (TIMP-2) mRNA expression in tumor cell lines and human tumor tissues. *J. Biol. Chem.* **265**: 13933–13938.

1198. **Stetler-Stevenson, W. G., H. C. Krutzsch, and L. A. Liotta.** 1989. Tissue inhibitor of metalloproteinase (TIMP-2). A new member of the metalloproteinase inhibitor family. *J. Biol. Chem.* **264**: 17374–17378.

1199. **Stetler-Stevenson, W. G., H. C. Krutzsch, and L. A. Liotta.** 1992. TIMP-2: identification and characterization of a new member of the metalloproteinase inhibitor family. *Matrix Suppl.* **1**: 299–306.

1200. **Stöhr, H., K. Roomp, U. Felbor, and B. H. Weber.**

1995. Genomic organization of the human tissue inhibitor of metalloproteinases-3 (TIMP3). *Genome Res.* **5**: 483–487.

1201. **Stricklin, G. P. and H. G. Welgus.** 1983. Human skin fibroblast collagenase inhibitor. Purification and biochemical characterization. *J. Biol. Chem.* **258**: 12252–12258.

1202. **Su, S., F. Dehnade, and M. Zafarullah.** 1996. Regulation of tissue inhibitor of metalloproteinases-3 gene expression by transforming growth factor-β and dexamethasone in bovine and human articular chondrocytes. *DNA Cell Biol.* **15**: 1039–1048.

1203. **Sun, Y., G. Hegamyer, and N. H. Colburn.** 1994. Molecular cloning of five messenger RNAs differentially expressed in preneoplastic or neoplastic JB6 mouse epidermal cells: one is homologous to human tissue inhibitor of metalloproteinases-3. *Cancer Res.* **54**: 1139–1144.

1204. **Sun, Y., G. Hegamyer, H. Kim, K. Sithanandam, H. Li, R. Watts, and N. H. Colburn.** 1995. Molecular cloning of mouse tissue inhibitor of metalloproteinases-3 and its promoter. Specific lack of expression in neoplastic JB6 cells may reflect altered gene methylation. *J. Biol. Chem.* **270**: 19312–19319.

1205. **Tanaka, T., N. Andoh, T. Takeya, and E. Sato.** 1992. Differential screening of ovarian cDNA libraries detected the expression of the porcine collagenase inhibitor gene in functional corpora lutea. *Mol. Cell. Endocrinol.* **83**: 65–71.

1206. **Uria, J. A., A. A. Ferrando, G. Velasco, J. M. Freije, and C. López-Otín.** 1994. Structure and expression in breast tumors of human TIMP-3, a new member of the metalloproteinase inhibitor family. *Cancer Res.* **54**: 2091–2094.

1207. **Van Ranst, M., K. Norga, S. Masure, P. Proost, F. Vandekerckhove, J. Auwerx, J. Van Damme, and G. Opdenakker.** 1991. The cytokine-protease connection: identification of a 96-kD THP-1 gelatinase and regulation by interleukin-1 and cytokine inducers. *Cytokine* **3**: 231–239.

1208. **Vigne, J. L., L. L. Halburnt, and M. K. Skinner.** 1994. Characterization of bovine ovarian surface epithelium and stromal cells: identification of secreted proteins. *Biol. Reprod.* **51**: 1213–1221.

1209. **Weber, B. H. F., G. Vogt, R. C. Pruett, H. Stöhr, and U. Felbor.** 1994. Mutations in the tissue inhibitor of metalloproteinases-3 (TIMP3) in patients with Sorsby's fundus dystrophy. *Nature Genet.* **8**: 352–356.

1210. **Wertheimer, S. J. and S. L. Katz.** 1995. Molecular cloning and characterization of rabbit TIMP2. *Inflamm. Res.* **44**(Suppl. 2): S121-S122

1211. **Wick, M., C. Bürger, S. Brüsselbach, F. C. Lucibello, and R. Müller.** 1994. A novel member of human tissue inhibitor of metalloproteinases (TIMP)

1212. **Wilde, C. G., P. R. Hawkins, R. T. Coleman, W. B. Levine, A. M. Delegeane, P. M. Okamoto, L. Y. Ito, R. W. Scott, and J. J. Seilhamer.** 1994. Cloning and characterization of human tissue inhibitor of metalloproteinases-3. *DNA Cell Biol.* **13**: 711–718.

1213. **Willenbrock, F. and G. Murphy.** 1994. Structure-function relationships in the tissue inhibitors of metalloproteinases. *Am. J. Respir. Crit. Care Med.* **150**: S165-S170

1214. **Williamson, R. A., F. A. O. Marston, S. Angal, P. Koklitis, M. Panico, H. R. Morris, A. F. Carne, B. J. Smith, T. J. R. Harris, and R. B. Freedman.** 1990. Disulphide bond assignment in human tissue inhibitor of metalloproteinases (TIMP). *Biochem. J.* **268**: 267–274.

1215. **Wilson, R., R. Ainscough, K. Anderson, C. Baynes, M. Berks, J. Bonfield, J. Burton, M. Connell, T. Copsey, J. Cooper,** *et al.* 1994. 2.2 Mb of contiguous nucleotide sequence from chromosome III of *C. elegans*. *Nature* **368**: 32–38.

1216. **Wu, I. and M. A. Moses.** 1996. Cloning and expression of the cDNA encoding rat tissue inhibitor of metalloproteinase 3 (TIMP-3). *Gene* **168**: 243–246.

1217. **Wu, I. M. and M. A. Moses.** 1998. Molecular cloning and expression analysis of the cDNA encoding rat tissue inhibitor of metalloproteinase-4. *Matrix Biol.* **16**: 339–342.

1218. **Yang, M. Z. and M. Kurkinen.** 1998. Cloning and developmental regulation of tissue inhibitor of metalloproteinases-3 (TIMP3) in *Xenopus laevis* early embryos. *Gene* **211**: 95–100.

See also review article: 38.

Specificity of MMPs

1219. **Abramson, S. R., G. E. Conner, H. Nagase, I. Neuhaus, and J. F. Woessner, Jr.** 1995. Characterization of rat uterine matrilysin and its cDNA. Relationship to human pump-1 and activation of procollagenases. *J. Biol. Chem.* **270**: 16016–16022.

1220. **Aimes, R. T. and J. P. Quigley.** 1995. Matrix metalloproteinase-2 is an interstitial collagenase. Inhibitor-free enzyme catalyzes the cleavage of collagen fibrils and soluble native type I collagen generating the specific 3/4-and 1/4-length fragments. *J. Biol. Chem.* **270**: 5872–5876.

1221. **Andrade-Gordon, P., M. J. Cox, P. J. Brynes, and C. W. Wu.** 1989. Immobilized fluorogenic substrates for proteinases. A method for visualization and quantitation of elastase release from human monocytes. *Cell. Mol. Biol.* **35**: 431–440.

1222. **Arbelaez, L. F., U. Bergmann, A. Tuuttila, V. P.**

Shanbhag, and T. Stigbrand. 1997. Interaction of matrix metalloproteinases-2 and -9 with pregnancy zone protein and α_2-macroglobulin. *Arch. Biochem. Biophys.* **347**: 62–68.

1223. **Azzo, W. and J. F. Woessner, Jr.** 1986. Purification and characterization of an acid metalloproteinase from human articular cartilage. *J. Biol. Chem.* **261**: 5434-5441.

1224. **Banda, M. J., E. J. Clark, S. Sinha, and J. Travis.** 1987. Interaction of mouse macrophage elastase with native and oxidized human α_1-proteinase inhibitor. *J. Clin. Invest.* **79**: 1314-1317.

1225. **Beekman, B., J. W. Drijfhout, W. Bloemhoff, H. K. Ronday, P. P. Tak, and J. M. te Koppele.** 1996. Convenient fluorometric assay for matrix metalloproteinase activity and its application in biological media. *FEBS Lett.* **390**: 221-225.

1226. **Bickett, D. M., M. D. Green, J. Berman, M. Dezube, A. S. Howe, P. J. Brown, J. T. Roth, and G. M. McGeehan.** 1993. A high throughput fluorogenic substrate for interstitial collagenase (MMP-1) and gelatinase (MMP-9). *Anal. Biochem.* **212**: 58-64.

1227. **Bini, A., Y. Itoh, B. J. Kudryk, and H. Nagase.** 1996. Degradation of cross-linked fibrin by matrix metalloproteinase 3 (stromelysin 1): hydrolysis of the γ Gly 404-Ala 405 peptide bond. *Biochemistry* **35**: 13056-13063.

1228. **Brownell, J., W. Earley, E. Kunec, B. A. Morgan, B. Olyslager, R. C. Wahl, and D. R. Houck.** 1994. Comparison of native matrix metalloproteinases and their recombinant catalytic domains using a novel radiometric assay. *Arch. Biochem. Biophys.* **314**: 120-125.

1229. **Butler, G. S., H. Will, S. J. Atkinson, and G. Murphy.** 1997. Membrane-type-2 matrix metalloproteinase can initiate the processing of progelatinase A and is regulated by the tissue inhibitors of metalloproteinases. *Eur. J. Biochem.* **244**: 653-657.

1230. **Cha, J., M. V. Pedersen, and D. S. Auld.** 1996. Metal and pH dependence of heptapeptide catalysis by human matrilysin. *Biochemistry* **35**: 15831-15838.

1231. **Clark, I. M. and T. E. Cawston.** 1989. Fragments of human fibroblast collagenase. Purification and characterization. *Biochem. J.* **263**: 201-206.

1232. **Crabbe, T., J. P. O'Connell, B. J. Smith, and A. J. P. Docherty.** 1994. Reciprocated matrix metalloproteinase activation: a process performed by interstitial collagenase and progelatinase A. *Biochemistry* **33**: 14419-14425.

1233. **Crabbe, T., B. Smith, J. O'Connell, and A. Docherty.** 1994. Human progelatinase A can be activated by matrilysin. *FEBS Lett.* **345**: 14-16.

1234. **Crabbe, T., F. Willenbrock, D. Eaton, P. Hynds, A. F. Carne, G. Murphy, and A. J. P. Docherty.** 1992. Biochemical characterization of matrilysin. Activation conforms to the stepwise mechanisms proposed for other matrix metalloproteinases. *Biochemistry* **31**: 8500-8507.

1235. **Desrochers, P. E., J. J. Jeffrey, and S. J. Weiss.** 1991. Interstitial collagenase (matrix metalloproteinase-1) expresses serpinase activity. *J. Clin. Invest.* **87**: 2258-2265.

1236. **Desrochers, P. E., K. Mookhtiar, H. E. Van Wart, K. A. Hasty, and S. J. Weiss.** 1992. Proteolytic inactivation of α_1-proteinase inhibitor and α_1-antichymotrypsin by oxidatively activated human neutrophil metalloproteinases. *J. Biol. Chem.* **267**: 5005–5012.

1237. **Diekmann, O. and H. Tschesche.** 1994. Degradation of kinins, angiotensins and substance P by polymorphonuclear matrix metalloproteinases MMP 8 and MMP 9. *Braz. J. Med. Biol. Res.* **27**: 1865-1876.

1238. **Dixit, S. N., C. L. Mainardi, J. M. Seyer, and A. H. Kang.** 1979. Covalent structure of collagen: amino acid sequence of α2-CB5 of chick skin collagen containing the animal collagenase cleavage site. *Biochemistry* **18**: 5416-5422.

1239. **Enghild, J. J., G. Salvesen, K. Brew, and H. Nagase.** 1989. Interaction of human rheumatoid synovial collagenase (matrix metalloproteinase 1) and stromelysin (matrix metalloproteinase 3) with human α_2-macroglobulin and chicken ovostatin. Binding kinetics and identification of matrix metalloproteinase cleavage sites. *J. Biol. Chem.* **264**: 8779–8785.

1240. **Fields, G. B., S. J. Netzel-Arnett, L. J. Windsor, J. A. Engler, H. Birkedal-Hansen, and H. E. Van Wart.** 1990. Proteolytic activities of human fibroblast collagenase: hydrolysis of a broad range of substrates at a single active site. *Biochemistry* **29**: 6670-6677.

1241. **Fields, G. B., H. E. Van Wart, and H. Birkedal-Hansen.** 1987. Sequence specificity of human skin fibroblast collagenase. Evidence for the role of collagen structure in determining the collagenase cleavage site. *J. Biol. Chem.* **262**: 6221–6226.

1242. **Fosang, A. J., K. Last, V. Knäuper, P. J. Neame, G. Murphy, T. E. Hardingham, H. Tschesche, and J. A. Hamilton.** 1993. Fibroblast and neutrophil collagenases cleave at two sites in the cartilage aggrecan interglobular domain. *Biochem. J.* **295**: 273–276.

1243. **Fosang, A. J., K. Last, P. J. Neame, G. Murphy, V. Knäuper, H. Tschesche, C. E. Hughes, B. Caterson, and T. E. Hardingham.** 1994. Neutrophil collagenase (MMP-8) cleaves at the aggrecanase site E373-A374 in the interglobular domain of cartilage aggrecan. *Biochem. J.* **304**: 347–351.

1244. **Fosang, A. J., P. J. Neame, K. Last, T. E. Hardingham, G. Murphy, and J. A. Hamilton.** 1992. The interglobular domain of cartilage aggrecan is cleaved by PUMP, gelatinases, and cathepsin B. *J. Biol. Chem.* **267**: 19470–19474.

1245. **Fowlkes, J. L., J. J. Enghild, K. Suzuki, and H. Nagase.** 1994. Matrix metalloproteinases degrade insulin-like growth factor-binding protein-3 in dermal fibroblast cultures. *J. Biol. Chem.* **269**: 25742-25746.

1246. Galloway, W. A., G. Murphy, J. D. Sandy, J. Gavrilovic, T. E. Cawston, and J. J. Reynolds. 1983. Purification and characterization of a rabbit bone metalloproteinase that degrades proteoglycan and other connective-tissue components. *Biochem. J.* **209**: 741–752.

1247. Gearing, A. J., P. Beckett, M. Christodoulou, M. Churchill, J. Clements, A. H. Davidson, A. H. Drummond, W. A. Galloway, R. Gilbert, J. L. Gordon, *et al.* 1994. Processing of tumour necrosis factor-α precursor by metalloproteinases. *Nature* **370**: 555–557.

1248. Giannelli, G., J. Falk-Marzillier, O. Schiraldi, W. G. Stetler-Stevenson, and V. Quaranta. 1997. Induction of cell migration by matrix metalloprotease-2 cleavage of laminin-5. *Science* **277**: 225–228.

1249. Gronski, T. J., Jr., R. L. Martin, D. K. Kobayashi, B. C. Walsh, M. C. Holman, M. Huber, H. E. Van Wart, and S. D. Shapiro. 1997. Hydrolysis of a broad spectrum of extracellular matrix proteins by human macrophage elastase. *J. Biol. Chem.* **272**: 12189–12194.

1250. Gross, J., E. Harper, E. D. Harris, Jr., P. A. McCroskery, J. H. Highberger, C. Corbett, and A. H. Kang. 1974. Animal collagenases: specificity of action, and structures of the substrate cleavage site. *Biochem. Biophys. Res. Commun.* **61**: 605–612.

1251. Harrison, R. K., B. Chang, L. Niedzwiecki, and R. L. Stein. 1992. Mechanistic studies on the human matrix metalloproteinase stromelysin. *Biochemistry* **31**: 10757–10762.

1252. Imai, K., A. Hiramatsu, D. Fukushima, M. D. Pierschbacher, and Y. Okada. 1997. Degradation of decorin by matrix metalloproteinases: identification of the cleavage sites, kinetic analyses and transforming growth factor-β1 release. *Biochem. J.* **322**: 809–814.

1253. Imai, K., Y. Yokohama, I. Nakanishi, E. Ohuchi, Y. Fujii, N. Nakai, and Y. Okada. 1995. Matrix metalloproteinase 7 (matrilysin) from human rectal carcinoma cells. Activation of the precursor, interaction with other matrix metalloproteinases and enzymic properties. *J. Biol. Chem.* **270**: 6691–6697.

1254. Imper, V. and H. E. Van Wart. 1998. Substrate specificity and mechanisms of substrate recognition of the matrix metalloproteinases. In *Matrix Metalloproteinases.* W. C. Parks and R. P. Mecham (eds), pp. 219–242. Academic Press, San Diego.

1255. Ito, A., A. Mukaiyama, Y. Itoh, H. Nagase, I. B. Thogersen, J. J. Enghild, Y. Sasaguri, and Y. Mori. 1996. Degradation of interleukin 1β by matrix metalloproteinases. *J. Biol. Chem.* **271**: 14657–14660.

1256. Kato, K. H., K. Takemoto, E. Kato, K. Miyazaki, H. Kobayashi, and S. Ikegami. 1998. Inhibition of sea urchin fertilization by jaspisin, a specific inhibitor of matrix metalloendoproteinase. *Devel. Growth Different.* **40**: 221–230.

1257. Kettner, C., E. Shaw, R. White, and A. Janoff. 1981. The specificity of macrophage elastase on the insulin B-chain. *Biochem. J.* **195**: 369–372.

1258. Kinoshita, T., H. Sato, T. Takino, M. Itoh, T. Akizawa, and M. Seiki. 1996. Processing of a precursor of 72-kilodalton type IV collagenase/gelatinase A by a recombinant membrane-type 1 matrix metalloproteinase. *Cancer Res.* **56**: 2535–2538.

1259. Knäuper, V., S. Cowell, B. Smith, C. López-Otín, M. O'Shea, H. Morris, L. Zardi, and G. Murphy. 1997. The role of the C-terminal domain of human collagenase-3 (MMP-13) in the activation of procollagenase-3, substrate specificity, and tissue inhibitor of metalloproteinase interaction. *J. Biol. Chem.* **272**: 7608–7616.

1260. Knäuper, V., C. López-Otín, B. Smith, G. Knight, and G. Murphy. 1996. Biochemical characterization of human collagenase-3. *J. Biol. Chem.* **271**: 1544–1550.

1261. Knäuper, V., G. Murphy, and H. Tschesche. 1996. Activation of human neutrophil procollagenase by stromelysin 2. *Eur. J. Biochem.* **235**: 187–191.

1262. Knäuper, V., A. Osthues, Y. A. DeClerck, K. E. Langley, J. Bläser, and H. Tschesche. 1993. Fragmentation of human polymorphonuclear-leucocyte collagenase. *Biochem. J.* **291**: 847–854.

1263. Knäuper, V., H. Reinke, and H. Tschesche. 1990. Inactivation of human plasma α1-proteinase inhibitor by human PMN leucocyte collagenase. *FEBS Lett.* **263**: 355–357.

1264. Knäuper, V., B. Smith, C. López-Otín, and G. Murphy. 1997. Activation of progelatinase B (proMMP-9) by active collagenase-3 (MMP-13). *Eur. J. Biochem.* **248**: 369–373.

1265. Knäuper, V., S. Triebel, H. Reinke, and H. Tschesche. 1991. Inactivation of human plasma C1-inhibitor by human PMN leucocyte matrix metalloproteinases. *FEBS Lett.* **290**: 99–102.

1266. Knäuper, V., S. M. Wilhelm, P. K. Seperack, Y. A. DeClerck, K. E. Langley, A. Osthues, and H. Tschesche. 1993. Direct activation of human neutrophil procollagenase by recombinant stromelysin. *Biochem. J.* **295**: 581–586.

1267. Knäuper, V., H. Will, C. López-Otín, B. Smith, S. J. Atkinson, H. Stanton, R. M. Hembry, and G. Murphy. 1996. Cellular mechanisms for human procollagenase-3 (MMP-13) activation. Evidence that MT1-MMP (MMP-14) and gelatinase A (MMP-2) are able to generate active enzyme. *J. Biol. Chem.* **271**: 17124–17131.

1268. Knight, C. G. 1995. Fluorimetric assays of proteolytic enzymes. *Methods Enzymol.* **248**: 18–34.

1269. Knight, C. G., F. Willenbrock, and G. Murphy. 1992. A novel coumarin-labelled peptide for sensitive continuous assays of the matrix metalloproteinases. *FEBS Lett.* **296**: 263–266.

1270. Kolkenbrock, H., H. M. Ali, A. Hecker-Kia, G.

Buchlow, H. Sörensen, R. W. Hauer, and N. Ulbrich. 1991. Characterization of a gelatinase from human rheumatoid synovial fluid cells. *Eur. J. Clin. Chem. Clin. Biochem.* **29**: 499–505.

1271. Lemaître, V., A. Jungbluth, and Y. Eeckhout. 1997. The recombinant catalytic domain of mouse collagenase-3 depolymerizes type I collagen by cleaving its aminotelopeptides. *Biochem. Biophys. Res. Commun.* **230**: 202–205.

1272. Levi, E., R. Fridman, H. Q. Miao, Y. S. Ma, A. Yayon, and I. Vlodavsky. 1996. Matrix metalloproteinase 2 releases active soluble ectodomain of fibroblast growth factor receptor 1. *Proc. Natl. Acad. Sci. USA* **93**: 7069–7074.

1273. Liko, Z., J. Botyanszki, J. Bodi, E. Vass, Z. Majer, M. Hollosi, and H. Suli-Vargha. 1996. Effect of Ca^{2+} on the secondary structure of linear and cyclic collagen sequence analogs. *Biochem. Biophys. Res. Commun.* **227**: 351–359.

1274. Llano, E., A. M. Pendás, V. Knäuper, T. Sorsa, T. Salo, E. Salido, G. Murphy, J. P. Simmer, J. D. Bartlett, and C. López-Otín. 1997. Identification and structural and functional characterization of human enamelysin (MMP-20). *Biochemistry* **36**: 15101–15108.

1275. Mallya, S. K., K. A. Mookhtiar, Y. Gao, K. Brew, M. Dioszegi, H. Birkedal-Hansen, and H. E. Van Wart. 1990. Characterization of 58-kilodalton human neutrophil collagenase: comparison with human fibroblast collagenase. *Biochemistry* **29**: 10628–10634.

1276. Mañes, S., E. Mira, M. M. Barbacid, A. Ciprés, P. Fernández-Resa, J. M. Buesa, I. Mérida, M. Aracil, G. Márquez, and C. Martínez. 1997. Identification of insulin-like growth factor-binding protein-1 as a potential physiological substrate for human stromelysin-3. *J. Biol. Chem.* **272**: 25706–25712.

1277. Marcotte, P. A., I. M. Kozan, S. A. Dorwin, and J. M. Ryan. 1992. The matrix metalloproteinase pump-1 catalyzes formation of low molecular weight (pro)urokinase in cultures of normal human kidney cells. *J. Biol. Chem.* **267**: 13803–13806.

1278. Mast, A. E., J. J. Enghild, H. Nagase, K. Suzuki, S. V. Pizzo, and G. Salvesen. 1991. Kinetics and physiologic relevance of the inactivation of α_1-proteinase inhibitor, α_1-antichymotrypsin, and antithrombin III by matrix metalloproteinases-1 (tissue collagenase), -2 (72-kDa gelatinase/type IV collagenase), and -3 (stromelysin). *J. Biol. Chem.* **266**: 15810–15816.

1279. Masui, Y., T. Takemoto, S. Sakakibara, H. Hori, and Y. Nagai. 1977. Synthetic substrates for vertebrate collagenase. *Biochem. Med.* **17**: 215–221.

1280. Mayer, U., K. Mann, R. Timpl, and G. Murphy. 1993. Sites of nidogen cleavage by proteases involved in tissue homeostasis and remodelling. *Eur. J. Biochem.* **217**: 877–884.

1281. McGeehan, G. M., D. M. Bickett, M. Green, D. Kassel, J. S. Wiseman, and J. Berman. 1994. Characterization of the peptide substrate specificities of interstitial collagenase and 92-kDa gelatinase. Implications for substrate optimization. *J. Biol. Chem.* **269**: 32814–32820.

1282. Mecham, R. P., T. J. Broekelmann, C. J. Fliszar, S. D. Shapiro, H. G. Welgus, and R. M. Senior. 1997. Elastin degradation by matrix metalloproteinases—cleavage site specificity and mechanisms of elastolysis. *J. Biol. Chem.* **272**: 18071–18076.

1283. Miller, E. J., E. D. Harris, Jr., E. Chung, J. E. Finch, Jr., P. A. McCroskery, and W. T. Butler. 1976. Cleavage of Type II and III collagens with mammalian collagenase: site of cleavage and primary structure at the NH_2-terminal portion of the smaller fragment released from both collagens. *Biochemistry* **15**: 787–792.

1284. Mitchell, P. G., H. A. Magna, L. M. Reeves, L. L. Lopresti-Morrow, S. A. Yocum, P. J. Rosner, K. F. Geoghegan, and J. E. Hambor. 1996. Cloning, expression, and type II collagenolytic activity of matrix metalloproteinase-13 from human osteoarthritic cartilage. *J. Clin. Invest.* **97**: 761–768.

1285. Miyazaki, K., M. Hasegawa, K. Funahashi, and M. Umeda. 1993. A metalloproteinase inhibitor domain in Alzheimer amyloid protein precursor. *Nature* **362**: 839–841.

1286. Moncrief, J. S., R. Obiso, Jr., L. A. Barroso, J. J. Kling, R. L. Wright, R. L. van Tassell, D. M. Lyerly, and T. D. Wilkins. 1995. The enterotoxin of *Bacteroides fragilis* is a metalloprotease. *Infect. Immun.* **63**: 175–181.

1287. Mucha, A., P. Cuniasse, R. Kannan, F. Beau, A. Yiotakis, P. Basset, and V. Dive. 1998. Membrane type-1 matrix metalloprotease and stromelysin-3 cleave more efficiently synthetic substrates containing unusual amino acids in their P_1' positions. *J. Biol. Chem.* **273**: 2763–2768.

1288. Nagase, H., J. J. Enghild, K. Suzuki, and G. Salvesen. 1990. Stepwise activation mechanisms of the precursor of matrix metalloproteinase 3 (stromelysin) by proteinases and (4-aminophenyl)mercuric acetate. *Biochemistry* **29**: 5783–5789.

1289. Nagase, H., C. G. Fields, and G. B. Fields. 1994. Design and characterization of a fluorogenic substrate selectively hydrolyzed by stromelysin 1 (matrix metalloproteinase-3). *J. Biol. Chem.* **269**: 20952–20957.

1290. Nakagawa, H. and H. Debuchi. 1992. Inactivation of substance P by granulation tissue-derived gelatinase. *Biochem. Pharmacol.* **44**: 1773–1777.

1291. Netzel-Arnett, S., G. B. Fields, H. Birkedal-Hansen, and H. E. Van Wart. 1991. Sequence specificities of human fibroblast and neutrophil collagenase. *J. Biol. Chem.* **266**: 6747–6755.

1292. Netzel-Arnett, S., S. K. Mallya, H. Nagase, H.

Birkedal-Hansen, and H. E. Van Wart. 1991. Continuously recording fluorescent assays optimized for five human matrix metalloproteinases. *Anal. Biochem.* **195**: 86–92.

1293. Netzel-Arnett, S., Q. X. Sang, W. G. Moore, M. Navre, H. Birkedal-Hansen, and H. E. Van Wart. 1993. Comparative sequence specificities of human 72- and 92-kDa gelatinases (type IV collagenases) and PUMP (matrilysin). *Biochemistry* **32**: 6427–6432.

1294. Nguyen, Q., G. Murphy, C. E. Hughes, J. S. Mort, and P. J. Roughley. 1993. Matrix metalloproteinases cleave at two distinct sites on human cartilage link protein. *Biochem. J.* **295**: 595–598.

1295. Niedzwiecki, L., J. Teahan, R. K. Harrison, and R. L. Stein. 1992. Substrate specificity of the human matrix metalloproteinase stromelysin and the development of continuous fluorometric assays. *Biochemistry* **31**: 12618–12623.

1296. Niyibizi, C., R. Chan, J. J. Wu, and D. Eyre. 1994. A 92 kDa gelatinase (MMP-9) cleavage site in native type V collagen. *Biochem. Biophys. Res. Commun.* **202**: 328–333.

1297. Nomura, K., H. Tanaka, Y. Kikkawa, M. Yamaguchi, and N. Suzuki. 1991. The specificity of sea urchin hatching enzyme (envelysin) places it in the mammalian matrix metalloproteinase family. *Biochemistry* **30**: 6115–6123.

1298. Ochieng, J., R. Fridman, P. Nangia-Makker, D. E. Kleiner, L. A. Liotta, W. G. Stetler-Stevenson, and A. Raz. 1994. Galectin-3 is a novel substrate for human matrix metalloproteinases-2 and -9. *Biochemistry* **33**: 14109–14114.

1299. Ogata, Y., J. J. Enghild, and H. Nagase. 1992. Matrix metalloproteinase 3 (stromelysin) activates the precursor for the human matrix metalloproteinase 9. *J. Biol. Chem.* **267**: 3581–3584.

1300. Patterson, B. C. and Q. X. A. Sang. 1997. Angiostatin-converting enzyme activities of human matrilysin (MMP-7) and gelatinase B type IV collagenase (MMP-9). *J. Biol. Chem.* **272**: 28823–28825.

1301. Pei, D., G. Majmudar, and S. J. Weiss. 1994. Hydrolytic inactivation of a breast carcinoma cell-derived serpin by human stromelysin-3. *J. Biol. Chem.* **269**: 25849–25855.

1302. Pendás, A. M., V. Knäuper, X. S. Puente, E. Llano, M. G. Mattei, S. Apte, G. Murphy, and C. López-Otín. 1997. Identification and characterization of a novel human matrix metalloproteinase with unique structural characteristics, chromosomal location, and tissue distribution. *J. Biol. Chem.* **272**: 4281–4286.

1303. Proost, P., J. Van Damme, and G. Opdenakker. 1993. Leukocyte gelatinase B cleavage releases encephalitogens from human myelin basic protein. *Biochem. Biophys. Res. Commun.* **192**: 1175–1181.

1304. Rantala-Ryhänen, S., L. Ryhänen, F. V. Nowak, and J. Uitto. 1983. Proteinases in human polymorphonuclear leukocytes. Purification and characterization of an enzyme which cleaves denatured collagen and a synthetic peptide with a Gly-Ile sequence. *Eur. J. Biochem.* **134**: 129–137.

1305. Roher, A. E., T. C. Kasunic, A. S. Woods, R. J. Cotter, M. J. Ball, and R. Fridman. 1994. Proteolysis of Aβ peptide from Alzheimer disease brain by gelatinase A. *Biochem. Biophys. Res. Commun.* **205**: 1755–1761.

1306. Sasaki, T., W. Göhring, K. Mann, P. Maurer, E. Hohenester, V. Knäuper, G. Murphy, and R. Timpl. 1997. Limited cleavage of extracellular matrix protein BM-40 by matrix metalloproteinases increases its affinity for collagens. *J. Biol. Chem.* **272**: 9237–9243.

1307. Sasaki, T., K. Mann, G. Murphy, M. L. Chu, and R. Timpl. 1996. Different susceptibilities of fibulin-1 and fibulin-2 to cleavage by matrix metalloproteinases and other tissue proteases. *Eur. J. Biochem.* **240**: 427–434.

1308. Seltzer, J. L., S. A. Adams, G. A. Grant, and A. Z. Eisen. 1981. Purification and properties of a gelatin-specific neutral protease from human skin. *J. Biol. Chem.* **256**: 4662–4668.

1309. Seltzer, J. L., K. T. Akers, H. Weingarten, G. A. Grant, D. W. McCourt, and A. Z. Eisen. 1990. Cleavage specificity of human skin type IV collagenase (gelatinase). Identification of cleavage sites in type I gelatin, with confirmation using synthetic peptides. *J. Biol. Chem.* **265**: 20409–20413.

1310. Seltzer, J. L., H. Weingarten, K. T. Akers, M. L. Eschbach, G. A. Grant, and A. Z. Eisen. 1989. Cleavage specificity of type IV collagenase (gelatinase) from human skin. Use of synthetic peptides as model substrates. *J. Biol. Chem.* **264**: 19583–19586.

1311. Sires, U. I., G. L. Griffin, T. J. Broekelmann, R. P. Mecham, G. Murphy, A. E. Chung, H. G. Welgus, and R. M. Senior. 1993. Degradation of entactin by matrix metalloproteinases. Susceptibility to matrilysin and identification of cleavage sites. *J. Biol. Chem.* **268**: 2069–2074.

1312. Smith, M. M., L. Shi, and M. Navre. 1995. Rapid identification of highly active and selective substrates for stromelysin and matrilysin using bacteriophage peptide display libraries. *J. Biol. Chem.* **270**: 6440–6449.

1313. Sottrup-Jensen, L. and H. Birkedal-Hansen. 1989. Human fibroblast collagenase-α-macroglobulin interactions. Localization of cleavage sites in the bait regions of five mammalian α-macroglobulins. *J. Biol. Chem.* **264**: 393–401.

1314. Sottrup-Jensen, L. and H. Birkedal-Hansen. 1992. Localization of cleavage sites for human fibroblast collagenase in the bait region of five mammalian α-macroglobulins. *Matrix Suppl.* **1**: 263–268.

1315. Stack, M. S. and R. D. Gray. 1989. Comparison of vertebrate collagenase and gelatinase using a new fluorogenic substrate peptide. *J. Biol. Chem.* **264**: 4277–4281.

1316. **Stack, M. S. and R. D. Gray.** 1990. The effect of pH, temperature, and D_2O on the activity of porcine synovial collagenase and gelatinase. *Arch. Biochem. Biophys.* **281**: 257–263.

1317. **Stein, R. L. and M. Izquierdo-Martin.** 1994. Thioester hydrolysis by matrix metalloproteinases. *Arch. Biochem. Biophys.* **308**: 274–277.

1318. **Suzuki, K., J. J. Enghild, T. Morodomi, G. Salvesen, and H. Nagase.** 1990. Mechanisms of activation of tissue procollagenase by matrix metalloproteinase 3 (stromelysin). *Biochemistry* **29**: 10261–10270.

1319. **Teahan, J., R. Harrison, M. Izquierdo, and R. L. Stein.** 1989. Substrate specificity of human fibroblast stromelysin. Hydrolysis of substance P and its analogues. *Biochemistry* **28**: 8497–8501.

1320. **Timár, F., J. Botyánszki, H. Süli-Vargha, I. Babó, J. Oláh, G. Pogány, and A. Jeney.** 1998. The antiproliferative action of a melphalan hexapeptide with collagenase-cleavable site. *Cancer Chemother. Pharmacol.* **41**: 292–298.

1321. **Triebel, S., J. Bläser, H. Reinke, V. Knäuper, and H. Tschesche.** 1992. Mercurial activation of human PMN leucocyte type IV procollagenase (gelatinase). *FEBS Lett.* **298**: 280–284.

1322. **Tschesche, H., J. Fedrowitz, U. Kohnert, J. Michaelis, and H. W. Macartney.** 1986. Matrix degrading proteinases from human granulocytes: type I, II, III collagenase, gelatinase and type IV, V-collagenase. A survey of recent findings and inhibition by gamma-anticollagenase. *Folia Histochem. Cytobiol.* **24**: 125–131.

1323. **Upadhye, S. and V. S. Ananthanarayanan.** 1995. Interaction of peptide substrates of fibroblast collagenase with divalent cations: Ca^{2+} binding by substrate as a suggested recognition signal for collagenase action. *Biochem. Biophys. Res. Commun.* **215**: 474–482.

1324. **Weingarten, H. and J. Feder.** 1985. Spectrophotometric assay for vertebrate collagenase. *Anal. Biochem.* **147**: 437–440.

1325. **Weingarten, H. and J. Feder.** 1986. Cleavage site specificity of vertebrate collagenases. *Biochem. Biophys. Res. Commun.* **139**: 1184–1187.

1326. **Weingarten, H., R. Martin, and J. Feder.** 1985. Synthetic substrates of vertebrate collagenase. *Biochemistry* **24**: 6730–6734.

1327. **Welch, A. R., C. M. Holman, M. Huber, M. C. Brenner, M. F. Browner, and H. E. Van Wart.** 1996. Understanding the P_1' specificity of the matrix metalloproteinases: effect of S_1' pocket mutations in matrilysin and stromelysin-1. *Biochemistry* **35**: 10103–10109.

1328. **Welgus, H. G., C. J. Fliszar, J. L. Seltzer, T. M. Schmid, and J. J. Jeffrey.** 1990. Differential susceptibility of type X collagen to cleavage by two mammalian interstitial collagenases and 72-kDa type IV collagenase. *J. Biol. Chem.* **265**: 13521–13527.

1329. **Welgus, H. G., G. A. Grant, J. C. Sacchettini, W. T. Roswit, and J. J. Jeffrey.** 1985. The gelatinolytic activity of rat uterus collagenase. *J. Biol. Chem.* **260**: 13601–13606.

1330. **Wilhelm, S. M., Z. H. Shao, T. J. Housley, P. K. Seperack, A. P. Baumann, Z. Gunja-Smith, and J. F. Woessner, Jr.** 1993. Matrix metalloproteinase-3 (stromelysin-1). Identification as the cartilage acid metalloprotease and effect of pH on catalytic properties and calcium affinity. *J. Biol. Chem.* **268**: 21906–21913.

1331. **Windsor, L. J., H. Grenett, B. Birkedal-Hansen, M. K. Bodden, J. A. Engler, and H. Birkedal-Hansen.** 1993. Cell type-specific regulation of SL-1 and SL-2 genes. Induction of the SL-2 gene but not the SL-1 gene by human keratinocytes in response to cytokines and phorbolesters. *J. Biol. Chem.* **268**: 17341–17347.

1332. **Woessner, J. F., Jr. and C. J. Taplin.** 1988. Purification and properties of a small latent matrix metalloproteinase of the rat uterus. *J. Biol. Chem.* **263**: 16918–16925.

1333. **Wu, J. J., M. W. Lark, L. E. Chun, and D. R. Eyre.** 1991. Sites of stromelysin cleavage in collagen types II, IX, X, and XI of cartilage. *J. Biol. Chem.* **266**: 5625–5628.

1334. **Xia, T., K. Akers, A. Z. Eisen, and J. L. Seltzer.** 1996. Comparison of cleavage site specificity of gelatinases A and B using collagenous peptides. *Biochim. Biophys. Acta* **1293**: 259–266.

1335. **Ye, Q. Z., L. L. Johnson, D. J. Hupe, and V. Baragi.** 1992. Purification and characterization of the human stromelysin catalytic domain expressed in *Escherichia coli*. *Biochemistry* **31**: 11231–11235.

1336. **Ye, Q. Z., L. L. Johnson, A. E. Yu, and D. Hupe.** 1995. Reconstructed 19 kDa catalytic domain of gelatinase A is an active proteinase. *Biochemistry* **34**: 4702–4708.

1337. **Zhang, Z., P. G. Winyard, K. Chidwick, G. Murphy, M. Wardell, R. W. Carrell, and D. R. Blake.** 1994. Proteolysis of human native and oxidised α_1-proteinase inhibitor by matrilysin and stromelysin. *Biochim. Biophys. Acta* **1199**: 224–228.

See also review article: 92.

Structure of MMPs

1338. **Allan, J. A., A. J. P. Docherty, P. J. Barker, N. S. Huskisson, J. J. Reynolds, and G. Murphy.** 1995. Binding of gelatinases A and B to type-I collagen and other matrix components. *Biochem. J.* **309**: 299–306.

1339. **Bányai, L. and L. Patthy.** 1991. Evidence for the involvement of type II domains in collagen binding by 72 kDa type IV procollagenase. *FEBS Lett.* **282**: 23–25.

1340. **Bányai, L., H. Tordai, and L. Patthy**. 1994. The gelatin-binding site of human 72 kDa type IV collagenase (gelatinase A). *Biochem. J.* **298**: 403–407.

1341. **Baumann, U.** 1994. Crystal structure of the 50 kDa metallo protease from *Serratia marcescens*. *J. Mol. Biol.* **242**: 244–251.

1342. **Baumann, U., S. Wu, K. M. Flaherty, and D. B. McKay**. 1993. Three-dimensional structure of the alkaline protease of *Pseudomonas aeruginosa*: A two-domain protein with a calcium binding parallel beta roll motif. *EMBO J.* **12**: 3357–3364.

1343. **Becker, J. W., A. I. Marcy, L. L. Rokosz, M. G. Axel, J. J. Burbaum, P. M. Fitzgerald, P. M. Cameron, C. K. Esser, W. K. Hagmann, J. D. Hermes, *et al.* 1995. Stromelysin-1: three-dimensional structure of the inhibited catalytic domain and of the C-truncated proenzyme. *Protein Sci.* **4**: 1966–1976.

1344. **Benbow, U., G. Buttice, H. Nagase, and M. Kurkinen**. 1996. Characterization of the 46-kDa intermediates of matrix metalloproteinase 3 (stromelysin 1) obtained by site-directed mutation of phenylalanine 83. *J. Biol. Chem.* **271**: 10715–10722.

1345. **Betz, M., P. Huxley, S. J. Davies, Y. Mushtaq, M. Pieper, H. Tschesche, W. Bode, and F. X. Gomis-Rüth**. 1997. 1.8-Å crystal structure of the catalytic domain of human neutrophil collagenase (matrix metalloproteinase-8) complexed with a peptidomimetic hydroxamate prime-side inhibitor with a distinct selectivity profile. *Eur. J. Biochem.* **247**: 356–363.

1346. **Bigg, H. F., I. M. Clark, and T. E. Cawston**. 1994. Fragments of human fibroblast collagenase: interaction with metalloproteinase inhibitors and substrates. *Biochim. Biophys. Acta* **1208**: 157–165.

1347. **Bode, W., F. X. Gomis-Rüth, R. Huber, R. Zwilling, and W. Stöcker**. 1992. Structure of astacin and implications for activation of astacins and zinc-ligation of collagenases. *Nature* **358**: 164–167.

1348. **Bode, W., P. Reinemer, R. Huber, T. Kleine, S. Schnierer, and H. Tschesche**. 1994. The X-ray crystal structure of the catalytic domain of human neutrophil collagenase inhibited by a substrate analogue reveals the essentials for catalysis and specificity. *EMBO J.* **13**: 1263–1269.

1349. **Borkakoti, N., F. K. Winkler, D. H. Williams, A. D'Arcy, M. J. Broadhurst, P. A. Brown, W. H. Johnson, and E. J. Murray**. 1994. Structure of the catalytic domain of human fibroblast collagenase complexed with an inhibitor. *Nature Struct. Biol.* **1**: 106–110.

1350. **Browner, M. F., W. W. Smith, and A. L. Castelhano**. 1995. Matrilysin-inhibitor complexes: common themes among metalloproteases. *Biochemistry* **34**: 6602–6610.

1351. **Cameron, P. M., A. I. Marcy, L. L. Rokosz, and J. D. Hermes**. 1995. Use of an active-site inhibitor of stromelysin to elucidate the mechanism of prostromelysin activation. *Bioorg. Chem.* **23**: 415–426.

1351a. **Carson, M.** 1997. Ribbons. *Methods Enzymol.* **277**: 493–505.

1352. **Clark, I. M. and T. E. Cawston**. 1989. Fragments of human fibroblast collagenase. Purification and characterization. *Biochem. J.* **263**: 201–206.

1353. **Collier, I. E., P. A. Krasnov, A. Y. Strongin, H. Birkedal-Hansen, and G. I. Goldberg**. 1992. Alanine scanning mutagenesis and functional analysis of the fibronectin-like collagen-binding domain from human 92-kDa type IV collagenase. *J. Biol. Chem.* **267**: 6776–6781.

1354. **Collier, I. E., S. M. Wilhelm, A. Z. Eisen, B. L. Marmer, G. A. Grant, J. L. Seltzer, A. Kronberger, C. S. He, E. A. Bauer, and G. I. Goldberg**. 1988. H-ras oncogene-transformed human bronchial epithelial cells (TBE-1) secrete a single metalloprotease capable of degrading basement membrane collagen. *J. Biol. Chem.* **263**: 6579–6587.

1355. **Constantine, K. L., M. Madrid, L. Bányai, M. Trexler, L. Patthy, and M. Llinas**. 1992. Refined solution structure and ligand-binding properties of PDC-109 domain b. A collagen-binding type II domain. *J. Mol. Biol.* **223**: 281–298.

1356. **Constantine, K. L., V. Ramesh, L. Bányai, M. Trexler, L. Patthy, and M. Llinas**. 1991. Sequence-specific ^{1}H NMR assignments and structural characterization of bovine seminal fluid protein PDC-109 domain b. *Biochemistry* **30**: 1663–1672.

1357. **Coulombe, R., P. Grochulski, J. Sivaraman, R. Menard, J. S. Mort, and M. Cygler**. 1996. Structure of human procathepsin L reveals the molecular basis of inhibition by the prosegment. *EMBO J.* **15**: 5492–5503.

1358. **Crabbe, T., S. Zucker, M. I. Cockett, F. Willenbrock, S. Tickle, J. P. O'Connell, J. M. Scothern, G. Murphy, and A. J. P. Docherty**. 1994. Mutation of the active site glutamic acid of human gelatinase A: effects on latency, catalysis, and the binding of tissue inhibitor of metalloproteinases-1. *Biochemistry* **33**: 6684–6690.

1359. **Cygler, M., J. Sivaraman, P. Grochulski, R. Coulombe, A. C. Storer, and J. S. Mort**. 1996. Structure of rat procathepsin B: Model for inhibition of cysteine protease activity by the proregion. *Structure* **4**: 405–416.

1360. **de Souza, S. J., H. M. Pereira, S. Jacchieri, and R. R. Brentani**. 1996. Collagen/collagenase interaction: does the enzyme mimic the conformation of its own substrate? *FASEB J.* **10**: 927–930.

1361. **Dhanaraj, V., Q. Z. Ye, L. L. Johnson, D. J. Hupe, D. F. Ortwine, J. B. Dunbar, Jr., J. R. Rubin, A. Pavlovsky, C. Humblet, and T. L. Blundell**. 1996. X-ray structure of a hydroxamate inhibitor complex of stromelysin catalytic domain and its comparison with members of the zinc metalloproteinase superfamily. *Structure* **4**: 375–386.

1362. **Esch, F. S., N. C. Ling, P. Bohlen, S. Y. Ying, and R. Guillemin.** 1983. Primary structure of PDC-109, a major protein constituent of bovine seminal plasma. *Biochem. Biophys. Res. Commun.* **113**: 861–867.

1363. **Esser, C. K., R. L. Bugianesi, C. G. Caldwell, K. T. Chapman, P. L. Durette, N. N. Girotra, I. E. Kopka, T. J. Lanza, D. A. Levorse, M. MacCoss,** *et al.* 1997. Inhibition of stromelysin-1 (MMP-3) by P_1'-biphenylylethyl carboxyalkyl dipeptides. *J. Med. Chem.* **40**: 1026–1040.

1363a. **Evans, S. V.** 1993. SETOR: hardware lighted three-dimensional solid model representations of macromolecules. *J. Mol. Graph.* **11**: 134–138.

1364. **Faber, H. R., C. R. Groom, H. M. Baker, W. T. Morgan, A. Smith, and E. N. Baker.** 1995. 1.8 Å crystal structure of the C-terminal domain of rabbit serum hemopexin. *Structure* **3**: 551–559.

1365. **Fernandez-Catalan, C., W. Bode, R. Huber, D. Turk, J. J. Calvete, A. Lichte, H. Tschesche, and K. Maskos.** 1998. Crystal structure of the complex formed by the membrane type 1-matrix metalloproteinase with the tissue inhibitor of metalloproteinases-2, the soluble progelatinase a receptor. *EMBO J.* **17**: 5238–5248.

1366. **Finzel, B. C., E. T. Baldwin, G. L. Bryant, G. F. Hess, J. W. Wilks, C. M. Trepod, J. E. Mott, V. P. Marshall, G. L. Petzold, R. A. Poorman,** *et al.* 1998. Structural characterizations of nonpeptidic thiadiazole inhibitors of matrix metalloproteinases reveal the basis for stromelysin selectivity. *Prot. Sci.* **7**: 2118–2126.

1367. **Gohlke, U., F. X. Gomis-Rüth, T. Crabbe, G. Murphy, A. J. P. Docherty, and W. Bode.** 1996. The C-terminal (haemopexin-like) domain structure of human gelatinase A (MMP2): structural implications for its function. *FEBS Lett.* **378**: 126–130.

1368. **Gomis-Rüth, F. X., U. Gohlke, M. Betz, V. Knäuper, G. Murphy, C. López-Otín, and W. Bode.** 1996. The helping hand of collagenase-3 (MMP-13): 2.7 Å crystal structure of its C-terminal haemopexin-like domain. *J. Mol. Biol.* **264**: 556–566.

1369. **Gomis-Rüth, F. X., L. F. Kress, and W. Bode.** 1993. First structure of a snake venom metalloproteinase: a prototype for matrix metalloproteinases/collagenases. *EMBO J.* **12**: 4151–4157.

1370. **Gomis-Rüth, F. X., L. F. Kress, J. Kellermann, I. Mayr, X. Lee, R. Huber, and W. Bode.** 1994. Refined 2.0 Å X-ray crystal structure of the snake venom zinc-endopeptidase adamalysin II. Primary and tertiary structure determination, refinement, molecular structure and comparison with astacin, collagenase and thermolysin. *J. Mol. Biol.* **239**: 513–544.

1371. **Gomis-Rüth, F. X., K. Maskos, M. Betz, A. Bergner, R. Huber, K. Suzuki, N. Yoshida, H. Nagase, K.**

Brew, G. P. Bourenkov, H. Bartunik, and W. Bode. 1997. Mechanism of inhibition of the human matrix metalloproteinase stromelysin-1 by TIMP-1. *Nature* **389**: 77–81.

1372. **Gooley, P. R., B. A. Johnson, A. I. Marcy, G. C. Cuca, S. P. Salowe, W. K. Hagmann, C. K. Esser, and J. P. Springer.** 1993. Secondary structure and zinc ligation of human recombinant short-form stromelysin by multidimensional heteronuclear NMR. *Biochemistry* **32**: 13098–13108.

1373. **Gooley, P. R., J. F. O'Connell, A. I. Marcy, G. C. Cuca, M. G. Axel, C. G. Caldwell, W. K. Hagmann, and J. W. Becker.** 1996. Comparison of the structure of human recombinant short form stromelysin by multidimensional heteronuclear NMR and X-ray crystallography. *J. Biomol. NMR* **7**: 8–28.

1374. **Gooley, P. R., J. F. O'Connell, A. I. Marcy, G. C. Cuca, S. P. Salowe, B. L. Bush, J. D. Hermes, C. K. Esser, W. K. Hagmann, J. P. Springer,** *et al.* 1994. The NMR structure of the inhibited catalytic domain of human stromelysin-1. *Nature Struct. Biol.* **1**: 111–118.

1375. **Grams, F., P. Reinemer, J. C. Powers, T. Kleine, M. Pieper, H. Tschesche, R. Huber, and W. Bode.** 1995. X-ray structures of human neutrophil collagenase complexed with peptide hydroxamate and peptide thiol inhibitors. Implications for substrate binding and rational drug design. *Eur. J. Biochem.* **228**: 830–841.

1376. **Hirose, T., C. Patterson, T. Pourmotabbed, C. L. Mainardi, and K. A. Hasty.** 1993. Structure-function relationship of human neutrophil collagenase: identification of regions responsible for substrate specificity and general proteinase activity. *Proc. Natl. Acad. Sci. U.S.A* **90**: 2569–2573.

1377. **Ishizaki, J., K. Hanasaki, K. Higashino, J. Kishino, N. Kikuchi, O. Ohara, and H. Arita.** 1994. Molecular cloning of pancreatic group I phospholipase A2 receptor. *J. Biol. Chem.* **269**: 5897–5904.

1378. **Knäuper, V., S. Cowell, B. Smith, C. López-Otín, M. O'Shea, H. Morris, L. Zardi, and G. Murphy.** 1997. The role of the C-terminal domain of human collagenase-3 (MMP-13) in the activation of procollagenase-3, substrate specificity, and tissue inhibitor of metalloproteinase interaction. *J. Biol. Chem.* **272**: 7608–7616.

1379. **Knäuper, V., A. J. P. Docherty, B. Smith, H. Tschesche, and G. Murphy.** 1997. Analysis of the contribution of the hinge region of human neutrophil collagenase (HNC, MMP-8) to stability and collagenolytic activity by alanine scanning mutagenesis. *FEBS Lett.* **405**: 60–64.

1380. **Knäuper, V., A. Osthues, Y. A. DeClerck, K. E. Langley, J. Bläser, and H. Tschesche.** 1993. Fragmentation of human polymorphonuclear-leucocyte collagenase. *Biochem. J.* **291**: 847–854.

1381. **Knäuper, V., S. M. Wilhelm, P. K. Seperack, Y. A. DeClerck, K. E. Langley, A. Osthues, and H. Tschesche.** 1993. Direct activation of human neutrophil procollagenase by recombinant stromelysin. *Biochem. J.* **295**: 581–586.

1382. **Li, J., P. Brick, M. C. O'Hare, T. Skarzynski, L. F. Lloyd, V. A. Curry, I. M. Clark, H. F. Bigg, B. L. Hazleman, T. E. Cawston,** *et al.* 1995. Structure of full-length porcine synovial collagenase reveals a C-terminal domain containing a calcium-linked, four-bladed beta-propeller. *Structure* **3**: 541–549.

1383. **Libson, A. M., A. G. Gittis, I. E. Collier, B. L. Marmer, G. I. Goldberg, and E. E. Lattman.** 1995. Crystal structure of the haemopexin-like C-terminal domain of gelatinase A. *Nature Struct. Biol.* **2**: 938–942

1384. **Lloyd, L. F., T. Skarzynski, A. J. Wonacott, T. E. Cawston, I. M. Clark, C. J. Mannix, and G. P. Harper.** 1989. Crystallization and preliminary X-ray analysis of porcine synovial collagenase. *J. Mol. Biol.* **210**: 237–238.

1385. **Lobel, P., N. M. Dahms, J. Breitmeyer, J. M. Chirgwin, and S. Kornfeld.** 1987. Cloning of the bovine 215-kDa cation-independent mannose 6-phosphate receptor. *Proc. Natl. Acad. Sci. USA* **84**: 2233–2237.

1386. **Lovejoy, B., A. Cleasby, A. M. Hassell, K. Longley, M. A. Luther, D. Weigl, G. McGeehan, A. B. McElroy, D. Drewry, M. H. Lambert,** *et al.* 1994. Structure of the catalytic domain of fibroblast collagenase complexed with an inhibitor. *Science* **263**: 375–377.

1387. **Lovejoy, B., A. M. Hassell, M. A. Luther, D. Weigl, and S. R. Jordan.** 1994. Crystal structures of recombinant 19-kDa human fibroblast collagenase complexed to itself. *Biochemistry* **33**: 8207–8217.

1388. **Massova, I., R. Fridman, and S. Mobashery.** 1997. Structural insights into the catalytic domains of human matrix metalloprotease-2 and human matrix metalloprotease-9: implications for substrate specificities. *J. Mol. Model.* **3**: 17–30.

1389. **Matthews, B. W.** 1988. Structural basis of the action of thermolysin and related zinc peptidases. *Acc. Chem. Res.* **21**: 333–340.

1390. **Matthews, B. W., J. N. Jansonius, P. M. Colman, B. P. Schoenborn, and D. Dupourque.** 1972. Three-dimensional structure of thermolysin. *Nature New Biol.* **238**: 37–41.

1391. **McCoy, M. A., M. J. Dellwo, D. M. Schneider, T. M. Banks, J. Falvo, K. J. Vavra, A. M. Mathiowetz, M. W. Qoronfleh, R. Ciccarelli, E. R. Cook,** *et al.* 1997. Assignments and structure determination of the catalytic domain of human fibroblast collagenase using 3D double and triple resonance NMR spectroscopy. *J. Biomol. NMR* **9**: 11–24.

1392. **McMullen, B. A. and K. Fujikawa.** 1985. Amino acid sequence of the heavy chain of human alpha-factor XIIa (activated Hageman factor). *J. Biol. Chem.* **260**: 5328–5341.

1393. **Miyazawa, K., T. Shimomura, A. Kitamura, J. Kondo, Y. Morimoto, and N. Kitamura.** 1993. Molecular cloning and sequence analysis of the cDNA for a human serine protease reponsible for activation of hepatocyte growth factor. Structural similarity of the protease precursor to blood coagulation factor XII. *J. Biol. Chem.* **268**: 10024–10028.

1394. **Morgan, D. O., J. C. Edman, D. N. Standring, V. A. Fried, M. C. Smith, R. A. Roth, and W. J. Rutter.** 1987. Insulin-like growth factor II receptor as a multifunctional binding protein. *Nature* **329**: 301–307.

1395. **Moy, F. J., P. K. Chanda, S. Cosmi, M. R. Pisano, C. Urbano, J. Wilhelm, and R. Powers.** 1998. High-resolution solution structure of the inhibitor-free catalytic fragment of human fibroblast collagenase determined by multidimensional NMR. *Biochemistry* **37**: 1495–1504.

1396. **Moy, F. J., M. R. Pisano, P. K. Chanda, C. Urbano, L. M. Killar, M. L. Sung, and R. Powers.** 1997. Assignments, secondary structure and dynamics of the inhibitor-free catalytic fragment of human fibroblast collagenase. *J. Biomol. NMR* **10**: 9–19.

1397. **Murphy, G., J. A. Allan, F. Willenbrock, M. I. Cockett, J. P. O'Connell, and A. J. P. Docherty.** 1992. The role of the C-terminal domain in collagenase and stromelysin specificity. *J. Biol. Chem.* **267**: 9612–9618.

1398. **Murphy, G., Q. Nguyen, M. I. Cockett, S. J. Atkinson, J. A. Allan, C. G. Knight, F. Willenbrock, and A. J. P. Docherty.** 1994. Assessment of the role of the fibronectin-like domain of gelatinase A by analysis of a deletion mutant. *J. Biol. Chem.* **269**: 6632–6636.

1399. **Muskett, F. W., T. A. Frenkiel, J. Feeney, R. B. Freedman, M. D. Carr, and R. A. Williamson.** 1998. High resolution structure of the N-terminal domain of tissue inhibitor of metalloproteinases-2 and characterization of its interaction site with matrix metalloproteinase-3. *J. Biol. Chem.* **273**: 21736–21743.

1400. **Nagase, H., J. J. Enghild, K. Suzuki, and G. Salvesen.** 1990. Stepwise activation mechanisms of the precursor of matrix metalloproteinase 3 (stromelysin) by proteinases and (4-aminophenyl)mercuric acetate. *Biochemistry* **29**: 5783–5789.

1401. **Nagase, H., C. G. Fields, and G. B. Fields.** 1994. Design and characterization of a fluorogenic substrate selectively hydrolyzed by stromelysin 1 (matrix metalloproteinase-3). *J. Biol. Chem.* **269**: 20952–20957.

1402. **Netzel-Arnett, S., Q. X. Sang, W. G. Moore, M. Navre, H. Birkedal-Hansen, and H. E. Van Wart.** 1993. Comparative sequence specificities of human 72- and 92-kDa gelatinases (type IV collagenases) and PUMP (matrilysin). *Biochemistry* **32**: 6427–6432.

1402a. **Nichols, A., K. A. Sharp, and B. Honing.** 1991. Protein folding and associaiton: insights from the interface and thermodynamic properties of hydrocarbons. *Proteins* **11**: 281–296.

1403. **Okada, Y., E. D. Harris, Jr., and H. Nagase.** 1988. The precursor of a metalloendopeptidase from human rheumatoid synovial fibroblasts. Purification and mechanisms of activation by endopeptidases and 4-aminophenylmercuric acetate. *Biochem. J.* **254**: 731–741.

1404. **Petersen, T. E., H. C. Thogersen, K. Skorstengaard, K. Vibe-Pedersen, P. Sahl, L. Sottrup-Jensen, and S. Magnusson.** 1983. Partial primary structure of bovine plasma fibronectin: three types of internal homology. *Proc. Natl. Acad. Sci. USA* **80**: 137–141.

1405. **Pickford, A. R., J. R. Potts, J. R. Bright, I. Phan, and I. D. Campbell.** 1997. Solution structure of a type 2 module from fibronectin: implications for the structure and function of the gelatin-binding domain. *Structure* **5**: 359–370.

1406. **Reinemer, P., F. Grams, R. Huber, T. Kleine, S. Schnierer, M. Piper, H. Tschesche, and W. Bode.** 1994. Structural implications for the role of the N terminus in the 'superactivation' of collagenases. A crystallographic study. *FEBS Lett.* **338**: 227–233.

1407. **Salowe, S. P., A. I. Marcy, G. C. Cuca, C. K. Smith, I. E. Kopka, W. K. Hagmann, and J. D. Hermes.** 1992. Characterization of zinc-binding sites in human stromelysin-1: stoichiometry of the catalytic domain and identification of a cysteine ligand in the proenzyme. *Biochemistry* **31**: 4535–4540.

1408. **Sanchez-Lopez, R., C. M. Alexander, O. Behrendtsen, R. Breathnach, and Z. Werb.** 1993. Role of zinc-binding- and hemopexin domain-encoded sequences in the substrate specificity of collagenase and stromelysin-2 as revealed by chimeric proteins. *J. Biol. Chem.* **268**: 7238–7247.

1409. **Seidah, N. G., P. Manjunath, J. Rochemont, M. R. Sairam, and M. Chretien.** 1987. Complete amino acid sequence of BSP-A3 from bovine seminal plasma. Homology to PDC-109 and to the collagen-binding domain of fibronectin. *Biochem. J.* **243**: 195–203.

1410. **Spurlino, J. C., A. M. Smallwood, D. D. Carlton, T. M. Banks, K. J. Vavra, J. S. Johnson, E. R. Cook, J. Falvo, R. C. Wahl, T. A. Pulvino,** *et al.* 1994. 1.56 Å structure of mature truncated human fibroblast collagenase. *Proteins* **19**: 98–109.

1411. **Stams, T., J. C. Spurlino, D. L. Smith, R. C. Wahl, T. F. Ho, M. W. Qoronfleh, T. M. Banks, and B. Rubin.** 1994. Structure of human neutrophil collagenase reveals large S1' specificity pocket. *Nature Struct. Biol.* **1**: 119–123.

1412. **Stöcker, W., F. Grams, U. Baumann, P. Reinemer, F. X. Gomis-Rüth, D. B. McKay, and W. Bode.** 1995. The metzincins–topological and sequential relations between the astacins, adamalysins, serralysins, and matrixins (collagenases) define a superfamily of zinc-peptidases. *Protein Sci.* **4**: 823–840.

1413. **Suzuki, K., J. J. Enghild, T. Morodomi, G. Salvesen, and H. Nagase.** 1990. Mechanisms of activation of tissue procollagenase by matrix metalloproteinase 3 (stromelysin). *Biochemistry* **29**: 10261–10270.

1414. **Taylor, M. E., J. T. Conary, M. R. Lennartz, P. D. Stahl, and K. Drickamer.** 1990. Primary structure of the mannose receptor contains multiple motifs resembling carbohydrate-recognition domains. *J. Biol. Chem.* **265**: 12156–12162.

1415. **Turk, D., M. Podobnik, R. Kuhelj, M. Dolinar, and V. Turk.** 1996. Crystal structures of human procathepsin B at 3.2 and 3.3 Å resolution reveal an interaction motif between a papain-like cysteine protease and its propeptide. *FEBS Lett.* **384**: 211–214.

1416. **Van Doren, S. R., A. V. Kurochkin, W. Hu, Q. Z. Ye, L. L. Johnson, D. J. Hupe, and E. R. Zuiderweg.** 1995. Solution structure of the catalytic domain of human stromelysin complexed with a hydrophobic inhibitor. *Protein Sci.* **4**: 2487–2498.

1417. **Van Doren, S. R., A. V. Kurochkin, Q. Z. Ye, L. L. Johnson, D. J. Hupe, and E. R. Zuiderweg.** 1993. Assignments for the main-chain nuclear magnetic resonances and delineation of the secondary structure of the catalytic domain of human stromelysin-1 as obtained from triple-resonance 3D NMR experiments. *Biochemistry* **32**: 13109–13122.

1418. **Welch, A. R., C. M. Holman, M. Huber, M. C. Brenner, M. F. Browner, and H. E. Van Wart.** 1996. Understanding the P_1' specificity of the matrix metalloproteinases: effect of S_1' pocket mutations in matrilysin and stromelysin-1. *Biochemistry* **35**: 10103–10109.

1419. **Wetmore, D. R. and K. D. Hardman.** 1996. Roles of the propeptide and metal ions in the folding and stability of the catalytic domain of stromelysin (matrix metalloproteinase 3). *Biochemistry* **35**: 6549–6558.

1420. **Wilhelm, S. M., I. E. Collier, B. L. Marmer, A. Z. Eisen, G. A. Grant, and G. I. Goldberg.** 1989. SV40-transformed human lung fibroblasts secrete a 92-kDa type IV collagenase which is identical to that secreted by normal human macrophages. *J. Biol. Chem.* **264**: 17213–17221.

1421. **Zhang, D., I. Botos, F. X. Gomis-Rüth, R. Doll, C. Blood, F. G. Njoroge, J. W. Fox, W. Bode, and E. F. Meyer.** 1994. Structural interaction of natural and synthetic inhibitors with the venom metalloproteinase, atrolysin C (form d). *Proc. Natl. Acad. Sci. USA* : 8447–8451.

See also review articles: 18, 19, 123.

Structure of the TIMPs

1422. **Arumugam, S., C. L. Hemme, N. Yoshida, K. Suzuki, H. Nagase, M. Bejanskii, B. Wu, and S. R. Van Doren.** 1998. TIMP-1 contact sites and perturbations of stromelysin 1 mapped by NMR and a paramagnetic surface probe. *Biochemistry* **37**: 9650–9657.

1423. **Bodden, M. K., G. J. Harber, B. Birkedal-Hansen, L. J. Windsor, N. C. Caterina, J. A. Engler, and H. Birkedal-Hansen.** 1994. Functional domains of human TIMP-1 (tissue inhibitor of metalloproteinases). *J. Biol. Chem.* **269**: 18943–18952.

1424. **Bodden, M. K., L. J. Windsor, N. C. Caterina, A. Yermovsky, B. Birkedal-Hansen, G. Galazka, J. A. Engler, and H. Birkedal-Hansen.** 1994. Analysis of the TIMP-1/FIB-CL complex. *Ann. N. Y. Acad. Sci.* **732**: 84–95.

1425. **Caterina, N. C. M., L. J. Windsor, A. E. Yermovsky, M. K. Bodden, K. B. Taylor, H. Birkedal-Hansen, and J. A. Engler.** 1997. Replacement of conserved cysteines in human tissue inhibitor of metalloproteinases-1. *J. Biol. Chem.* **272**: 32141–32149.

1426. **Docherty, A. J. P., A. Lyons, B. J. Smith, E. M. Wright, P. E. Stephens, T. J. R. Harris, G. Murphy, and J. J. Reynolds.** 1985. Sequence of human tissue inhibitor of metalloproteinases and its identity to erythroid-potentiating activity. *Nature* **318**: 66–69.

1427. **Fernandez-Catalan, C., W. Bode, R. Huber, D. Turk, J. J. Calvete, A. Lichte, H. Tschesche, and K. Maskos.** 1998. Crystal structure of the complex formed by the membrane type 1-matrix metalloproteinase with the tissue inhibitor of metalloproteinases-2, the soluble progelatinase a receptor. *EMBO J.* **17**: 5238–5248.

1428. **Gomis-Rüth, F. X., K. Maskos, M. Betz, A. Bergner, R. Huber, K. Suzuki, N. Yoshida, H. Nagase, K. Brew, G. P. Bourenkov, H. Bartunik, and W. Bode.** 1997. Mechanism of inhibition of the human matrix metalloproteinase stromelysin-1 by TIMP-1. *Nature* **389**: 77–81.

1429. **Hanglow, A. C., A. Lugo, R. Walsky, M. Visnick, J. W. Coffey, and N. Fotouhi.** 1994. Inhibition of human stromelysin by peptides based on the N-terminal domain of tissue inhibitor of metalloproteinases-1. *Biochem. Biophys. Res. Commun.* **205**: 1156–1163.

1430. **Huang, W., Q. Meng, K. Suzuki, H. Nagase, and K. Brew.** 1997. Mutational study of the amino-terminal domain of human tissue inhibitor of metalloproteinases I (TIMP-1) locates an inhibitory region for matrix metalloproteinases. *J. Biol. Chem.* **272**: 22086–22091.

1431. **Murzin, A. G.** 1993. OB(oligonucleotide/oligosaccharide binding)-fold: common structural and functional solution for non-homologous sequences. *EMBO J.* **12**: 861–867.

1432. **Muskett, F. W., T. A. Frenkiel, J. Feeney, R. B. Freedman, M. D. Carr, and R. A. Williamson.** 1998. High resolution structure of the N-terminal domain of tissue inhibitor of metalloproteinases-2 and characterization of its interaction site with matrix metalloproteinase-3. *J. Biol. Chem.* **273**: 21736–21743.

1433. **Nagase, H., K. Suzuki, T. E. Cawston, and K. Brew.** 1997. Involvement of a region near valine-69 of tissue inhibitor of metalloproteinases (TIMP)-1 in the interaction with matrix metalloproteinase 3 (stromelysin 1). *Biochem. J.* **325**: 163–167.

1434. **O'Shea, M., F. Willenbrock, R. A. Williamson, M. I. Cockett, R. B. Freedman, J. J. Reynolds, A. J. P. Docherty, and G. Murphy.** 1992. Site-directed mutations that alter the inhibitory activity of the tissue inhibitor of metalloproteinases-1: importance of the N-terminal region between cysteine 3 and cysteine 13. *Biochemistry* **31**: 10146–10152.

1435. **Tolley, S., G. Murphy, M. O'Shea, R. Ward, A. J. P. Docherty, M. Cockett, A. Rawas, and G. Davies.** 1993. Crystallization and preliminary X-ray analysis of a truncated tissue metalloproteinase inhibitor delta 128–194 TIMP-2. *J. Mol. Biol.* **229**: 1163–1164.

1436. **Tolley, S. P., G. J. Davies, M. O'Shea, M. I. Cockett, A. J. P. Docherty, and G. Murphy.** 1993. Crystallization and preliminary X-ray analysis of nonglycosylated tissue inhibitor of metalloproteinases-1, N30QN78Q TIMP-1. *Proteins* **17**: 435–437.

1437. **Williamson, R. A., M. D. Carr, T. A. Frenkiel, J. Feeney, and R. B. Freedman.** 1997. Mapping the binding site for matrix metalloproteinase on the N-terminal domain of the tissue inhibitor of metalloproteinases-2 by NMR chemical shift perturbation. *Biochemistry* **36**: 13882–13889.

1438. **Williamson, R. A., F. A. O. Marston, S. Angal, P. Koklitis, M. Panico, H. R. Morris, A. F. Carne, B. J. Smith, T. J. R. Harris, and R. B. Freedman.** 1990. Disulphide bond assignment in human tissue inhibitor of metalloproteinases (TIMP). *Biochem. J.* **268**: 267–274.

1439. **Williamson, R. A., G. Martorell, M. D. Carr, G. Murphy, A. J. P. Docherty, R. B. Freedman, and J. Feeney.** 1994. Solution structure of the active domain of tissue inhibitor of metalloproteinases-2. A new member of the OB fold protein family. *Biochemistry* **33**: 11745–11759.

1440. **Williamson, R. A., B. J. Smith, S. Angal, and R. B. Freedman.** 1993. Chemical modification of tissue inhibitor of metalloproteinases-1 and its inactivation by diethyl pyrocarbonate. *Biochim. Biophys. Acta* **1203**: 147–154.

Substrates of MMPs

1441. **Abramson, S. R., G. E. Conner, H. Nagase, I. Neuhaus, and J. F. Woessner, Jr.** 1995. Characterization of rat uterine matrilysin and its cDNA.

Relationship to human pump-1 and activation of procollagenases. *J. Biol. Chem.* **270**: 16016–16022.

1442. **Aimes, R. T. and J. P. Quigley.** 1995. Matrix metalloproteinase-2 is an interstitial collagenase. Inhibitor-free enzyme catalyzes the cleavage of collagen fibrils and soluble native type I collagen generating the specific 3/4- and 1/4-length fragments. *J. Biol. Chem.* **270**: 5872–5876.

1443. **Allan, J. A., A. J. P. Docherty, P. J. Barker, N. S. Huskisson, J. J. Reynolds, and G. Murphy.** 1995. Binding of gelatinases A and B to type-I collagen and other matrix components. *Biochem. J.* **309**: 299–306.

1444. **Allan, J. A., A. J. P. Docherty, and G. Murphy.** 1994. The binding of gelatinases A and B to type I collagen yields both high and low affinity sites. *Ann. N. Y. Acad. Sci.* **732**: 365–366.

1445. **Arner, E. C., C. P. Decicco, R. Cherney, and M. D. Tortorella.** 1997. Cleavage of native cartilage aggrecan by neutrophil collagenase (MMP-8) is distinct from endogenous cleavage by aggrecanase. *J. Biol. Chem.* **272**: 9294–9299.

1446. **Arthur, M. J. P., S. L. Friedman, F. J. Roll, and D. M. Bissell.** 1989. Lipocytes from normal rat liver release a neutral metalloproteinase that degrades basement membrane (type IV) collagen. *J. Clin. Invest.* **84**: 1076–1085.

1447. **Azzo, W. and J. F. Woessner, Jr.** 1986. Purification and characterization of an acid metalloproteinase from human articular cartilage. *J. Biol. Chem.* **261**: 5434–5441.

1448. **Backstrom, J. R., G. P. Lim, M. J. Cullen, and Z. A. Tökés.** 1996. Matrix metalloproteinase-9 (MMP-9) is synthesized in neurons of the human hippocampus and is capable of degrading the amyloid-β peptide (1–40). *J. Neurosci.* **16**: 7910–7919.

1449. **Backstrom, J. R. and Z. A. Tökés.** 1995. The 84-kDa form of human matrix metalloproteinase-9 degrades substance P and gelatin. *J. Neurochem.* **64**: 1312–1318.

1450. **Baici, A., P. Salgam, G. Cohen, K. Fehr, and A. Böni.** 1982. Action of collagenase and elastase from human polymorphonuclear leukocytes on human articular cartilage. *Rheumatol. Int.* **2**: 11–16.

1451. **Banda, M. J., E. J. Clark, S. Sinha, and J. Travis.** 1987. Interaction of mouse macrophage elastase with native and oxidized human α_1-proteinase inhibitor. *J. Clin. Invest.* **79**: 1314–1317.

1452. **Banda, M. J., E. J. Clark, and Z. Werb.** 1980. Limited proteolysis by macrophage elastase inactivates human α_1-proteinase inhibitor. *J. Exp. Med.* **152**: 1563–1570.

1453. **Banda, M. J., E. J. Clark, and Z. Werb.** 1983. Selective proteolysis of immunoglobulins by mouse macrophage elastase. *J. Exp. Med.* **157**: 1184–1196.

1454. **Banda, M. J., A. G. Rice, G. L. Griffin, and R. M. Senior.** 1988. α_1-Proteinase inhibitor is a neutrophil chemoattractant after proteolytic inactivation by macrophage elastase. *J. Biol. Chem.* **263**: 4481–4484.

1455. **Banda, M. J. and Z. Werb.** 1981. Mouse macrophage elastase. Purification and characterization as a metalloproteinase. *Biochem. J.* **193**: 589–605.

1456. **Bauer, E. A., J. J. Jeffrey, and A. Z. Eisen.** 1971. Preparation of three vertebrate collagenases in pure form. *Biochem. Biophys. Res. Commun.* **44**: 813–818.

1457. **Bejarano, P. A., M. E. Noelken, K. Suzuki, B. G. Hudson, and H. Nagase.** 1988. Degradation of basement membranes by human matrix metalloproteinase 3 (stromelysin). *Biochem. J.* **256**: 413–419.

1458. **Bicsak, T. A. and E. Harper.** 1984. Purification and characterization of tadpole back-skin collagenase with low gelatinase activity. *J. Biol. Chem.* **259**: 13145–13150.

1459. **Billinghurst, R. C., L. Dahlberg, M. Ionescu, A. Reiner, R. Bourne, C. Rorabeck, P. Mitchell, J. Hambor, O. Diekmann, H. Tschesche, *et al.* 1997. Enhanced cleavage of type II collagen by collagenases in osteoarthritic articular cartilage. *J. Clin. Invest.* **99**: 1534–1545.

1460. **Bini,** A.1997.Fibrin(ogen) degradation and clot lysis by fibrinolytic matrix metalloproteinase. PCT Int. Appl WO 96 36,277, 1–57

1461. **Bini, A., Y. Itoh, B. J. Kudryk, and H. Nagase.** 1996. Degradation of cross-linked fibrin by matrix metalloproteinase 3 (stromelysin 1): hydrolysis of the γ Gly 404-Ala 405 peptide bond. *Biochemistry* **35**: 13056–13063.

1462. **Birkedal-Hansen, H., R. E. Taylor, A. S. Bhown, J. Katz, H. Y. Lin, and B. R. Wells.** 1985. Cleavage of bovine skin type III collagen by proteolytic enzymes. Relative resistance of the fibrillar form. *J. Biol. Chem.* **260**: 16411–16417.

1463. **Black, R. A., F. H. Durie, C. Otten-Evans, R. Miller, J. L. Slack, D. H. Lynch, B. Castner, K. M. Mohler, M. Gerhart, R. S. Johnson, *et al.* 1996. Relaxed specificity of matrix metalloproteinases (MMPS) and TIMP insensitivity of tumor necrosis factor-α (TNF-α) production suggest the major TNF-α converting enzyme is not an MMP. *Biochem. Biophys. Res. Commun.* **225**: 400–405.

1464. **Black, R. A., C. T. Rauch, C. J. Kozlosky, J. J. Peschon, J. L. Slack, M. F. Wolfson, B. J. Castner, K. L. Stocking, P. Reddy, S. Srinivasan, *et al.* 1997. A metalloproteinase disintegrin that releases tumour-necrosis factor-α from cells. *Nature* **385**: 729–733.

1464a. **Brown, D. J., P. Bishop, H. Hamdi, and M. C. Kenney.** 1996. Cleavage of structural components of mammalian vitreous by endogenous matrix metalloproteinase-2.. *Curr. Eye Res.* **15**: 439–445.

1465. **Bruns, R. R. and J. Gross.** 1973. Band pattern of the segment-long-spacing form of collagen. Its use in the analysis of primary structure. *Biochemistry* **12** 808–815.

1466. **Bruns, R. R. and J. Gross.** 1974. High-resolution

analysis of the modified quarter-stagger model of the collagen fibril. *Biopolymers* **13**: 931–941.

1467. **Burleigh, M. C., Z. Werb, and J. J. Reynolds.** 1977. Evidence that species specificity and rate of collagen degradation are properties of collagen, not collagenase. *Biochim. Biophys. Acta* **494**: 198–208.

1468. **Buttner, F. H., S. Chubinskaya, D. Margerie, K. Huch, J. Flechtenmacher, A. A. Cole, K. E. Kuettner, and E. Bartnik.** 1997. Expression of membrane type 1 matrix metalloproteinase in human articular cartilage. *Arthritis Rheum.* **40**: 704–709.

1469. **Byers, P. H., E. M. Click, E. Harper, and P. Bornstein.** 1975. Interchain disulfide bonds in procollagen are located in a large nontriple-helical COOH-terminal domain. *Proc. Natl. Acad. Sci. USA* **72**: 3009–3013.

1470. **Campbell, I. K., P. J. Roughley, and J. S. Mort.** 1986. The action of human articular-cartilage metalloproteinase on proteoglycan and link protein. Similarities between products of degradation *in situ* and *in vitro. Biochem. J.* **237**: 117–122.

1471. **Cawston, T. E.** 1978. Connective tissue catabolism in relation to the cornea. *Trans. Ophthalmol. Soc. U. K.* **98**: 329–334.

1472. **Chandler, S., R. Coates, A. Gearing, J. Lury, G. Wells, and E. Bone.** 1995. Matrix metalloproteinases degrade myelin basic protein. *Neurosci. Lett.* **201**: 223–226.

1473. **Chandler, S., J. Cossins, J. Lury, and G. Wells.** 1996. Macrophage metalloelastase degrades matrix and myelin proteins and processes a tumour necrosis factor-α fusion protein. *Biochem. Biophys. Res. Commun.* **228**: 421–429.

1474. **Chin, J. R., G. Murphy, and Z. Werb.** 1985. Stromelysin, a connective tissue-degrading metalloendopeptidase secreted by stimulated rabbit synovial fibroblasts in parallel with collagenase. Biosynthesis, isolation, characterization, and substrates. *J. Biol. Chem.* **260**: 12367–12376.

1475. **Cole, A. A., T. Boyd, L. Luchene, K. E. Kuettner, and T. M. Schmid.** 1993. Type X collagen degradation in long-term serum-free culture of the embryonic chick tibia following production of active collagenase and gelatinase. *Dev. Biol.* **159**: 528–534.

1476. **Cole, T.-C., J. Melrose, and P. Ghosh.** 1989. Isolation and characterisation of a neutral proteinase from the canine intervertebral disc. *Biochim. Biophys. Acta* **990**: 254–262.

1477. **Collier, I. E., S. M. Wilhelm, A. Z. Eisen, B. L. Marmer, G. A. Grant, J. L. Seltzer, A. Kronberger, C. S. He, E. A. Bauer, and G. I. Goldberg.** 1988. H-ras oncogene-transformed human bronchial epithelial cells (TBE-1) secrete a single metalloprotease capable of degrading basement membrane collagen. *J. Biol. Chem.* **263**: 6579–6587.

1478. **Crabbe, T., F. Willenbrock, D. Eaton, P. Hynds, A. F. Carne, G. Murphy, and A. J. P. Docherty.** 1992. Biochemical characterization of matrilysin. Activation conforms to the stepwise mechanisms proposed for other matrix metalloproteinases. *Biochemistry* **31**: 8500–8507.

1479. **d'Ortho, M. P., H. Will, S. Atkinson, G. Butler, A. Messent, J. Gavrilovic, B. Smith, R. Timpl, L. Zardi, and G. Murphy.** 1997. Membrane-type matrix metalloproteinases 1 and 2 exhibit broad-spectrum proteolytic capacities comparable to many matrix metalloproteinases. *Eur. J. Biochem.* **250**: 751–757.

1480. **Davies, M., G. J. Thomas, J. Martin, and D. H. Lovett.** 1988. The purification and characterization of a glomerular-basement-membrane-degrading neutral proteinase from rat mesangial cells. *Biochem. J.* **251**: 419–425.

1481. **Davison, P. F. and M. Berman.** 1973. Corneal collagenase: specific cleavage of types $(\alpha 1)_2 \alpha 2$ and $(\alpha 1)_3$ collagens. *Connect. Tissue Res.* **2**: 57–64.

1482. **Delshammar, M. and K. Ohlsson.** 1978. Granulocyte collagenase and elastase and the plasma protease inhibitors in human pus. *Surgery* **83**: 323–327.

1483. **Desrochers, P. E., J. J. Jeffrey, and S. J. Weiss.** 1991. Interstitial collagenase (matrix metalloproteinase-1) expresses serpinase activity. *J. Clin. Invest.* **87**: 2258–2265.

1484. **Desrochers, P. E., K. Mookhtiar, H. E. Van Wart, K. A. Hasty, and S. J. Weiss.** 1992. Proteolytic inactivation of α_1-proteinase inhibitor and α_1-antichymotrypsin by oxidatively activated human neutrophil metalloproteinases. *J. Biol. Chem.* **267**: 5005–5012.

1485. **Desrochers, P. E. and S. J. Weiss.** 1988. Proteolytic inactivation of α_1-proteinase inhibitor by a neutrophil metalloproteinase. *J. Clin. Invest.* **81**: 1646–1650.

1486. **Diekmann, O. and H. Tschesche.** 1994. Degradation of kinins, angiotensins and substance P by polymorphonuclear matrix metalloproteinases MMP 8 and MMP 9. *Braz. J. Med. Biol. Res.* **27**: 1865–1876.

1487. **Dong, Z., R. Kumar, X. Yang, and I. J. Fidler.** 1997. Macrophage-derived metalloelastase is responsible for the generation of angiostatin in Lewis lung carcinoma. *Cell* **88**: 801–810.

1488. **Eble, J. A., A. Ries, A. Lichy, K. Mann, H. Stanton, J. Gavrilovic, G. Murphy, and K. Kühn.** 1996. The recognition sites of the integrins $\alpha_1 \beta_1$ and $\alpha_2 \beta_1$ within collagen IV are protected against gelatinase A attack in the native protein. *J. Biol. Chem.* **271**: 30964–30970.

1489. **Eeckhout, Y., H. Riccomi, C. Cambiaso, G. Vaes, and P. Masson.** 1976. Studies on properties common to collagen and C1q. *Arch. Int. Physiol. Biochim.* **84**: 611–612.

1490. **Eisen, A. Z., J. J. Jeffrey, and J. Gross.** 1968. Human skin collagenase. Isolation and mechanism of attack on the collagen molecule. *Biochim. Biophys. Acta* **151**: 637–645.

1491. **Enghild, J. J., G. Salvesen, K. Brew, and H. Nagase.** 1989. Interaction of human rheumatoid synovial collagenase (matrix metalloproteinase 1) and stromelysin (matrix metalloproteinase 3) with human α_2-macroglobulin and chicken ovostatin. Binding kinetics and identification of matrix metalloproteinase cleavage sites. *J. Biol. Chem.* **264**: 8779–8785.

1492. **Eyre, D. R., J. J. Wu, and P. E. Woods.** 1991. The cartilage collagens: structural and metabolic studies. *J. Rheumatol. Suppl.* **27**: 49–51.

1493. **Eyre, D. R., J. J. Wu, and D. E. Woolley.** 1984. All three chains of $\alpha 1 \alpha 2 \alpha 3$ collagen from hyaline cartilage resist human collagenase. *Biochem. Biophys. Res. Commun.* **118**: 724–729.

1494. **Fessler, L. I., K. G. Duncan, J. H. Fessler, T. Salo, and K. Tryggvason.** 1984. Characterization of the procollagen IV cleavage products produced by a specific tumor collagenase. *J. Biol. Chem.* **259**: 9783–9789.

1495. **Fields, G. B.** 1991. A model for interstitial collagen catabolism by mammalian collagenases. *J. Theor. Biol.* **153**: 585–602.

1496. **Fletcher, D. S., H. R. Williams, and T.-Y. Lin.** 1978. Effects of human polymorphonuclear leukocyte collagenase on sub-component C1q of the first component of human complement. *Biochim. Biophys. Acta* **540**: 270–277.

1497. **Fosang, A. J., K. Last, P. Gardiner, D. C. Jackson, and L. Brown.** 1995. Development of a cleavage-site-specific monoclonal antibody for detecting metalloproteinase-derived aggrecan fragments: detection of fragments in human synovial fluids. *Biochem. J.* **310**: 337–343.

1498. **Fosang, A. J., K. Last, V. Knäuper, G. Murphy, and P. J. Neame.** 1996. Degradation of cartilage aggrecan by collagenase-3 (MMP-13). *FEBS Lett.* **380**: 17–20.

1499. **Fosang, A. J., K. Last, V. Knäuper, P. J. Neame, G. Murphy, T. E. Hardingham, H. Tschesche, and J. A. Hamilton.** 1993. Fibroblast and neutrophil collagenases cleave at two sites in the cartilage aggrecan interglobular domain. *Biochem. J.* **295**: 273–276.

1500. **Fosang, A. J., K. Last, and R. A. Maciewicz.** 1996. Aggrecan is degraded by matrix metalloproteinases in human arthritis. Evidence that matrix metalloproteinase and aggrecanase activities can be independent. *J. Clin. Invest.* **98**: 2292–2299.

1501. **Fosang, A. J., K. Last, P. J. Neame, G. Murphy, V. Knäuper, H. Tschesche, C. E. Hughes, B. Caterson, and T. E. Hardingham.** 1994. Neutrophil collagenase (MMP-8) cleaves at the aggrecanase site E373-A374 in the interglobular domain of cartilage aggrecan. *Biochem. J.* **304**: 347–351.

1502. **Fosang, A. J., P. J. Neame, T. E. Hardingham, G. Murphy, and J. A. Hamilton.** 1991. Cleavage of cartilage proteoglycan between G1 and G2 domains by stromelysins. *J. Biol. Chem.* **266**: 15579–15582.

1503. **Fosang, A. J., P. J. Neame, K. Last, T. E.**

1504. **Fowlkes, J. L., J. J. Enghild, K. Suzuki, and H. Nagase.** 1994. Matrix metalloproteinases degrade insulin-like growth factor-binding protein-3 in dermal fibroblast cultures. *J. Biol. Chem.* **269**: 25742–25746.

1505. **Fowlkes, J. L., K. Suzuki, H. Nagase, and K. M. Thrailkill.** 1994. Proteolysis of insulin-like growth factor binding protein-3 during rat pregnancy: a role for matrix metalloproteinases. *Endocrinology* **135**: 2810–2813.

1506. **Fujihashi, H. and T. Shimojima.** 1992. Localization of type VI collagen in drug-induced gingival hyperplasia. *Meikai Daigaku Shigaku Zasshi* **21**: 459–475.

1507. **Fukai, F., M. Ohtaki, N. Fujii, H. Yajima, T. Ishii, Y. Nishizawa, K. Miyazaki, and T. Katayama.** 1995. Release of biological activities from quiescent fibronectin by a conformational change and limited proteolysis by matrix metalloproteinases. *Biochemistry* **34**: 11453–11459.

1508. **Fullmer, H. M., W. A. Gibson, G. S. Lazarus, H. A. Bladen, and K. A. Whedon.** 1969. Origin of collagenase in peridontal tissues of man. *J. Dent. Res.* **48**: 646–651.

1509. **Gadek, J. E., G. A. Fells, D. G. Wright, and R. G. Crystal.** 1980. Human neutrophil elastase functions as a type III collagen 'collagenase'. *Biochem. Biophys. Res. Commun.* **95**: 1815–1822.

1510. **Gadher, S. J., D. R. Eyre, V. C. Duance, S. F. Wotton, L. W. Heck, T. M. Schmid, and D. E. Woolley.** 1988. Susceptibility of cartilage collagens type II, IX, X, and XI to human synovial collagenase and neutrophil elastase. *Eur. J. Biochem.* **175**: 1–7.

1511. **Gadher, S. J., D. R. Eyre, S. F. Wotton, T. M. Schmid, and D. E. Woolley.** 1990. Degradation of cartilage collagens type II, IX, X and XI by enzymes derived from human articular chondrocytes. *Matrix* **10**: 154–163.

1512. **Gadher, S. J., T. M. Schmid, L. W. Heck, and D. E. Woolley.** 1989. Cleavage of collagen type X by human synovial collagenase and neutrophil elastase. *Matrix* **9**: 109–115.

1513. **Galloway, W. A., G. Murphy, J. D. Sandy, J. Gavrilovic, T. E. Cawston, and J. J. Reynolds.** 1983. Purification and characterization of a rabbit bone metalloproteinase that degrades proteoglycan and other connective-tissue components. *Biochem. J.* **209**: 741–752.

1514. **Gearing, A. J., P. Beckett, M. Christodoulou, M. Churchill, J. Clements, A. H. Davidson, A. H. Drummond, W. A. Galloway, R. Gilbert, J. L. Gordon,** *et al.* 1994. Processing of tumour necrosis factor-α precursor by metalloproteinases. *Nature* **370**: 555–557.

1515. **Gearing, A. J., P. Beckett, M. Christodoulou, M.**

Churchill, J. M. Clements, M. Crimmin, A. H. Davidson, A. H. Drummond, W. A. Galloway, R. Gilbert, *et al.* 1995. Matrix metalloproteinases and processing of pro-TNF-α. *J. Leukoc. Biol.* **57**: 774–777.

1516. Giannelli, G., J. Falkmarzillier, O. Schiraldi, W. G. Stetler-Stevenson, and V. Quaranta. 1997. Induction of cell migration by matrix metalloprotease-2 cleavage of laminin-5. *Science* **277**: 225–228.

1517. Gijbels, K., P. Proost, S. Masure, H. Carton, A. Billiau, and G. Opdenakker. 1993. Gelatinase B is present in the cerebrospinal fluid during experimental autoimmune encephalomyelitis and cleaves myelin basic protein. *J. Neurosci. Res.* **36**: 432–440.

1518. Gohji, K., M. Nomi, I. Hara, S. Arakawa, and S. Kamidono. 1998. Influence of cytokines and growth factors on matrix metalloproteinase-2 production and invasion of human renal cancer. *Urol. Res.* **26**: 33–37.

1519. Gresham, H. D., I. L. Graham, G. L. Griffin, J. C. Hsieh, L. J. Dong, A. E. Chung, and R. M. Senior. 1996. Domain-specific interactions between entactin and neutrophil integrins. G2 domain ligation of integrin $\alpha_3\beta_1$ and E domain ligation of the leukocyte response integrin signal for different responses. *J. Biol. Chem.* **271**: 30587–30594.

1520. Gronski, T. J., Jr., R. L. Martin, D. K. Kobayashi, B. C. Walsh, M. C. Holman, M. Huber, H. E. Van Wart, and S. D. Shapiro. 1997. Hydrolysis of a broad spectrum of extracellular matrix proteins by human macrophage elastase. *J. Biol. Chem.* **272**: 12189–12194.

1521. Gross, J., E. Harper, E. D. Harris, Jr., P. A. McCroskery, J. H. Highberger, C. Corbett, and A. H. Kang. 1974. Animal collagenases: specificity of action, and structures of the substrate cleavage site. *Biochem. Biophys. Res. Commun.* **61**: 605–612.

1522. Gross, J., J. H. Highberger, B. Johnson-Wint, and C. Biswas. 1980. Mode of action and regulation of tissue collagenases. In *Collagenase in Normal and Pathological Connective Tissues.* D. E. Woolley and J. M. Evanson (eds), pp. 11–35. Wiley, Chichester,England.

1523. Gross, J. and C. M. Lapière. 1962. Collagenolytic activity in amphibian tissues: a tissue culture assay. *Proc. Natl. Acad. Sci. USA* **48**: 1014–1022.

1524. Gross, J. and Y. Nagai. 1965. Specific degradation of the collagen molecule by tadpole collagenolytic enzyme. *Proc. Natl. Acad. Sci. USA* **54**: 1197–1204.

1525. Gunja-Smith, Z., H. Nagase, and J. F. Woessner, Jr. 1989. Purification of the neutral proteoglycan-degrading metalloproteinase from human articular cartilage tissue and its identification as stromelysin matrix metalloproteinase-3. *Biochem. J.* **258**: 115–119.

1526. Harris, E. D., Jr. and M. E. Farrell. 1972. Resistance to collagenase: a characteristic of collagen fibrils cross-linked by formaldehyde. *Biochim. Biophys. Acta* **278**: 133–141.

1527. Hasty, K. A., J. J. Jeffrey, M. S. Hibbs, and H. G. Welgus. 1987. The collagen substrate specificity of human neutrophil collagenase. *J. Biol. Chem.* **262**: 10048–10052.

1528. Hasty, K. A., H. Wu, M. Byrne, M. B. Goldring, J. M. Seyer, R. Jaenisch, S. M. Krane, and C. L. Mainardi. 1993. Susceptibility of type I collagen containing mutated $\alpha1(1)$ chains to cleavage by human neutrophil collagenase. *Matrix* **13**: 181–186.

1529. Hayashi, T., T. Nakamura, H. Hori, and Y. Nagai. 1980. Degradation rates of type II and III collagens by tadpole collagenase are modulated by mutual presence. *J. Biochem. (Tokyo)* **87**: 993–995.

1530. Hazuda, D. J., J. Strickler, F. Kueppers, P. L. Simon, and P. R. Young. 1990. Processing of precursor interleukin 1β and inflammatory disease. *J. Biol. Chem.* **265**: 6318–6322.

1531. Highberger, J. H., C. Corbett, and J. Gross. 1979. Isolation and characterization of a peptide containing the site of cleavage of the chick skin collagen alpha 1[I] chain by animal collagenases. *Biochem. Biophys. Res. Commun.* **89**: 202–208.

1532. Hiro, D., A. Ito, K. Matsuta, and Y. Mori. 1986. Hyaluronic acid is an endogenous inducer of interleukin-1 production by human monocytes and rabbit macrophages. *Biochem. Biophys. Res. Commun.* **140**: 715–722.

1533. Hirose, T., R. A. Reife, G. N. Smith, Jr., R. M. Stevens, C. L. Mainardi, and K. A. Hasty. 1992. Characterization of type V collagenase (gelatinase) in synovial fluid of patients with inflammatory arthritis. *J. Rheumatol.* **19**: 593–599.

1534. Hori, H. and Y. Nagai. 1979. Purification of tadpole collagenase and characterization using collagen and synthetic substrates. *Biochim. Biophys. Acta* **566**: 211–221.

1535. Horwitz, A. L., A. J. Hance, and R. G. Crystal. 1977. Granulocyte collagenase: selective digestion of type I relative to type III collagen. *Proc. Natl. Acad. Sci. USA* **74**: 897–901.

1536. Hughes, C., G. Murphy, and T. E. Hardingham. 1991. Metalloproteinase digestion of cartilage proteoglycan. Pattern of cleavage by stromelysin and susceptibility to collagenase. *Biochem. J.* **279**: 733–739.

1537. Hughes, C. E., B. Caterson, A. J. Fosang, P. J. Roughley, and J. S. Mort. 1995. Monoclonal antibodies that specifically recognize neoepitope sequences generated by 'aggrecanase' and matrix metalloproteinase cleavage of aggrecan: application to catabolism in situ and in vitro. *Biochem. J.* **305**: 799–804.

1538. Ilic, M. Z., C. J. Handley, H. C. Robinson, and M. T. Mok. 1992. Mechanism of catabolism of aggrecan by articular cartilage. *Arch. Biochem. Biophys.* **294**: 115–122.

1539. Imai, K., A. Hiramatsu, D. Fukushima, M. D. Pierschbacher, and Y. Okada. 1997. Degradation of decorin by matrix metalloproteinases: identification of the cleavage sites, kinetic analyses and transforming growth factor-β1 release. *Biochem. J.* **322**: 809–814.

1540. **Imai, K., M. Kusakabe, T. Sakakura, I. Nakanishi, and Y. Okada.** 1994. Susceptibility of tenascin to degradation by matrix metalloproteinases and serine proteinases. *FEBS Lett.* **352**: 216–218.

1541. **Imai, K., E. Ohuchi, T. Aoki, H. Nomura, Y. Fujii, H. Sato, M. Seiki, and Y. Okada.** 1996. Membrane-type matrix metalloproteinase 1 is a gelatinolytic enzyme and is secreted in a complex with tissue inhibitor of metalloproteinases 2. *Cancer Res.* **56**: 2707–2710.

1542. **Imai, K., H. Shikata, and Y. Okada.** 1995. Degradation of vitronectin by matrix metalloproteinases-1, -2, -3, -7 and -9. *FEBS Lett.* **369**: 249–251.

1543. **Ishibashi, M., A. Ito, K. Sakyo, and Y. Mori.** 1987. Procollagenase activator produced by rabbit uterine cervical fibroblasts. *Biochem. J.* **241**: 527–534.

1544. **Ito, A.** 1995. [Processing of interleukin 1 precursor at inflammatory sites.] [Japanese] *Yakugaku Kenkyu no Shinpo* **12**: 35–44.

1545. **Ito, A., A. Mukaiyama, Y. Itoh, H. Nagase, I. B. Thogersen, J. J. Enghild, Y. Sasaguri, and Y. Mori.** 1996. Degradation of interleukin 1β by matrix metalloproteinases. *J. Biol. Chem.* **271**: 14657–14660.

1546. **Ito, A., Y. Ojima, Y. Mori, and H. Nagase.** 1990. [Action mechanism of interleukin-1 on the production of collagenase in fibroblasts][Japanese]. *Connect. Tissue* **21**: 184–187.

1547. **Itoh, M., K. Masuda, Y. Ito, T. Akizawa, M. Yoshioka, K. Imai, Y. Okada, H. Sato, and M. Seiki.** 1996. Purification and refolding of recombinant human proMMP-7 (pro-matrilysin) expressed in *Escherichia coli* and its characterization. *J. Biochem. (Tokyo)* **119**: 667–673.

1548. **Jeffrey, J. J., H. G. Welgus, R. E. Burgeson, and A. Z. Eisen.** 1983. Studies on the activation energy and deuterium isotope effect of human skin collagenase on homologous collagen substrates. *J. Biol. Chem.* **258**: 11123–11127.

1549. **Katsuda, S., Y. Okada, K. Imai, and I. Nakanishi.** 1994. Matrix metalloproteinase-9 (92-kd gelatinase/type IV collagenase equals gelatinase B) can degrade arterial elastin. *Am. J. Pathol.* **145**: 1208–1218.

1550. **Kawaguchi, Y., H. Tanaka, T. Okada, H. Konishi, M. Takahashi, M. Ito, and J. Asai.** 1996. The effects of ultraviolet A and reactive oxygen species on the mRNA expression of 72-kDa type IV collagenase and its tissue inhibitor in cultured human dermal fibroblasts. *Arch. Dermatol. Res.* **288**: 39–44.

1551. **Kielty, C. M., M. Lees, C. A. Shuttleworth, and D. Woolley.** 1993. Catabolism of intact type VI collagen microfibrils: susceptibility to degradation by serine proteinases. *Biochem. Biophys. Res. Commun.* **191**: 1230–1236.

1552. **Knäuper, V., S. Cowell, B. Smith, C. López-Otín, M. O'Shea, H. Morris, L. Zardi, and G. Murphy.** 1997. The role of the C-terminal domain of human collagenase-3 (MMP-13) in the activation of procollagenase-3, substrate specificity, and tissue inhibitor of metalloproteinase interaction. *J. Biol. Chem.* **272**: 7608–7616.

1553. **Knäuper, V., C. López-Otín, B. Smith, G. Knight, and G. Murphy.** 1996. Biochemical characterization of human collagenase-3. *J. Biol. Chem.* **271**: 1544–1550.

1554. **Knäuper, V., H. Reinke, and H. Tschesche.** 1990. Inactivation of human plasma α_1-proteinase inhibitor by human PMN leucocyte collagenase. *FEBS Lett.* **263**: 355–357.

1555. **Knäuper, V., S. Triebel, H. Reinke, and H. Tschesche.** 1991. Inactivation of human plasma C1-inhibitor by human PMN leucocyte matrix metalloproteinases. *FEBS Lett.* **290**: 99–102.

1556. **Kobayashi, S. and Y. Nagai.** 1978. Human leucocyte neutral proteases, with special reference to collagen metabolism. *J. Biochem. (Tokyo)* **84**: 559–567.

1557. **Koita, H., K. Nabeshima, T. Inoue, and M. Koono.** 1991. Sequential degradation of interstitial collagen by metalloproteinases extracted from tumors of murine ascites hepatomas. *Clin. Exp. Metastasis* **9**: 441–456.

1558. **Kolkenbrock, H., H. M. Ali, A. Hecker-Kia, G. Buchlow, H. Sörensen, R. W. Hauer, and N. Ulbrich.** 1991. Characterization of a gelatinase from human rheumatoid synovial fluid cells. *Eur. J. Clin. Chem. Clin. Biochem.* **29**: 499–505.

1559. **Krane, S. M.** 1995. Is collagenase (matrix metalloproteinase-1) necessary for bone and other connective tissue remodeling? *Clin. Orthop.* **313**: 47–53.

1560. **Krane, S. M., M. H. Byrne, V. Lemaître, P. Henriet, J. J. Jeffrey, J. P. Witter, X. Liu, H. Wu, R. Jaenisch, and Y. Eeckhout.** 1996. Different collagenase gene products have different roles in degradation of type I collagen. *J. Biol. Chem.* **271**: 28509–28515.

1561. **Kryshtalskyj, E., J. Sodek, and J. M. Ferrier.** 1986. Correlation of collagenolytic enzymes and inhibitors in gingival crevicular fluid with clinical and microscopic changes in experimental periodontitis in the dog. *Arch. Oral Biol.* **31**: 21–31.

1562. **Kudo, K., A. Saito, K. Sudo, M. Adachi, A. Ikai, Y. Ofuji, Y. Tadeuchi, H. Yoshie, and K. Hara.** 1988. [The inhibitory effects of chicken ovomacroglobulin on collagenolytic activity in Bacteroides gingivalis culture supernatant, human PMN and human gingival crevicular fluid][Japanese]. *Nippon Shishubyo Gakkai Kaishi* **30**: 1061–1069.

1563. **Lark, M. W., E. K. Bayne, J. Flanagan, C. F. Harper, L. A. Hoerrner, N. I. Hutchinson, I. I. Singer, S. A. Donatelli, J. R. Weidner, H. R. Williams,** *et al.* 1997. Aggrecan degradation in human cartilage—evidence for both matrix metalloproteinase and aggrecanase activity in normal, osteoarthritic, and rheumatoid joints. *J. Clin. Invest.* **100**: 93–106.

1564. **Leibovich, S. J. and J. B. Weiss.** 1973. Elucidation of the exact sites of cleavage of tropocollagen by rheumatoid synovial collagenase: correlation of cleavage sites with fibril structure. *Connect. Tissue Res.* **2**: 11–19.

1565. **Lemaître, V., A. Jungbluth, and Y. Eeckhout.** 1997. The recombinant catalytic domain of mouse collagenase-3 depolymerizes type I collagen by cleaving its aminotelopeptides. *Biochem. Biophys. Res. Commun.* **230**: 202–205.

1566. **LePage, R. N., A. J. Fosang, S. J. Fuller, G. Murphy, G. Evin, K. Beyreuther, C. L. Masters, and D. H. Small.** 1995. Gelatinase A possesses a β-secretase-like activity in cleaving the amyloid protein precursor of Alzheimer's disease. *FEBS Lett.* **377**: 267–270.

1567. **Levi, E., R. Fridman, H. Q. Miao, Y. S. Ma, A. Yayon, and I. Vlodavsky.** 1996. Matrix metalloproteinase 2 releases active soluble ectodomain of fibroblast growth factor receptor 1. *Proc. Natl. Acad. Sci. USA* **93**: 7069–7074.

1568. **Liotta, L. A., T. Kalebic, C. A. Reese, and R. Mayne.** 1982. Protease susceptibilities of HMW, 1α, 2α, but not 3α cartilage collagens are similar to type V collagen. *Biochem. Biophys. Res. Commun.* **104**: 500–506.

1569. **Liotta, L. A., W. L. Lanzer, and S. Garbisa.** 1981. Identification of a type V collagenolytic enzyme. *Biochem. Biophys. Res. Commun.* **98**: 184–190.

1570. **Liotta, L. A., K. Tryggvason, S. Garbisa, P. G. Robey, and S. Abe.** 1981. Partial purification and characterization of a neutral protease which cleaves type IV collagen. *Biochemistry* **20**: 100–104.

1571. **Liu, J., J. D. Cassidy, A. Allan, P. J. Neame, J. S. Mort, and P. J. Roughley.** 1992. Link protein shows species variation in its susceptibility to proteolysis. *J. Orthop. Res.* **10**: 621–630.

1572. **Liu, X., H. Wu, M. Byrne, J. Jeffrey, S. Krane, and R. Jaenisch.** 1995. A targeted mutation at the known collagenase cleavage site in mouse type I collagen impairs tissue remodeling. *J. Cell Biol.* **130**: 227–237.

1573. **Llano, E., A. M. Pendás, V. Knäuper, T. Sorsa, T. Salo, E. Salido, G. Murphy, J. P. Simmer, J. D. Bartlett, and C. López-Otín.** 1997. Identification and structural and functional characterization of human enamelysin (MMP-20). *Biochemistry* **36**: 15101–15108.

1574. **Loulakis, P., A. Shrikhande, G. Davis, and C. A. Maniglia.** 1992. N-terminal sequence of proteoglycan fragments isolated from medium of interleukin-1-treated articular-cartilage cultures. Putative site(s) of enzymic cleavage. *Biochem. J.* **284**: 589–593.

1575. **Lyons, J. G., B. Birkedal-Hansen, W. G. Moore, R. L. O'Grady, and H. Birkedal-Hansen.** 1991. Characteristics of a 95-kDa matrix metalloproteinase produced by mammary carcinoma cells. *Biochemistry* **30**: 1449–1456.

1576. **Mackay, A. R., D. E. Gomez, A. M. Nason, and U. P. Thorgeirsson.** 1994. Studies on the effects of laminin, E-8 fragment of laminin and synthetic laminin peptides PA22–2 and YIGSR on matrix metalloproteinases and tissue inhibitor of metalloproteinase expression. *Lab. Invest.* **70**: 800–806.

1577. **Mackay, A. R., J. L. Hartzler, M. D. Pelina, and U. P. Thorgeirsson.** 1990. Studies on the ability of 65-kDa and 92-kDa tumor cell gelatinases to degrade type IV collagen. *J. Biol. Chem.* **265**: 21929–21934.

1578. **Mañes, S., E. Mira, M. M. Barbacid, A. Ciprés, P. Fernández-Resa, J. M. Buesa, I. Mérida, M. Aracil, G. Márquez, and C. Martínez.** 1997. Identification of insulin-like growth factor-binding protein-1 as a potential physiological substrate for human stromelysin-3. *J. Biol. Chem.* **272**: 25706–25712.

1579. **Marcotte, P. A., D. Dudlak, M. L. Leski, J. Ryan, and J. Henkin.** 1992. Characterization of a metalloprotease which cleaves with high site-specificity the Glu(143)-Leu(144) bond of urokinase. *Fibrinolysis* **6**(Suppl. 1): 57–62.

1580. **Marcy, A. I., L. L. Eiberger, R. Harrison, H. K. Chan, N. I. Hutchinson, W. K. Hagmann, P. M. Cameron, D. A. Boulton, and J. D. Hermes.** 1991. Human fibroblast stromelysin catalytic domain: expression, purification, and characterization of a C-terminally truncated form. *Biochemistry* **30**: 6476–6483.

1581. **Mast, A. E., J. J. Enghild, H. Nagase, K. Suzuki, S. V. Pizzo, and G. Salvesen.** 1991. Kinetics and physiologic relevance of the inactivation of α_1-proteinase inhibitor, α_1-antichymotrypsin, and antithrombin III by matrix metalloproteinases-1 (tissue collagenase), -2 (72-kDa gelatinase/type IV collagenase), and -3 (stromelysin). *J. Biol. Chem.* **266**: 15810–15816.

1582. **Masui, Y., T. Takemoto, S. Sakakibara, H. Hori, and Y. Nagai.** 1977. Synthetic substrates for vertebrate collagenase. *Biochem. Med.* **17**: 215–221.

1583. **Maurer, P., W. Göhring, T. Sasaki, K. Mann, R. Timpl, and R. Nischt.** 1997. Recombinant and tissue-derived mouse BM-40 bind to several collagen types and have increased affinities after proteolytic activation. *Cell. Mol. Life Sci.* **53**: 478–484.

1584. **Mayer, U., K. Mann, R. Timpl, and G. Murphy.** 1993. Sites of nidogen cleavage by proteases involved in tissue homeostasis and remodelling. *Eur. J. Biochem.* **217**: 877–884.

1585. **Mayne, R., P. M. Mayne, Z. X. Ren, M. A. Accavitti, S. Gurusiddappa, and P. G. Scott.** 1994. Monoclonal antibody to the aminotelopeptide of type II collagen: loss of the epitope after stromelysin digestion. *Connect. Tissue Res.* **31**: 11–21.

1586. **McCroskery, P. A., S. Wood, Jr., and E. D. Harris, Jr.**

1973. Gelatin: a poor substrate for a mammalian collagenase. *Science* **182**: 70–71.

1587. **Mecham, R. P., T. J. Broekelmann, C. J. Fliszar, S. D. Shapiro, H. G. Welgus, and R. M. Senior.** 1997. Elastin degradation by matrix metalloproteinases—cleavage site specificity and mechanisms of elastolysis. *J. Biol. Chem.* **272**: 18071–18076.

1588. **Mehul, B., S. Bawumia, S. R. Martin, and R. C. Hughes.** 1994. Structure of baby hamster kidney carbohydrate-binding protein CBP30, an S-type animal lectin. *J. Biol. Chem.* **269**: 18250–18258.

1589. **Menzel, E. J. and J. S. Smolen.** 1978. [Degradation of C1q, the first subcomponent of the complement sequence, by synovial collagenase from patients with rheumatoid arthritis][German]. *Wien. Klin. Wochenschr.* **90**: 727–730.

1590. **Menzel, J. and W. Borth.** 1983. Influence of plasma fibronectin on collagen cleavage by collagenase. *Coll. Relat. Res.* **3**: 217–230.

1591. **Miller, E. J., J. E. Finch, Jr., E. Chung, W. T. Butler, and P. B. Robertson.** 1976. Specific cleavage of the native type III collagen molecule with trypsin. Similarity of the cleavage products to collagenase-produced fragments and primary structure at the cleavage site. *Arch. Biochem. Biophys.* **173**: 631–637.

1592. **Miller, E. J., E. D. Harris, Jr., E. Chung, J. E. Finch, Jr., P. A. McCroskery, and W. T. Butler.** 1976. Cleavage of Type II and III collagens with mammalian collagenase: site of cleavage and primary structure at the NH$_2$-terminal portion of the smaller fragment released from both collagens. *Biochemistry* **15**: 787–792.

1593. **Mitchell, T. I., J. J. Jeffrey, R. D. Palmiter, and C. E. Brinckerhoff.** 1993. The acute phase reactant serum amyloid A (SAA3) is a novel substrate for degradation by the metalloproteinases collagenase and stromelysin. *Biochim. Biophys. Acta* **1156**: 245–254.

1594. **Miyazaki, K., K. Funahashi, M. Umeda, and A. Nakano.** 1994. Gelatinase A and APP. *Nature* **368**: 695–696.

1595. **Miyazaki, K., Y. Hattori, F. Umenishi, H. Yasumitsu, and M. Umeda.** 1990. Purification and characterization of extracellular matrix-degrading metalloproteinase, matrin (pump-1), secreted from human rectal carcinoma cell line. *Cancer Res.* **50**: 7758–7764.

1596. **Moll, U. M., G. L. Youngleib, K. B. Rosinski, and J. P. Quigley.** 1990. Tumor promoter-stimulated Mr 92,000 gelatinase secreted by normal and malignant human cells: isolation and characterization of the enzyme from HT1080 tumor cells. *Cancer Res.* **50**: 6162–6170.

1597. **Morodomi, T., Y. Ogata, Y. Sasaguri, M. Morimatsu, and H. Nagase.** 1992. Purification and characterization of matrix metalloproteinase 9 from U937 monocytic leukaemia and HT1080 fibrosarcoma cells. *Biochem. J.* **285**: 603–611.

1598. **Moss, M. L., S. L. C. Jin, M. E. Milla, W. Burkhart, H. L. Carter, W. J. Chen, W. C. Clay, J. R. Didsbury, D. Hassler, C. R. Hoffman, et al.** 1997. Cloning of a disintegrin metalloproteinase that processes precursor tumour-necrosis factor-α. *Nature* **385**: 733–736.

1599. **Mott, J. D., R. G. Khalifah, H. Nagase, C. F. Shield, J. K. Hudson, and B. G. Hudson.** 1997. Nonenzymatic glycation of type IV collagen and matrix metalloproteinase susceptibility. *Kidney Int.* **52**: 1302–1312.

1600. **Muir, D. and M. Manthorpe.** 1992. Stromelysin generates a fibronectin fragment that inhibits Schwann cell proliferation. *J. Cell Biol.* **116**: 177–185.

1601. **Murphy, G., J. A. Allan, F. Willenbrock, M. I. Cockett, J. P. O'Connell, and A. J. P. Docherty.** 1992. The role of the C-terminal domain in collagenase and stromelysin specificity. *J. Biol. Chem.* **267**: 9612–9618.

1602. **Murphy, G., E. C. Cartwright, A. Sellers, and J. J. Reynolds.** 1977. The detection and characterisation of collagenase inhibitors from rabbit tissues in culture. *Biochim. Biophys. Acta* **483**: 493–498.

1603. **Murphy, G., T. E. Cawston, W. A. Galloway, M. J. Barnes, R. A. D. Bunning, E. Mercer, J. J. Reynolds, and R. E. Burgeson.** 1981. Metalloproteinases from rabbit bone culture medium degrade types IV and V collagens, laminin and fibronectin. *Biochem. J.* **199**: 807–811.

1604. **Murphy, G., M. I. Cockett, R. V. Ward, and A. J. P. Docherty.** 1991. Matrix metalloproteinase degradation of elastin, type IV collagen and proteoglycan. A quantitative comparison of the activities of 95 kDa and 72 kDa gelatinases, stromelysins-1 and -2 and punctuated metalloproteinase (PUMP). *Biochem. J.* **277**: 277–279.

1605. **Murphy, G., C. G. McAlpine, C. T. Poll, and J. J. Reynolds.** 1985. Purification and characterization of a bone metalloproteinase that degrades gelatin and types IV and V collagen. *Biochim. Biophys. Acta* **831**: 49–58.

1606. **Murphy, G., J. J. Reynolds, U. Bretz, and M. Baggiolini.** 1982. Partial purification of collagenase and gelatinase from human polymorphonuclear leucocytes. Analysis of their actions on soluble and insoluble collagens. *Biochem. J.* **203**: 209–221.

1607. **Murphy, G., J. P. Segain, M. O'Shea, M. Cockett, C. Ioannou, O. Lefebvre, P. Chambon, and P. Basset.** 1993. The 28-kDa N-terminal domain of mouse stromelysin-3 has the general properties of a weak metalloproteinase. *J. Biol. Chem.* **268**: 15435–15441.

1608. **Murphy, G., R. Ward, R. M. Hembry, J. J. Reynolds, K. Kuhn, and K. Tryggvason.** 1989. Characterization of gelatinase from pig polymorphonuclear leucocytes. A metalloproteinase resembling tumour type IV collagenase. *Biochem. J.* **258**: 463–472.

1609. **Myint, E., D. J. Brown, A. V. Ljubimov, M. Kyaw, and M. C. Kenney.** 1996. Cleavage of human corneal type VI collagen α3 chain by matrix metalloproteinase-2. *Cornea* **15**: 490–496.

1610. **Nagase, H.** 1995. Human stromelysins 1 and 2. *Methods Enzymol.* **248**: 449–470.

1611. **Nakagawa, H. and H. Debuchi.** 1992. Inactivation of substance P by granulation tissue-derived gelatinase. *Biochem. Pharmacol.* **44**: 1773–1777.

1612. **Narayanan, A. S., D. F. Meyers, R. C. Page, and H. G. Welgus.** 1984. Action of mammalian collagenases on type I trimer collagen. *Coll. Relat. Res.* **4**: 289–296.

1613. **Nethery, A. and R. L. O'Grady.** 1989. Identification, partial purification and characterization of high-molecular-weight gelatin-degrading metalloproteinases produced by a rat mammary carcinoma cell line. *Biochim. Biophys. Acta* **993**: 42–47.

1614. **Nethery, A. and R. L. O'Grady.** 1991. Interstitial collagenase from rat mammary carcinoma cells: interaction with substrates and inhibitors. *Invasion Metastasis* **11**: 241–248.

1615. **Nguyen, Q., J. Liu, P. J. Roughley, and J. S. Mort.** 1991. Link protein as a monitor in situ of endogenous proteolysis in adult human articular cartilage. *Biochem. J.* **278**: 143–147.

1616. **Nguyen, Q., G. Murphy, C. E. Hughes, J. S. Mort, and P. J. Roughley.** 1993. Matrix metalloproteinases cleave at two distinct sites on human cartilage link protein. *Biochem. J.* **295**: 595–598.

1617. **Nguyen, Q., G. Murphy, P. J. Roughley, and J. S. Mort.** 1989. Degradation of proteoglycan aggregate by a cartilage metalloproteinase. Evidence for the involvement of stromelysin in the generation of link protein heterogeneity in situ. *Biochem. J.* **259**: 61–67.

1618. **Nicholson, R., G. Murphy, and R. Breathnach.** 1989. Human and rat malignant-tumor-associated mRNAs encode stromelysin-like metalloproteinases. *Biochemistry* **28**: 5195–5203.

1619. **Noël, A., M. Santavicca, I. Stoll, C. L'Hoir, A. Staub, G. Murphy, M. C. Rio, and P. Basset.** 1995. Identification of structural determinants controlling human and mouse stromelysin-3 proteolytic activities. *J. Biol. Chem.* **270**: 22866–22872.

1620. **Ochieng, J. and B. Green.** 1996. The interactions of alpha 2HS glycoprotein with metalloproteinases. *Biochem. Mol. Biol. Int.* **40**: 13–20.

1621. **Ochieng, J., B. Green, S. Evans, O. James, and P. Warfield.** 1998. Modulation of the biological functions of galectin-3 by matrix metalloproteinases. *Biochim. Biophys. Acta—Gen. Subj.* **1379**: 97–106.

1622. **Ohhama, Y.** 1996. [Vitronectin distribution in oral tissue and its degradation by matrix metalloproteinases][Japanese]. *Meikai Daigaku Shigaku Zasshi* **25**: 176–186.

1623. **Ohuchi, E., K. Imai, Y. Fujii, H. Sato, M. Seiki, and Y. Okada.** 1997. Membrane type 1 matrix metalloproteinase digests interstitial collagens and other extracellular matrix macromolecules. *J. Biol. Chem.* **272**: 2446–2451.

1624. **Ohyama, H. and K. Hashimoto.** 1977. Collagenase of human skin basal cell epithelioma. *J. Biochem. (Tokyo)* **82**: 175–183.

1625. **Okada, Y., H. Konomi, T. Yada, K. Kimata, and H. Nagase.** 1989. Degradation of type IX collagen by matrix metalloproteinase 3 (stromelysin) from human rheumatoid synovial cells. *FEBS Lett.* **244**: 473–476.

1626. **Okada, Y., T. Morodomi, J. J. Enghild, K. Suzuki, A. Yasui, I. Nakanishi, G. Salvesen, and H. Nagase.** 1990. Matrix metalloproteinase 2 from human rheumatoid synovial fibroblasts. Purification and activation of the precursor and enzymic properties. *Eur. J. Biochem.* **194**: 721–730.

1627. **Okada, Y., H. Nagase, and E. D. Harris, Jr.** 1986. A metalloproteinase from human rheumatoid synovial fibroblasts that digests connective tissue matrix components. Purification and characterization. *J. Biol. Chem.* **261**: 14245–14255.

1628. **Okada, Y., H. Nagase, and E. D. Harris, Jr.** 1986. A multi-substrate metalloproteinase from rheumatoid synovial cells. *Trans. Assoc. Am. Physicians.* **99**: 143–153.

1629. **Okada, Y., K. Naka, T. Minamoto, Y. Ueda, Y. Oda, I. Nakanishi, and R. Timpl.** 1990. Localization of type VI collagen in the lining cell layer of normal and rheumatoid synovium. *Lab. Invest.* **63**: 647–656.

1630. **Olsen, B. R.** 1964. Electron microscope studies on collagen. III. Tryptic digestion of tropocollagen macromolecules. *Z. Zellforsch.* **61**: 913–919.

1631. **Pardo, A., H. Soto, I. Montfort, and R. P. Tamayo.** 1980. Collagen-bound collagenase. *Connect. Tissue Res.* **7**: 253–261.

1632. **Pasternak, R. D., S. J. Hubbs, R. G. Caccese, R. L. Marks, J. M. Conaty, and G. DiPasquale.** 1986. Interleukin-1 stimulates the secretion of proteoglycan- and collagen-degrading proteases by rabbit articular chondrocytes. *Clin. Immunol. Immunopathol.* **41**: 351–367.

1633. **Patterson, B. C. and Q. X. A. Sang.** 1997. Angiostatin-converting enzyme activities of human matrilysin (MMP-7) and gelatinase B type IV collagenase (MMP-9). *J. Biol. Chem.* **272**: 28823–28825.

1634. **Pei, D., G. Majmudar, and S. J. Weiss.** 1994. Hydrolytic inactivation of a breast carcinoma cell-derived serpin by human stromelysin-3. *J. Biol. Chem.* **269**: 25849–25855.

1635. **Pelletier, J.-P., J. Martel-Pelletier, D. S. Howell, L. Ghandur-Mnaymneh, J. E. Enis, and J. F. Woessner, Jr.** 1983. Collagenase and collagenolytic activity in human osteoarthritic cartilage. *Arthritis Rheum.* **26**: 63–68.

1636. **Perejda, A. J., E. J. Zaragoza, E. Eriksen, and J. Uitto.** 1984. Nonenzymatic glucosylation of lysyl and hydroxylysyl residues in type I and type II collagens. *Coll. Relat. Res.* **4**: 427–440.

1637. **Perides, G., R. A. Asher, M. W. Lark, W. S. Lane, R. A. Robinson, and A. Bignami.** 1995. Glial hyaluronate-binding protein: a product of metalloproteinase digestion of versican? *Biochem. J.* **312**: 377–384.

1638. **Pourmotabbed, T.** 1994. Relation between substrate specificity and domain structure of 92-kDa type IV collagenase. *Ann. N. Y. Acad. Sci.* **732**: 372–374.

1639. **Pourmotabbed, T., T. L. Solomon, K. A. Hasty, and C. L. Mainardi.** 1994. Characteristics of 92 kDa type IV collagenase/gelatinase produced by granulocytic leukemia cells: structure, expression of cDNA in *E. coli* and enzymic properties. *Biochim. Biophys. Acta* **1204**: 97–107.

1640. **Proost, P., J. Van Damme, and G. Opdenakker.** 1993. Leukocyte gelatinase B cleavage releases encephalitogens from human myelin basic protein. *Biochem. Biophys. Res. Commun.* **192**: 1175–1181.

1641. **Quantin, B., G. Murphy, and R. Breathnach.** 1989. Pump-1 cDNA codes for a protein with characteristics similar to those of classical collagenase family members. *Biochemistry* **28**: 5327–5334.

1642. **Rajah, R., S. E. Nunn, D. J. Herrick, M. M. Grunstein, and P. Cohen.** 1996. Leukotriene D_4 induces MMP-1, which functions as an IGFBP protease in human airway smooth muscle cells. *Am. J. Physiol.* **271**: L1014-L1022.

1643. **Rantala-Ryhänen, S., L. Ryhänen, F. V. Nowak, and J. Uitto.** 1983. Proteinases in human polymorphonuclear leukocytes. Purification and characterization of an enzyme which cleaves denatured collagen and a synthetic peptide with a Gly-Ile sequence. *Eur. J. Biochem.* **134**: 129–137.

1644. **Reboul, P., J. P. Pelletier, G. Tardif, J. M. Cloutier, and J. Martel-Pelletier.** 1996. The new collagenase, collagenase-3, is expressed and synthesized by human chondrocytes but not by synoviocytes. A role in osteoarthritis. *J. Clin. Invest.* **97**: 2011–2019.

1645. **Robertson, P. B. and E. J. Miller.** 1972. Cartilage collagen: inability to serve as a substrate for collagenases active against skin and bone collagen. *Biochim. Biophys. Acta* **289**: 247–250.

1646. **Roher, A. E., T. C. Kasunic, A. S. Woods, R. J. Cotter, M. J. Ball, and R. Fridman.** 1994. Proteolysis of Aβ peptide from Alzheimer disease brain by gelatinase A. *Biochem. Biophys. Res. Commun.* **205**: 1755–1761.

1647. **Roughley, P. J., Q. Nguyen, and J. S. Mort.** 1991. Mechanisms of proteoglycan degradation in human articular cartilage. *J. Rheumatol. Suppl.* **27**: 52–54.

1648. **Rucklidge, G. J., M. Lund-Johansen, G. Milne, and R. Bjerkvig.** 1990. Isolation and characterization of a metalloproteinase secreted by rat glioma cells in serum-free culture. *Biochem. Biophys. Res. Commun.* **172**: 544–550.

1649. **Sage, H., G. Balian, A. M. Vogel, and P. Bornstein.** 1984. Type VIII collagen. Synthesis by normal and malignant cells in culture. *Lab. Invest.* **50**: 219–231.

1650. **Sage, H., B. Trueb, and P. Bornstein.** 1983. Biosynthetic and structural properties of endothelial cell type VIII collagen. *J. Biol. Chem.* **258**: 13391–13401.

1651. **Sakai, T. and J. Gross.** 1967. Some properties of the products of reaction of tadpole collagenase with collagen. *Biochemistry* **6**: 518–528.

1652. **Sakamoto, W., K. Fujie, M. Kaga, H. Handa, K. Gotoh, J. Nishihira, J. Kishi, T. Hayakawa, and Y. Okada.** 1996. Degradation of T-kininogen by cathepsin D and matrix metalloproteinases. *Immunopharmacology* **32**: 73–75.

1653. **Sakata, K., K. Hoshino, and H. Nakagawa.** 1989. Purification and characterization of exudate gelatinases in the chronic-phase of carrageenin-induced inflammation in rats. *J. Biochem. (Tokyo)* **105**: 384–389.

1654. **Salo, T., L. A. Liotta, and K. Tryggvason.** 1983. Purification and characterization of a murine basement membrane collagen-degrading enzyme secreted by metastatic tumor cells. *J. Biol. Chem.* **258**: 3058–3063.

1655. **Sanchez-Lopez, R., C. M. Alexander, O. Behrendtsen, R. Breathnach, and Z. Werb.** 1993. Role of zinc-binding- and hemopexin domain-encoded sequences in the substrate specificity of collagenase and stromelysin-2 as revealed by chimeric proteins. *J. Biol. Chem.* **268**: 7238–7247.

1656. **Sandy, J. D., P. J. Neame, R. E. Boynton, and C. R. Flannery.** 1991. Catabolism of aggrecan in cartilage explants. Identification of a major cleavage site within the interglobular domain. *J. Biol. Chem.* **266**: 8683–8685.

1657. **Sasaki, T., W. Göhring, K. Mann, P. Maurer, E. Hohenester, V. Knäuper, G. Murphy, and R. Timpl.** 1997. Limited cleavage of extracellular matrix protein BM-40 by matrix metalloproteinases increases its affinity for collagens. *J. Biol. Chem.* **272**: 9237–9243.

1658. **Sasaki, T., K. Mann, G. Murphy, M. L. Chu, and R. Timpl.** 1996. Different susceptibilities of fibulin-1 and fibulin-2 to cleavage by matrix metalloproteinases and other tissue proteases. *Eur. J. Biochem.* **240**: 427–434.

1659. **Sawamura, D., T. Sugawara, I. Hashimoto, L. Bruckner-Tuderman, D. Fujimoto, Y. Okada, N. Utsumi, and H. Shikata.** 1991. Increased gene expression of matrix metalloproteinase-3 (stromelysin) in skin fibroblasts from patients with severe recessive dystrophic epidermolysis bullosa. *Biochem. Biophys. Res. Commun.* **174**: 1003–1008.

1660. **Schmid, T. M., R. Mayne, J. J. Jeffrey, and T. F. Linsenmayer.** 1986. Type X collagen contains two cleavage sites for a vertebrate collagenase. *J. Biol. Chem.* **261**: 4184–4189.

1661. **Sellers, A. and J. F. Woessner, Jr.** 1980. The extraction of a neutral metalloproteinase from the involuting rat uterus, and its action on cartilage proteoglycan. *Biochem. J.* **189**: 521–531.

1662. **Seltzer, J. L., S. A. Adams, G. A. Grant, and A. Z. Eisen.** 1981. Purification and properties of a gelatin-specific neutral protease from human skin. *J. Biol. Chem.* **256**: 4662–4668.

1663. **Seltzer, J. L., A. Z. Eisen, E. A. Bauer, N. P. Morris, R. W. Glanville, and R. E. Burgeson.** 1989. Cleavage of type VII collagen by interstitial collagenase and type IV collagenase (gelatinase) derived from human skin. *J. Biol. Chem.* **264**: 3822–3826.

1664. **Senior, R. M., G. L. Griffin, C. J. Fliszar, S. D. Shapiro, G. I. Goldberg, and H. G. Welgus.** 1991. Human 92- and 72-kilodalton type IV collagenases are elastases. *J. Biol. Chem.* **266**: 7870–7875.

1665. **Shapiro, S. D., G. L. Griffin, D. J. Gilbert, N. A. Jenkins, N. G. Copeland, H. G. Welgus, R. M. Senior, and T. J. Ley.** 1992. Molecular cloning, chromosomal localization, and bacterial expression of a murine macrophage metalloelastase. *J. Biol. Chem.* **267**: 4664–4671.

1666. **Shapiro, S. D., D. K. Kobayashi, and T. J. Ley.** 1993. Cloning and characterization of a unique elastolytic metalloproteinase produced by human alveolar macrophages. *J. Biol. Chem.* **268**: 23824–23829.

1667. **Shlopov, B. V., W. R. Lie, C. L. Mainardi, A. A. Cole, S. Chubinskaya, and K. A. Hasty.** 1997. Osteoarthritic lesions: involvement of three different collagenases. *Arthritis Rheum.* **40**: 2065–2074.

1668. **Shofuda, K., H. Yasumitsu, A. Nishihashi, K. Miki, and K. Miyazaki.** 1997. Expression of three membrane-type matrix metalloproteinases (MT-MMPs) in rat vascular smooth muscle cells and characterization of MT3-MMPs with and without transmembrane domain. *J. Biol. Chem.* **272**: 9749–9754.

1669. **Singer, I. I., D. W. Kawka, E. K. Bayne, S. A. Donatelli, J. R. Weidner, H. R. Williams, J. M. Ayala, R. A. Mumford, M. W. Lark, T. T. Glant,** *et al.* 1995. VDIPEN, a metalloproteinase-generated neoepitope, is induced and immunolocalized in articular cartilage during inflammatory arthritis. *J. Clin. Invest.* **95**: 2178–2186.

1670. **Sires, U. I., B. Dublet, E. Aubert-Foucher, M. van der Rest, and H. G. Welgus.** 1995. Degradation of the COL1 domain of type XIV collagen by 92-kDa gelatinase. *J. Biol. Chem.* **270**: 1062–1067.

1671. **Sires, U. I., G. L. Griffin, T. J. Broekelmann, R. P. Mecham, G. Murphy, A. E. Chung, H. G. Welgus, and R. M. Senior.** 1993. Degradation of entactin by matrix metalloproteinases. Susceptibility to matrilysin

and identification of cleavage sites. *J. Biol. Chem.* **268**: 2069–2074.

1672. **Sires, U. I., G. Murphy, V. M. Baragi, C. J. Fliszar, H. G. Welgus, and R. M. Senior.** 1994. Matrilysin is much more efficient than other matrix metalloproteinases in the proteolytic inactivation of α_1-antitrypsin. *Biochem. Biophys. Res. Commun.* **204**: 613–620.

1673. **Sires, U. I., T. M. Schmid, C. J. Fliszar, Z. Q. Wang, S. L. Gluck, and H. G. Welgus.** 1995. Complete degradation of type X collagen requires the combined action of interstitial collagenase and osteoclast-derived cathepsin-B. *J. Clin. Invest.* **95**: 2089–2095.

1674. **Siri, A., V. Knäuper, N. Veirana, F. Caocci, G. Murphy, and L. Zardi.** 1995. Different susceptibility of small and large human tenascin-C isoforms to degradation by matrix metalloproteinases. *J. Biol. Chem.* **270**: 8650–8654.

1675. **Smith, G. N., Jr., K. A. Hasty, L. P. Yu, Jr., K. S. Lamberson, E. A. Mickler, and K. D. Brandt.** 1991. Cleavage of type XI collagen fibers by gelatinase and by extracts of osteoarthritic canine cartilage. *Matrix* **11**: 36–42.

1676. **Sopata, I. and A. M. Dancewicz.** 1974. Presence of a gelatin-specific proteinase and its latent form in human leucocytes. *Biochim. Biophys. Acta* **370**: 510–523.

1677. **Sottrup-Jensen, L. and H. Birkedal-Hansen.** 1989. Human fibroblast collagenase-α-macroglobulin interactions. Localization of cleavage sites in the bait regions of five mammalian α-macroglobulins. *J. Biol. Chem.* **264**: 393–401.

1678. **Sottrup-Jensen, L. and H. Birkedal-Hansen.** 1992. Localization of cleavage sites for human fibroblast collagenase in the bait region of five mammalian α-macroglobulins. *Matrix Suppl.* **1**: 263–268.

1679. **Stack, M. S., C. G. Emberts, and R. D. Gray.** 1991. Application of N-carboxyalkyl peptides to the inhibition and affinity purification of the porcine matrix metalloproteinases collagenase, gelatinase, and stromelysin. *Arch. Biochem. Biophys.* **287**: 240–249.

1680. **Stack, M. S. and R. D. Gray.** 1990. The effect of pH, temperature, and D_2O on the activity of porcine synovial collagenase and gelatinase. *Arch. Biochem. Biophys.* **281**: 257–263.

1681. **Stolow, M. A., D. D. Bauzon, J. Li, T. Sedgwick, V. C. Liang, Q. A. Sang, and Y. B. Shi.** 1996. Identification and characterization of a novel collagenase in *Xenopus laevis*: possible roles during frog development. *Mol. Biol. Cell.* **7**: 1471–1483.

1682. **Sudbeck, B. D., J. J. Jeffrey, H. G. Welgus, R. P. Mecham, D. McCourt, and W. C. Parks.** 1992. Purification and characterization of bovine interstitial collagenase and tissue inhibitor of metalloproteinases. *Arch. Biochem. Biophys.* **293**: 370–376.

1683. **Suzuki, M., G. Raab, M. A. Moses, C. A. Fernandez,

and M. Klagsbrun. 1997. Matrix metalloproteinase-3 releases active heparin-binding EGF-like growth factor by cleavage at a specific juxtamembrane site. *J. Biol. Chem.* **272**: 31730–31737.

1684. **Templeton, N. S., P. D. Brown, A. T. Levy, I. M. Margulies, L. A. Liotta, and W. G. Stetler-Stevenson.** 1990. Cloning and characterization of human tumor cell interstitial collagenase. *Cancer Res.* **50**: 5431–5437.

1685. **Thrailkill, K. M., L. D. Quarles, H. Nagase, K. Suzuki, D. M. Serra, and J. L. Fowlkes.** 1995. Characterization of insulin-like growth factor-binding protein 5-degrading proteases produced throughout murine osteoblast differentiation. *Endocrinology* **136**: 3527–3533.

1686. **Tschesche, H., J. Fedrowitz, U. Kohnert, J. Michaelis, and H. W. Macartney.** 1986. Matrix degrading proteinases from human granulocytes: type I, II, III collagenase, gelatinase and type IV, V-collagenase. A survey of recent findings and inhibition by γ-anticollagenase. *Folia Histochem. Cytobiol.* **24**: 125–131.

1687. **Tyler, J. A. and T. E. Cawston.** 1980. Properties of pig synovial collagenase. *Biochem. J.* **189**: 349–357.

1688. **Umenishi, F., H. Yasumitsu, Y. Ashida, J. Yamauti, M. Umeda, and K. Miyazaki.** 1990. Purification and properties of extracellular matrix-degrading metallo-proteinase overproduced by Rous sarcoma virus-transformed rat liver cell line, and its identification as transin. *J. Biochem. (Tokyo)* **108**: 537–543.

1689. **Vater, C. A., C. L. Mainardi, and E. D. Harris, Jr.** 1978. Binding of latent rheumatoid synovial collagenase to collagen fibrils. *Biochim. Biophys. Acta* **539**: 238–247.

1690. **Vater, C. A., H. Nagase, and E. D. Harris, Jr.** 1986. Proactivator-dependent activation of procollagenase induced by treatment with EGTA. *Biochem. J.* **237**: 853–858.

1691. **Vissers, M. C., P. M. George, I. C. Bathurst, S. O. Brennan, and C. C. Winterbourn.** 1988. Cleavage and inactivation of α_1-antitrypsin by metalloproteinases released from neutrophils. *J. Clin. Invest.* **82**: 706–711.

1692. **von Bredow, D. C., R. B. Nagle, G. T. Bowden, and A. E. Cress.** 1995. Degradation of fibronectin fibrils by matrilysin and characterization of the degradation products. *Exp. Cell Res.* **221**: 83–91.

1693. **von Bredow, D. C., R. B. Nagle, G. T. Bowden, and A. E. Cress.** 1997. Cleavage of β4 integrin by matrilysin. *Exp. Cell. Res.* **236**: 341–345.

1694. **Walsh, D. M., C. H. Williams, H. E. Kennedy, D. Allsop, and G. Murphy.** 1994. Gelatinase A not alpha-secretase? *Nature* **367**: 27–28.

1695. **Watanabe, H., I. Nakanishi, K. Yamashita, T. Hayakawa, and Y. Okada.** 1993. Matrix metalloproteinase-9 (92 kDa gelatinase/type IV collagenase) from U937 monoblastoid cells: correlation with cellular invasion. *J. Cell Sci.* **104**: 991–999.

1696. **Welgus, H. G., R. E. Burgeson, J. A. Wootton, R. R. Minor, C. Fliszar, and J. J. Jeffrey.** 1985. Degradation of monomeric and fibrillar type III collagens by human skin collagenase. Kinetic constants using different animal substrates. *J. Biol. Chem.* **260**: 1052–1059.

1697. **Welgus, H. G., E. J. Campbell, Z. Bar-Shavit, R. M. Senior, and S. L. Teitelbaum.** 1985. Human alveolar macrophages produce a fibroblast-like collagenase and collagenase inhibitor. *J. Clin. Invest.* **76**: 219–224.

1698. **Welgus, H. G., C. J. Fliszar, J. L. Seltzer, T. M. Schmid, and J. J. Jeffrey.** 1990. Differential susceptibility of type X collagen to cleavage by two mammalian interstitial collagenases and 72-kDa type IV collagenase. *J. Biol. Chem.* **265**: 13521–13527.

1699. **Welgus, H. G., G. A. Grant, J. C. Sacchettini, W. T. Roswit, and J. J. Jeffrey.** 1985. The gelatinolytic activity of rat uterus collagenase. *J. Biol. Chem.* **260**: 13601–13606.

1700. **Welgus, H. G., J. J. Jeffrey, and A. Z. Eisen.** 1981. The collagen substrate specificity of human skin fibroblast collagenase. *J. Biol. Chem.* **256**: 9511–9515.

1701. **Welgus, H. G., J. J. Jeffrey, A. Z. Eisen, W. T. Roswit, and G. P. Stricklin.** 1985. Human skin fibroblast collagenase: interaction with substrate and inhibitor. *Coll. Relat. Res.* **5**: 167–179.

1702. **Welgus, H. G., J. J. Jeffrey, G. P. Stricklin, and A. Z. Eisen.** 1982. The gelatinolytic activity of human skin fibroblast collagenase. *J. Biol. Chem.* **257**: 11534–11539.

1703. **Welgus, H. G., D. K. Kobayashi, and J. J. Jeffrey.** 1983. The collagen substrate specificity of rat uterus collagenase. *J. Biol. Chem.* **258**: 14162–14165.

1704. **Werb, Z., M. J. Banda, and P. A. Jones.** 1980. Degradation of connective tissue matrices by macrophages. I. Proteolysis of elastin, glycoproteins, and collagen by proteinases isolated from macrophages. *J. Exp. Med.* **152**: 1340–1357.

1705. **Whitelock, J. M., A. D. Murdoch, R. V. Iozzo, and P. A. Underwood.** 1996. The degradation of human endothelial cell-derived perlecan and release of bound basic fibroblast growth factor by stromelysin, collagenase, plasmin, and heparanases. *J. Biol. Chem.* **271**: 10079–10086.

1706. **Wilhelm, S. M., I. E. Collier, A. Kronberger, A. Z. Eisen, B. L. Marmer, G. A. Grant, E. A. Bauer, and G. I. Goldberg.** 1987. Human skin fibroblast stromelysin: structure, glycosylation, substrate specificity, and differential expression in normal and tumorigenic cells. *Proc. Natl. Acad. Sci. USA* **84**: 6725–6729.

1707. **Wilhelm, S. M., I. E. Collier, B. L. Marmer, A. Z. Eisen, G. A. Grant, and G. I. Goldberg.** 1989. SV40-transformed human lung fibroblasts secrete a 92-kDa type IV collagenase which is identical to that secreted by normal human macrophages. *J. Biol. Chem.* **264**: 17213–17221.

1708. **Wilhelm, S. M., T. Javed, and R. L. Miller.** 1984. Human gingival fibroblast collagenase: purification and properties of precursor and active forms. *Coll. Relat. Res.* **4**: 129–152.

1709. **Wilhelm, S. M., Z. H. Shao, T. J. Housley, P. K. Seperack, A. P. Baumann, Z. Gunja-Smith, and J. F. Woessner, Jr.** 1993. Matrix metalloproteinase-3 (stromelysin-1). Identification as the cartilage acid metalloprotease and effect of pH on catalytic properties and calcium affinity. *J. Biol. Chem.* **268**: 21906–21913.

1710. **Winyard, P. G., Z. Zhang, K. Chidwick, D. R. Blake, R. W. Carrell, and G. Murphy.** 1991. Proteolytic inactivation of human α_1 antitrypsin by human stromelysin. *FEBS Lett.* **279**: 91–94.

1711. **Wirl, G.** 1977. Extractable collagenase and carcinogenesis of the mouse skin. *Connect. Tissue Res.* **5**: 171–178.

1712. **Woessner, J. F., Jr. and M. G. Selzer.** 1984. Two latent metalloproteases of human articular cartilage that digest proteoglycan. *J. Biol. Chem.* **259**: 3633–3638.

1713. **Woessner, J. F., Jr. and C. J. Taplin.** 1988. Purification and properties of a small latent matrix metalloproteinase of the rat uterus. *J. Biol. Chem.* **263**: 16918–16925.

1714. **Wojtecka-Lukasik, E., J. Kaczanowska, Z. Tomczak, I. Sopata, and M. Kopec.** 1984. Effects of neutral proteases from human leukocytes on plasma fibronectin. *Thromb. Res.* **33**: 471–476.

1715. **Woolley, D. E., R. W. Glanville, M. J. Crossley, and J. M. Evanson.** 1975. Purification of rheumatoid synovial collagenase and its action on soluble and insoluble collagen. *Eur. J. Biochem.* **54**: 611–622.

1716. **Woolley, D. E., R. W. Glanville, K. A. Lindberg, A. J. Bailey, and J. M. Evanson.** 1973. Action of human skin collagenase on cartilage collagen. *FEBS Lett.* **34**: 267–269.

1717. **Wu, J. J. and D. R. Eyre.** 1995. Structural analysis of cross-linking domains in cartilage type XI collagen. Insights on polymeric assembly. *J. Biol. Chem.* **270**: 18865–18870.

1718. **Wu, J. J., M. W. Lark, L. E. Chun, and D. R. Eyre.** 1991. Sites of stromelysin cleavage in collagen types II, IX, X, and XI of cartilage. *J. Biol. Chem.* **266**: 5625–5628.

1719. **Yasumitsu, H., K. Miyazaki, F. Umenishi, N. Koshikawa, and M. Umeda.** 1992. Comparison of extracellular matrix-degrading activities between 64-kDa and 90-kDa gelatinases purified in inhibitor-free forms from human schwannoma cells. *J. Biochem. (Tokyo)* **111**: 74–80.

1720. **Yu, L. P., Jr., G. N. Smith, Jr., K. D. Brandt, and W. Capello.** 1990. Type XI collagen-degrading activity in human osteoarthritic cartilage. *Arthritis Rheum.* **33**: 1626–1633.

1721. **Yu, L. P., Jr., G. N. Smith, Jr., K. A. Hasty, and K. D. Brandt.** 1991. Doxycycline inhibits type XI collagenolytic activity of extracts from human osteoarthritic cartilage and of gelatinase. *J. Rheumatol.* **18**: 1450–1452.

1722. **Zhang, Z., P. G. Winyard, K. Chidwick, G. Murphy, M. Wardell, R. W. Carrell, and D. R. Blake.** 1994. Proteolysis of human native and oxidised α_1-proteinase inhibitor by matrilysin and stromelysin. *Biochim. Biophys. Acta* **1199**: 224–228.

See also review articles: 30, 43, 64, 93, 147.

Protein sequence alignments

Matrix metalloproteinase alignments

```
                                                SIGNAL PEPTIDE————————\/————PROPEPTIDE —

A1  COG1_BOVIN  P28053    ..........10 ..........20 ..........30 ..........40 .MPRLPLLLL 50 LLWGTGSHGF 60 PA........ 70 .......... 80 ......AT.. 90 ...SETQ... 100
A2  COG1_HUMAN  P03956    .......... .......... .......... .......... MHSFPPLLLL LFWGVVSHSF P......... .......... ......AT.. ...LETQ...
A3  COG1_RABIT  P13943    .......... .......... .......... .......... .MPGLPLLLL LLWGVGSHGF P......... .......... ......AA.. ...SETQ...
A4  COG1_RANCA  Q11133    .......... .......... .......... .......... MLSGLWSSIL ALLGVFLQSV GE........ .......... ......FR.. ...AETQ...
A5  COG1_PIG    P21692    .......... .......... .......... .......... MFS.LLLLLL LLCNTGSHGF PA........ .......... ......AT.. ...SETQ...
A6  COG2_CHICK  Q90611    .......... .......... .......... ......MK   THSVFGFFFK VLLIQVYLFN KTLA...... .......... ......APSP IIKFPGDSTP
A7  COG2_HUMAN  P08253    .......... .......... .......... ....MEALM  ARGALTGPLR ALCLLGCLLS HAAA...... .......... ......APSP IIKFPGDVAP
A8  COG2_MOUSE  P33434    .......... .......... .......... ....MEARV  AWGALAGPLR VLCVLCCLLG RAIA...... .......... ......APSP IIKFPGDVAP
A9  COG2_RABIT  P50757    .......... .......... .......... ....MEALG  ARGALAGFLR ALCVLGCLLG RATA...... .......... ......PPSP VIKFPGDVAP
A10 COG2_RAT    P33436    .......... .......... .......... ....MEARL  VWGVLVGPLR VLCVLCCLLG HAIA...... .......... ......APSP IIKFPGDVAP
A11 G257061     G257061   .......... .......... .......... .......... .......... .......... .......... .......... .......... ..........
A12 Q98857      Q98857    .......... .......... .......... .......... ...MKSLSLL LLCVVHTYAF PAV....... .......... ......PA.. ...TEDR...
A13 Q98858      Q98858    .......... .......... .......... .......... ...MKILSLL LLCAAGAYAV QEA....... .......... ......PV.. ...HEED...
A14 Q28397      Q28397    .......... .......... .......... .......... ..MKNLPILL LLCVAACSAY PLDR...... .......... ......SA.. ...RDED...
A15 COG3_HUMAN  P08254    .......... .......... .......... .......... ..MKSLPILL LLCVAVCSAY PLDG...... .......... ......AA.. ...RGED...
A16 COG3_MOUSE  P28862    .......... .......... .......... .......... ..MKGLPVLL WLCVVVCSSY PLHD...... .......... ......SA.. ...RDDD...
A17 COG3_RABIT  P28863    .......... .......... .......... .......... ..MKTLPTLL LLCVALCSAY PLDG...... .......... ......AS.. ...RDAD...
A18 COG3_RAT    P03957    .......... .......... .......... .......... ..MKGLPVLL WLCTAVCSSY PLHG...... .......... ......SE.. ...ED...
A19 COG7_FELCA  P55032    .......... .......... .......... .......... .......LCV LCLLPQSPAL PLPR...... .......... ......EAGG HS......
A20 COG7_HUMAN  P09237    .......... .......... .......... .......... ..MRLTVLCA VCLLPGSLAL PLPQ...... .......... ......EAGG MS......
A21 COG7_MOUSE  Q10738    .......... .......... .......... .......... ..MQLTLFCF VCLLPGHLAL PLSQ...... .......... ......EAGD VS......
A22 COG7_RAT    P50280    .......... .......... .......... .....M     AAMRLTLFRI VCLLPGCLAL PLSQ...... .......... ......EAGE VT......
A23 COG8_HUMAN  P22894    .......... .......... .......... .....M     FSLKTLPFLL LLHVQISKAF PV........ .......... ......S... ...SK...
A24 O70138      O70138    .......... .......... .......... .....M     FRLKTLPLLI FLHTQLANAF PV........ .......... ......P... ...EHLE...
A25 O88733      O88733    .......... .......... .......... .....M     FRLKTLPLLI FLHTQLANAF PV........ .......... ......P... ...EHLE...
A26 O88766      O88766    .......... .......... .......... .....M     LHLKTLPFLF FFHTQLATAL PV........ .......... ......PP.. ...EHLE...
A27 COG9_BOVIN  P52176    .......... .......... .......... .......... MSPLQPLVLA LLVLACCSAV PRRR...... .......... ......QP.T VVVFPGEPRT
A28 COG9_CANFA  O18733    .......... .......... .......... .......... MSPRQPLVLV FLVLGCCSAA PRPH...... .......... ......KP.T VVVFPGDLRT
A29 Q95166      Q95166    .......... .......... .......... .......... .......... .......... .......... .......... .......... ..........
A30 O19130      O19130    .......... .......... .......... .......... MSPRQPLVLV FLVLGCCSAA PRPH...... .......... ......KP.T VVVFPGDLRT
A31 Q98856      Q98856    .......... .......... .......... .......... MKPQLALLAL GLLALGCRAA PLQS...... .......... ...QPQV    RVTFPGELVS
A32 COG9_HUMAN  P14780    .......... .......... .......... .......... MSLWQPLVLV LLVLGCCFAA PRQR...... .......... ......QS.T LVLFPGDLRT
A33 COG9_MOUSE  P41245    .......... .......... .......... .......... MSPWQPLLLA LLAFGCSSAA PYQR...... .......... ......QP.T FVVFPKDLKT
A34 COG9_RABIT  P41246    .......... .......... .......... .......... MSPRQPLVLA LLVLGCCSAA PRRR...... .......... ......QP.T LVVFPGELRT
A35 COG9_RAT    P50282    .......... .......... .......... .......... MSPWQPLLLV LLALGYSFAA PHQR...... .......... ......QP.T YVVFPRDLKT
A36 COGX_HUMAN  P09238    .......... .......... .......... .......... ..MMHLAFLV LLCLPVCSAY PLSG...... .......... ......AA.. ...KEED...
A37 COGX_MOUSE  O55123    .......... .......... .......... .......... ..MEPLAILA LLSLPICSAY PLHG...... .......... ......AV.. ...T.QG...
A38 COGX_RAT    P07152    .......... .......... .......... .......... ..MEPLAILV LLCFPICSAY PLHG...... .......... ......AV.. ...R.QD...
A39 COGY_HUMAN  P24347    .......... .......... .......... .......... ...MAPAAWL RSAAARALLP PM........ .......... .......... ..........
A40 COGY_MOUSE  Q02853    .......... .......... .......... .......... ...MARAACL LRAISRVLLL PL........ .......... .......... ..........
A41 P97568      P97568    .......... .......... .......... .......... .......... .......... .......... .......... .......... ..........
A42 COGY_XENLA  Q11005    .......... .......... .......... .......... .......... ...MHLLILL P......... .......... .......... ..........
A43 COGM_HUMAN  P39900    .......... .......... .......... .......... ...MKFLLIL LLQATASGAL PLNS...... .......... ......ST.. ...SLEK...
A44 Q99745      Q99745    .......... .......... .......... .......... .......... .......... .......... .......... .......... ..........
A45 COGM_MOUSE  P34960    .......... .......... .......... .......... ..MKFLMMIV FLQVSACGAA PMN....... .......... .......... ..........
A46 COGM_RABIT  P79227    .......... .......... .......... .......... ..MKFLLLIL TLWVTSSGAD PLK....... .......... .......... ...E...
A47 Q63341      Q63341    .......... .......... .......... .......M   KFLLVLVLLV SLQVSACGAA PMN....... .......... .......... ..........
A48 O77656      O77656    .......... .......... .......... .......... MHPRVLAGFL FFSWTACWSL PLPS...... .......... ......DG.. ...DSED.LS
A49 O77632      O77632    .......... .......... .......... .......... .......... .......... .......... .......... .......... ..........
A50 Q98859      Q98859    .......... .......... .......... .......M   MPSVLSAAIF FLSLAFGLPV PVP....... .......... ......H... ...ERDSDVT
A51 O18927      O18927    .......... .......... .......... .......... MHPGVLAAFL FLSWTRCWSL PVPN...... .......... ......DD.. ...DDDDDMS
A52 COGZ_HUMAN  P45452    .......... .......... .......... .......... MHPGVLAAFL FLSWTHCRAL PLPS...... .......... ......GG.. ...DEDD.LS
A53 COGZ_MOUSE  P33435    .......... .......... .......... .......... MHSAILATFF LLSWTPCWSL PLPY...... .......... ......GD.. ...DDDDDLS
A54 O73923      O73923    .......... .......... .......... .......... .......... .......... .......... .......... .......... ..........
A55 O62806      O62806    .......... .......... .......... .......... MQPGVLAACL LLSWTHCWSL PLLN...... .......... ......SN.. ...EDDD.LS
A56 COGZ_RAT    P23097    .......... .......... .......... .......... .....ATFF  LLSWTHCWSL PLPY...... .......... ......GD.. ...DDDDDLS
A57 O77631      O77631    .......... .......... .......... .......... .......... .......... .......... .......... .......... ..........
A58 COGZ_XENLA  Q10835    .......... .......... .......... .......... ..SSLSVLVL SLSFAYCLSA PVP....... .......... ......Q... ...DEDSELT
A59 CGZA_XENLA  Q10833    .......... .......... .......... .......M   APSSLSVFVL SLSFTYCLSA PVS....... .......... ......Q... ...DEDSELT
A60 COGT_HUMAN  P50281    .......... .......... .......... ......MSP  APRPSRCLLL PLLTLGTALA SLGS...... .......... ......AQ.. ...
A61 COGT_MOUSE  P53690    .......... .......... .......... ......MSP  APRPSRSLLL PLLTLGTALA SLGW...... .......... ......AQ.. ...
A62 O08645      O08645    .......... .......... .......... ......MSP  APRPSRSLLL PLLTLGTALA SLGW...... .......... ......AQ.. ...
A63 COGT_RABIT  Q95220    .......... .......... .......... ......MSP  APRPSRRLLL PLLTLGTALA SLGS...... .......... ......AQ.. ...
A64 COGT_RAT    Q10739    .......... .......... .......... ......MSP  APRPSRSLLL PLLTLGTTLA SLGW...... .......... ......AQ.. ...
A65 COGU_HUMAN  P51511    .......... ........MG SDPSAPGRPG WTGSLLGDRE EAARPRLLPL LLVLLGCLGL GVAA...... .......... ......ED.  ..........
A66 Q14111      Q14111    .......... .......... .......... .......... .......... .......... .......... .......... .......... ..........
A67 D1023087    D1023087  .......... .......... .......... .......... .......... .......... .......... .......... .......... ..........
A68 O54732      O54732    .......... ........MG SDRSALGRPG CTGSCLSSR. ....ASLLPL LLVLLDCLGH GTAS...... .......... ......KD.. ...
A69 Q98947      Q98947    .......... .......... .......... .........M IVLALSTGSR FLQTCVWILC ATVC...... .......... ......KA.. ..........
A70 COGV_HUMAN  P51512    .......... .......... .......... .........M ILLTFSTGRR LDFVHHSGVF FLQTLLWILC ATVC...... ......GT.. ..........
A71 Q14824      Q14824    .......... .......... .......... .........M ILLTFSTGRR LDFVHHSGVF FLQTLLWILC ATVC...... ......GT.. ..........
A72 COGV_RAT    O35548    .......... .......... .......... .........M ILLAFSSGRR LDFVHRSGVF FFQTLLWILC ATVC...... ......GT.. ..........
A73 O35541      O35541    .......... .......... .......... .........M ILLAFSSGRR LDFVHRSGVF FFQTLLWILC ATVC...... ......GT.. ..........
A74 Q14850      Q14850    .......... .......... .......... .......... .......... .......... .......... .......... .......... ..........
A75 O13065      O13065    .......... .......... .......... .......... ..MNSLLLKL LLCVAITAAF PAD....... .......... ......KQ.. ...DEPP...
A76 Q99542      Q99542    .......... .......... .......... .......... .......... .......... .......... .MNCQQLWLG FLLPMTVSGR
A77 Q99580      Q99580    .......... .......... .......... .......... .......... .......... .......... .MNCQQLWLG FLLPMTVSGR
A78 O15278      O15278    .......... .......... .......... .......... .......... .......... .......... .......... .......... ..........
A79 O18767      O18767    .......... .......... .......... .........M LPASGLAVLL VTALKFSTAA PSLP...... .......... ......AA.. ...SPRTS..
A80 O60882      O60882    .......... .......... .......... .......MKV LPASGLAVFL IMALKFSTAA PSLV...... .......... ......AA.. ...SPRTW..
A81 P79287      P79287    .......... .......... .......... .......MKV LPASGLAVLL VTALKFSAAA PSLF...... .......... ......AA.. ...TPRTS..
A82 O04529      O04529    .......... .......... .......... .......MRF CVFGFLSLFL IVSPASAWFF PN........ .STAVPPS.L RNTTRVFWDA FSNFTGCHHG
A83 O23507      O23507    .......... .......... .......... ......MSRN LIYRRNRALC FVLILFCFPY RFGARNTPEA EQSTAKATQI IHVSNSTWHD FSRLVDVQIG
A84 O48680      O48680    .......... .......... .......... .......... ...MVFLLFF APSPVSAGFY TN........ .SSAIPPQLL RNATGNPWNS FLNFTGCHAG
A85 O65340      O65340    .......... .......... .......... .......MHH HHHPCNRKPF TTIFSFFLLY LN........ ...LHNQQI  IEARNPSQFT TNPSPDVSIP
A86 O16901      O16901    .......... .......... ...MNHHTYF HNFCPNCPNS FLSLYVCVYM NVTIYRKMFT NSS....... .ILVKIVYDC GNSSPSSSSS SSSFSNSRKP
A87 O61264      O61264    .......... .......... .......... .......... MRLIYVIAIL LVSTCQAGFF SS........ ..LVSRFTGG GNSSPSSSSS SSSFSNSRKP
A88 O61266      O61266    .......... .......... .......... .......... MFTGLHDILI ILFLLVTLKI AQN....... .......... .......... ..........
A89 O61265      O61265    .......... .......... .......... .......... .......... .......... .......... .......... .......... ..........
A90 O44836      O44836    .......... .......... .......... .......... .......... .......... .......... ....MKNDKI FIISTCYHQT LKIFFSIQKF
A91 O55761      O55761    .......... .......... .......... .......... .......... .......... .......... .......... .......... ..........
A92 MEP1_SOYBN  P29136    .......... .......... .......... .......... ..MTLRNHQ  ELLVALATLY FL........ .ATSLPSVSA HGPYAWDGEA TYKFTTYHPG
A93 HE_HEMPU    P91953    MANFCLIFAA VFLTRLTTVL NTPISVTFG. .PTQLTDITK LVSETGGDFG LTTPRSAILT TVSEDDSDDD DGGESVEDTT ILQTTTSSEI VISGIVVDED
A94 HE_PARLI    P22757    MANSGLILLV MFMIHVTTVH NVPLPSTAPS IITQLSDITT SIIEE.DAFG LTTPTGLLT  PVSENDSDDD .GD..DITT  IQTTTSSQT  LVKDMLSAQQ
A95 O93470      O93470    .......... .......... .......... ......MPS  IKLLVWCCLC VISPRLCHSE KLFHSRDRSD LQPSAIEQAE LVKDMLSAQQ FLAKYGWTQP
A96 P89294      P89294    .......... .......... .......... .......... .......... .......... .......... .......... .......... ..........
```

PROPEPTIDE ──────────────────────┐
                                                                                                          Cys switch

ID	110	120	130	140	150	160	170	180	190	200
A1	E..QDVETVK	KY.LENYYNL	N.SNGKKVER	......QRNG	GLITEKLKQM	QKFFGLRVTG	KPDAETLNVM	KQPRCGVPDV	AP........	....
A2	E..QDVDLVQ	KY.LEKYYNL	K.NDGRQVEK	.....RRNS	GPVVEKLKQM	QEFFGLKVTG	KPDAETLKVM	KQPRCGVPDV	AQ........	
A3	E..QDVEMVQ	KY.LENYYNL	K.DEWRKIPK	.....QRGN	GLAVEKLKQM	QEFFGLKVTG	KPDAETLKMM	KQPRCGVPDV	AQ........	
A4	E..QDVEIVQ	KY.LKNYYNS	...E......	.....KRNS	GLVVEILK..	.QFFGLKVTG	KPDAETL.VM	KQSTCGVPDV	GE........	
A5	E..QDVEIVQ	KY.LKNYYNL	N.SDGVPVEK	.....KRNS	GLVVEKLKQM	QQFFGLKVTG	KPDAETLNVM	KQPRCGVPDV	AE........	
A6	K..TDKELAV	QY.LNKYYGC	PKDNCN....	L	FVLKDTLKKM	QKFFGLPETG	DLDQNTIETM	KKPRCGNPDV	AN........	
A7	K..TDKELAV	QY.LNTFYGC	PKESCN....	L	FVLKDTLKKM	QKFFGLPQTG	DLDQNTIETM	RKPRCGNPDV	AN........	
A8	K..TDKELAV	QY.LNTFYGC	PKESCN....	L	FVLKDTLKKM	QKFFGLPQTG	DLDQNTIETM	RKPRCGNPDV	AN........	
A9	K..TDKELAV	QY.LNTFYGC	PKDSCN....	L	FVLKDTLKKM	QKFFGLPQTG	ELDQSTIETM	RKPRCGNPDV	AN........	
A10	K..TDKELAV	QY.LNTFYGC	PKESCN....	L	FVLKDTLKKM	QKFFGLPQTG	DLDQNTIETM	RKPRCGNPDV	AN........	
A11	....	....	....	....	....	....	....	....	....	
A12	G..ENEQLAE	EY.LKKFYNL	N.EDGTPITR	.......KKH	SPFSEKLQEM	QAFFGLEVTG	KLDSNTLEMM	HKPRCGVADV	AE........	
A13	D..TIRQDVE	EY.LKKYYGL	N.SDKTPDLR	.......KAA	SPLAEKIREM	QKFCGLQVTG	KVDSNTLEVM	QQPRCGVSDV	AA........	
A14	...SNMDLLQ	DY.LEKYYDL	G.KEMRQYVR	......RKDS	GPIVKKIQEM	QKFLGLKVTG	KLDSDTVEVM	HKSRCGVPDV	GH........	
A15	...TSMNLVQ	KY.LENYYDL	K.KDVKQFVR	......RKDS	GPVVKKIREM	QKFLGLEVTG	KLDSDTLEVM	RKPRCGVPDV	GH........	
A16	...AGMELLQ	KY.LENYYGL	A.KDVKQFIK	......KKDS	SLIVKKIQEM	QKFLGLEMTG	KLDSNTMELM	HKPRCGVPDV	GG........	
A17	T..TNMDLLQ	QY.LENYYNL	E.KDVKQFVK	......RKDS	SPVVKKIQEM	QKFLGLEVTG	KLDSNTLEVI	RKPRCGVPDV	GH........	
A18	...AGMEVLQ	KY.LENYYGL	E.KDVKQFTK	......KKDS	SPVVKKIQEM	QKFLGLKMTG	KLDSNTMELM	HKPRCGVPDV	GG........	
A19	..ESQWKQAQ	EY.LKRFYPS	DAKSRD....	.........A	DSFGAQLKEM	QKFFRLPVTG	MLDSRVIVVM	QQPRCGLPDT	GE........	
A20	.ELQWEQAQ	DY.LKRFYLY	DSETKN....	.........A	NSLEAKLKEM	QKFFGLPITG	MLNSRVIEIM	QKPRCGVPDV	AE........	
A21	.AHQWEQAQ	NY.LRKFYPH	DSKTKK....	V	NSLVDNLKEM	QKFFGLPMTG	KLSPYIMEIM	QKPRCGVPDV	AE........	
A22	.ALQWEQAQ	NY.LRKFYLH	DSKTKK....	.........A	TSAVDKLREM	QKFFGLPETG	KLSPRVMEIM	QKPRCGVPDV	AE........	
A23	E..KNTKTVQ	DY.LEKFYQL	P.SNQYQSTR	.....KNGT	NVIVEKLKEM	QRFFGLNVTG	KPNEETLDMM	KKPRCGVPDS	GG........	
A24	E..KNIKTAE	NY.LRKFYNL	P.SNQFRSSR	.....N..A	TMVAEKLKEM	QRFFSLAETG	KLDAATMGIM	EMPRCGVPDS	GD........	
A25	E..KNIKTAE	NY.LRKFYNL	P.SNQFRSSR	.....N..A	TMVAEKLKEM	QRFFSLAETG	KLDAATMGIM	EMPRCGVPDS	GD........	
A26	E..KNMKTAE	NY.LRKFYHL	P.SNQFRSAR	.....N..A	TMIAEKLKEM	QRFFGLPETG	KPDAATIEIM	EKPRCGVPDS	GD........	
A27	N.LTNRQLAE	EY.LYRYGYT	PGAELSE...	........DG	QSLQRALLRF	QRRLSLPETG	ELDSTTLNAM	RAPRCGVPDV	GR........	
A28	N.LTDKQLAE	EY.LFRYGYT	QVAELSN...	........DK	QSLSRGLRLL	QRRLALPETG	ELDKTTLEAM	RAPRWAFPDL	GK........	
A29	....	....	....	....	....	....	....	....	....	
A30	N.LTDKQLAE	EY.LFRYGYT	QVAELSN...	........DK	QSLSRGLRLL	QRRLALPETG	ELDKTTLEAM	RAPRCGVPDL	GK........	
A31	G.ISDDELAE	SY.LERFGYI	SKRARSS...	........TH	VSLSKALLQM	QKKLGLNETG	ELDQSTLEAM	KTPRCGVPDV	GN........	
A32	N.LTDRQLAE	EY.LYRYGYT	RVAEMRG...	........ES	KSLGPALLLL	QKQLSLPETG	ELDSATLKAM	RTPRCGVPDL	GR........	
A33	SNLTDTQLAE	AY.LYRYGYT	RAAQMMG...	........EK	QSLRPALLML	QKQLSLPQTG	ELDSQTLKAI	RTPRCGVPDV	GR........	
A34	R.LTDRQLAE	EY.LFRYGYT	RVASMHG...	........DS	QSLRLPLLLL	QKHLSLPETG	ELDNATLEAM	RAPRCGVPDV	GK........	
A35	SNLTDTQLAE	DY.LYRYGYT	RAAQMMG...	........EK	QSLRPALLML	QKQLSLPQTG	ELDSETLKAI	RSPRCGVPDV	GK........	
A36	...SNKDLAQ	QY.LEKYYNL	E.KDVKQFR.	......RKDS	NLIVKKIQGM	QKFLGLEVTG	KLDTDTLEVM	RKPRCGVPDV	GH........	
A37	H..PSMDLAQ	QY.LENYYNF	K.KNEKQIFK	......RKDS	SPVVKKIQEM	QKFLGLEMTG	KLDSNTMELM	HKPRCGVPDV	GG........	
A38	H..STMDLAQ	QY.LENYYNF	R.KNEKQFFK	......RKDS	SPVVKKIEEM	QKFLGLEMTG	KLDSNTVEMM	HKPRCGVPDV	GG........	
A39	...LLLLLQ.	....PPPLLA	RALPPDVHHL	......HAER	RGPQPWHAAL	PSSPAPAPAT	QEAPRPASSL	RPPRCGVPD.	..PSDGLSAR	N.
A40	...PLLLLLL	LL.LPSPLMA	RARPPESHRH	......HPVK	KGPRLLHAAL	PNTLTSVPAS	HWVPSPAGSS	RPLRCGVPD.	..LPDVLNAR	N.
A41		.......MA	RARPPENHRH	......RPVK	RVPQLLPAAL	PNSLSPSVPAS	HWVPGPASSS	RPLRCGVPD.	..PPDVLNAR	N.
A42	...ALCVLG.	....AHSA	PLS..YTYLQ	......HRIQ	EKPQKDHGRL	QFNSLHYPHI	KGLLNAHGSW	NPPRCGVPDV	PAPPDSSSGR	N.
A43	...NNVLFGE	RY.LFKFYGI	FTNKIPVTKM	......KYSG	NLMKEKIQEM	QHFLQLKVTG	QLDTDTLDMM	HAFRCGVPDV	HH........	
A44	....	....	....	....	....	....	....	....	....	
A45	...DSEFAE	WY.LSRFYDY	GKDRIPMTKT	......KTNR	NFLKEKLQEM	QQFFGLEATG	QLDNSTLAIM	HIPRCGVPDV	QH........	
A46	...NDMLFAE	NY.LENFYGL	KVERIPMTKM	......KTNR	NFIEEKVQEM	QQFLGLNVTG	QLDTSTLEMM	HKPRCGVPDV	YH........	
A47	...ESEFAE	WY.LSRFFDY	QGDRIPMTKT	......KTNR	NLLEEKLQEM	QQFFGLEVTG	QLDTSTLKIM	HTSRCGVPDV	QH........	
A48	E..EDFQFAE	SY.LKSYYYP	Q..NPAGILK	......KTAA	SSVIDRLREM	QSFFGLEVTG	RLDDNTLDIM	KKPRCGVPDV	GE........	
A49	....	....	....	....	....	....	....	....	....	
A50	E..QELRLAE	KY.LKTFYVA	S..DHAGIMT	......KKGG	NALASKLREM	QSFFDLEVTG	KLDEDTLEVM	KQPRCGVPDV	GE........	
A51	E..EDFQLAE	RY.LKSYYYP	L..NPAGILK	......KTAA	NSVVDRLREM	QSFFGLEVTG	KLDDNTLDIM	KKPRCGVPDV	GE........	
A52	E..EDLQFAE	RY.LRSYYHP	T..NLAGILK	......ENAA	SSMTERLREM	QSFFGLEVTG	KLDDNTLDVM	KKPRCGVPDV	GE........	
A53	E..EDLVFAE	HY.LKSYYHP	A..TLAGILK	......KSTV	TSTVDRLREM	QSFFGLEVTG	KLDDPTLDIM	RKPRCGVPDV	GE........	
A54	....	....	....	....	....	....	....	....	....	
A55	E..EDFQFAE	SY.LRSYYHP	L..NPAGILK	......KNAA	GSMVDRLREM	QSFFGLEVTG	KLDDNTLAIM	KQPRCGVPDV	GE........	
A56	E..EDLEFAE	HY.LKSYYHP	V..TLAGILK	......KSTV	TSTVDRLREM	QSFFGLDVTG	KLDDPTLDIM	RKPRCGVPDV	GV........	
A57	....	....	....	....	....	....	....	....	....	
A58	P..GDLQLAE	HY.LNRLYSS	SS.NFVGMLR	......MKNV	NSIETKLKEM	QSFFGLEVTG	KLNEDTLDIM	KQPRCGVPDV	GQ........	
A59	P..GALQLAE	HY.LNRLYSS	SS.NFAGMLR	......MKDV	NSVETKLKEM	QSFFGLEVTG	KLNEDTLDIM	KQPRCGVPDV	GQ........	
A60	...SSSFSPE	AW.LQQYGYL	PPGDLRTHTQ	......RSP	QSLSAAIAAM	QKFYGLQVTG	KADADTMKAM	RRPRCGVPDK	FGAEIKANVR	R.
A61	...GSNFSPE	AW.LQQFGYL	PRGDLRTHTQ	......RSP	QTLSVDIAAI	QKFYGLYVTG	KAYSETMKAM	RRPRCGVPDK	FGTEIKANVR	R.
A62	...GSNFSPE	AW.LQQYGYL	PPGDLRTHTQ	......RSP	QSLSAAIAAM	QKFYGLQVTG	KADLATMMAM	RRPRCGVPDK	FGTEIKANVR	R.
A63	...SNSFSPE	AW.LQQYGYL	PPGDLRTHTQ	......RSP	QSLSAAIAAM	QRFYGLRVTG	KADTDTMKAM	RRPRCGVPDK	FGAEIKANVR	R.
A64	...SSNFSPE	AW.LQQYGYL	PPGDLRTHTQ	......RSP	QSLSAAIAAI	QRFYGLQVTG	KADSDTMKAM	RRPRCGVPDK	FGTEIKANVR	R.
A65	...AEVHAEN	.W.LRLYGYL	PQPSRHMSTM	......RSA	QILASALAEM	QRFYGIPVTG	VLDEETKEWM	KRPRCGVPDQ	FGVRVKANLR	R.
A66	....	....	....	....	....	....	..........M	KRPRCGVPDQ	FGVRVKANLR	R.
A67	....	....	....	....	....	....	....	....	....	
A68	...AEVYAAE	NW.LRLYGYL	PQPSRHMSTM	......RSA	QILASALAEM	QSFYGIPVTG	VLDEETKTWM	KRPRCGVPDQ	FGVHVKANLR	R.
A69	...EQYFNVE	VW.LQKYGYL	PPTDPRMSVL	......RSA	ETMQSAIAAM	QQFYGINMTG	KVDRNTIDWM	KKPRCGVPDQ	TRGSSKFNIR	R.
A70	...EQYFNVE	VW.LQKYGYL	PPTDPRMSVL	......RSA	ETMQSALAAM	QQFYGINMTG	KVDRNTIDWM	KKPRCGVPDQ	TRGSSKFHIR	R.
A71	...EQYFNVE	VW.LQKYGYL	PPTDPRMSVL	......RSA	ETMQSALAAM	QQFYGINMTG	KVDRNTIDWM	KKPRCGVPDQ	TRGSSKFHIR	R.
A72	...EQYFNVE	VW.LQKYGYL	PPTDPRMSVL	......RSA	ETMQSALAAM	QQFYGINMTG	KVDRNTIDWM	KKPRCGVPDQ	TRGSSKFNIR	R.
A73	...EQYFNVE	VW.LQKYGYL	PPTDPRMSVL	......RSA	ETMQSALAAM	QQFYGINMTG	KVDRNTIDWM	KKPRCGVPDQ	TRGSSKFNIR	R.
A74	....	....	....	....	...MQQFGGL	EATGILDEAT	LALM.....K	TP.RCSLPD.	..LPVLTQAR	R.
A75	A..TKEEMAE	NY.LKRFYSL	G.TDGSPVGR	......KKHI	QPFTEKLEQM	QKFFGLKVTG	TLDPKTVEVM	EKPRCGVYDV	GQ........	
A76	VLGLAEVAPV	DY.LSQYGYL	QKPLEJSNNF	K........P	EDITEALRAF	QEASELPVSG	QLDDATRARM	RQPRCGLEDP	FN........	
A77	VLGLAEVAPV	DY.LSQYGYL	QKPLEJSNNF	K........P	EDITEALRAF	QEASELPVSG	QLDDATRARM	RQPRCGLEDP	FN........	
A78	....	....	....	....	....	....	....	....	....	
A79	R..NNYRLAQ	AY.LDKYYTK	KGGPQIGEMV	......ARGG	NSTVKKIKEL	QEFFGLRVTG	KLDRATMDVI	KRPRCGVPDV	AN........	
A80	R..NNYRLAQ	AY.LDKYYTN	KEGHQIGEMV	......ARGS	NSMIRKIKEL	QAFFGLQVTG	KLDQTTMNVI	KKPRCGVPDV	AN........	
A81	R..NNYHLAQ	AY.LDKYYTK	KGGHQVGEMV	......AKGG	NSMVKKIKEL	QAFFGLRVTG	KLDRTTMDVI	KRPRCGVPDV	AN........	
A82	QNVDGLYRIK	KY.FQRFGYI	PET.FSGNFT	......DDFD	DILKAAVELY	QTNFNLNVTG	ELDALTIQHI	VIPRCGNPDV	VNGTSLMHGG	RRKTFEV...
A83	SHVSGVSELK	RY.LHRFGYV	NDGSE..IFS	......DVFD	GPLESAISLY	QENLGLPITG	RLDTSTVTLM	SLPRCGVSDT	HMTINNDFL.	..HTTA....
A84	KKYDGLYMLK	QY.FQHFGYI	TETNLSGNFT	......DDFD	DILKNAVEMY	QRNFQLNVTG	VLDELTLKHV	VIPRCGNPDV	VNGTSMHSG	R.KTFEV...
A85	.......EIK	RH.LQQYGYL	PQNK......	......ESDD	VSFEQALVRY	QKNLGLPITG	KPDSDTLSQI	LLPRCGFPD.	DVEPKTAPF.	..HTGK....
A86	SLSD..EKAR	SY.LQTFGYV	PPSNSLQSRN	GMAGDIQSAE	QVFKSAIRKF	QEFAGIAKTG	FLDAATKAKM	ALSRCGVTDA	PLALTS....	
A87	SLSD..EKAR	SY.LQTFGYV	PPSNSLQSRN	GMAGDIQSAE	QVFKSAIRKF	QEFAGIAKTG	FLDAATKAKM	ALSRCGVTDA	PLALTS....	
A88	.......VDH	TKFLQKYGYL	TSGDNQ..LS	.........S	ESLSDALKNM	QRMAGLEETG	ELDERTIQMM	ERPRCGHPDV	EDHQKSRG..	
A89	....	....	....	....	......VEQTG	NVDEMTVEAA	SKPRCTQTDV	RQE.QTKRTK	RFSSLKTSQV	
A90	LPEKIFFFLT	GQ.MEKYGYL	KGID......	......HSSP	QEFRQALMFF	QEVLEVEQTG	NVDEMTVEAA	SKPRCTQTDV	RQE.QTKRTK	RFTLSKRAKW
A91	....	....	....	....	....	....	....	....	....	
A92	QNYKGLSNVK	NY.FHHLGYI	PNA...PHFD	......DNFD	DTLVSAIKTY	QKNYNLNVTG	KFDINTLKQI	MTPRCGVPDI	I....INTN.	KTTSPG....
A93	IDESKVVKLK	AN.LEQFGYV	PLGS....TF	......GEAN	INYTSAILEY	QQNGGINQTG	ILDAETAALL	DTPRCGVPDI	LP........	
A94	VHESNVEILK	AH.LEKFGYT	PPGS....TF	......GEAN	LNYTSAILDF	QEHGGINQTG	ILDADTAELL	STPRCGVPDV	LP........	
A95	VIWDPSSTNE	NEPLKDFSLM	QEGVCNPRQE	...VAEPTKS	PQFIDALKKF	QKLNNLPVTG	TLDDATINAM	NKPRCGVPDN	QMAKKETEKP	TAAQSLENKT
A96	....	....	....	....	....	....	....	....	....	

┌─ CATALYTICS DOMAIN ─┐

```
                           210         220         230         240         250        260        270         280         290         300
A1  COG1_BOVIN    ..........  .....FVLTP  GKSCWENTN.  ..LTYRIENY  TPDL......  ..........  ..........  SRADVDQ...  AIEKAFQLWS  NVTPLTFTKV
A2  COG1_HUMAN    ..........  .....FVLTE  GNPRWEQTH.  ..LTYRIENY  TPDL......  ..........  ..........  PRADVDH...  AIEKAFQLWS  NVTPLTFTKV
A3  COG1_RABIT    ..........  .....FVLTP  GNPRWEQTH.  ..LTYRIENY  TPDL......  ..........  ..........  SRADVDN...  AIEKAFQLWS  NVTPLTFTKV
A4  COG1_RANCA    ..........  .....YVLTP  GNPRWENTH.  ..LTYRIENY  TPDL......  ..........  ..........  ..........  ..........  .VSPLTFTKV
A5  COG1_PIG      ..........  .....FVLTP  GNPRWENTH.  ..LTYRIENY  TPDL......  ..........  ..........  SREDVDR...  AIEKAFQLWS  NVSPLTFTKV
A6  COG2_CHICK    ..........  .....YNFFP  RKPKWEKNH.  ..ITYRIIGY  TPDL......  ..........  ..........  DPETVDD...  AFARAFKVWS  DVTPLRFNRI
A7  COG2_HUMAN    ..........  .....YNFFP  RKPKWDKNQ.  ..ITYRIIGY  TPDL......  ..........  ..........  DPETVDD...  AFARAFQVWS  DVTPLRFSRI
A8  COG2_MOUSE    ..........  .....YNFFP  RKPKWDKNQ.  ..ITYRIIGY  TPDL......  ..........  ..........  DPETVDD...  AFARALKVWS  DVTPLRFSRI
A9  COG2_RABIT    ..........  .....YNFFP  RKPKWDKNQ.  ..ITYRIIGY  TPDL......  ..........  ..........  DPETVDD...  AFARAFQVWS  NVTPLRFSRI
A10 COG2_RAT      ..........  .....YNFFP  RKPKWDKNQ.  ..ITYRIIGY  TPDL......  ..........  ..........  DPETVDD...  AFARALKVWS  DVTPLRFSRI
A11 G257061       ..........  .....FTTFP  GMPKWRKXH.  ..LTYRIMNY  XP........  ..........  ..........  ..........  ..........  ..........
A12 Q98857        ..........  .....YSHFG  GRPTWRTTS.  ..LTYRILGY  TPDM......  ..........  ..........  AEADVDT...  AIRRAFKVWS  DVTPLTFSRI
A13 Q98858        ..........  .....YSTFP  GRPAWRTHA.  ..LTYRILNY  TPDM......  ..........  ..........  ARADVDT...  AIQKAFKVWS  DVTPLTFTQI
A14 Q28397        ..........  .....FTTFP  GMPKWSKTH.  ..LTYRIVNY  TQDL......  ..........  ..........  PRDAVDS...  DVEKALKIWE  EVTPLTFSRI
A15 COG3_HUMAN    ..........  .....FRTFP  GIPKWRKTH.  ..LTYRIVNY  TPDL......  ..........  ..........  PKDAVDS...  AVEKALKVWE  EVTPLTFSRL
A16 COG3_MOUSE    ..........  .....FSTFP  GSPKWRKSH.  ..ITYRIVNY  TPDL......  ..........  ..........  PRQSVDS...  AIEKALKVWE  EVTPLTFSRI
A17 COG3_RABIT    ..........  .....FSTFP  GTPKWTKTH.  ..LTYRIVNY  TPDL......  ..........  ..........  PRDAVDA...  AIEKALKVWE  EVTPLTFSRK
A18 COG3_RAT      ..........  .....FSTFP  GSPKWRKNH.  ..ISYRIVNY  TLDL......  ..........  ..........  PRESVDS...  AIERALKVWE  EVTPLTFSRI
A19 COG7_FELCA    ..........  .....DLPSR  NRPKWISKV.  ..VTYRIISY  TRDL......  ..........  ..........  PRVTVDH...  LVAKALNMWS  KEIPLSFRRV
A20 COG7_HUMAN    ..........  .....YSLFP  NSPKWTSKV.  ..VTYRIVSY  TRDL......  ..........  ..........  PHITVDR...  LVSKALNMWG  KEIPLHFRKV
A21 COG7_MOUSE    ..........  .....YSLMP  NSPKWHSRI.  ..VTYRIVSY  TSDL......  ..........  ..........  PRIVVDQ...  IVKKALRMWS  MQIPLNFKRV
A22 COG7_RAT      ..........  .....FSLMP  NSPKWHSRT.  ..VTYRIVSY  TTDL......  ..........  ..........  PRFLVDQ...  IVKRALRMWS  MQIPLNFKRV
A23 COG8_HUMAN    ..........  .....FMLTP  GNPKWERTN.  ..LTYRIRNY  TPQL......  ..........  ..........  SEAEVER...  AIKDAFELWS  VASPLIFTRI
A24 O70138        ..........  .....FLLTP  GSPKWTHTN.  ..LTYWIINH  TPQL......  ..........  ..........  SRAEVKT...  AIEKAFHVWS  VASPLTFTEI
A25 O88733        ..........  .....FLLTP  GSPKWTHTN.  ..LTYRIINH  TPQL......  ..........  ..........  SRAEVKT...  AIEKAFHVWS  VASPLTFTEI
A26 O88766        ..........  .....FLLTP  GSPKWTHTN.  ..LTYRIINH  TPQM......  ..........  ..........  SKAEVKT...  EIEKAFKIWS  VPSTLTFTET
A27 COG9_BOVIN    ..........  .....FQTFE  GELKWHHHN.  ..ITYWIQNY  SEDL......  ..........  ..........  PRAVIDD...  AFARAFALWS  AVTPLTFTRV
A28 COG9_CANFA    ..........  .....FQTFE  GDLKWHHND.  ..ITYWIQNY  SEDL......  ..........  ..........  PRDVIDD...  AFARAFAVWS  AVTPLTFTRV
A29 Q95166        ..........  ..........  ..........  ..........  ..........  ..........  ..........  ..........  ..........  ..........
A30 O19130        ..........  .....FQTFE  GDLKWHHND.  ..ITYWIQNY  SEEL......  ..........  ..........  PRDVIDD...  AFARAFAVWS  AVTPLTFTRV
A31 Q98856        ..........  .....FQTFD  GDLKWDHND.  ..ITFRVLNY  SPDL......  ..........  ..........  DGDVIED...  AFRRAFKVWS  DVSPLTFTQI
A32 COG9_HUMAN    ..........  .....FQTFE  GDLKWHHHN.  ..ITYWIQNY  SEDL......  ..........  ..........  PRAVIDD...  AFARAFALWS  AVTPLTFTRV
A33 COG9_MOUSE    ..........  .....FQTFK  .GLKWDHHN.  ..ITYWIQNY  SEDL......  ..........  ..........  PRDMIDD...  AFARAFAVWG  EVAPLTFTRV
A34 COG9_RABIT    ..........  .....FQTFE  GDLKWHHHN.  ..ITYWIQNY  SEDL......  ..........  ..........  PRDVIDD...  AFARAFALWS  AVTPLTFTRV
A35 COG9_RAT      ..........  .....FQTFD  GDLKWHHHN.  ..ITYWIQSY  TEDL......  ..........  ..........  PRDVIDD...  SFARAFAVWS  AVTPLTFTRV
A36 COGX_HUMAN    ..........  .....FSSFP  GMPKWRKTH.  ..LTYRIVNY  TPDL......  ..........  ..........  PRDAVDS...  AIEKALKVWE  EVTPLTFSRL
A37 COGX_MOUSE    ..........  .....FSTFP  GSPKWRKSH.  ..ITYRIVNY  TPDL......  ..........  ..........  PRQSVDS...  AIEKALKVWE  EVTPLTFSRI
A38 COGX_RAT      ..........  .....FSTFP  GSPKWRKNH.  ..ISYRIVNY  TLDL......  ..........  ..........  PRESVDS...  AIERALKVWE  EVTPLTFSRI
A39 COGY_HUMAN    ..........  ...RQKRFVL  SGGRWEKTD.  ..LTYRILRF  PWQ.......  ..........  ........LV  .QEQVRQ...  TMAEALKVWS  DVTPLTFTEV
A40 COGY_MOUSE    ..........  ...RQKRFVL  SGGRWEKTD.  ..LTYRILRF  PWQ.......  ..........  ........LV  .REQVRQ...  TVAEALQVWS  EVTPLTFTEV
A41 P97568        ..........  ...RQKRFVL  SGGRWEKTD.  ..LTYRILRF  PWQ.......  ..........  ........LV  .REQVRQ...  TVAEALRVWS  EVTPLTFTEV
A42 COGY_XENLA    ..........  ...RQKRFVL  SGGRWDKTN.  ..LTYKIIRF  PWQ.......  ..........  ........LS  .KVKVRR...  TIAEALKVWS  EVTPLTFTEV
A43 COGM_HUMAN    ..........  .....FREMP  GGPVWRKHY.  ..ITYRINNY  TPDM......  ..........  ..........  NREDVDY...  AIRKAFQVWS  NVTPLKFSKI
A44 Q99745        ..........  ..........  ..........  ..........  ..........  ..........  ..........  ..........  ..........  ..........
A45 COGM_MOUSE    ..........  .....LRAVP  QRSRWMKRY.  ..LTYRIYNY  TPDM......  ..........  ..........  KREDVDY...  IFQKAFQVWS  DVTPLRFRKL
A46 COGM_RABIT    ..........  .....FKTMP  GRPVWRKHY.  ..ITYRIKNY  TPDM......  ..........  ..........  KREDVEY...  AIQKAFQVWS  DVTPLKFRKI
A47 Q63341        ..........  .....LRAVP  QRSRWMKRY.  ..LTYRIYNY  TPDM......  ..........  ..........  KRADVDY...  IFQKAFQVWS  DVTPLRFRKI
A48 O77656        ..........  .....YNVFP  RTLKWSKMN.  ..LTYRIVNY  TPDL......  ..........  ..........  THSEVEK...  AFRKAFKVWS  DVTPLNFTRI
A49 O77632        ..........  ..........  ..........  ..........  ..........  ..........  ..........  ..........  ..........  ..........
A50 Q98859        ..........  .....YNVFP  RSLKWPRFN.  ..LTYRIENY  TPDM......  ..........  ..........  THAEVDR...  AIKKAFRVWS  EVTPLNFTRL
A51 O18927        ..........  .....YNVFP  RTLKWPKMN.  ..LTYRIVNY  TPDL......  ..........  ..........  THSEVEK...  AFKKAFKVWS  DVTPLNFTRL
A52 COGZ_HUMAN    ..........  .....YNVFP  RTLKWSKMN.  ..LTYRIVNY  TPDM......  ..........  ..........  THSEVEK...  AFKKAFKVWS  DVTPLNFTRL
A53 COGZ_MOUSE    ..........  .....YNVFP  RTLKWSQTN.  ..LTYRIVNY  TPDM......  ..........  ..........  SHSEVEK...  AFRKAFKVWS  DVTPLNFTRI
A54 O73923        ..........  ..........  ..........  ..........  ..........  ..........  ..........  ..........  ..........  ...PLRFTRI
A55 O62806        ..........  .....YNVFP  RTLKWSQTN.  ..LTYRIVNY  TPDL......  ..........  ..........  THSEVEK...  AFKKAFKVWS  DVTPLNFTRI
A56 COGZ_RAT      ..........  .....YNVFP  RTLKWSQTN.  ..LTYRIVNY  TPDI......  ..........  ..........  SHSEVEK...  AFRKAFKVWS  DVTPLNFTRI
A57 O77631        ..........  ..........  ..........  ..........  ..........  ..........  ..........  ..........  ..........  ..........
A58 COGZ_XENLA    ..........  .....YNFFP  RKLKWPRNN.  ..LTYRIVNY  TPDL......  ..........  ..........  STSEVDR...  AIKKALKVWS  DVTPLNFTRL
A59 CGZA_XENLA    ..........  .....YNFFP  RKLKWPRNN.  ..LTYRIVNY  TPDL......  ..........  ..........  STSDVDR...  AIKKALKVWS  DVTPLNFTRL
A60 COGT_HUMAN    ..........  .....KRYAI  QGLKWQHNE.  ..ITFCIQNY  TPKV......  ..........  ..........  GEYATYE...  AIRKAFRVWE  SATPLRFREV
A61 COGT_MOUSE    ..........  .....KRYAI  QGLKWQHNE.  ..ITFCIQNY  TPKV......  ..........  ..........  GEYATFE...  AIRKAFRVWE  SATPLRFREV
A62 O08645        ..........  .....KRYAI  QGLKWQHNE.  ..ITFCIQNY  TPKV......  ..........  ..........  GEYATFE...  AIRKAFRVWE  SATPLRFREV
A63 COGT_RABIT    ..........  .....KRYAI  QGLKWQHNE.  ..ITFCIQNY  TPKV......  ..........  ..........  GEYATFE...  AIRKAFRVWE  SATPLRFREV
A64 COGT_RAT      ..........  .....KRYAI  QGLKWQHNE.  ..ITFCIQNY  TPKV......  ..........  ..........  GEYATFE...  AIRKAFRVWE  SATPLRFREV
A65 COGU_HUMAN    ..........  ...RRKRYAL  TGRKWNNHH.  ..LTFSIQNY  TEKL......  ..........  ..........  GWYHSME...  AVRRAFRVWE  QATPLVFQEV
A66 Q14111        ..........  ...RRKRYAL  TGRKWNNHH.  ..LTFSIQNY  TEKL......  ..........  ..........  GWYHSME...  AVRRAFRVWE  QATPLVFQEV
A67 D1023087      ..........  ..........  ..........  ..........  ..........  ..........  ..........  ..........  AVRRAFRVWE  QATPLVFQEV
A68 O54732        ..........  ...RRKRYTL  TGKAWNNYH.  ..LTFSIQNY  TEKL......  ..........  ..........  GWYNSME...  AVRRAFQVWE  QVTPLVFQEV
A69 Q98947        ..........  .....KRYAL  TGQKWQHKH.  ..IAYSIKNV  TPKV......  ..........  ..........  GDAETRK...  AIRRAFDVWQ  NVTPLTFEEV
A70 COGV_HUMAN    ..........  .....KRYAL  TGQKWQHKH.  ..ITYSIKNV  TPKV......  ..........  ..........  GDPETRK...  AIRRAFDVWQ  NVTPLTFEEV
A71 Q14824        ..........  .....KRYAL  TGQKWQHKH.  ..ITYSIKNV  TPKV......  ..........  ..........  GDPETRK...  AIRRAFDVWQ  NVTPLTFEEV
A72 COGV_RAT      ..........  .....KRYAL  TGQKWQHKH.  ..ITYSIKNV  TPKV......  ..........  ..........  GDPETRK...  AIRRAFDVWQ  NVTPLTFEEV
A73 O35541        ..........  .....KRYAL  TGQKWQHKH.  ..ITYSIKNV  TPKV......  ..........  ..........  GDPETRR...  AIRRAFDVWQ  NVTPLTFEEV
A74 Q14850        ..........  ....R.RQAP  APTKWNKRN.  ..LSWRVRTF  PRDS......  ..........  .......PLG  .HDTVRA...  LMYYALKVWS  DIAPLNFHEV
A75 O13065        ..........  .....YSTVA  RKSSAWQKKD  ..LTYRILNF  TPDL......  ..........  ..........  PQADVET...  AIQRAFKVWS  DVTPLTFTRI
A76 Q99542        ..........  ...QKTLKYL  LLGRWRKKH.  ..LTFRILNL  PSTL......  ..........  ..........  PPHTARA...  ALRQAFQDWS  NVAPLTFQEV
A77 Q99580        ..........  ...QKTLKYL  LLGRWRKKH.  ..LTFRILNL  PSTL......  ..........  ..........  PPHTARA...  ALRQAFQDWS  NVAPLTFQEV
A78 O15278        ..........  ..........  ..........  ..........  ..........  ..........  ..........  ..........  ..........  ..........
A79 O18767        ..........  .....YRLFP  GEPKWKKNT.  ..LTYRISKY  TPSM......  ..........  ..........  TPAEVDR...  AMEMALRAWS  SAVPLNFVRI
A80 O60882        ..........  .....YRLFP  GEPKWKKNT.  ..LTYRISKY  TPSM......  ..........  ..........  SSVEVDR...  AVEMALQAWS  SAVPLSFVRI
A81 P79287        ..........  .....YRLFP  GEPKWKKNT.  ..LTYRISKY  TPSM......  ..........  ..........  TPAEVDK...  AMEMALQAWS  SAVPLSFVRV
A82 O04529        .NFSR..THL  HAVKRYTLFP  GEPRWPRNRR  .DLTYAFDPK  ..........  ..........  ........NP  LTEEVKS...  VFSRAFGRWS  DVTALNFTLS
A83 O23507        .HYT......  ......Y..FN  GKPKWNRDT.  ..LTYAISKT  HKLD......  ..........  ........YL  TSEDVKT...  VFRRAFSQWS  SVIPVSFEEV
A84 O48680        .SFAGRGQRF  HAVKHYSFFP  ..........  .DLTYAFDPR  ..........  ..........  ........NA  LTEEVKS...  VFSRAFTRWE  EVTPLTFTRV
A85 O65340        ..KYV.....  ......Y..FP  GRPRWTRDVP  LKLTYAFSQE  NLTP......  ..........  ........YL  APTDIR....  VFRRAFGEWA  SVIPVSFIET
A86 O16901        ..........  .......GS  SQFKWSKTR.  ..LTYSIESW  SS........  ..........  ........DL  SKDDVRR...  AISEAYGLWS  KVTPLEFSEV
A87 O61264        ..........  .......GS  SQFKWSKTR.  ..LTYSIESW  SS........  ..........  ........DL  SKDDVRR...  AISEAYGLWS  KVTPLEFSEV
A88 O61266        ..........  ....KRYAP  PQFKWKEKI.  ..ITYGCKAV  GTST......  ..........  ..........  .RISLDDLRR  TMHQAASQWS  ELADVEIVES
A89 O61265        G.........  AR....SGQS  VTLKWYISDY  TSDIDRLETR  ..........  ..........  ........KV  VEKAFKL...  WSSQSYIKNE  KKVTLTFQEA
A90 O44836        A.........  HA....SGQS  VTLKWYISDY  TSDIDRLETR  ..........  ..........  ........KV  VEKAFKL...  WSSQSYIKNE  KKVTLTFQEA
A91 O55761        ..........  ..........  ..........  ..MKINTT  ..........  ..........  ..........  .DVRT...  AIYHSLQQWQ  SVSLLTFKEE
A92 MEP1_SOYBN    .MIS......  ....DYTFFK  DMPRWQAGTT  .QLTYAFSPE  SN........  ..........  ........PR  LDDTFKS...  AIARAFSKWT  PVVNIAFQET
A93 HE_HEMPU      ..........  .......YVT  GGIAWPRNVA  ..VTYSPGTL  SN........  ..........  ........DL  SQTAIKN...  ELRRAFQVWD  DVSSLTFREV
A94 HE_PARLI      ..........  ......FVT  SSITWSRNQP  ..VTYSFGAL  TS........  ..........  ........DL  NQNDVKD...  EIRRAFRVWD  DVSGLSFREV
A95 O93470        K.........  DSENVTQQNP  DPPKIRRKRF  LDMLMYSNKY  REFQEALQKS  TGKVFTKKLL  KWRMIGEGYS  NQLSINEQRY  VFRLAFRMWS  EVMPLDFEED
A96 P89294        ..........  ..........  ..........  ..........  ..........  ..........  ..........  ..........  ..........  ..........
```

			Zr.				Zn Ca	Fibronectin insert repeat 1		
A1	S........E 310	GQADIMISFV 320	.RGDHRDNSP 330	FDGPGG.... 340	.....NLAHA 350	FQPGAGI... 360	GGDAHFDDDE 370	WWTSNF.... 380	 390	 400
A2	S........E	GQADIMISFV	.RGDHRDNSP	FDGPGG....	.....NLAHA	FQPGPGI...	GGDAHFDEDE	RWTNNF....	....	....
A3	S........K	GQADIMISFV	.RGDHRDNSP	FDGPEG....	.....QLAHA	FQPGLGI...	GGDVHFDEDD	RWTKDF....	....	....
A4	S........E	GQADIMISFV	.RGDHRDKYP	FDGPGG....	.....NLAHA	SQPGPGI...	GGDAHFDEYE	RWTKNF....	....	....
A5	S........E	GQADIMISFV	.RGDHRDNSP	FDGPGG....	.....NLAHA	FQPGPGI...	GGDAHFDEDE	RWTKNF....	....	....
A6	N........D	GEADIMINFG	.RWEHGDGYP	FDGKDG....	.....LLAHA	FAPGPGI...	GGDSHFDDDE	LWTLGEGQVV	RVKYGNADGE	YCKFPFWFNG
A7	H........D	GEADIMINFG	.RWEHGDGYP	FDGKDG....	.....LLAHA	FAPGTGV...	GGDSHFDDDE	LWTLGEGQVV	RVKYGNADGE	YCKFPFLFNG
A8	H........D	GEADIMINFG	.RWEHGDGYP	FDGKDG....	.....LLAHA	FAPGTGV...	GGDSHFDDDE	LWTLGEGQVV	RVKYGNADGE	YCKFPFLFNG
A9	H........D	GEADIMINFG	.RWEHGDGYP	FDGKDG....	.....LLAHA	FAPGTGV...	GGDSHFDDDE	LWTLGEGQVV	RVKYGNADGE	YCKFPFLFNG
A10	H........D	GEADIMINFG	.RWEHGDGYP	FDGKDG....	.....LLAHA	FAPGTGV...	GGDSHFDDDE	LWTLGEGQVV	RVKYGNADGE	YCKFPFLFNG
A11	....	....	....	....	....	....	....	....	....	....
A12	Y........E	GTADIQISFG	.AGVHGDFYP	FDGPHG....	.....TLAHA	FAPGNSI...	GGDAHFDEDE	TWTAGS....	....	....
A13	Y........Y	GTADIQISFG	.AREHGDFNP	FDGPYG....	.....TLAHA	FAPGTGI...	GGDAHFDEDE	KWSKVS....	....	....
A14	Y........E	GEADIMITFA	.VREHGDFFP	FDGPGK....	.....VLAHA	YPPGPGM...	NGDAHFDDDE	HWTKDA....	....	....
A15	Y........E	GEADIMISFA	.VREHGDFYP	FDGPGN....	.....VLAHA	YAPGPGI...	NGDAHFDDDE	QWTKDT....	....	....
A16	S........E	GEADIMISFA	.VGEHGDFVP	FDGPGT....	.....VLAHA	YAPGPGI...	NGDAHFDDDE	RWTEDV....	....	....
A17	Y........E	GEADIMISFG	.VREHGDFIP	FDGPGN....	.....VLAHA	YAPGPGI...	NGDAHFDDDE	QWTKDT....	....	....
A18	S........E	GEADIMISFA	.VEEHGDFIP	FDGPGM....	.....VLAHA	YAPGPGT...	NGDAHFDDDE	RWTDDV....	....	....
A19	V........L	GIPDIVIGFA	.RGAHGDFYP	FDGPGG....	.....TLAHA	YEPGPGL...	GGDAHFDEDE	RWADGR....	....	....
A20	V........W	GTADIMIGFA	.RGAHGDSYP	FDGPGN....	.....TLAHA	FAPGTGL...	GGDAHFDEDE	RWTDGS....	....	....
A21	S........W	GTADIIIGFA	.RRDHGDSFP	FDGPGN....	.....TLGHA	FAPGPGL...	GGDAHFDKDE	YWTDGE....	....	....
A22	S........W	GTADIIIGFA	.RGDHGDNFP	FDGPGN....	.....TLGHA	FAPGPGL...	GGDAHFDKDE	YWTDGE....	....	....
A23	S........Q	GEADINIAFY	.QRDHGDNSP	FDGPNG....	.....ILAHA	FQPGQGI...	GGDAHFDAEE	TWTNTS....	....	....
A24	L........Q	GEADINIAFV	.SRDHGDNSP	FDGPNG....	.....ILAHA	FQPGQGI...	GGDAHFDSEE	TWTQDS....	....	....
A25	L........Q	GEADINIAFV	.SRDHGDNSP	FDGPNG....	.....ILAHA	FQPGQGI...	GGDAHFDSEE	TWTQDS....	....	....
A26	L........E	GEADINIAFV	.SRDHGDNSP	FDGPNG....	.....ILAHA	FQPGRGI...	GGDAHFDSEE	TWTQDS....	....	....
A27	Y........G	PEADIVIQFG	.VREHGDGYP	FDGKNG....	.....LLAHA	FPPGKGI...	QGDAHFDDEE	LWSLGKGVVI	PTYFGNAKGA	ACHFPFTFEG
A28	Y........G	PEADIIIQFG	.VREHGDGYP	FDGKNG....	.....LLAHA	FAPGPAI...	QGDAHFDDEE	LWTLGKGVVV	PTHFGNADGA	PCHFPFTFEG
A29	....	....	....	....	....	....	....	....	....	....
A30	Y........G	PEADIIIQFG	.VREHGDGYP	FDGKNG....	.....LLAHA	FPPGPGI...	QGDAHFDDEE	LWTLGKGVVV	PTHFGNADGA	PCHFPFTFEG
A31	Y........S	GEADIMILFG	.SDDHGDPYP	FDGKDG....	.....LLAHA	YPPGEGV...	QGDAHFDDDE	FWTLGTGVVV	KTRFGNANGA	ACKFPFKFNG
A32	Y........S	RDADIVIQFG	.VAEHGDGYP	FDGKDG....	.....LLAHA	FPPGPGI...	QGDAHFDDDE	LWSLGKGVVV	PTRFGNADGA	ACHFPFIFEG
A33	Y........G	PEADIVIQFG	.VAEHGDGYP	FDGKDG....	.....LLAHA	FPPGAGV...	QGDAHFDDDE	LWSLGKGVVI	PTYYGNSNGA	PCHFPFTFEG
A34	Y........S	RDADIVIQFG	.VAEHGDGYP	FDGKDG....	.....LLAHA	FPPGPGI...	QGDAHFDDEE	LWSLGKGVVV	PTYFGNADGA	PCHFPFTFEG
A35	Y........G	LEADIVIQFG	.VAEHGDGYP	FDGKDG....	.....LLAHA	FPPGPGI...	QGDAHFDDDE	LWSLGKGAVV	PTYFGNANGA	PCHFPFTFEG
A36	Y........E	GEADIMISFA	.VKEHGDFYS	FDGPGH....	.....SLAHA	YPPGPGL...	YGDIHFDDDE	KWTEDA....	....	....
A37	S........E	GEADIMISFA	.VGEHGDFYP	FDGPGQ....	.....SLAHA	YPPGPGF...	YGDVHFDDDE	KWTLAP....	....	....
A38	S........E	GEADIMISFA	.VGEHGDFYP	FDGVGQ....	.....SLAHA	YPPGPGF...	YGDAHFDDDE	KWSLGP....	....	....
A39	HE........	GRADIMIDFA	.RYWDGDDLP	FDGPGG....	.....ILAHA	FFPKTHR...	EGDVHFDYDE	TWTIGDD...	....	....
A40	HE........	GRADIMIDFA	.RYWHGDNLP	FDGPGG....	.....ILAHA	FFPKTHR...	EGDVHFDYDE	TWTIGDN...	....	....
A41	HE........	GRADIMIDFT	.RYWHGDNLP	FDGPGG....	.....ILAHA	FFPKTHR...	EGDVHFDYDE	TWTIGD....	....	....
A42	HE........	GRSDIIIDFT	.RYWHGDNLP	FDGPGG....	.....ILAHA	FFPKTHR...	EGDVHFDYDE	AWTIGNN...	....	....
A43	N........I	GMADILVVFA	.RGAHGDFHA	FDGKGG....	.....ILAHA	FGPGSGI...	GGDAHFDEDE	FWTTHS....	....	....
A44	....	....	....	....	....	....	....	....	....	....
A45	H........K	DEADIMILFA	.FGAHGDFNY	FDGKGG....	.....TLAHV	FYPGPGI...	QGDAHFDEAE	TWTKSF....	....	....
A46	T........T	GKADIMILFA	.SGAHGDYGA	FDGRGG....	.....VIAHA	FGPGPGI...	GGDTHFDEDE	IWSKSY....	....	....
A47	H........K	GEADITILFA	.FGDHGDFYD	FDGKGG....	.....TLAHA	FYPGPGI...	QGDAHFDEAE	TWTKSF....	....	....
A48	H........N	GTADIMISFG	.TKEHGDFYP	FDGPSG....	.....LLAHA	FPPGPNY...	GGDAHFDDDE	TWTSSS....	....	....
A49	....	....	....	....	....	....	..DAHFDDDE	TWTSSS....	....	....
A50	R........S	GTADIMISFG	.TKEHGDFYP	FDGPNG....	.....LLAHA	FPPGQRI...	GGDTHFDDDE	TFTSGS....	....	....
A51	Y........N	GTADIMISFG	.TKEHGDFYP	FDGPSG....	.....LLAHA	FPPGPNY...	GGDAHFDDDE	TWTSSS....	....	....
A52	H........D	GIADIMISFG	.IKEHGDFYP	FDGPSG....	.....LLAHA	FPPGPNY...	GGDAHFDDDE	TWTSSS....	....	....
A53	Y........D	GTADIMISFG	.TKEHGDFYP	FDGPSG....	.....LLAHA	FPPGPNY...	GGDAHFDDDE	TWTSSS....	....	....
A54	Y........S	DIADIMISFG	.TRNHGDSYP	FDGPDG....	.....TLAHA	FSPGADI...	GGDAHFDDDE	SFSFSSTRGY	NLFLVAVPAL	RHYLPA....
A55	H........N	GTADIMISFG	.TKEHGDFYP	FDGPSG....	.....LLAHA	FPPGPNY...	GGDAHFDDDE	TWTSSS....	....	....
A56	H........D	GTADIMISFG	.TKEHGDFYP	FDGPSG....	.....LLAHA	FPPGPNL...	GGDAHFDDDE	TWTSSS....	....	....
A57	....	....	....	....	....	....	..DPHFDDDE	TWTSSS....	....	....
A58	R........T	GTADIMVSFG	.KKEHGDYYP	FDGPDG....	.....LLAHA	FPPGEKL...	GGDTHFDDDE	MFSTDN....	....	....
A59	R........T	GTADIMVAFG	.KKEHGDYYP	FDGPDG....	.....LLAHA	FPPGEKI...	GGDTHFDDDE	MFSTDN....	....	....
A60	PYAYIREGHE	KQADIMIFFA	.EGFHGDSTP	FDGEGG....	.....FLAHA	YFPGPNI...	GGDTHFDSAE	PWTVRN....	....	....
A61	PYAYIREGHE	KQADIMILFP	.EGLHGDSTP	FDGEGG....	.....FLAHA	YFPGPNI...	GGDTHFDSAE	PWTVQN....	....	....
A62	PYAYIREGHE	KQADIMILFA	.EGFHGDSTP	FDGEGG....	.....FLAHA	YFPGPNI...	GGDTHFDSAE	PWTVQN....	....	....
A63	HYAYIRDGRE	KQADIMIFFA	.EGFHGDSTP	FDGEGG....	.....FLAHA	YFPGPNI...	GGDTHFDSAE	PWTVRN....	....	....
A64	PYAYIREGHE	KQADIMILFA	.EGFHGDSTP	FDGEGG....	.....FLAHA	YFPGPNI...	GGDTHFDSAE	PWTVQN....	....	....
A65	PYEDIRLRRQ	KEADIMVLFA	.SGFHGDSSP	FDGTGG....	.....FLAHA	YFPGPGL...	GGDTHFDADE	PWTFSS....	....	....
A66	PYEDIRLRRQ	KEADIMVLFA	.SGFHGDSSP	FDGTGG....	.....FLAHA	YFPGPGL...	GGDTHFDADE	PWTFSS....	....	....
A67	PYEDIRLRRQ	KEADIMVLFA	.SGFHGDSSP	FDGTGG....	.....FLAHA	YFPGPGL...	GGDTHFDADE	PWTFSS....	....	....
A68	SYDDIRLRRR	AEADIMVLFA	.SGFHGDSSP	FDGVGG....	.....FLAHA	YFPGPGL...	GGDTHFDADE	PWTFSS....	....	....
A69	PYIELENGK.	RDVDITIIFA	.SGFHGDSSP	FDGEGG....	.....FLAHA	YFPGPGI...	GGDTHFDSDE	PWTLGN....	....	....
A70	PYSELENGK.	RDVDITIIFA	.SGFHGDSSP	FDGEGG....	.....FLAHA	YFPGPGI...	GGDTHFDSDE	PWTLGN....	....	....
A71	PYSELENGK.	RDVDITIIFA	.SGFHGDSSP	FDGEGG....	.....FLAHA	YFPGPGI...	GGDTHFDSDE	PWTLGN....	....	....
A72	PYSELENGK.	RDVDITIIFA	.SGFHGDRSP	FDGEGG....	.....FLAHA	YFPGPGI...	GGDTHFDSDE	PWTLGN....	....	....
A73	PYSELENGK.	RDVDITIIFA	.SGFHGDRSP	FDGEGG....	.....FLAHA	YFPGPGI...	GGDTHFDSDE	PWTLGN....	....	....
A74	AG........	STADIQIDFS	.KADHNDGYP	FDGPGG....	.....TVAHA	FFPGHHH..T	AGDTHFDDDE	AWTFRSSD..	....	....
A75	Y........N	EVSDIEISFT	.AGDHKDNSP	FDGSGG....	.....ILAHA	FQPGNGI...	GGDAHFDEDE	TWTKTS....	....	....
A76	Q........A	GAADIRLSFH	GRQSSYCSNT	FDGPGR....	.....VLAHA	DIP......E	LGSVHFDEDE	FWTEGT....	....	....
A77	Q........A	GAADIRLSFH	GRQSSYCSNT	FDGPGR....	.....VLAHA	DIP......E	LGSVHFDEDE	FWTEGT....	....	....
A78	....	....	....	....	....	....	....	....	....	....
A79	N........A	GEADIMISFE	.TGDHGDSYP	FDGPRG....	.....TLAHA	FAPGEGL...	GGDTHFDNAE	KWTMGT....	....	....
A80	N........S	GEADIMISFE	.NGDHGDSYP	FDGPRG....	.....TLAHA	FAPGEGL...	GGDTHFDNPE	KWTMGT....	....	....
A81	N........A	GEADIMISFE	.TGDHGDSYP	FDGPRG....	.....TLAHA	FAPGEGL...	GGDTHFDNAE	KWTMGM....	....	....
A82	ES.......F	STSDITIGFY	.TGDHGDGEP	FDGVLG....	.....TLAHA	FSP.....P.	SGKFHLDADE	NWVVSG....	....	....
A83	DD.......F	TTADLKIGFY	.AGDHGDGLP	FDGVLG....	.....TLAHA	FAP.....E.	NGRLHLDAAE	TWIVD.....	....	....
A84	ER.......F	STSDISIGFY	.SGEHGDGEP	FDGPMR....	.....TLAHA	FSP.....P.	TGHFHLDGEE	NWIVSGE...	....	....
A85	ED.......Y	VIADIKIGFF	.NGDHGDGEP	FDGVLG....	.....VLAHT	FSP.....E.	NGRLHLDKAE	TWAVD.....	....	....
A86	PA.......G	STSDIKIRFG	.VRNHNDPWP	FDGEGG....	.....VLAHA	TMP.....E.	SGMFHFDDDE	NWTYK.....	....	....
A87	PA.......G	STSDIKIRFG	.VRNHNDPWP	FDGEGG....	.....VLAHA	TMP.....E.	SGMFHFDDDE	NWTYK.....	....	....
A88	S........V	KNPMMVISAG	RENHYPCTVR	FDTKTL....	.....A..HA	FFP.T.....	NGQIHINDRV	QFAMTN....	....	....
A89	SS.......K	DEADINILWA	.EGNHGDEHD	FDGANGKIEG	NKKENVLAHT	FFPGYARPL.	NGDIHFDDAE	DWEID.....	....	....
A90	SS.......K	DEADINILWA	.EGNHGDEHD	FDGANGKIEG	NKKENVLAHT	FFPGYARPL.	NGDIHFDDAE	DWEID.....	....	....
A91	YY.......D	ENADIKVSFV	.KGKHNDGWD	FDGKGR....	.....ILAHA	FFPSGSM...	RGVVHLDYDE	DWDFAS....	....	....
A92	TS.......Y	ETANIKILFA	.SKNHGDFYP	FDGPGG....	.....ILGHA	FAP.....T.	DGRCHFDADE	YWVASG....	....	....
A93	VD.......S	SSVDIRIKFG	.SYEHGDGIS	FDGQGG....	.....VLAHA	FLP.....R.	NGDAHFDDSE	RWTIG.....	....	....
A94	PD.......T	TSVDIRIKFG	.SYDHGDGIS	FDGRGG....	.....VLAHA	FLP.....R.	NGDAHFDDSE	TWTEG.....	....	....
A95	N.....TSPL	SQIDIKLGFG	RGRHLGCSRA	FDGSGQ....	.....EFAHA	WFL.......	.GDIHFDDDE	HFTAPS....	....	....
A96	....	....	....	....	....	....	.....FDAAE	DWRLL.....	....	....

```
                                                          ⌐ Repeat 2 →                                                          ⌐ Repeat 3
                    .......... 410 .......... 420 .......... 430 ......V... 440 .......... 450 .......... 460 .......... 470 .......... 480 .......... 490 ...V.. 500
A1  COG1_BOVIN ..........  ..........  ..........  ..........  ..........  ..........  ..........  ..........  ..........
A2  COG1_HUMAN ..........  ..........  ..........  ..........  ..........  ..........  ..........  ..........  ..........
A3  COG1_RABIT ..........  ..........  ..........  ..........  ..........  ..........  ..........  ..........  ..........
A4  COG1_RANCA ..........  ..........  ..........  ..........  ..........  ..........  ..........  ..........  ..........
A5  COG1_PIG   ..........  ..........  ..........  ..........  ..........  ..........  ..........  ..........  ..........
A6  COG2_CHICK KEYNSCTDAG  RNDGFLWCST  TKDFDADGKY  GFCPHESLFT  MGGNGDGQPC  KFPFKFQGQS  YDQCTTEGRT  DGYRWCGTTE  DYDRDKKYGF  CPET.AMSTV
A7  COG2_HUMAN KEYNSCTDTG  RSDGFLWCST  TYNFEKDGKY  GFCPHEALFT  MGGNAEGQPC  KFPFRFQGTS  YDSCTTEGRT  DGYRWCGTTE  DYDRDKKYGF  CPET.AMSTV
A8  COG2_MOUSE REYSSCTDTG  RSDGFLWCST  TYNFEKDGKY  GFCPHEALFT  MGGNADGQPC  KFPFRFQGTS  YNSCTTEGRT  DGYRWCGTTE  DYDRDKKYGF  CPET.AMSTV
A9  COG2_RABIT KEYTSCTDTG  RSDGFLWCST  TYNFEKDGKY  GFCPHEALFT  MGGNADGQPC  KFPFRFQGTS  YSSCTTEGRT  DGYRWCGTTE  DYDRDKKYGF  CPET.AMSTI
A10 COG2_RAT   REYSSCTDTG  RSDGFLWCST  TYNFEKDGKY  GFCPHEALFT  MGGNADGQPC  KFPFRFQGTS  YNSCTTEGRT  DGYRWCGTTE  DYDRDKKYGF  CPET.AMSTV
A11 G257061    ..........  ..........  ..........  ..........  ..........  ..........  ..........  ..........  ..........
A12 Q98857     ..........  ..........  ..........  ..........  ..........  ..........  ..........  ..........  ..........
A13 Q98858     ..........  ..........  ..........  ..........  ..........  ..........  ..........  ..........  ..........
A14 Q28397     ..........  ..........  ..........  ..........  ..........  ..........  ..........  ..........  ..........
A15 COG3_HUMAN ..........  ..........  ..........  ..........  ..........  ..........  ..........  ..........  ..........
A16 COG3_MOUSE ..........  ..........  ..........  ..........  ..........  ..........  ..........  ..........  ..........
A17 COG3_RABIT ..........  ..........  ..........  ..........  ..........  ..........  ..........  ..........  ..........
A18 COG3_RAT   ..........  ..........  ..........  ..........  ..........  ..........  ..........  ..........  ..........
A19 COG7_FELCA ..........  ..........  ..........  ..........  ..........  ..........  ..........  ..........  ..........
A20 COG7_HUMAN ..........  ..........  ..........  ..........  ..........  ..........  ..........  ..........  ..........
A21 COG7_MOUSE ..........  ..........  ..........  ..........  ..........  ..........  ..........  ..........  ..........
A22 COG7_RAT   ..........  ..........  ..........  ..........  ..........  ..........  ..........  ..........  ..........
A23 COG8_HUMAN ..........  ..........  ..........  ..........  ..........  ..........  ..........  ..........  ..........
A24 O70138     ..........  ..........  ..........  ..........  ..........  ..........  ..........  ..........  ..........
A25 O88733     ..........  ..........  ..........  ..........  ..........  ..........  ..........  ..........  ..........
A26 O88766     ..........  ..........  ..........  ..........  ..........  ..........  ..........  ..........  ..........
A27 COG9_BOVIN RSYSACTTDG  RSDDMLWCST  TADYDADRQF  GFCPSERLYT  QDGNADGKPC  VFPFTFQGRT  YSACTSDGRS  DGYRWCATTA  NYDQDKLYGF  CPTRVDATVT
A28 COG9_CANFA RSYSACTTDG  RSDDTPWCST  TADYDTDRRF  GFCPSEKLYT  QDGNGDGKPC  VFPFTFEGRS  YSTCTTDGRS  DGYRWCSTTA  DYDQDKLYGF  CPTRVDSAVT
A29 Q95166     ..........  ..........  ..........  ..........  ..........  ........RS  YSPCTTEGRS  DGYRWCATTA  NYDQDKLYGF  CPTRADATVV
A30 O19130     RSYSACTTDG  RSDDTPWCST  TADYDTDRRF  GFCPSEKFYT  QDGNGDGKPC  VFPFTFEGRS  YSTCTTDGRS  DGYRWCSTTA  DYDQDKLYGF  CPTRVDSAVT
A31 Q98856     NSYSSCTSEG  RTDGLLWCST  TTDYDKDKKY  GFCPSELLYT  YGGNSDGDKC  VFPFIFDGDS  YDACTKEGRS  DGYRWCATTD  NFDKDKKYGF  CPNR.DTAVI
A32 COG9_HUMAN RSYSACTTDG  RSDGLPWCST  TANYDTDDRF  GFCPSERLYT  RDGNADGKPC  QFPFIFQGQS  YSACTTDGRS  DGYRWCATTA  NYDRDKLFGF  CPTRADSTVM
A33 COG9_MOUSE RSYSACTTDG  RNDGTPWCST  TADYDKDGKF  GFCPSERLYT  EHGNGEGKPC  VFPFIFEGRS  YSACTTKGRS  DGYRWCATTA  NYDQDKLYGF  CPTRVDATVV
A34 COG9_RABIT RSYTACTTDG  RSDGMAWCST  TADYDTDRRF  GFCPSERLYT  QDGNADGKPC  EFPFIFQGRT  YSACTTDGRS  DGHRWCATTA  SYDKDKLYGF  CPTRADSTVV
A35 COG9_RAT   RSYLSCTTDG  RNDGKPWCGT  TADYDTDRKY  GFCPSENLYT  EHGNGDGKPC  VFPFIFEGHS  YSACTTKGRS  DGYRWCATTA  NYDQDKADGF  CPTRADVTVT
A36 COGX_HUMAN ..........  ..........  ..........  ..........  ..........  ..........  ..........  ..........  ..........
A37 COGX_MOUSE ..........  ..........  ..........  ..........  ..........  ..........  ..........  ..........  ..........
A38 COGX_RAT   ..........  ..........  ..........  ..........  ..........  ..........  ..........  ..........  ..........
A39 COGY_HUMAN ..........  ..........  ..........  ..........  ..........  ..........  ..........  ..........  ..........
A40 COGY_MOUSE ..........  ..........  ..........  ..........  ..........  ..........  ..........  ..........  ..........
A41 P97568     ..........  ..........  ..........  ..........  ..........  ..........  ..........  ..........  ..........
A42 COGY_XENLA ..........  ..........  ..........  ..........  ..........  ..........  ..........  ..........  ..........
A43 COGM_HUMAN ..........  ..........  ..........  ..........  ..........  ..........  ..........  ..........  ..........
A44 Q99745     ..........  ..........  ..........  ..........  ..........  ..........  ..........  ..........  ..........
A45 COGM_MOUSE ..........  ..........  ..........  ..........  ..........  ..........  ..........  ..........  ..........
A46 COGM_RABIT ..........  ..........  ..........  ..........  ..........  ..........  ..........  ..........  ..........
A47 Q63341     ..........  ..........  ..........  ..........  ..........  ..........  ..........  ..........  ..........
A48 O77656     ..........  ..........  ..........  ..........  ..........  ..........  ..........  ..........  ..........
A49 O77632     ..........  ..........  ..........  ..........  ..........  ..........  ..........  ..........  ..........
A50 Q98859     ..........  ..........  ..........  ..........  ..........  ..........  ..........  ..........  ..........
A51 O18927     ..........  ..........  ..........  ..........  ..........  ..........  ..........  ..........  ..........
A52 COGZ_HUMAN ..........  ..........  ..........  ..........  ..........  ..........  ..........  ..........  ..........
A53 COGZ_MOUSE ..........  ..........  ..........  ..........  ..........  ..........  ..........  ..........  ..........
A54 O73923     ..........  ..........  ..........  ..........  ..........  ..........  ..........  ..........  ..........
A55 O62806     ..........  ..........  ..........  ..........  ..........  ..........  ..........  ..........  ..........
A56 COGZ_RAT   ..........  ..........  ..........  ..........  ..........  ..........  ..........  ..........  ..........
A57 O77631     ..........  ..........  ..........  ..........  ..........  ..........  ..........  ..........  ..........
A58 COGZ_XENLA ..........  ..........  ..........  ..........  ..........  ..........  ..........  ..........  ..........
A59 CGZA_XENLA ..........  ..........  ..........  ..........  ..........  ..........  ..........  ..........  ..........
A60 COGT_HUMAN ..........  ..........  ..........  ..........  ..........  ..........  ..........  ..........  ..........
A61 COGT_MOUSE ..........  ..........  ..........  ..........  ..........  ..........  ..........  ..........  ..........
A62 O08645     ..........  ..........  ..........  ..........  ..........  ..........  ..........  ..........  ..........
A63 COGT_RABIT ..........  ..........  ..........  ..........  ..........  ..........  ..........  ..........  ..........
A64 COGT_RAT   ..........  ..........  ..........  ..........  ..........  ..........  ..........  ..........  ..........
A65 COGU_HUMAN ..........  ..........  ..........  ..........  ..........  ..........  ..........  ..........  ..........
A66 Q14111     ..........  ..........  ..........  ..........  ..........  ..........  ..........  ..........  ..........
A67 D1023087   ..........  ..........  ..........  ..........  ..........  ..........  ..........  ..........  ..........
A68 O54732     ..........  ..........  ..........  ..........  ..........  ..........  ..........  ..........  ..........
A69 Q98947     ..........  ..........  ..........  ..........  ..........  ..........  ..........  ..........  ..........
A70 COGV_HUMAN ..........  ..........  ..........  ..........  ..........  ..........  ..........  ..........  ..........
A71 Q14824     ..........  ..........  ..........  ..........  ..........  ..........  ..........  ..........  ..........
A72 COGV_RAT   ..........  ..........  ..........  ..........  ..........  ..........  ..........  ..........  ..........
A73 O35541     ..........  ..........  ..........  ..........  ..........  ..........  ..........  ..........  ..........
A74 Q14850     ..........  ..........  ..........  ..........  ..........  ..........  ..........  ..........  ..........
A75 O13065     ..........  ..........  ..........  ..........  ..........  ..........  ..........  ..........  ..........
A76 Q99542     ..........  ..........  ..........  ..........  ..........  ..........  ..........  ..........  ..........
A77 Q99580     ..........  ..........  ..........  ..........  ..........  ..........  ..........  ..........  ..........
A78 O15278     ..........  ..........  ..........  ..........  ..........  ..........  ..........  ..........  ..........
A79 O18767     ..........  ..........  ..........  ..........  ..........  ..........  ..........  ..........  ..........
A80 O60882     ..........  ..........  ..........  ..........  ..........  ..........  ..........  ..........  ..........
A81 P79287     ..........  ..........  ..........  ..........  ..........  ..........  ..........  ..........  ..........
A82 O04529     ..........  ..........  ..........  ..........  ..........  ..........  ..........  ..........  ..........
A83 O23507     ..........  ..........  ..........  ..........  ..........  ..........  ..........  ..........  ..........
A84 O48680     ..........  ..........  ..........  ..........  ..........  ..........  ..........  ..........  ..........
A85 O65340     ..........  ..........  ..........  ..........  ..........  ..........  ..........  ..........  ..........
A86 O16901     ..........  ..........  ..........  ..........  ..........  ..........  ..........  ..........  ..........
A87 O61264     ..........  ..........  ..........  ..........  ..........  ..........  ..........  ..........  ..........
A88 O61266     ..........  ..........  ..........  ..........  ..........  ..........  ..........  ..........  ..........
A89 O61265     ..........  ..........  ..........  ..........  ..........  ..........  ..........  ..........  ..........
A90 O44836     ..........  ..........  ..........  ..........  ..........  ..........  ..........  ..........  ..........
A91 O55761     ..........  ..........  ..........  ..........  ..........  ..........  ..........  ..........  ..........
A92 MEP1_SOYBN ..........  ..........  ..........  ..........  ..........  ..........  ..........  ..........  ..........
A93 HE_HEMPU   ..........  ..........  ..........  ..........  ..........  ..........  ..........  ..........  ..........
A94 HE_PARLI   ..........  ..........  ..........  ..........  ..........  ..........  ..........  ..........  ..........
A95 O93470     ..........  ..........  ..........  ..........  ..........  ..........  ..........  ..........  ..........
A96 P89294     ..........  ..........  ..........  ..........  ..........  ..........  ..........  ..........  ..........
```

Repeat 3 begins at column 560 (".. QDYNLYRV"); the **Zinc binding** motif spans column 570; the **Turn** spans column 590.

Seq	510	520	530	540	550	560	570 (Zinc binding)	580	590 (Turn)	600
A1	....	....	....	....	....	..QDYNLYRV	AAHEFGHSLG	LAHSTDIGAL	MYPSYTF...	S..GDVQLS.
A2	....	....	....	....	....	..REYNLHRV	AAHELGHSLG	LSHSTDIGAL	MYPSYTF...	S..GDVQLA.
A3	....	....	....	....	....	..RNYNLYRV	AAHELGHSLG	LSHSTDIGAL	MYPNYMF...	S..GDVQLA.
A4	....	....	....	....	....	..QDYNLYRV	AAHELGHSLG	LSHSTDIGAL	MYPTYLR...	...GDVQLS.
A5	....	....	....	....	....	..RDYNLYRV	AAHELGHSLG	LSHSTDIGAL	MYPNYIY...	T..GDVQLS.
A6	GGNSEGAPCV	FPFIFLGNKY	DSCTSAGRND	GKLWCASTSS	YDDDRKWGFC	PDQGYSLFLV	AAHEFGHAMG	LEHSEDPGAL	MAPIYTY...	TK..NFRLS.
A7	GGNSEGAPCV	FPFTFLGNKY	ESCTSAGRSD	GKMWCATTAN	YDDDRKWGFC	PDQGYSLFLV	AAHEFGHAMG	LEHSQDPGAL	MAPIYTY...	TK..NFRLS.
A8	GGNSEGAPCV	FPFTFLGNKY	ESCTSAGRND	GKVWCATTTN	YDDDRKWGFC	PDQGYSLFLV	AAHEFGHAMG	LEHSQDPGAL	MAPIYTY...	TK..NFRLS.
A9	GGNSEGAPCV	FPFTFLGNKY	ESCTSAGRSD	GKMWCATSTN	YDDDRKWGFC	PDQGYSLFLV	AAHEFGHAMG	LEHSQDPGAL	MAPIYTY...	TK..NFRLS.
A10	GGNSEGAPCV	FPFTFLGNKY	ESCTSAGRND	GKVWCATTTN	YDDDRKWGFC	PDQGYSLFLV	AAHEFGHAMG	LEHSQDPGAL	MAPIYTY...	TK..NFRLS.
A11	....	....	....	....	....	....	....	....	....	....
A12	....	....	....	....	....	..AGYNLFLV	AAHEFGHSLG	LSHSGDRSAL	MYPTYSY...	IDPARFRLP.
A13	....	....	....	....	....	..TGTNLFLV	AAHEFGHSLG	LSHSNDRNAL	MFPTYSY...	TDPARFRLP.
A14	....	....	....	....	....	..SGINFLLV	AAHEFGHSLG	LYHSTNTEAL	MYPLYNT...	LKGPARVRLS
A15	....	....	....	....	....	..TGTNLFLV	AAHEIGHSLG	LFHSANTEAL	MYPLYHS...	LTDLTRFRLS
A16	....	....	....	....	....	..TGTNLFLV	AAHELGHSLG	LYHSAKAEAL	MYPVYKS...	STDLSRFHLS
A17	....	....	....	....	....	..TGTNLFLV	AAHELGHSLG	LFHSANPEAL	MYPVYNA...	FTDLARFRLS
A18	....	....	....	....	....	..TGTNLFLV	AAHELGHSLG	LFHSANAEAL	MYPVYKS...	STDLARFHLS
A19	....	....	....	....	....	.GLGINFLAV	ATHELGHSLG	LRHSSDPDSV	MYPTYGA...	RDSENFKLS.
A20	....	....	....	....	....	.SLGINFLYA	ATHELGHSLG	MGHSSDPNAV	MYPTYGN...	GDPQNFKLS.
A21	....	....	....	....	....	.DAGVNFLFA	ATHEFGHSLG	LSHSSVPGTV	MYPTYQR...	DYSEDFSLT.
A22	....	....	....	....	....	.DSGVNFLFV	ATHELGHSLG	LGHSSVPSSV	MYPTYQG...	DHSEDFSLT.
A23	....	....	....	....	....	..ANYNLFLV	AAHEFGHSLG	LAHSSDPGAL	MYPNYAF...	RETSNYSLP.
A24	....	....	....	....	....	..KNYNLFLV	AAHEFGHSLG	LSHSTDPGAL	MYPNYAY...	REPSTYSLP.
A25	....	....	....	....	....	..KNYNLFLV	AAHEFGHSLG	LSHSTDPGAL	MYPNYAY...	REPSTYSLP.
A26	....	....	....	....	....	..KNYNLFLV	AAHEFGHSLG	LSHSTDPGAL	MYPNYAY...	REPSTYSLP.
A27	GGNAAGELCV	FPFTFLGKEY	SACTREGRND	GHLWCATTSN	FDKDKKWGFC	PDQGYSLFLV	AAHEFGHALG	LDHTSVPEAL	MYPMYRF...	TE..EHPLH.
A28	GGNSAGEPCV	FPFIFLGKQY	STCTREGRGD	GHLWCATTSN	FDRDKKWGFC	PDQGYSLFLV	AAHEFGHALG	LDHSSVPEAL	MYPMYSF...	TE..GPPLH.
A29	GGNSAGELCV	FPFVFLGKEY	STCTSDGRRD	GRLWCATTSN	FDTDKKWGFC	PDQGYSLFLV	AAHEFGHAL.	....	MYPMYSF...	....
A30	GGNSAGEPCV	FPFIFLGKQY	STCTREGRGD	GHLWCATTSN	FDRDKKWGFC	PDQGYSLFLV	AAHEFGHALG	LDHSSVPEAL	MYPMYSF...	TE..GPPLH.
A31	GGNSQGDPCV	FPFVFLEKTY	HSCTSDGRGD	RKLWCATTSS	YDSDRKWGFC	PDQGYSLFLV	GAHEFGHAIG	LEHSTVRDAL	MYPMFRY...	IE..GPQLH.
A32	GGNSAGELCV	FPFTFLGKEY	STCTSEGRGD	GRLWCATTSN	FDSDKKWGFC	PDQGYSLFLV	AAHEFGHALG	LDHSSVPEAL	MYPMYRF...	TE..GPPLH.
A33	GGNSAGELCV	FPFVFLGKQY	SSCTSDGRRD	GRLWCATTSN	FDTDKKWGFC	PDQGYSLFLV	AAHEFGHALG	LDHSSVPEAL	MYPLYSY...	LE..GPFLN.
A34	GGNSAGELCV	FPFVFLGKEY	SSCTSEGRRD	GRLWCATTSN	FDSDKKWGFC	PDKGYSLFLV	AAHEFGHALG	LDHSSVPERL	MYPMYRY...	LE..GSPLH.
A35	GGNSAGEMCV	FPFVFLGKQY	STCTSEGRSD	GRLWCATTSN	FDADKKWGFC	PDQGYSLFLV	AAHEFGHALG	LDHSSVPEAL	MYPMYHY...	HE..DSPLH.
A36	....	....	....	....	....	..SGTNLFLV	AAHELGHSLG	LFHSANTEAL	MYPLYNS...	FTELAQFRLS
A37	....	....	....	....	....	..SGTNLFLV	AAHELGHSLG	LFHSDKKESL	MYPVYRF...	STSPANFHLS
A38	....	....	....	....	....	..SGTNLFLV	AAHELGHSLG	LFHSNNKESL	MYPVYRF...	STSQANIRLS
A39	....	....	....	....	....	..QGTDLLQV	AAHEFGHVLG	LQHTTAAKAL	MSAFYT....	.FRYPLSLSP
A40	....	....	....	....	....	..QGTDLLQV	AAHEFGHVLG	LQHTTAAKAL	MSPFYT....	.FRYPLSLSP
A41	....	....	....	....	....	..KGTDLLQV	AAHEFGHVLG	LQHTTAAKAL	MSPFYT....	.FRYPLSLSP
A42	....	....	....	....	....	..IGTDLLQV	AAHEFGHMLG	LQHSSISKSL	MSPFYT....	.FRYPLSLSA
A43	....	....	....	....	....	..GGTNLFLT	AVHEIGHSLG	LGIIGQDPKAV	MFPTYK....	YVDINTFRLS
A44	....	....	....	....	....	....	....	.....DPKAV	MFPTYK....	YVDINTFRLS
A45	....	....	....	....	....	..QGTNLFLV	AVHELGHSLG	LQHSNNPKSI	MYPTYR....	YLNPSTFRLS
A46	....	....	....	....	....	..KGTNLFLV	AVHELGHALG	LDHSNDPKAI	MFPTYG....	YIDLNTFHLS
A47	....	....	....	....	....	..QGTNLFLV	AVHELGHSLG	LRHSNNPKSI	MYPTYR....	YLHPNTFRLS
A48	....	....	....	....	....	..KGYNLFLV	AAHEFGHSLG	LDHSKDPGAL	MFPIYTY...	TGKSHFMLP.
A49	....	....	....	....	....	..KGYNLFLV	AAHEFGHSLG	LDHSKDPGAL	MFPIYTY...	TGKSHFMLP.
A50	....	....	....	....	....	..NGYNLFIV	AAHEFGHALG	LDHSRDPGSL	MYPVYSY...	TEPSRFLLP.
A51	....	....	....	....	....	..KGYNLFLV	AAHEFGHSLG	LDHSKDPGAL	MFPIYTY...	TGKSHFVLP.
A52	....	....	....	....	....	..KGYNLFLV	AAHEFGHSLG	LDHSKDPGAL	MFPIYTY...	TGKSHFMLP.
A53	....	....	....	....	....	..KGYNLFIV	AAHELGHSLG	LDHSKDPGAL	MFPIYTY...	TGKSHFMLP.
A54	....	....	....	....	....	....	....	....	....	....
A55	....	....	....	....	....	..KGYNLFLV	AAHEFGHSLG	LDHSKDPGAL	MFPIYTY...	TGKSHFMLP.
A56	....	....	....	....	....	..KGYNLFIV	AAHELGHSLG	LDHSKDPGAL	MFPIYTY...	TGKSHFMLP.
A57	....	....	....	....	....	..KGYNLFLV	AAHEFGHSLG	LDHSKDPGAL	MFPIYTY...	TGKSHFMLP.
A58	....	....	....	....	....	..KGYNLFVV	AAHEFGHALG	LDHSRDPGSL	MFPVYTY...	TETSRFVLP.
A59	....	....	....	....	....	..KGYNLFVV	AAHEFGHALG	LDHSRDPGSL	MFPVYTY...	TETSRFVLP.
A60	....	....	.....E..D.	....	....	.LNGNDIFLV	AVHELGHALG	LEHSSDPSAI	MAPFYQW...	MDTENFVLP.
A61	....	....	.....E..D.	....	....	.LNGNDIFLV	AVHELGHALG	LEHSNDPSDI	MSPFYQW...	MDTENFVLP.
A62	....	....	.....E..D.	....	....	.LNGNDIFLV	AVHELGHALG	LEHSNDPSAI	MSPFYQW...	MDTENFVLP.
A63	....	....	.....E..D.	....	....	.LNGNDIFLV	AVHELGHALG	LEHSNDPSAI	MAPFYQW...	MDTEKFLLP.
A64	....	....	.....E..D.	....	....	.LNGNDIFLV	AVHELGHALG	LEHSNDPSDI	MAPFYQW...	MDTENFVLP.
A65	....	....	.....T..D.	....	....	.LHGNNLFLV	AVHELGHALG	LEHSSNPNAI	MAPFYQW...	KDVDNFKLP.
A66	....	....	.....T..D.	....	....	.LHGNNLFLV	AVHELGHALG	LEHSSNPNAI	MAPFYQW...	KDVDNFKLP.
A67	....	....	.....T..D.	....	....	.LHGNNLFLV	AVHELGHALG	LEHSSNPNAI	MAPFYQW...	KDVDNFKLP.
A68	....	....	.....T..D.	....	....	.LHGISLFLV	AVHELGHALG	LEHSSNPSAI	MAPFYQW...	MDTDNFQLP.
A69	....	....	.....P..N.	....	....	.HDGNDLFLV	AVHELGHALG	LEHSNDPTAI	MAPFYQY...	METDNFKLP.
A70	....	....	.....P..N.	....	....	.HDGNDLFLV	AVHELGHALG	LEHSNDPTAI	MAPFYQY...	METDNFKLP.
A71	....	....	.....P..N.	....	....	.HDGNDLFLV	AVHELGHALG	LEHSNDPTAI	MAPFYQY...	METDNFKLP.
A72	....	....	.....P..N.	....	....	.HDGNDLFLV	AVHELGHALG	LEHSNDPTAI	MAPFYQY...	METDNFKLP.
A73	....	....	.....P..N.	....	....	.HDGNDLFLV	AVHELGHALG	LEHSNDPTAI	MAPFYQY...	METDNFKLP.
A74	....	....	....	....	....	.AHGMDLFAV	AVHEFGHAIG	LSHVAAAHSI	MRPYYQG...	PVGDPLRYGL
A75	....	....	....	....	....	..EIYNLFLV	AAHEFGHSLG	LSHSTDQGAL	MYPTYSN...	TDPKTFQLP.
A76	....	....	....	....	....	.YRGVNLRII	AAHEVGHALG	LGHSRYSQAL	MAPVYEG...	...YRPHFKL
A77	....	....	....	....	....	.YRGVNLRII	AAHEVGHALG	LGHSRYSQAL	MAPVYEG...	...YRPHFKL
A78	....	....	....	....	....	....	....	....	....	....
A79	....	....	....	....	....	..NGFNLFTV	AAHEFGHALG	LAHSTDPSAL	MFPTYKY...	QNPYGFRLP.
A80	....	....	....	....	....	..NGFNLFTV	AAHEFGHALG	LAHSTDPSAL	MYPTYKY...	KNPYGFHLP.
A81	....	....	....	....	....	..NGFNLFTV	AAHEFGHALG	LAHSTDPSAL	MYPTYKY...	CNPYGFHLP.
A82	....	....	....	....	...DLDSFLS	VTAAVDLESV	AVHEIGHLLG	LGHSSVEESI	MYPTITT.GK	RK.....VDL
A83	....	....	....	....	....DDLKGS	SEVAVDLESV	ATHEIGHLLG	LGHSSQESAV	MYPSLRP.RT	KK.....VDL
A84	....	....	....	....	...GGDGFIS	VSEAVDLESV	AVHEIGHLLG	LGHSSVEGSI	MYPTIRT.GR	RK.....VDL
A85	....	....	....	....	....FDEEKS	S.VAVDLESV	AVHEIGHVLG	LGHSSVKDAA	MYPTLKP.RS	KK.....VNL
A86	....	....	....	....	....DARKIH	NNEATDLLAV	AIHEGGHTLG	LEHSRDENAI	MAPFYQKTTD	SSGNYVYPNL
A87	....	....	....	....	....DARKIH	NNEATDLLAV	AIHEGGHTLG	LEHSRDENAI	MAPFYQKTTD	SSGNYVYPNL
A88	....	....	....	....	.......YTE	RMGANSLYSV	VAHEMGHALG	FSHSPDIDSV	MFAYDTP...	.....RKWKF
A89	....	....	....	....	....VDQVGH	GSNKRFFPYV	LAHEIGHALG	LDHSQKADAL	MHPYYKNVPI	NE.....IQL
A90	....	....	....	....	....VDQVGH	GSNKRFFPYV	LAHEIGHALG	LDHSQKADAL	MHPYYKNVPI	NE.....IQL
A91	....	....	....	....	....	......LKQV	LLHELGHTFG	LGHSSDNKSI	MFPWNSS...	.S..VDNVN.
A92	....	....	....	....	....DVTKSP	VTSAFDLESV	AVHEIGHLLG	LGHSSDLRAI	MYPSIPP.RT	RK.....VNL
A93	....	....	....	....	....	TNSGTNLFQV	AAHEFGHSLG	LYHSDVQSAL	MYPYYRG.YN	PN.....FNL
A94	....	....	....	....	....	TRSGTNLFQV	AAHEFGHSLG	LYHSTVRSAL	MYPYYQG.YV	PN.....FRL
A95	....	....	....	....	....	SEHGISLLKV	AAHEIGHVLG	LSHIHRVGSI	MQPNYIP...	.QDSGFELD.
A96	....	....	....	....	....DDDHKI	PEDGISLYLV	AAHEIGHALG	LHHTSVRDAI	MYWYYNNEKT	GL.....HQD

```
                CATALYTIC DOMAIN──┐ ┌─HINGE→                                                        HINGE──┐ ┌─HEMOPEXIN DOMAIN

A1   COG1_BOVIN  .QDDIDGIQA 610 IYGPSQNPT. 620 .......... 630 .......... 640 .......... 650 ......QPV 660 GP........ 670 .......... 680 .......... 690 .QTPEVCDSK 700
A2   COG1_HUMAN  .QDDIDGIQA     IYGRSQNPV.     ..........     ..........     ..........     ......QPI     GP........     ..........     ..........     .QTPKACDSK
A3   COG1_RABIT  .QDDIDGIQA     IYGPSQNPS.     ..........     ..........     ..........     ......QPV     GP........     ..........     ..........     .QTPKVCDSK
A4   COG1_RANCA  .QDDID....     .GPSGNPV.      ..........     ..........     ..........     ......QPR     GP........     ..........     ..........     .QTPQVCDSK
A5   COG1_PIG    .QDDIDGIQA     IYGPSENPV.     ..........     ..........     ..........     ......QPS     GP........     ..........     ..........     .QTPQVCDSK
A6   COG2_CHICK  .QDDIKGIQE     LYE..VSPDV     EPG......      ..........     ..........     PGPGPGPGPR     ..........     ..........     ...PTLG.       PVTPELCKHD
A7   COG2_HUMAN  .QDDIKGIQE     LYG..ASP..     ..........     ..........     ..........     .DIDLGTGPT     ..........     ..........     ...PTLG.       PVTPEICKQD
A8   COG2_MOUSE  .HDDIKGIQE     LYG..PSPDA     ..........     ..........     ..........     .DTDTGTGPT     ..........     ..........     ...PTLG.       PVTPEICKQD
A9   COG2_RABIT  .HDDIKGIQE     LYG..ASPDA     ..........     ..........     ..........     .GTDAGTGPT     ..........     ..........     ...PTLG.       PVTPEICTQD
A10  COG2_RAT    .HDDIKGIQE     LYG..PSPDA     ..........     ..........     ..........     .DTDTGTGPT     ..........     ..........     ...PTLG.       PVTPEICKQD
A11  G257061     ..........     ..........     ..........     ..........     ..........     ..........     ..........     ..........     ..........     ..........
A12  Q98857      .QDDVDGIQA     LYGASPNPVP     ..........     ..........     ..........     ....TTPQAT     TPTTTVST..     ..........     ...TTTTTSS     PINPSICDPT
A13  Q98858      .KDDINGIQA     IYGPSRKPSP     ..........     ..........     ..........     ....QTPPPT     KP........     ..........     ..........     .ALQSYCDPA
A14  Q28397      .QDDVTGIQS     LYGPPPASPD     SPVEP.....     ..........     ..........     ....SEPEPP     APG.......     ..........     ..........     ..TLAMCDPA
A15  COG3_HUMAN  .QDDINGIQS     LYGPPPDSPE     TPLVP.....     ..........     ..........     ....TEPVPP     EPG.......     ..........     ..........     ..TPANCDPA
A16  COG3_MOUSE  .QDDVDGIQS     LYGTPTASPD     VLVVP.....     ..........     ..........     ....TKSNSL     EPE.......     ..........     ..........     ..TSPMCSST
A17  COG3_RABIT  .QDDVDGIQS     LYGPAPASPD     NSGVP.....     ..........     ..........     ....MEPVPP     GSG.......     ..........     ..........     ..TPVMCDPD
A18  COG3_RAT    .QDDVDGIQS     LYGPPTESPD     VLVVP.....     ..........     ..........     ....TKSNSL     DPE.......     ..........     ..........     ..TLPMCSSA
A19  COG7_FELCA  .PGDIREIQE     LYGKRSKSRK     K.........     ..........     ..........     ..........     ..........     ..........     ..........     ..........
A20  COG7_HUMAN  .QDDIKGIQK     LYGKRSNSRK     K.........     ..........     ..........     ..........     ..........     ..........     ..........     ..........
A21  COG7_MOUSE  .KDDIAGIQK     LYGKRNTL..     ..........     ..........     ..........     ..........     ..........     ..........     ..........     ..........
A22  COG7_RAT    .KDDIAGIQK     LYGKRNKL..     ..........     ..........     ..........     ..........     ..........     ..........     ..........     ..........
A23  COG8_HUMAN  .QDDINGIQA     IYGLSSNPI.     ..........     ..........     ..........     ......QPT     GP........     ..........     ..........     .STPKPCDPS
A24  O70138      .QDDINGIQT     IYGPSDNPI.     ..........     ..........     ..........     ......QPT     GP........     ..........     ..........     .STPKACDPH
A25  O88733      .QDDINGIQT     IYGPSDNPI.     ..........     ..........     ..........     ......QPT     GP........     ..........     ..........     .STPKACDPH
A26  O88766      .QDDINGIQT     IYGPSDNPV.     ..........     ..........     ..........     ......QPT     GP........     ..........     ..........     .STPTACDPH
A27  COG9_BOVIN  .RDDVQGIQH     LYGPRPEPEP     RPPTTTTTTT     TEPQPTAPPT     VCVTGPPTAR     PSEGPTTGPT     GPPAAG....     ..........     ..PTGPPTAG     PSAAPTESPD
A28  COG9_CANFA  .EDDVRGIQH     LYGPRPEPEP     QPP.......     TAP.PTAPPT     VCATGPPTTR     PSERPTAGPT     GPPAAG....     ..........     ..PTGPPTAG     PSEAPTVPVD
A29  Q95166      ..........     ..........     ..........     ..........     ..........     ..........     ..........     ..........     ..........     ..........
A30  O19130      .EDDVRGIQH     LYGPRPEPEP     ..........     TAP.PTAPPT     VCATGPPTTR     PSERPTAGPT     GPPAAG....     ..........     ..PTGPPTAG     PSEAPTVPVD
A31  Q98856      .QDDIEGVQY     LYGSGTGPHP     SPP.....TT     ....MPTTKS     P.D..VSGKT     ..........     ..........     ..........     ..TTTVTTSP     TTTTELVPVD
A32  COG9_HUMAN  .KDDVNGIRH     LYG..PRPEP     EPRPP...TT     TTPQPTAPPT     VCPTGPPTVH     PSERPTAGPT     GPPSAG....     ..........     ..PTGPPTAG     PSTATTVPLS
A33  COG9_MOUSE  .KDDIDGIQY     LYGRGSKPDP     RPPAT...TT     TEPQPTAPPT     MCPTIPPTAY     PTVGPTVGPT     GAPSPGPTSS     PSPGPTGAPS     PGPTAAPTAG     SSEASTESLS
A34  COG9_RABIT  .EDDVRGIQH     LYG..PNPNP     QPPAT...TT     PEPQPTAPPT     ACPTWPATVR     PSEHPTTSPT     GAPSAG....     ..........     ..PTGPPTAS     PSAAPTASLD
A35  COG9_RAT    .EDDIKGIHH     LYGRGSKPDP     RPPAT...TA     AEPQPTAPPT     MCSTAPPMAY     PTGGPTVAPT     GAPSPG....     ..........     ..PTGPPTAG     PSEAPTESST
A36  COGX_HUMAN  .QDDVNGIQS     LYGPPPASTE     EPLVP.....     ..........     ..........     ....TKSVPS     GSE.......     ..........     ..........     ..MPAKCDPA
A37  COGX_MOUSE  .QDDIEGIQS     LYGAGPSS.D     ATVVP.....     ..........     ..........     ....VLSVSP     RPE.......     ..........     ..........     ..TPDKCDPA
A38  COGX_RAT    .QDDIEGIQS     LYGARPSS.D     ATVVP.....     ..........     ..........     ....VPSVSP     KPE.......     ..........     ..........     ..TPVKCDPA
A39  COGY_HUMAN  ..DDCRGVQH     LYGQPWPT..     ..........     ..........     ..........     ...VTSRTPA     LGP.......     ..........     ......QAGI     DTNEIAPLEP
A40  COGY_MOUSE  ..DDRRGIQH     LYGRPQMA..     ..........     ..........     ..........     ...PTSPAPT     LSS.......     ..........     ......QAGT     DTNEIALLEP
A41  P97568      ..DDRRGIQH     LYGRPQLT..     ..........     ..........     ..........     ...PTSPTPT     LSS.......     ..........     ......QAGT     DTNEIALQEP
A42  COGY_XENLA  ..DDKHGIQF     LYGAPRPP..     ..........     ..........     ..........     ...TPSPTPR     VEVN......     ..........     ......QVEN     ESNEIPAAEP
A43  COGM_HUMAN  .ADDIRGIQS     LYGDPKEN..     ...Q......     ..........     ..........     ....RLPNPD     NS........     ..........     ..........     ..EPALCDPN
A44  Q99745      .ADDIRGIQS     LYGDPKEN..     ...Q......     ..........     ..........     ....RLPNPD     NS........     ..........     ..........     ..XPALCDPN
A45  COGM_MOUSE  .ADDIRNIQS     LYGAPVKP..     ...P......     ..........     ..........     ....SLTKPS     SP........     ..........     ..........     .PSTFCHQS
A46  COGM_RABIT  .ADDIHSIQS     LYGGPEQH..     ...Q......     ..........     ..........     ....PMPKPD     NP........     ..........     ..........     ..EPTACDHN
A47  Q63341      .ADDIHSIQS     LYGAPVKN..     ...P......     ..........     ..........     ....SLTNPG     SP........     ..........     ..........     .PSTVCHQS
A48  O77656      .DDDVQGIQS     LYGPGDED..     ..........     ..........     ..........     ......PYSK     HP........     ..........     ..........     .KTPDKCDPS
A49  O77632      .DDDVQGIQS     LYGPGDED..     ..........     ..........     ..........     ......PN..     ..........     ..........     ..........     ..........
A50  Q98859      .DDDVQGIQS     LYGPGNRD..     ..........     ..........     ..........     ......PNPK     HP........     ..........     ..........     .KTPEKCDPE
A51  O18927      .DDDVQGIQY     LYGPGDED..     ..........     ..........     ..........     ......PNPK     HP........     ..........     ..........     .KTPDKCDPS
A52  COGZ_HUMAN  .DDDVQGIQS     LYGPGDED..     ..........     ..........     ..........     ......PNPK     HP........     ..........     ..........     .KTPDKCDPS
A53  COGZ_MOUSE  .DDDVQGIQF     LYGPGDED..     ..........     ..........     ..........     ......PNPK     HP........     ..........     ..........     .KTPEKCDPA
A54  O73923      ..........     ..........     ..........     ..........     ..........     ..........     ..........     ..........     ..........     ..........
A55  O62806      .DDDVQGIQS     LYGPGDED..     ..........     ..........     ..........     ......PNPK     HP........     ..........     ..........     .KTPDKCDPS
A56  COGZ_RAT    .DDDVQGIQS     LYGPGDED..     ..........     ..........     ..........     ......PNPK     HP........     ..........     ..........     .KTPEKCDPA
A57  O77631      .DDDVQGIQS     LYGPGDED..     ..........     ..........     ..........     ......PN..     ..........     ..........     ..........     ..........
A58  COGZ_XENLA  .DDDVQGIQV     LYGPGNRD..     ..........     ..........     ..........     ......PHPK     HP........     ..........     ..........     .KTPEKCDPD
A59  CGZA_XENLA  .DDDVQGIQA     LYGSGNRD..     ..........     ..........     ..........     ......PHPK     HP........     ..........     ..........     .KTPEKCDPD
A60  COGT_HUMAN  .DDDRRGIQQ     LYGG......     ..........     .ESGFPTKMP     PQPRTT....     SRPSVPDKPK     NP........     ..........     ..........     TYGPNICDGN
A61  COGT_MOUSE  .DDDRRGIQQ     LYGS......     ..........     .KSGSPTKMP     PQPRTT....     SRPSVPDKPK     NP........     ..........     ..........     AYGPNICDGN
A62  O08645      .DDDRRGIQQ     LYGS......     ..........     .KSGSPTKMP     PQPRTT....     SRPSVPDKPK     NP........     ..........     ..........     AYGPNICDGN
A63  COGT_RABIT  .DDERRGIQQ     LYGS......     ..........     .QSGSPTRCL     LNPGQP....     SGLLFRISPG     NP........     ..........     ..........     TYGPKICDGN
A64  COGT_RAT    .DDDRRGIQQ     LYGS......     ..........     .KSGSPTKMP     PQPRTT....     SRPSVPDKPR     NP........     ..........     ..........     TYGPNICDGN
A65  COGU_HUMAN  .EDDLRGIQQ     LYGTPDGQPQ     PTQPLPTVTP     RRPGRPDHRP     PRPPQPPPPG     GKPERPPKPG     PPVQPR....     ..........     ...ATERPD     QYGPNICDGD
A66  Q14111      .EDDLRGIQQ     LYGTPDGQPQ     PTQPLPTVTP     RRPGRPDHRP     PRPPQPPPPG     GKPERPPKPG     PPVQPR....     ..........     ...ATERPD     QYGPNICDGD
A67  D1023087    .EDDLRGIQQ     LYGTPDGQPQ     PTQPLPTVTP     RRPGRPDHRP     PRPPQPPPPG     GKPERPPKPG     PPVQPR....     ..........     ...ATERPD     QYGPNICDGD
A68  O54722      .EDDLRGIQQ     LYGSPDGKPQ     PTRPLPTVRP     RRPGRPDHQP     PRPPQPPHPG     GKPERPPKPG     PPPQPR....     ..........     ...ATERPD     QYGPNICDGN
A69  Q98947      .NDDLQGIQK     IYGPPD....     ......RIPP     PTRPLPTVPP     HRSIPPADPR     KNDRQPKPPR     PPTGD.....     ..........     ....KPSYP     GAKPNICDGN
A70  COGV_HUMAN  .NDDLQGIQK     IYGPPD....     ......KIPP     PTRPLPTVPP     HRSIPPADPR     KNDR.PKPPR     PPTG......     ..........     ....RPSYP     GAKPNICDGN
A71  Q14824      .NDDLQGIQK     IYGPPD....     ......KIPP     PTRPLPTVPP     HRSIPPADPR     KNDR.PKPPR     PPTG......     ..........     ....RPSYP     GAKPNICDGN
A72  COGV_RAT    .NDDLQGIQK     IYGPPD....     ......KIPP     PTRPLPTVPP     HRSVPPADPR     KNDR.PKPPR     PPTG......     ..........     ....RPSYP     GAKPNICDGN
A73  O35541      .NDDLQGIQK     IYGPPD....     ......KIPP     PTRPLPTVPP     HRSVPPADPR     KNDR.PKPPR     PPTG......     ..........     ....RPSYP     GAKPNICDGN
A74  Q14850      PYEDKVRVWQ     LYGVRESV..     ..........     ..........     ..........     ...SPTAQPE     EPPL......     ..........     .....LPEP     PDNRSSAPPR
A75  O13065      .QDDINAIQY     LYGKSSNP..     ..........     ..........     ..........     ......VQPT     GP........     ..........     ..........     .STPSRCDPN
A76  Q99542      HPDDVAGIQA     LYGKKSPVIR     DEEEE.....     ..........     .....ETEL     PTVPPVPTEP     SP........     ..........     ..........     ..MPDPCSSE
A77  Q99580      HPDDVAGIQA     LYGKKSPVIR     DEEEE.....     ..........     .....ETEL     PTVPPVPTEP     SP........     ..........     ..........     ..MPDPCSSE
A78  O15278      ..........     ..........     ..........     ..........     ..........     ..........     ..........     ..........     ..........     ....MGVTWD
A79  O18767      .KDDVKGIQA     LYGPRRAFSG     ..........     ..........     ..........     ....KPTAPH     GPPHN.....     ..........     ..........     PSIPDLCDSN
A80  O60882      .KDDVKGIQA     LYGPRKVFLG     ..........     ..........     ..........     ....KPTLPH     APHHK.....     ..........     ..........     PSIPDLCDSS
A81  P79287      .KDDVKGIQA     LYGPRKTFTG     ..........     ..........     ..........     ....KPTVPH     GPPHN.....     ..........     ..........     PSLPDICDSS
A82  O04529      TNDDVEGIQY     LYGA......     ..........     ..........     ..........     ..........     ..........     ..........     ..........     ..NPNFNGTT
A83  O23507      TVDDVAGVLK     LYGP......     ..........     ..........     ..........     ..........     ..........     ..........     ..........     ..NPKLRLDS
A84  O48680      TTDDVEGVQY     LYGA......     ..........     ..........     ..........     ..........     ..........     ..........     ..........     ..NPNFNGSR
A85  O65340      NMDDVVGVQS     LYGT......     ..........     ..........     ..........     ..........     ..........     ..........     ..........     ..NPNFTLNS
A86  O16901      KSDDISAIQA     IYGAGSGRSS     ..........     ..........     ..........     ..........     ..........     ..........     ..........     .SGSDFGGSS
A87  O61264      KSDDISAIQA     IYGAGSGRSS     ..........     ..........     ..........     ..........     ..........     ..........     ..........     .SGSDFGGSS
A88  O61266      TSMDKYNMRS     YYGAKASKKE     NEEEE.....     ..........     ....RKTENE     DKRRKTEKDR     GRTREHE...     ..........     .......SD     DIRPNECRVE
A89  O61265      DIDDKCGVIW     NYGG......     ..........     ..........     ..........     ..........     ..........     ..........     ..........     ..ASDFCLYV
A90  O44836      DIDDKCGVIW     NYGG......     ..........     ..........     ..........     ..........     ..........     ..........     ..........     ..ASDFCLYV
A91  O55761      .QDDKNGIEW     LYGLKN....     ..........     ..........     ..........     ..........     ..........     ..........     ..........     ..........
A92  MEP1_SOYBN  AQDDIDGIRK     LYGI......     ..........     ..........     ..........     ..........     ..........     ..........     ..........     ..NP......
A93  HE_HEMPU    DRDDIAGITS     LYGR......     ..........     ..........     ..........     ..........     ..........     ..........     ..........     ..NTGS.TTT
A94  HE_PARLI    DNDDIAGIRS     LYGS......     ..........     ..........     ..........     ..........     ..........     ..........     ..........     ..NSGSGTTT
A95  O93470      .LSDRRAIQN     LYGSCEGP..     ..........     ..........     ..........     ..........     ..........     ..........     .........F     DTAFDWIYKE
A96  P89294      DANGMSQLYA     DNPF......     ..........     ..........     ..........     ..........     ..........     ..........     ..........     ..........
```

Blade 1 spans columns 710–750; Blade 2 spans columns 760–790.

		710	720	730	740	750	760	770	780	790
A1	L.......T	FDAITTIRGE	VMFFKDRFYM	RTNPLY.P.E	VELN.FISVF	WPQLP...NG	LQAAYEVADR	DEVRFFKGNK	YWAVK.....	.GQDVLRGYP
A2	L.......T	FDAITTIRGE	VMFFKDRFYM	RTNPFY.P.E	VELN.FISVF	WPQLP...NG	LEAAYEFADR	DEVRFFKGNK	YWAVQ.....	.GQNVLHGYP
A3	L.......T	FDAITTIRGE	IMFFKDRFYM	RANPYY.S.E	VELN.FISVF	WPHLP...NG	LQAAYEVAHR	DEILFFKGNK	YWTVQ.....	.GQNELPGYP
A4	L.......T	FDAITTVRGE	LMFFK....M	RTNRFY.P.E	VEL.......	.........G	LQAAYEMADR	DEVRFFKGNK	YWAVS.....	.GQDVLYGYP
A5	L.......T	FDAITTLRGE	LMFFKDRFYM	RTNSFY.P.E	VELN.FISVF	WPQVP...NG	LQAAYEIADR	DEVRFFKGNK	YWAVR.....	.GQDVLYGYP
A6	......I.V	FDGVAQIRGE	IFFFKDRFMW	RTVNPR.G.K	PTGPLLVATF	WPDLP...EK	IDAVYESPQD	EKAVFFAGNE	YWVYT.....	.ASNLDRGYP
A7	......I.V	FDGIAQIRGE	IFFFKDRFIW	RTVTPR.D.K	PMGPLLVATF	WPELP...EK	IDAVYEAPQE	EKAVFFAGNE	YWIYS.....	.ASTLERGYP
A8	......I.V	FDGIAQIRGE	IFFFKDRFIW	RTVTPR.D.K	PTGPLLVATF	WPELP...EK	IDAVYEAPQE	EKAVFFAGNE	YWVYS.....	.ASTLERGYP
A9	......I.V	FDGIAQIRGE	IFFFKDRFIW	RTVTPG.D.K	PMGPLLVATF	WPELP...EK	IDAVYEAPQE	EKAVFFAGNE	YWVYS.....	.ASTLERGYP
A10	......I.V	FDGIAQIRGE	IFFFKDRFIW	RTVTPR.D.K	PTGPLLVATF	WPELP...EK	IDAVYEAPQE	EKAVFFAGNE	YWVYS.....	.ASTLERGYP
A11	....	....	....	....	....	....	....	....	....	....
A12	L.......V	FDAITTLRGE	ILFFKDSSFW	REVPTI.K.E	VYNY.PISTS	WPSLP...SG	IQAAYENPET	DQIFLFKGSK	YWALQ.....	.GFDILPNYP
A13	I.......R	WDAITTLRNE	ILFFNGRTFL	RSMPHT.G.R	IISY.TISAV	WPSLP...SG	IHAAYENQQK	DQVLLFKGNK	YWAMK.....	.GYQMLPNYP
A14	L.......S	FDAISTLRGE	ILFFKDRYFW	RKTFRT.L.V	PEFH.PISSF	WPSLP...SG	IDAAYEVTSR	DSVFIFKGNK	FWAIR.....	.GNEEQAGYP
A15	L.......S	FDAVSTLRGE	ILIFKDRHFW	RKSLRK.L.E	PELH.LISSF	WPSLP...SG	VDAAYEVTSK	DLVFIFKGNQ	FWAIR.....	.GNEVRAGYP
A16	L.......F	FDAVSTLRGE	VLFFKDRHFW	RKSLRT.P.E	PEFY.LISSF	WPSLP...SN	MDAAYEVTNR	DTVFIFKGNQ	FWAIR.....	.GHEELAGYP
A17	L.......S	FDAISTLRGE	ILFFKDRYFW	RKSLRI.L.E	PEFH.LISSF	WPSLP...SA	VDAAYEVISR	DTVFIFKGTQ	FWAIR.....	.GNEVQAGYP
A18	L.......S	FDAVSTLRGE	VLFFKDRHFW	RKSLRT.P.E	PGFY.LISSF	WPSLP...SN	MDAAYEVTNR	DTVFILKGNQ	IWAIR.....	.GHEELAGYP
A19	....	....	....	....	....	....	....	....	....	....
A20	....	....	....	....	....	....	....	....	....	....
A21	....	....	....	....	....	....	....	....	....	....
A22	....	....	....	....	....	....	....	....	....	....
A23	L.......T	FDAITTLRGE	ILFFKDRYFW	RRHPQL.Q.R	VEMN.FISLF	WPSLP...TG	IQAAYEDFDR	DLIFLFKGNQ	YWALS.....	.GYDILQGYP
A24	L.......R	FDATTTLRGE	IYFFKEKYFW	RRHPQL.R.T	VDLN.FISLF	WPGLP...NG	LQAAYEDFDR	DLVFLFKGRQ	YWALS.....	.GYDLQQGYP
A25	L.......R	FDATTTLRGE	IYFFKDKYFW	RRHPQL.R.T	VDLN.FISLF	WPFLP...NG	LQAAYEDFDR	DLVFLFKGRQ	YWALS.....	.GYDLQQGYP
A26	L.......R	FDAATTLRGE	IYFFKDKYFW	RRHPQL.R.T	VDLN.FISLF	WPFLP...NG	LQAAYEDFDR	DLVFLFKGRQ	YWALS.....	.AYDLQQGYP
A27	PAEDVCNVDI	FDAIAEIRNR	LHFFKAGKYW	RLSEGG.GRR	VQGPFLVKSK	WPALP...RK	LDSAFEDPLT	KKIFFFSGRQ	VWVYT.....	.GASLLG..P
A28	PAEDICKVNI	FDAIAEIRNY	LHFFKEGKYW	RFSKGK.GRR	VQGPFLSPST	WPALP...RK	LDSAFEDGLT	KKTFFFSGRQ	VWVYT.....	.GTSVVG..P
A29	....	....	....	....	....	....	....	....	....	....
A30	PAEDICKVNI	FDAIAEIRNY	LHFFKEGKYW	RFSKGK.GRR	VQGPFLITDT	WPALP...RK	LDSAFEDGLT	KKTFFFSGRQ	VWVYT.....	.GTSVVG..P
A31	PTTDACKVKA	FDAITEIEGQ	LHFFKDGKYW	KASSAR.PGA	IKGPVKISDT	WPALP...AI	IDSAFEDLLT	KKIFFFSGRR	FWVYT.....	.GTTVLG..P
A32	PVDDACNVNI	FDAIAEIGNQ	LYLFKDGKYW	RFSEGR.GSR	PQGPFLIADK	WPALP...RK	LDSVFEEPLS	KKLFFFSGRQ	VWVYT.....	.GASVLG..P
A33	PADNPCNVDV	FDAIAEIQGA	LHFFKDGWYW	KFLNHR.GSP	LQGPFLTART	WPALP...AT	LDSAFEDPQT	KRVFFFSGRQ	MWVYT.....	.GKTVLG..P
A34	PAEDVCNVNV	FDAIAEIGNK	LHVFKDGRYW	RFSEGS.GRR	PQGPFLIADT	WPALP...AK	LDSAFEEPLT	KKLFFFSGRQ	VWVYT.....	.GASVLG..P
A35	PDDNPCNVDV	FDAIADIQGA	LHFFKDGKYW	KFSNHG.GNQ	LQGPFLIART	WPAFP...SK	LNSAFEDPQP	KKIFFFLWAQ	MWVYT.....	.GQSVLG..P
A36	L.......S	FDAISTLRGE	YLFFKDRYFW	RRSHWN.P.E	PEFH.LISAF	WPSLP...SY	LDAAYEVNSR	DTVFIFKGNE	FWAIR.....	.GNEVQAGYP
A37	L.......S	FDSVSTLRGE	VLFFKDRYFW	RRSHWN.P.E	PEFH.LISAF	WPTLP...SD	LDAAYEAHNT	DSVLIFKGSQ	FWAVR.....	.GNEVQAGYP
A38	L.......S	FDAVTMLRGE	FLFFKDRHFW	RRTQWN.P.E	PEFH.LISAF	WPSLP...SG	LDAAYEANNK	DRVLIFKGSQ	FWAVR.....	.GNEVQAGYP
A39	DAPPDACEAS	FDAVSTIRGE	LFFFKAGFVW	RLRGGQ..LQ	PGYPALASRH	WQGLP...SP	VDAAFEDAQG	H.IWFFGQAQ	YWVY......	.DGEKPVLGP
A40	ETPPDVCETS	FDAVSTIRGE	LFFFKAGFVW	RLRSGR..LQ	PGYPALASRH	WQGLP...SP	VDAAFEDAQG	Q.IWFFGQAQ	YWVY......	.DGEKPVLGP
A41	EVPPEVCETS	FDAVSTIRGE	LFFFKAGFVW	RLRSGQ..LQ	PGYPALASRH	WQGLP...SP	VDAAFEDAQG	Q.IWFFGQAQ	YWVY......	.DGEKPVLGP
A42	....DACKTN	FDAVSTIRGE	LFFFKSGYVW	RLRGGK..LQ	NGYPALASRH	WRGIP...DT	VDAAFEDSVG	N.IWFFYGSQ	FWVF......	.DGKLQASGP
A43	L.......S	FDAVTTVGNK	IFFFKDRFFW	LKVSER.P.K	TSVN.LISSL	WPTLP...SG	IEAAYEIEAR	NQVFLFKDDK	YWLIS.....	.NLRPEPNYP
A44	L.......S	FDAVTTVGNK	IFFFKDRFFW	LKVSER.P.K	TSVN.LISSL	WPTLP...SG	IEAAYEIEAR	NQVFLFKDDK	YWLIS.....	.NLRPEPNYP
A45	L.......S	FDAVTTVGEK	ILFFKDWFFW	WKLPGS.P.A	TNIT.SISSI	WPSIP...SA	IQAAYEIESR	NQLFLFKDEK	YWLIN.....	.NLVPEPHYP
A46	L.......K	FDAVTTVGNK	IFFFKDSFFW	WKIPKS.S.T	TSVR.LISSL	WPTLP...SG	IEAAYEIGDR	HQVFLFKGDK	FWLIS.....	.HLRLQPNYP
A47	L.......S	FDAVTTVGDK	IFFFKDWFFW	WRLPGS.P.A	TNIT.SISSM	WPTIP...SG	IQAAYEIGGR	NQLFLFKGDK	YWLIN.....	.NLVPEPHYP
A48	L.......S	LDAITSLRGE	TLIFKDRFFW	RLHPQQ.V.E	AELF.LTKSF	GPELP...NR	IDAAYEHPSH	DLIFIFRGRK	FWALS.....	.GYDILEDYP
A49	....	....	....	....	....	....	....	....	....	....
A50	L.......S	LDAITEMRGE	KLIFKDRFFW	RQHPQM.T.D	VELV.LIRNF	WPELP...SK	IDAAYEVPFK	DLIVIFRGRK	FWALN.....	.GYDILADYP
A51	L.......S	LDAITSLRGE	TMVFKDRFFW	RLHPQL.V.D	AELF.LTKSF	WPELP...NR	IDAAYEHPSK	DLIFIFRGRK	FWALN.....	.GYDILEGYP
A52	L.......S	LDAITSLRGE	TMIFKDRFFW	RLHPQQ.V.D	AELF.LTKSF	WPELP...NR	IDAAYEHPSH	DLIFIFRGRK	FWALN.....	.GYDILEGYP
A53	L.......S	LDAITSLRGE	TMIFKDRFFW	RLHPQQ.V.E	AELF.LTKSF	WPELP...NH	VDAAYEHPSR	DLMFIFRGRK	FWALN.....	.GYDILEGYP
A54	....	....	....	....	....	....	....	....	....	....
A55	L.......S	LDAITSLRGE	TMIFKDRFFW	RLHPQQ.V.D	AELF.LTKSF	WPELP...NR	IDAAYEHPAR	DLIFIFRGKK	FWAPN.....	.GYDILEGYP
A56	L.......S	LDAITSLRGE	TMIFKDRFFW	RLHPQQ.V.E	PELF.LTKSF	WPELP...NH	VDAAYEHPSR	DLMFIFRGRK	FWALN.....	.GYDIMEGYP
A57	....	....	....	....	....	....	....	....	....	....
A58	L.......S	IDAITELRGE	KMIFKDRFFW	RVHPQM.T.D	AELV.LIKSF	WPELP...NK	LDAAYEHPAK	DLSYLFRGKK	FWALN.....	.GYDIVEDYP
A59	L.......T	IDAITELRGE	KMIFKDRFFW	RVHPQM.T.D	AELV.LIKSF	WPELP...NK	IDAAYEHPAK	DLIYIFRGKK	FWALN.....	.GYDFVEDYP
A60	....	FDTVAMLRGE	MFVFKKRWFW	RVFNN..QVM	DGYPMPIGQF	WRGLP...AS	INTAYERKD.	GKFVFFKGDK	HWVFD.....	.EASLEPGYP
A61	....	FDTVAMLRGE	MFVFKERWLW	RVFNN..QVM	DGYPMPIGQF	WRGLP...AS	INTAYERKD.	GTFVFFKGDK	HWVCV.....	.EASLEPGYA
A62	....	FDTVAMLRGE	MFVFKERWFW	RVRNN..QVM	DGYPMPIGQF	WRGLP...AS	INTAYERKD.	GKFVFFKGDK	HWVFD.....	.EASLEPGYP
A63	....	FDTVAVFRGE	MFVFKERWFW	RVRNN..QVM	DGYPMPIGQL	WRGLP...AS	INTAYERKD.	GKFVFFKGDK	HWVFD.....	.EASLEPGYP
A64	....	FDTVAMLRGE	MFVFKERWFW	RVRNN..QVM	DGYPMPIGQF	WRGLP...AS	INTAYERKD.	GKFVFFKGDK	HWVFD.....	.EASLEPGYP
A65	....	FDTVAMLRGE	MFVFKGRWFW	RVRHN..RVL	DNYPMPIGHF	WRGLP...GD	ISAAYERQD.	GRFVFFKGDR	YWLFR.....	.EANLEPGYP
A66	....	FDTVAMLRGE	MFVFKGRWFW	RVRHN..RVL	DNYPMPIGHF	WRGLP...GD	ISAAYERQD.	GRFVFFKGDR	YWLFR.....	.EANLEPGYP
A67	....	FDTVAMLRGE	MFVFKGRWFW	RVRHN..RVL	DNYPMPIGHF	WRGLP...GD	ISAAYERQD.	GRFVFFKGDR	YWLFR.....	.EANLEPGYP
A68	....	FDTVAVLRGE	MFVFKGRWFW	RVRHN..RVL	DNYPMPIGHF	WRGLP...GN	ISAAYERQD.	GHFVFFKGNR	YWLFR.....	.EANLEPGYP
A69	....	FNTLVILRRE	MFVFKDQWFW	RVRNN..RVM	DGYPMQITYF	WRGLP...PS	IDAVYENSE.	GNFVFFKSNK	FWVFK.....	.DTTLQPGYP
A70	....	FNTLAILRRE	MFVFKDQWFW	RVRNN..RVM	DGYPMQITYF	WRGLP...PS	IDAVYENSD.	GNFVFFKGNK	YWVFK.....	.DTTLQPGYP
A71	....	FNTLAILRRE	MFVFKDQWFW	RVRNN..RVM	DGYPMQITYF	WRGLP...PS	IDAVYENSD.	GNFVFFK...	..V.K.....	.GDTLS....
A72	....	FNTLAILRRE	MFVFKDQWFW	RVRNN..RVM	DGYPMQITYF	WRGLP...PS	IDAVYENSD.	GNFVFFKGNK	YWVFK.....	.DTTLQPGYP
A73	....	FNTLAILRRE	MFVFKDQWFW	RVRNN..RVM	DGYPMQITYF	WRGLP...PS	IDAVYENSD.	GNFVFFKGNK	YWVFK.....	.DTTLQPGYP
A74	KDVPHRCSTH	FDAVAQIRGE	AFFFKGKYFW	RLTRDR.HLV	SLQPAQMHRF	WRGLPLHLDS	VDAVYERTSD	HKIVFFKGDR	YWVFK.....	.DNNVEEGYP
A75	V.......V	FNAVTTMRGE	LIFFVKRFLW	RKHPQA.S.E	AELM.FVQAF	WPSLP...TN	IDAAYENPIT	EQILVFKGSK	YTALD.....	.GFDVVQGYP
A76	LD........	.AMMLGPRGK	TYAFKGDYVW	TVSDS....G	PGPLFRVSAL	WEGLP...GN	LDAAVYSPRT	QWIHFFKGDK	VWRYI.....	.NFKMSPGFP
A77	LD........	.AMMLGPRGK	TYAFKGDYVW	TVSDS....G	PGPLFRVSAL	WEGLP...GN	LDAAVYSPRT	QWIHFFKGDK	VWRYI.....	.NFKMSPGFP
A78	FS........	.MSNGGPRGK	TYAFKGDYVW	TVSDS....G	PGPLFRVSAL	WEGLP...GN	LDAAVYSPRT	QWIHFFKGDK	VWRYI.....	.NFKMSPGFP
A79	L.......S	FDAVTMLGKE	LLLFRDRIFW	RRQVHL.M.S	GIRPSTITSS	FPQLM...SN	VDAAYEVAER	GTAYFFKGPH	YWITR.....	.GFQMQ.GPP
A80	S.......S	FDAVTMLGKE	LLLFKDRIFW	RRQVHL.R.T	GIRPSTITSS	FPQLM...SN	VDAAYEVAER	GTAYFFKGPH	YWITR.....	.GFQMQ.GPP
A81	S.......S	FDAVTMLGKE	LLFFRDRIFW	RRQVHL.M.S	GIRPSTITSS	FPQLM...SN	VDAAYEVADR	GMAYFFKGPH	YWITR.....	.GFQMQ.GPP
A82	S.........	PPSTTKHQRD	TGGFSAAWRI	DGSS....RS	TIVSLLLSTV	GLVLWFLP..	....	....	....	....
A83	L.........	TQSEDS..IK	NGTVSHRFLS	GNFI....GY	VLLVVGLILF	L.........	....	....	....	....
A84	S.........	PPPSTQ.QRD	TGDSGAFGRS	DGS.....RS	VLTNLLQYYF	WIIFGLFLYL	V.........	....	....	....
A85	L.........	LASETS..TN	LADGS.RIRS	QGMI....YS	ALSTVIALCF	LNW.......	....	....	....	....
A86	G.........	GGSRTTARPT	TTTRSWFGRF	FGDDDD.DVR	SRTTTRRTTL	WPTTQSPFSG	DDWGSGSGSS	GRGGSSSGSS	GGGCP.....	.SHIDAYTPS
A87	G.........	GGSRTTARPT	TTTRSWFGRF	FGDDDD.DVR	SRTTTRRTTL	WPTTQSPFSG	DDWGSGSGSS	GRGGSSSGSS	GGGCP.....	.SHIDAYTPS
A88	N.........	.PIVVQYRGE	YLIFKSQVVW	RVSSDWKR.L	IIKAVPINQL	FPGLP...NP	IDAAVTVG..	HNLWVFVGEM	IYVIY.....	.GNHMVHAP
A89	W.........	LMSQIVEAHN	SSAQNNDGVG	SITS....SR	TNKKSFK...	....	....	....	.DSKI.....	.PKCSSNNSS
A90	W.........	LMSQIVEAHN	SSAQNNHGVG	SITS....SR	TNKKSPKSEG	FFLFQLKFPH	STLTHTDDVV	MREKDKRSYR	GDSKI.....	.PKCSSNNSS
A91	.........K	WAKLSPPVTK	PPVTKPPVTK	PILYTV.PRF	PVPTTQVPII	QPRFG.....	NPTTHSPTTR	HPLQKIPIQP	RFGHS.....	.NPPIRYNPP
A92	....	....	....	....	....	....	....	....	....	....
A93	T.........	TRRPTITR.T	TTRRTTTRRT	TTQL....AT	TQTTTIRPPT	YPTPPRQ...	ACTGSFDAVI	KDNSDRIYAL	AGRYY.....	.WRLDQASPS
A94	T.........	TRRPTTTRAT	TTRRTTTTRA	TTTR....AT	TTTTTS..PS	RPSPPRR...	ACSGSFDAVV	RDSSNRIYAL	TGPYF.....	.WQLDQPSPS
A95	K.......N	QYGELVVRYN	TYFFRNSWYW	MYENRSNRTR	YGDPLAIANG	WHGIPVQNID	AFVHVWTWTR	DASYFFKGTQ	YWRYDSENDK	AYAEDAQGKS
A96	....	....	....	....	....	....	....	....	....	....

                           Blade 3   Blade 4
A1 COG1_BOVIN RDIYRSFGFP 810 RTVKSIDAAV 820 SEEDTGKT.. 830 YFFVANKCWR 840 YDEYK..QSM 850 DAGYPKMIAE 860 DFPGIGNK.. 870 ....VDAVFQ 880 KGG..FFYFF 890 HGRRQYKFDP 900
A2 COG1_HUMAN KDIYSSFGFP   RTVKHIDAAL   SEENTGKT..   YFFVANKYWR   YDEYK..RSM   DPGYPKMIAH   DFPGIGHK..   ....VDAVFM   KDG..FFYFF   HGTRQYKFDP
A3 COG1_RABIT KDIHSSFGFP   RSVNHIDAAV   SEEDTGKT..   YFFVANKYWR   YDEYK..RSM   DAGYPKMIEY   DFPGIGNK..   ....VDAVFK   KDG..FFYFF   HGTRQYKFDP
A4 COG1_RANCA KDIHSSFGFP   ....TG....   ..VAHECWS   YDEYK..QSM   DTGY....AD   EFPG......   ....DAVFQ   K.......FF   HGTRQYQFDL
A5 COG1_PIG   KDIHRSFGFP   STVKNIDAAV   FEEDTGKT..   YFFVAHECWR   YDEYK..QSM   DTGYPKMIAE   EFPGIGNK..   ....VDAVFQ   KDG..FLYFF   HGTRQYQFDF
A6 COG2_CHICK KK.LTSLGLP   PDVQRIDAAF   NWGRNKKT..   YIFSGDRYWK   YNEEK..KKM   ELATPKFIAD   SWNGVPDN..   ....LDAVLG   LTDSGYTYFF   KDQYYLQMED
A7 COG2_HUMAN KP.LTSLGLP   PDVQRVDAAF   NWSKNKKT..   YIFAGDKFWR   YNEVK..KKM   DPGFPKLIAD   AWNAIPDN..   ....LDAVVD   LQGGGHSYFF   KGAYYLKLEN
A8 COG2_MOUSE KP.LTSLGLP   PDVQQVDAAF   NWSKNKKT..   YIFAGDKFWR   YNEVK..KKM   DPGFPKLIAD   SWNAIPDN..   ....LDAVVD   LQGSGHSYFF   KGTYYLKLEN
A9 COG2_RABIT KP.LTSLGLP   PDVQQVDAAF   NWSKNKKT..   YIFAGDKFWR   YNEVK..KKM   DPGFPKLIAD   AWNAIPDH..   ....LDAVVD   LQGSGHSYFF   KGTYYLKLEN
A10 COG2_RAT   KP.LTSLGLP   PDVQQVDAAF   NWSKNKKT..   YIFAGDKFWR   YNEVK..KKM   DPGFPKLIAD   SWNAIPDN..   ....LDAVVD   LQGGGHSYFF   KGAYYLKLEN
A11 G257061   ....
A12 Q98857   KN.IDKLGFP   RTVKHINAAV   YLQSTGKT..   YFFAGEQYWS   YDEAR..KTM   DKESPRRIED   DFPGIGKK..   ....VHAVFE   DNG..LLYFF   SGHKQFEFNM
A13 Q98858   QN.IYTLGFP   RTVTRIDAAV   YHPDTRKT..   YYFVNDKYWS   FDEAL..QVM   DKDSPQQIVT   TFPRIGTK..   ....VDAVFY   AKG..LLYFF   NGQHQFEFNM
A14 Q28397   RG.IHTLGFP   PTVRKIDAAI   FDKEKQKT..   YFFVEDKYWR   FDEKR..QSM   EPGYPKQIAE   DFPGIDSK..   ....LDAAFE   SFG..FFYFF   SGSSQFEFDP
A15 COG3_HUMAN RG.IHTLGFP   PTVRKIDAAI   SDKEKNKT..   YFFVEDKYWR   FDEKR..NSM   EPGFPKQIAE   DFPGIDSK..   ....IDAVFE   EFG..FFYFF   TGSSQLEFDP
A16 COG3_MOUSE KS.IHTLGFP   ATVKKIDAAV   SNKEKRKT..   YFFVEDKYWR   FDEKR..QSM   EPGFPRKIAE   DFPGVDSR..   ....VDAVFE   AFG..FLYFF   SGSSQLEFDP
A17 COG3_RABIT RS.IHTLGFP   STIRKIDAAI   SDKERKKT..   YFFVEDKYWR   FDEKR..QSL   EPGFPRHIAE   DFPGINPK..   ....IDAVFE   AFG..FFYFF   SGSSQSEFDP
A18 COG3_RAT   KS.IHTLGLP   ETVQKIDAAI   SLKDQKKT..   YFFVEDKFWR   FDEKK..QSM   DPEFPRKIAE   NFPGIGTK..   ....VDAVFE   AFG..FLYFF   SGSSQLEFDP
A19 COG7_FELCA ....
A20 COG7_HUMAN ....
A21 COG7_MOUSE ....
A22 COG7_RAT   ....
A23 COG8_HUMAN KD.ISNYGFP   SSVQAIDAAV   FYR..SKT..   YFFVNDQFWR   YDNQF..QFM   EPGYPKSISG   AFPGIESK..   ....VDAVFQ   QEH..FFHVF   SGPRYYAFDL
A24 O70138   RD.ISNYGFP   RSVQAIDAAV   SYN..GKT..   YFFINNQCWR   YDNER..RSM   DPGYPKSIPS   MFPGVNCR..   ....VDAVFL   QDS..FFLFF   SGPQYFAFNF
A25 O88733   RD.ISNYGFP   RSVQAIDAAV   SYN..GKT..   YFFINNQCWR   YDNQR..RSM   DPGYPKSIPS   MFPGVNCR..   ....VDAVFL   QDS..FFLFF   SGPQYFAFNF
A26 O88766   RD.ISNYGFP   RSVQAIDAAV   SYN..GKT..   YFFVNNQCWR   YDNQR..RSM   DPGYPTSIAS   VFPGVNCR..   ....IDAVFQ   QDS..FFLFF   SGPQYFAFNL
A27 COG9_BOVIN RR.LDKLGLG   PEVAQVTGAL   PRPEGK.V..   LLFSGQSFWR   FDVKT..QKV   DPQSVTPVDQ   MFPGVPIS..   ....THDIFQ   YQEK..AYFC   QDHFYWRVSS
A28 COG9_CANFA RR.LDKLGLG   PEVTQVTGAL   PQAGGK.V..   LLFSRQRFWS   FDVKT..QTV   DPRSAGSVEQ   MYPGVPLN..   ....THDIFQ   YQEK..AYFC   QDRFYWRVNS
A29 Q95166   ....
A30 O19130   RR.LDKLGLG   PEVTQVTGAL   PQGGGK.V..   LLFSRQRFWS   FDVKT..QTV   DPGSAGSVEQ   MYPGVPLN..   ....THDIFQ   YQEK..AYFC   QDRFYWRVNS
A31 Q99856   RG.LEKLGIG   KDVEKIVGSL   QRGRGK.V..   LLFNGDKYWR   LDVKT..QVV   DKGYPRDTED   AFAGVPIN..   ....AHDVFL   YQEN..IHFC   KDQFYWRMTP
A32 COG9_HUMAN RR.LDKLGLG   ADVAQVTGAL   RSGRGK.M..   LLFSGRRLWR   FDVKA..QMV   DPRSASEVDR   MFPGVPLD..   ....THDVFQ   YREK..AYFC   QDRFYWRVSS
A33 COG9_MOUSE RS.LDKLGLG   PEVTHVSGLL   PRRPGK.A..   LLFSKGRVWR   FDLKS..QKV   DPQSVIRVDK   EFSGVPWN..   ....SHDIFQ   YQDK..AYFC   HGKFFWRVSF
A34 COG9_RABIT RR.LDKLGLG   PEVPHVTGAL   PRAGGK.A..   LLFGAQRFWR   FDVKT..QTV   DSRSGAPVDQ   MFPGVPLN..   ....THDVFQ   YREK..AYFC   QDRFWRVST
A35 COG9_RAT   RS.LDKLGLG   SEVTLVTGLL   PRRGGK.A..   LLISRERIWK   FDLKS..QKV   DPQSVTRLDN   EFSGVPWN..   ....SHNVFQ   YQDK..AYFC   HDKYFWRVSF
A36 COGX_HUMAN RG.IHTLGFP   PTIRKIDAAV   SDKEKKKT..   YFFAADKYWR   FDENS..QSM   EQGFPRLIAD   DFPGVEPK..   ....VDAVLQ   AFG..FFYFF   SGSSQFEFDP
A37 COGX_MOUSE RG.IHTLGFP   PTVKKIDAAV   FEKEKKKT..   YFFVGDKYWR   FDETR..HVM   DKGFPRQITD   DFPGIEPQ..   ....VDAVLH   EFG..FFYFF   RGSSQFEFDP
A38 COGX_RAT   KR.IHTLGFP   PTVKKIDAAV   FEKEKKKT..   YFFVGDKYWR   FDETR..QLM   DKGFPRLITD   DFPGIEPQ..   ....VDAVLH   AFG..FFYFF   CGSSQFEFDP
A39 COGY_HUMAN AP.LTELGLV   RF..PVHAAL   VWGPEKNK.I   YFFRGRDYWR   FHPST..RRV   DSPVPRR.AT   DWRGVPSE..   ....IDAAFQ   DADG.YAYFL   RGRLYWKFDP
A40 COGY_MOUSE AP.LSKLGLQ   GS..PVHAAL   VWGPEKNK.I   YFFRGGDYWR   FHPRT..QRV   DNPVPRR.ST   DWRGVPSE..   ....IDAAFQ   DAEG.YAYFL   RGHLYWKFDP
A41 P97568   AP.LSKLGLQ   GS..PVHAAL   VWGPEKNK.I   YFFRGGDYWR   FHPRT..QRV   DNPVPRR.TT   DWRGVPSE..   ....IDAAFQ   DAEG.YAYFL   RGHLYWKFDP
A42 COGY_XENLA FP.ITDIGIS   VT..QIQAAF   VWGTEKNKKT   YLFRGGEYWR   FNPET..RRV   ESRHSRR.IG   DWRGVPKG..   ....IDAAFQ   DEQG.YAYFV   KGRQYWKFDP
A43 COGM_HUMAN KS.IHSFGFP   NFVKKIDAAV   FNPRFYRT..   YFFVDNQYWR   YDERR..QMM   DPGYPKLITK   NFQGIGPK..   ....IDAVFY   SKNK.YYYFF   QGSNQFEYDF
A44 Q99745   KS.IHSFGFP   NFVKKIDAAV   FNPRFYRT..   YFFVDNQYWR   YDERR..QMM   DPGYPKLITK   NFQGIGPK..   ....IDAVFY   SKNK.YYYFF   QGSNQFEYDF
A45 COGM_MOUSE RS.IYSLGFS   ASVKKVDAAV   FDPLRQKV..   YFFVDKHYWR   YDVRQ..ELM   DPAYPKLIST   HFPGIKPK..   ....IDAVLY   FKR..HYYIF   QGAYQLEYDP
A46 COGM_RABIT KS.IHSLGFP   DFVKKIDAAV   FNPSLRKT..   YFFVDNLYWR   YDERR..EVM   DAGYPKLITK   HFPGIGPK..   ....IDAVFY   FQR..YYYFF   QGPNQLEYDT
A47 Q63341   RS.IHSLGFP   ASVKKIDAAV   FDPLRQKV..   YFFVDKQYWR   YDVRQ..ELM   DAAYPKLIST   HFPGIRPK..   ....IDAVLY   FKR..HYYIF   QGAYQLEYDP
A48 O77656   KK.ISELGFP   KHVKKISAAL   HFEDSGKT..   LFFSENQVWS   YDDTN..HVM   DKDYPRLIEE   VFPGIGDK..   ....VDAVYQ   KNG..YIYFF   NGPIQFEYSI
A49 O77632   ....
A50 Q98859   KK.IQELGFP   KSLRTIDAAV   YNRAMGKT..   LFFTGEKYWS   FDEEK..QTV   EKGYPRFIAD   DFPGIGET..   ....VDAAYQ   RNG..YIYFF   SGSLQFEYST
A51 O18927   QK.ISELGFP   KDVKKISAAV   HFEDTGKT..   LFFSGNQVWR   YDDTN..RMM   DKDYPRLIEE   DFPGIGDK..   ....VDAVYE   KNG..YIYFF   NGPIQFEYSI
A52 COGZ_HUMAN KK.ISELGLP   KEVKKISAAV   HFEDTGKT..   LLFSGNQVWR   YDDTN..HIM   DKDYPRLIEE   DFPGIGDK..   ....VDAVYE   KNG..YIYFF   NGPIQFEYSI
A53 COGZ_MOUSE RK.ISDLGFP   KEVKRLSAAV   HFENTGKT..   LFFSENHVWS   YDDVN..QTM   DKDYPRLIEE   EFPGIGNK..   ....VDAVYE   KNG..YIYFF   NGPIQFEYSI
A54 O73923   ....
A55 O62806   QK.LSELGFP   REVKKISAAV   HFEDTGKT..   LFFSGNQVWS   YDDTN..HTM   DQDYPRLIEE   EFPGIGGK..   ....VDAVYE   KNG..YIYFF   NGPIQFEYSI
A56 COGZ_RAT   RK.ISDLGFP   KEVKRLSAAV   HFEDTGKT..   LFFSGNHVWS   YDDAN..QTM   DKDYPRLIEE   EFPGIGDK..   ....VDAVYE   KNG..YIYFF   NGPIQFEYSI
A57 O77631   ....
A58 COGZ_XENLA KK.LHELGFP   KTLKAIDAAV   YNKDTGKT..   FFFTEDSYWS   FDEEA..RTL   DKGFPRLISE   DFPGIGEK..   ....VDAAYQ   RNG..YLYFF   NGALQFEYSI
A59 CGZA_XENLA KK.LHELGFP   KTLKAIDAAV   YNKAIGKT..   LFFAEDSYWS   FDEEA..RTM   DKGFPRLISE   DFPGIGEK..   ....VDAAYQ   RNG..YIYFF   NGALQFEYSI
A60 COGT_HUMAN KH.IKELGRG   LPTDKIDAAL   FWMPNGKT..   YFFRGNKYYR   FNEEL..RAV   DSEYPKN.IK   VWEGIPES..   ....PRGSFM   GSDEVFTYFY   KGNKYWKFNN
A61 COGT_MOUSE NH.IKELVRG   LPSDKIDTAL   FWMPNGKT..   YFFRGNKYYR   FNEEF..RAV   DSEYPKN.IK   VWEGIPES..   ....PRGSFM   GSDEVFTYFY   KGNKYWKFNN
A62 O08645   KH.IKELGRG   LPTDKIDAAL   FWMPNGKT..   YFFRGNKYYR   FNEEF..RAV   DSEYPKN.IK   VWEGIPES..   ....PRGSFM   GSDEVFTYFY   KGNKYWKFNN
A63 COGT_RABIT KH.IKELGRG   LPTDKIDAAL   FWMPNGKT..   YFFRGNKYYR   FNEEL..RAV   DSEYPKN.IK   VWEGIPES..   ....PRGSFM   GSDEVFTYFY   KGNKYWKFNN
A64 COGT_RAT   KH.IKELGRG   LPTDKIDAAL   FWMPNGKT..   YFFRGNKYYR   FNEEF..RAV   DSEYPKN.IK   VWEGIPES..   ....PRGSFM   GSDEVFTYFY   KGNKYWKFNN
A65 COGU_HUMAN QP.LTSYGLG   IPYDRIDTAI   WWEPTGHT..   FFFQEDRYWR   FNEET..QRG   DPGYPKP.IS   VWQGIPAS..   ....PKGAFL   SNDAAYTYFY   KGTKYWKFDN
A66 Q14111   QP.LTSYGLG   IPYDRIDTAI   WWEPTGHT..   FFFQEDRYWR   FNEET..QRG   DPGYPKP.IS   VWQGIPAS..   ....PKGAFL   SNDAAYTYFY   KGTKYWKFDN
A67 D1023087   QP.LTSYGLG   IPYDRIDTAI   WWEPTGHT..   FFFQEDRYWR   FNEET..QRG   DPGYPKP.IS   VWQGIPAS..   ....PKGAFL   SNDAAYTYFY   KGTKYWKFDN
A68 O54732   QP.LSSYGTD   IPYDRIDTAI   WWEPTGHT..   FFFQADRYWR   FNEET..QHG   DPGYPKP.IS   VWQGIPTS..   ....PKGAFL   SNDAAYTYFY   KGTKYWKFNN
A69 Q98947   HD.LITLGSG   IPSHGIDSAI   WWEDVGKT..   YFFKGDRYWR   YGEEM..RSM   DPGYPKP.IT   IWKGIPES..   ....PQGAFV   HKENGFTYFI   KEK.STGNST
A70 COGV_HUMAN HD.LITLGSG   IPPHGIDSAI   WWEDVGKT..   YFFKGDRYWR   YSEEM..KTM   DPGYPKP.IT   VWKGIPES..   ....PQGAFV   HKENGFTYFY   KGKEYWKFNN
A71 Q14824   ....VIQDG..   WLYK..YHWK   WILEQ..RQS   VPVLSRQ.TE   KHKTYEEL..   ....SSITY.   ....
A72 COGV_RAT   HD.LITLGNG   IPPHGIDSAI   WWEDVGKT..   YFFKGDRYWR   YSEEM..KTM   DPGYPKP.IT   IWKGIPES..   ....PQGAFV   HKENGFTYFY   KGKEYWKFNN
A73 O35541   HD.LITLGNG   IPPHGIDSAI   WWEDVGKT..   YFFKGDRYWR   YSEEM..KTM   DPGYPKP.IT   IWKGIPES..   ....PQGAFV   HKENGFTYFY   KGKEYWKFNN
A74 Q14850   RP.VSDFSLP   PG..GIDAAF   SWAHNDRT..   YFFKDQLYWR   YDDHT..RHM   DPGYPAQ.SP   LWRGVPST..   ....LDDAMR   WSDG.ASYFF   RGQEYWKVLD
A75 O13065   RN.IYSLGFP   KTVKRIDAAV   HIEQLGKT..   YFFAAKKYWS   YDEDK..KQM   DKGFPKQISN   DFPGIPDK..   ....IDAAFY   YRG..RLYFF   IGRSQFEYNI
A76 Q99542   KK....LNRV   EP..NLDAAL   YWPLNQKV..   FLFKGSGYWQ   WDELA..RTD   FSSYPKPIKG   LFTGVPNQ..   ....PSAAMS   WQDG.RVYFF   KGKVYWRLNQ
A77 Q99580   KK....LNRS   EP..NLDAAL   YWPLNQKV..   FLFKGSGYWQ   WDELA..RTD   FSSYPKPIKG   LFTGVPNQ..   ....PSAAMS   WQDG.RVYFF   KGKVYWRLNQ
A78 O15278   KK....LNRV   EP..NLDAAL   YWPLNQKV..   FLFKGSGYWQ   WDELA..RTD   FSSYPKPIKG   LFTGVPNQ..   ....PSAAMS   WQDG.RVYFF   KGKVYWRLNQ
A79 O18767   RT.IYDFGFP   RYVQRIDAAV   YLKDAQKT..   LFFVGDEYYS   YDERK..RKM   EKDYPKSTEE   EFSGVNGQ..   ....IDAAVE   LNG..YIYFF   SGPKAYKSDT
A80 O60882   RT.IYDFGFP   RHVQQIDAAV   YLREPQKT..   LFFVGDEYYS   YDERK..RKM   EKDYPKNTEE   EFSGVNGQ..   ....IDAAVE   LNG..YIYFF   SGPKTYKYDT
A81 P79287   RT.IYDFGFP   RYVQRIDAAV   HLKDTQKT..   LFFVGDEYYS   YDEPK..RKM   DKDYPKNTEE   EFSGVNGQ..   ....IDAAVE   LNG..YIYFF   SGPKAYKYDT
A82 O04529   ....
A83 O23507   ....
A84 O48680   ....
A85 O65340   ....
A86 O16901   SS..FSYAFS   GSQVYTISGT   KVTKVQSIHD   LFPSAPTPVN   AALWN..PIS   GSMLLFSSNR   VYSYYFSN..   ....IRQIFQ   MDSGFPKTLP   SDLGFSVSGA
A87 O61264   SS..FSYAFS   GSQVYTISGT   KVTKVQSIHD   LFPSAPTPVN   AALWN..PIS   GSMLLFSSNR   VYSYYFSN..   ....IRQIFQ   MDSGFPKTLP   SDLGFSVSGA
A88 O61266   LR.LSDIGIN   ..EKYVDLAY   EWHYFNPPAV   YIWKGSRYWK   LDEKMYHRRV   DERYPKDTDL   NWARVPKG..   ....VHSAFT   YEK..EIHLL   RGNQVFRMNS
A89 O61265   ....QRT   LAEKKLTLGL   HLSEADAKRY   AEMVCNFLAG   LHMWR..TNP   NHHASESLEK   EYKGVSQE..   ....MGTFSG   KSIAVRRLIR   HAEHQKERSE
A90 O44836   ....QRT   LAEKKLTLGL   HLSEADAKRY   TEMVCNFLAG   LHMWR..TNP   NHHASESLEK   EYKGVSQE..   ....MGTFSG   KSIAVRRLIR   HAEHQKERSE
A91 O55761   IRYNPKHPLS   PPFKSIPNSL   LNVVNSSITN   LYVYFPNTWN   SSIIK..FN.   ....
A92 MEP1_SOYBN ....
A93 HE_HEMPU   ....WGV   VRNRFGFDLP   ENVDASFQNG   IFSYFFSGCY   YYYQT..STR   RRFPRTPPNR   RWVGLPCD..   ....IDAVYK   SGDSGTTYFF   KGRFVYKFSS
A94 HE_PARLI   ....WGL   VSNRFGFGLP   QNIDASFQRG   VVTYFFSECY   YYYQT..STQ   RNFPRIPVNR   KWVGLPCN..   ....IDAVYR   SS.RGPTYFF   KDSFVYKFNS
A95 O93470   YPRLISEGFP   GIPSPINAAY   FDR..RRQYI   YPFRDSQVFA   FDINR..NRV   APDFPKRILD   FFPAVAANNH   PKGNIDVAYY   SYTYSSLFLF   KGKEFWKVVS
A96 P89294   ....

←—— HEMOPEXIN DOMAIN ——→ ┬— LINKER →   ←— LINKER —┬— TRANSMEMBRANE DOMAIN —→

```
                          910        920        930        940        950        960        970        980        990       1000
A1   QT.....KRI  LTLLKANSWF  NCRKN.....  ..........  ..........  ..........  ..........  ..........  ..........  ..........
A2   KT.....KRI  LTLQKANSWF  NCRKN.....  ..........  ..........  ..........  ..........  ..........  ..........  ..........
A3   KT.....KRI  LTLQKANSWF  NCRKN.....  ..........  ..........  ..........  ..........  ..........  ..........  ..........
A4   KT.....KRI  LTLQKANSWF  NCRKN.....  ..........  ..........  ..........  ..........  ..........  ..........  ..........
A5   KT.....KRI  LTLQKANSWF  NCRKN.....  ..........  ..........  ..........  ..........  ..........  ..........  ..........
A6   KS....LKIV  KIGKISSDWL  GC........  ..........  ..........  ..........  ..........  ..........  ..........  ..........
A7   QS....LKSV  KFGSIKSDWL  GC........  ..........  ..........  ..........  ..........  ..........  ..........  ..........
A8   QS....LKSV  KFGSIKSDWL  GC........  ..........  ..........  ..........  ..........  ..........  ..........  ..........
A9   QS....LKSV  KVGSIKTDWL  GC........  ..........  ..........  ..........  ..........  ..........  ..........  ..........
A10  QS....LKSV  KFGSIKSDWL  GC........  ..........  ..........  ..........  ..........  ..........  ..........  ..........
A11  ..........  ..........  ..........  ..........  ..........  ..........  ..........  ..........  ..........  ..........
A12  KS.....KKV  TRTLKNTSWL  GC........  ..........  ..........  ..........  ..........  ..........  ..........  ..........
A13  RL.....KKV  TRVLKKSSWF  SC........  ..........  ..........  ..........  ..........  ..........  ..........  ..........
A14  NA.....KKV  THVLKSNSWF  NC........  ..........  ..........  ..........  ..........  ..........  ..........  ..........
A15  NA.....KKV  THTLKSNSWL  NC........  ..........  ..........  ..........  ..........  ..........  ..........  ..........
A16  NA.....KKV  THILKSNSWF  NC........  ..........  ..........  ..........  ..........  ..........  ..........  ..........
A17  NA.....KKV  THVLKSNSWF  QC........  ..........  ..........  ..........  ..........  ..........  ..........  ..........
A18  NA.....GKV  THILKSNSWF  NC........  ..........  ..........  ..........  ..........  ..........  ..........  ..........
A19  ..........  ..........  ..........  ..........  ..........  ..........  ..........  ..........  ..........  ..........
A20  ..........  ..........  ..........  ..........  ..........  ..........  ..........  ..........  ..........  ..........
A21  ..........  ..........  ..........  ..........  ..........  ..........  ..........  ..........  ..........  ..........
A22  ..........  ..........  ..........  ..........  ..........  ..........  ..........  ..........  ..........  ..........
A23  IA.....QRV  TRVARGNKWL  NCRYG.....  ..........  ..........  ..........  ..........  ..........  ..........  ..........
A24  VS.....HRV  TRVARSNLWL  NCS.......  ..........  ..........  ..........  ..........  ..........  ..........  ..........
A25  VS.....HRV  TRVARSNLWL  NCS.......  ..........  ..........  ..........  ..........  ..........  ..........  ..........
A26  VS.....RRV  TRVARSNLWL  NCP.......  ..........  ..........  ..........  ..........  ..........  ..........  ..........
A27  QN..EVNQVD  YVGYVTFDLL  KCPED.....  ..........  ..........  ..........  ..........  ..........  ..........  ..........
A28  RN..EVNQVD  EVGYVTFDIL  QCPED.....  ..........  ..........  ..........  ..........  ..........  ..........  ..........
A29  ..........  ..........  ..........  ..........  ..........  ..........  ..........  ..........  ..........  ..........
A30  RN..EVNQVD  EVGYVTFDIL  QCPED.....  ..........  ..........  ..........  ..........  ..........  ..........  ..........
A31  R.....RQVD  QVGYVKYDIL  NCPEN.....  ..........  ..........  ..........  ..........  ..........  ..........  ..........
A32  RS..ELNQVD  QVGYVTYDIL  QCPED.....  ..........  ..........  ..........  ..........  ..........  ..........  ..........
A33  QN..EVNKVD  PEVNQVDDVG  YVTYDLLQCP  ..........  ..........  ..........  ..........  ..........  ..........  ..........
A34  RN..EVNLVD  QVGYVSFDIL  HCPED.....  ..........  ..........  ..........  ..........  ..........  ..........  ..........
A35  HN..RVNQVD  HVAYVTYDLL  QCP.......  ..........  ..........  ..........  ..........  ..........  ..........  ..........
A36  NA.....RMV  THILKSNSWL  HC........  ..........  ..........  ..........  ..........  ..........  ..........  ..........
A37  NA.....RTV  THILKSNSWL  LC........  ..........  ..........  ..........  ..........  ..........  ..........  ..........
A38  NA.....RTV  THTLKSNSWL  LC........  ..........  ..........  ..........  ..........  ..........  ..........  ..........
A39  VK..VKALEG  FPRLVGPDFF  GCAEP...AN  TFL.......  ..........  ..........  ..........  ..........  ..........  ..........
A40  VK..VKVLEG  FPRPVGPDFF  DCAEP...AN  TFP.......  ..........  ..........  ..........  ..........  ..........  ..........
A41  VK..VKVLES  FPRPIGPDFF  DCAEP...AN  TFP.......  ..........  ..........  ..........  ..........  ..........  ..........
A42  FK..VRVMDG  YPHLISQDFF  NCQASSTFVN  SLP.......  ..........  ..........  ..........  ..........  ..........  ..........
A43  LL.....QRI  TKTLKSNSWF  GC........  ..........  ..........  ..........  ..........  ..........  ..........  ..........
A44  LL.....QRI  TKTLKSNSWF  GC........  ..........  ..........  ..........  ..........  ..........  ..........  ..........
A45  LF.....RRV  TKTLKSTSWF  GC........  ..........  ..........  ..........  ..........  ..........  ..........  ..........
A46  FS.....SRV  TKKLKSNSWF  DC........  ..........  ..........  ..........  ..........  ..........  ..........  ..........
A47  LL.....DRV  TKTLSSTSWF  GC........  ..........  ..........  ..........  ..........  ..........  ..........  ..........
A48  WS.....NRI  VRVMTTNSLL  WC........  ..........  ..........  ..........  ..........  ..........  ..........  ..........
A49  ..........  ..........  ..........  ..........  ..........  ..........  ..........  ..........  ..........  ..........
A50  WS.....NKV  IRVLKTNGIL  WC........  ..........  ..........  ..........  ..........  ..........  ..........  ..........
A51  WS.....NRI  VRVMPTNSLL  WC........  ..........  ..........  ..........  ..........  ..........  ..........  ..........
A52  WS.....NRI  VRVMPANSIL  WC........  ..........  ..........  ..........  ..........  ..........  ..........  ..........
A53  WS.....NRI  VRVMPTNSIL  WC........  ..........  ..........  ..........  ..........  ..........  ..........  ..........
A54  ..........  ..........  ..........  ..........  ..........  ..........  ..........  ..........  ..........  ..........
A55  WS.....KRI  VRVMPTNSLL  WC........  ..........  ..........  ..........  ..........  ..........  ..........  ..........
A56  WS.....NRI  VRVMPTNSLL  WC........  ..........  ..........  ..........  ..........  ..........  ..........  ..........
A57  ..........  ..........  ..........  ..........  ..........  ..........  ..........  ..........  ..........  ..........
A58  WS.....QRI  TRILKTNFVL  MC........  ..........  ..........  ..........  ..........  ..........  ..........  ..........
A59  WS.....KRI  TRILKTNFVL  MC........  ..........  ..........  ..........  ..........  ..........  ..........  ..........
A60  QK..LKVEPG  YPKSALRDWM  GCPSGGRPD.  ..........  ......EGTE  EETEVIIIEV  D.........  ..........  EEGGGAVSAA  AVVLPVLLLL
A61  QK..LKVEPG  YPKSALRDWM  GCPSGGRPD.  ..........  ......EGTE  EETEVIIIEV  D.........  ..........  EEGSGAVSAA  AVVLPVLLLL
A62  QK..LKVEPG  YPKSALRDWM  GCPSGRRPD.  ..........  ......EGTE  EETEVIIIEV  D.........  ..........  EEGSGAVSAA  AVVLPVLLLL
A63  QK..LKVEPG  YPKSALRDWM  GCPAGGRPD.  ..........  ......EGTE  EETEVIIIEV  D.........  ..........  EEGSGAVSAA  AVVLPVLLLL
A64  QK..LKVEPG  YPKSALRDWM  GCPSGGRPD.  ..........  ......EGTE  EETEVIIIEV  D.........  ..........  EEGSGAVSAA  AVVLPVLLLL
A65  ER..LRMEPG  YPKSILRDFM  GCQEHVEPGP  RWPDVARPPF  NPHGGAEPGA  DSAEGDVGDG  DGDFGAGVNK  DGGSRVVVQM  EEVARTVNVV  MVLVPLLLLL
A66  ER..LRMEPG  YPKSILRDFM  GCQEHVEPGP  RWPDVARPPF  NPHGGAEPGA  DSAEGDVGDG  DGDFGAGVNK  DRGSRVVVQM  EEVARTVNVV  MVLVPLLLLL
A67  ER..LRMEPG  YPKSILRDFM  GCQEHVEPGP  RWPDVARPPF  NPHGGAEPGA  DSAEGDVGDG  DGDFGAGVNK  DGGSRVVVQM  EEVARTVNVV  MVLVPLLLLL
A68  ER..LRMEPG  HPKSILRDFM  GCQEHVEPRS  RWPDVARPPF  NPNGGAEPEA  DGDSKEENAG  D........K  DEGSRVVVQM  EEVVRTVNVV  MVLVPLLLLL
A69  KM..LRVEPG  YPRSILKDFM  GCDGPTDRDK  ..........  ......DRHS  PQDDVDIIIK  L.........  ..........  DNTASTVKAI  AIVIPCILAL
A70  QI..LKVEPG  YPRSILKDFM  GCDGPTDRVK  ..........  ......EGHS  PPDDVDIVIK  L.........  ..........  DNTASTVKAI  AIVIPCILAL
A71  ..........  ..........  ..........  ..........  ..........  ..........  ..........  ..........  ..........  ..........
A72  QI..LKVEPG  YPRSILKDFM  GCDGPTDRDK  ..........  ......EGLS  PPDDVDIVIK  L.........  ..........  DNTASTVKAI  AIVIPCILAL
A73  QI..LKVEPG  YPRSILKDFM  GCDGPTDRDK  ..........  ......EGLS  PPP.......  ..........  ..........  ..........  ..........
A74  GE..LEVAPG  YPQSTARDWL  VCGDSQ..AD  GSVAAGVDAA  EGPRAPPGQH  DQSRSEDGYE  VCSCTSGASS  PPGAPGPLVA  ATMLLLLPPL  SPGALWTAAQ
A75  NS.....KRI  VQVLRSNSWL  GC........  ..........  ..........  ..........  ..........  ..........  ..........  ..........
A76  QL..R.VEKG  YPRNISHNWM  HCRPRTIDTT  PSGGNTTPSG  TGITLDTTLS  ATETTFEY..  ..........  ..........  ..........  ..........
A77  QL..R.VEKG  YPRNISHNWM  HCRPRTIDTT  PSGGNTTPSG  TGITLDTTLS  ATETTFEY..  ..........  ..........  ..........  ..........
A78  QL..R.VEKG  YPRNISHNWM  HCRPRTIDTT  PSGGNTTPSG  TGITLDTTLS  ATETTFEY..  ..........  ..........  ..........  ..........
A79  EK.....EDV  VSELKSSSWI  GC........  ..........  ..........  ..........  ..........  ..........  ..........  ..........
A80  EK.....EDV  VSVVKSSSWI  GC........  ..........  ..........  ..........  ..........  ..........  ..........  ..........
A81  EK.....EDV  VSVLKSNSWI  GC........  ..........  ..........  ..........  ..........  ..........  ..........  ..........
A82  ..........  ..........  ..........  ..........  ..........  ..........  ..........  ..........  ..........  ..........
A83  ..........  ..........  ..........  ..........  ..........  ..........  ..........  ..........  ..........  ..........
A84  ..........  ..........  ..........  ..........  ..........  ..........  ..........  ..........  ..........  ..........
A85  ..........  ..........  ..........  ..........  ..........  ..........  ..........  ..........  ..........  ..........
A86  LR..WINGHQ  ILMS.SGDEF  AVYDEFWNQV  TLKNRISSYF  PNLPRGVKGV  ESPAGSVITA  FTSNQVFEYN  SRTKSIGRQS  GFSSYIAC..  ..........
A87  LR..WINGHQ  ILMS.SGDEF  AVYDEFWNQV  TLKNRISSYF  PNLPRGVKGV  ESPAGSVITA  FTSNQVFEYN  SRTKSIGRQS  GFSSYIAC..  ..........
A88  SRSVFDIADG  YPQPLQSFFG  FCPRNEKLVL  NSSSSHFSLI  YATITILILI  F.........  ..........  ..........  ..........  ..........
A89  KG..PLDPDY  FDDDFFENFF  MEYSK.....  ..........  ..........  ..........  ..........  ..........  ..........  ..........
A90  KG..PLG...  ..........  ..........  ..........  ..........  ..........  ..........  ..........  ..........  ..........
A91  ..........  ..........  ..........  ..........  ..........  ..........  ..........  ..........  ..........  ..........
A92  ..........  ..........  ..........  ..........  ..........  ..........  ..........  ..........  ..........  ..........
A93  SN..QLQRRS  PISSYFRNTP  YALRDGVEAV  VRVDDVYLHF  YRDGRYYRMI  ESTKQFVNFP  NGLSYRDVID  TLIPQCRSLN  LSVEIESCSN  SSE.......
A94  NN..RLQRRT  RISSLFNDVP  SALHDGVEAV  VRADRNYIHF  YRDGRYYRMT  DYGRQFVNFP  NGLPYSDVIE  SVIPQCRGRS  LSYESEGCSN  SSE.......
A95  DK....DRRQ  NPSLPYNGLF  PRRAISQQWF  DICNVHPSLL  KI........  ..........  ..........  ..........  ..........  ..........
A96  ..........  ..........  ..........  ..........  ..........  ..........  ..........  ..........  ..........  ..........
```

                        TRANSMEMBRANE─┐ ┌─CYTOPLASMIC DOMAIN→

A1   COG1_BOVIN .... 1010 .. .. 1020 .... 1030 ..
A2   COG1_HUMAN ....
A3   COG1_RABIT ....
A4   COG1_RANCA ....
A5   COG1_PIG   ....
A6   COG2_CHICK ....
A7   COG2_HUMAN ....
A8   COG2_MOUSE ....
A9   COG2_RABIT ....
A10 COG2_RAT   ....
A11 G257061   ....
A12 Q98857   ....
A13 Q98858   ....
A14 Q28397   ....
A15 COG3_HUMAN ....
A16 COG3_MOUSE ....
A17 COG3_RABIT ....
A18 COG3_RAT   ....
A19 COG7_FELCA ....
A20 COG7_HUMAN ....
A21 COG7_MOUSE ....
A22 COG7_RAT   ....
A23 COG8_HUMAN ....
A24 O70138   ....
A25 O88733   ....
A26 O88766   ....
A27 COG9_BOVIN ....
A28 COG9_CANFA ....
A29 Q95166   ....
A30 O19130   ....
A31 Q98856   ....
A32 COG9_HUMAN ....
A33 COG9_MOUSE ....
A34 COG9_RABIT ....
A35 COG9_RAT   ....
A36 COGX_HUMAN ....
A37 COGX_MOUSE ....
A38 COGX_RAT   ....
A39 COGY_HUMAN ....
A40 COGY_MOUSE ....
A41 P97568   ....
A42 COGY_XENLA ....
A43 COGM_HUMAN ....
A44 Q99745   ....
A45 COGM_MOUSE ....
A46 COGM_RABIT ....
A47 Q63341   ....
A48 O77656   ....
A49 O77632   ....
A50 Q98859   ....
A51 O18927   ....
A52 COGZ_HUMAN ....
A53 COGZ_MOUSE ....
A54 O73923   ....
A55 O62806   ....
A56 COGZ_RAT   ....
A57 O77631   ....
A58 COGZ_XENLA ....
A59 CGZA_XENLA ....
A60 COGT_HUMAN LVLAVGLAVF   FFRRHGTPRR   LLYCQRSLLD   KV
A61 COGT_MOUSE LVLAVGLAVF   FFRRHGTPKR   LLYCQRSLLD   KV
A62 O08645   LVLAVGLAVF   FFRRHGTPKR   LLYCQRSLLD   KV
A63 COGT_RABIT LVLAVGLAVF   FFRRHGTPKR   LLYCQRSLLD   KV
A64 COGT_RAT   LVLAVGLAVF   FFRRHGTPKR   LLYCQRSLLD   KV
A65 COGU_HUMAN CVLGLTYALV   QMQRKGAPRV   LLYCKRSLQE   WV
A66 Q14111   CVLGLTYALV   QMQRKGAPRV   LLYCKRSLQE   WV
A67 D1023087   CVLGLTYALV   QMQRKGAPRV   LLYCKRSLQE   WV
A68 O54732   CILGLAFALV   QMQRKGAPRM   LLYCKRSLQE   WV
A69 Q98947   CLLVLVYTVF   QFKREGTPRH   ILYCKRSMQE   WV
A70 COGV_HUMAN CLLVLVYTVF   QFKRKGTPRH   ILYCKRSMQE   WV
A71 Q14824   ....
A72 COGV_RAT   CLLVLVYTVF   QFKRKGTPRH   ILYCKRSMQE   WV
A73 O35541   ....
A74 Q14850   ALTL......   ....
A75 O13065   ....
A76 Q99542   ....
A77 Q99580   ....
A78 O15278   ....
A79 O18767   ....
A80 O60882   ....
A81 P79287   ....
A82 O04529   ....
A83 O23507   ....
A84 O48680   ....
A85 O65340   ....
A86 O16901   ....
A87 O61264   ....
A88 O61266   ....
A89 O61265   ....
A90 O44836   ....
A91 O55761   ....
A92 MEP1_SOYBN ....
A93 HE_HEMPU   ....
A94 HE_PARLI   ....
A95 O93470   ....
A96 P89294   ....

Tissue inhibitors of MMPs alignments

```
                                    SIGNAL PEPTIDE─┐ ┌1       ┌2              ┌3
C1  TIM1_BOVIN   P20414   ......MAPF 10  APMASGILLL 20  LWLTAPSRAC 30  TCVPPHPQTA 40  FCNSDVVIRA 50  KF...VGTAE 60  VNE....... 70  TALYQRYEIK 80  MTKMFKGFSA 90  LRDAPDIRFI 100
C2  TIM1_CANFA   P81546   ......MAPF     APLASCILLL     LWLTAPSRAC     TCAPPHPQTA     FCNSQIVIRA     KF...VGTAE     VNQ.......     TDLNRRYEIK     MTKMFKGFSA     LGNASDIRFV
C3  TIM1_HORSE   O02722   ......MAPF     APLVSGILLL     LWLTAPSRAC     TCVPPHPQTA     FCSSEFVIRA     KF...VGTSE     VNQ.......     TTLQRRYEIK     MTKMFKGFSA     LGDAPDTWFV
C4  TIM1_HUMAN   P01033   ......MAPF     EPLASGILLL     LWLIAPSRAC     TCVPPHPQTA     FCNSDLVIRA     KF...VGTPE     VNQ.......     TTLYQRYEIK     MTKMYKGFQA     LGDAADIRFV
C5  TIM1_MOUSE   P12032   .....MMAPF     ASLASGILLL     LSLIASSKAC     SCAPPHPQTA     FCNSDLVIRA     KF...MGSPE     INE.......     TTLYQRYKIK     MTKMLKGFKA     VGNAADIRYA
C6  TIM1_RABIT   P20614   ......MAPL     AALASSMLLL     LWLVAPSRAC     TCVPPHPQTA     FCNSDLVIRA     KF...VGAPE     VNH.......     TTLYQRYEIK     TTKMFKGFDA     LGHATDIRFV
C7  TIM1_SHEEP   P50122   ......MALF     APTVSGILLL     LWLTAPSRAC     TCVPPHPQTA     FCNSEVVIRA     KF...VGTAE     VNE.......     TALYQRYEIK     MTKMFKGFSA     LRDAPDIRFV
C8  TIM1_PAPCY   P49061   ......MAPF     EPLASGILLL     LWLIAPSRAC     TCVPPHPQTA     FCNSDLVIRA     KF...VGTPE     VNQ.......     TTLYQRYEIK     MTKMYKGFQA     LGDAADIRFV
C9  TIM1_RAT     P30120   ......MAPF     ASLASGILLL     LSLIASSKAC     SCAPTHPQTA     FCNSDLVIRA     KF...MGSPE     IIE.......     TTLYQRYEIK     MTKMLKGFDA     VGNATGFRFA
C10 TIM1-PIG     P35624   ......MSPF     APLASGILLL     LWLTAPSRAC     TCVPPHPQTA     FCSSDLVIRA     KF...VGAPE     FNQ.......     TASYQRYEIK     MTKMFKGFNA     LGDAPDIRFI
C11 TIM1_BOVINE  P16368   ...MGAAARS     LPLAFCLLLL     GTLLPRADAC     SCSPVHPQQA     FCNADIVIRA     KA...VNKKE     VDSGNDIYGN     PIKRIQYEIK     QIKMFKGPD.     ....QDIEFI
C12 Q21265       Q21265   .........M     QNLSLSLVIL     SVLIAVTLAC     KCREQSTKES     FCNAHWVSHV     KVKVRVGKQG     LPEGS..ERK     GLNNLRYTVQ     HVEVFKKPSN     MT..TLPDEI
C13 TIM2_CRILO   Q60453   ..........     ..........     .......RAC     SCSPVHPQQA     FCNADVVIRA     KA...VSEKE     VDSGNDIYGN     PIKRIQYEIK     QIKMFKGPD.     ....KDIEFI
C14 TIM2_CHICK   O42146   ...MPGAALP     SLLAWLAVLL     LGRARPADAC     SCSPIHPQQA     FCNADVVIRA     KR...VSAKE     VDSGNDIYGN     PIKRIQYEVK     QIKMFKGPD.     ....QDIEFI
C15 TIM2_HUMAN   P16035   ...MGAAART     LRLALGLLLL     ATLLRPADAC     SCSPVHPQQA     FCNADVVIRA     KA...VSEKE     VDSGNDIYGN     PIKRIQYEIK     QIKMFKGPE.     ....KDIEFI
C16 TIM2_MOUSE   P25785   ...MGAAARS     LRLALGLLLL     ASLVRPADAC     SCSPVHPQQA     FCNADVVIRA     KA...VSEKE     VDSGNDIYGN     PIKRIQYEIK     QIKMFKGPD.     ....KDIEFI
C17 G1477929     G1477929 ..........     ..........     .........C     SCSPVHPQQA     FCNADIVIRA     KA...VNKKE     VDSGNDIYGN     PIKRIQYEIK     QIKMFKGPD.     ....QDIEFI
C18 TIM2_RAT     P30121   ...MGAAARS     LRLALGLLLL     ATLLRPADAC     SCSPVHPQQA     FCNADVVIRA   · KA...VSEKE     VDSGNDIYGN     PIKRIQYEIK     QIKMFKGPD.     ....KDIEFI
C19 TIM3_BOVINE  P79121   .....MTPWL     G.LVVLLGSW     SLGDWGAEAC     TCSPSHPQDA     FCNSDIVIRA     KV...VGKKL     LKE......G     PFGTMVYTIK     QMKMYRGFTK     M...PHVQYI
C20 TIM3_CHICK   P26652   .....MTAWL     GFLAVFLCSW     SLRDLVAEAC     TCVPIHPQDA     FCNSDIVIRA     KV...VGKKL     MKD......G     PFGTMRYTVK     QMKMYRGFQI     M...PHVQYI
C21 TIM3_HUMAN   P35625   .....MTPWL     G.LIVLLGSW     SLGDWGAEAC     TCSPSHPQDA     FCNSDIVIRA     KV...VGKKL     VKE......G     PFGTLVYTIK     QMKMYRGFTK     M...PHVQYI
C22 TIM3_MOUSE   P39876   .....MTPWL     G.LVVLLSCW     SLGHWGAEAC     TCSPSHPQDA     FCNSDIVIRA     KV...VGKKL     VKE......G     PFGTLVYTIK     QMKMYRGFSK     M...PHVQYI
C23 TIM3_RAT     P48032   .....MTPWL     G.LVVLLSCW     SLGHWGTEAC     TCSPSHPQDA     FCNSDIVIRA     KV...VGKKL     VKE......G     PFGTLVYTIK     QMKMYRGFSK     M...PHVQYI
C24 TIM3_XENLA   O73746   ...MSVCALT     LILGCFLLFL     GDISKPAEGC     TCAPSHPQDA     FCNSDIVIRA     KV...VGKKL     MKD......G     PFGTMRYTVK     QMKMYRGFNK     M...PQVQYI
C25 TIM4_HUMAN   Q99727   MPGSPRPAPS     WVLLLRLLAL     LRPPGLGEAC     SCAPAHPQQH     ICHSALVIRA     KI...SSEKV     VPASA.DPAD     TEKMLRYEIK     QIKMFKGFEK     V...KDVQYI
```

```
                                                                                      Domains join
             1┐                          2┐                                          3┐ ┌4            5┐            6┐
C1   TIM1_BOVIN  P20414   YTPAMESVCG 110  YFHRSQNRSE 120  EFLIAGQLS. 130  NGHLHITTCS 140  FVAPWNSMSS 150  AQRRGFTKTY 160  AAGCEECTVF 170  PCSSIPCKLQ 180  SDTHCLWTDQ 190  LLTGSDKGFQ 200
C2   TIM1_CANFA  P81546   DTPALESVCG      YLHRSQNRSE      EFLVAGNLR.      DGHLQINTCS      FVAPWSSLST      AQRRGFTKTY      AAGCEGCTVF      TCSSIPCKLQ      SDTHCLWTDQ      FLTGSDKGFQ
C3   TIM1_HORSE  O02722   YTPAMESLCG      YFHRSENRSE      EFLIAGQLL.      DEKLYITTCS      FVAPWNSLSS      AQRQGFTKTY      AAGCGECSVF      PCSSIPCKLQ      SDTDCLWTDQ      LLTGSDKGFQ
C4   TIM1_HUMAN  P01033   YTPAMESVCG      YFHRSHNRSE      EFLIAGKLQ.      DGLLHITTCS      FVAPWNSLSL      AQRRGFTKTY      TVGCEECTVF      PCLSIPCKLQ      SGTHCLWTDQ      LLQGSEKGFQ
C5   TIM1_MOUSE  P12032   YTPVMESLCG      YAHKSQNRSE      EFLITGRLR.      NGNLHISACS      FLVPWRTLSP      AQQRAFSKTY      SAGCGVCTVF      PCLSIPCKLE      SDTHCLWTDQ      VLVGSED.YQ
C6   TIM1_RABIT  P20614   YTPAMESVCG      YSHKSQNRSE      EFLIAGQLR.      NGLLHITTCS      FVVPWNSLSF      SQRSGFTKTY      AAGCDMCTVF      ACASIPCHLE      SDTHCLWTDS      SLG.SDKGFQ
C7   TIM1_SHEEP  P50122   YTPAMESVCG      YFHRSQNRSE      EFLIAGQLS.      NGHLHITTCS      FVAPWNSMSS      AQRRGFTKTY      AAGCEECTVF      PCSSIPCKLQ      SDTHCLWTDQ      LLTGSDKGFQ
C8   TIM1_PAPCY  P49061   YTPAMESVCG      YFHRSHNRSE      EFLIAGKLQ.      DGLLHITTCS      FVAPWNSLSL      AQRRGPTKTY      TVGCEECTVF      PCLSIPCKLQ      SGTHCLWTDQ      LLQGSEKGFQ
C9   TIM1_RAT    P30120   YTPAMESLCG      YVHKSQNRSE      EFLIAGRLR.      NGNLHITACS      FLVPWHNLSP      AQQKAFVKTY      SAGCGVCTVF      PCSAIPCKLE      SDSHCLWTDQ      ILMGSEKGYQ
C10  TIM1-PIG    P35624   YTPAMESVCG      YFHRSQNRSQ      EFLIAGQLW.      NGHLHITTCS      FVAPWNSLSS      AQRQGPTEIY      AAGCEECTVF      PCTSIPCKLQ      SDTHCLWTDQ      LLTGSDKGFQ
C11  TIM1_BOVINE P16368   YTAPAAAVCG      VSLDIGG.KK      EYLIAGKAEG      NGNMHITLCD      FIVPWDTLSA      TQKKSLNHRY      QMGCE.CKIT      RCPMIPCYIS      SPDECLWMDW      VTEKNINGHQ
C12  Q21265      Q21265   FTPSEAPACG      LKIAAG...H      EYLLAGRVEG      PNALYTVLCG      QVLPDD..RS      .Q.TSF....      ...EN...VL      EWKNVPQTLQ      S.........      .........Q
C13  TIM2_CRILO  Q60453   YTAPSSAVCG      VSLDVGG.KK      EYLIAGKAEG      DGKMHITLCD      FIVPWDTLST      TQKKSLNHRY      QMGCE.CKIT      RCPMIPCYIS      SPDECLWMDW      VTEKSINGHQ
C14  TIM2_CHICK  O42146   YTAPSTEVCG      QPLDTGG.KK      EYLIAGKSEG      DGKMHITLCD      LVATWDSVSP      TQKKSLNQRY      QMGCE.CKIS      RCLSIPCFVS      SSDECLWTDW      AMEKIVGGRQ
C15  TIM2_HUMAN  P16035   YTAPSSAVCG      VSLDVGG.KK      EYLIAGKAEG      DGKMHITLCD      FIVPWDTLST      TQKKSLNHRY      QMGCE.CKIT      RCPMIPCYIS      SPDECLWMDW      VTEKNINGHQ
C16  TIM2_MOUSE  P25785   YTAPSSAVCG      VSLDVGG.KK      EYLIAGKAEG      DGKMHITLCD      FIVPWDTLSI      TQKKSLNHRY      QMGCE.CKIT      RCPMIPCYIS      SPDECLWMDW      VTEKSINGHQ
C17  G1477929    G1477929 YTAPSSAVCG      VSLDIGG.KK      EYLIAGKAEG      NGNMHITLCD      FIVPWDTLSA      TQKKSLNHRY      QMGCE.CKIT      RCPMIPCYIS      SPDECLWMDW      VTEKNINRHQ
C18  TIM2_RAT    P30121   YTAPSSAVCG      VSLDVGG.KK      EYLIAGKAEG      DGKMHITLCD      FIVPWDTLSI      TQKKSLNHRY      QMGCE.CKIT      RCPMIPCYIS      SPDECLWMDW      VTEKSINGHQ
C19  TIM3_BOVINE P79121   HTEASESLCG      LKLEVN..KY      QYLLTGRVY.      DGKMYTGLCN      FVERWDQLTL      SQRKGLNYRY      HLGCN.CKIK      SCYYLPCFVT      SKNECLWTDM      FSNFGYPGYQ
C20  TIM3_CHICK  P26652   YTEASESLCG      VKLEVN..KY      QYLITGRVY.      EGKVYTGLCN      WYEKWDRLTL      SQRKGLNHRY      HLGCG.CKIR      PCYYLPCFAT      SKNECIWTDM      LSNFGHSGHQ
C21  TIM3_HUMAN  P35625   HTEASESLCG      LKLEVN..KY      QYLLTGRVY.      DGKMYTGLCN      FVERWDQLTL      SQRKGLNYRY      HLGCN.CKIK      SCYYLPCFVT      SKNECLWTDM      LSNFGYPGYQ
C22  TIM3_MOUSE  P39876   HTEASESLCG      LKLEVN..KY      QYLLTGRVY.      EGKMYTGLCN      FVERWDHLTL      SQRKGLNYRY      HLGCN.CKIK      SCYYLPCFVT      SKNECLWTDM      LSNFGYPGYQ
C23  TIM3_RAT    P48032   HTEASESLCG      LKLEVN..KY      QYLLTGRVY.      EGKMYTGLCN      FVERWDHLTL      SQRKGLNYRY      HLGCN.CKIK      SCYYLPCFVT      SKKECLWTDM      LSNFGYPGYQ
C24  TIM3_XENLA  O73746   YTEASESLCG      VKLEVN..KY      QYLITGRVY.      EGKVYTGLCN      LIERWEKLTF      AQRKGLNHRY      PLGCT.CKIK      PCYYLPCFIT      SKNECLWTDM      LSNFGYPGYQ
C25  TIM4_HUMAN  Q99727   YTPFDSSLCG      VKLEANS.QK      QYLLTGQVLS      DGKVFIHLCN      YIEPWEDLSL      VQRESLNHHY      HLNCG.CQIT      TCYTVPCTIS      APNECLWTDW      LLERKLYGYQ
```

```
                                    6⌐                4⌐
C1  TIM1_BOVIN  P20414    SRHLACLPRE 210  PGLCTWQSLR 220  AQMA...... 230  ....
C2  TIM1_CANFA  P81546    SRHLACLPRE      PGICTWQSL.      PRMA......      ....
C3  TIM1_HORSE  O02722    SRYLACLPRE      PGLCTWQSLR      PRTA......      ....
C4  TIM1_HUMAN  P01033    SRHLACLPRE      PGLCTWQSLR      SQIA......      ....
C5  TIM1_MOUSE  P12032    SRHFACLPRN      PGLCTWRSLG      AR........      ....
C6  TIM1_RABIT  P20614    SRHLACLPQE      PGLCAWESLR      PRKD......      ....
C7  TIM1_SHEEP  P50122    SRHLACLPRE      PGMCTWQSLR      PRGA......      ....
C8  TIM1_PAPCY  P49061    SRHLACLPRE      PGLCTWQSLR      TRIA......      ....
C9  TIM1_RAT    P30120    SDHFACLPRN      PDLCTWQYLG      VSMTRSLPLA      KAEA
C10 TIM1-PIG    P35624    SRHLACMPRE      PGMCTWQSLR      PRVA......      ....
C11 TIM1_BOVINE P16368    AKFFACIKRS      DGSCAWYRGA      APPKQEFLDI      EDP.
C12 Q21265      Q21265    VKSIKC....      ..........      ..........      ....
C13 TIM2_CRILO  Q60453    AKFFACIKRS      DGSCAWYRGA      APPKQEFLDI      EDP.
C14 TIM2_CHICK  O42146    AKHYACIKRS      DGSCAWYRGM      APPKQEFLDI      EDP.
C15 TIM2_HUMAN  P16035    AKFFACIKRS      DGSCAWYRGA      APPKQEFLDI      EDP.
C16 TIM2_MOUSE  P25785    AKFFACIKRS      DGSCAWYRGA      APPKQEFLDI      EDP.
C17 G1477929    G1477929  AKFFACIKRS      DGSCAWYRGA      APPKQEFLDI      EDP.
C18 TIM2_RAT    P30121    AKFFACIKRS      DGSCAWYRGA      APPKQEFLDI      EDP.
C19 TIM3_BOVINE P79121    SKHYACIRQK      GGYCSWYRGW      APPDKSIINA      TDP.
C20 TIM3_CHICK  P26652    AKHYACIQRV      EGYCSWYRGW      APPDKTIINA      TDP.
C21 TIM3_HUMAN  P35625    SKHYACIRQK      GGYCSWYRGW      APPDKSIINA      TDP.
C22 TIM3_MOUSE  P39876    SKHYACIRQK      GGYCSWYRGW      APPDKSISNA      TDP.
C23 TIM3_RAT    P48032    SKHYACIRQK      GGYCSWYRGW      APPDKSISNA      TDP.
C24 TIM3_XENLA  O73746    SKNYACIKQK      EGYCSWYRGW      APPDKTTINT      TDP.
C25 TIM4_HUMAN  Q99727    AQHYVCMKHV      DGTCSWYRGH      LPLRKEFVDI      VQP.
```

The manufacturer's authorised representative in the EU for product safety is
Oxford University Press España S.A. of el Parque Empresarial San Fernando de
Henares, Avenida de Castilla, 2 – 28830 Madrid (www.oup.es/en or product.
safety@oup.com). OUP España S.A. also acts as importer into Spain of products
made by the manufacturer.